COPPEROPOLIS

LANDSCAPES OF THE EARLY INDUSTRIAL PERIOD IN SWANSEA

Sailing ships engaged in the two-way copper-ore and coal trade moored in Swansea's then tidal harbour in the early 1840s. The wharves and stores associated with the Landore Copperworks are visible on the right (east). (National Maritime Museum, Greenwich).

Royal Commission on the Ancient and Historical Monuments of Wales

COPPEROPOLIS

LANDSCAPES OF THE EARLY INDUSTRIAL PERIOD IN SWANSEA

Stephen Hughes

Royal Commission on the Ancient and Historical Monuments of Wales

ISBN 1-871184-17-7
British Library Cataloguing in Publication Data.
A catalogue record for this book is available from the British Library.

The Royal Commission has a national role in the management of the archaeological, built and maritime heritage of Wales as the originator, curator and supplier of authoritative information for individual, corporate and governmental decision-makers, researchers and the general public.

To this end it:
Surveys, Interprets and Records the man-made environment of Wales;
Compiles, Maintains and Curates the National Monuments Record of Wales;
Promotes an understanding of this information by all appropriate means.

Printed in Wales by
Cambrian Printers Limited
Llanbadarn Road, Aberystwyth, Ceredigion, SY23 3TN

(Photograph opposite) It has been possible to locate and clear side-passages inside Smith's Canal Tunnel which ran directly into the main White Rock Copperworks smelting-house to facilitate the delivery of coal from canal boats. (831008)

Contents

Acknowledgements

A work of this nature owes a great deal to many individuals whose contributions are appropriately acknowledged throughout. The text was researched and compiled by Stephen Hughes and has been edited and significantly improved by Paul Reynolds. Further editing and copy-editing have been undertaken by Patricia Moore, Ann Collis and David Browne. Part of its industrial archaeological content is based on an M.Phil. thesis by Stephen Hughes submitted to the University of Birmingham in 1983-4, jointly supervised by the late Gordon Tucker, Fellow in the History of Technology at the University and latterly a Royal Commissioner, and the late Professor John H. Harris. Much primary research on the lower Swansea Valley has been ably undertaken by Paul Reynolds of the Library of the University of Wales, Swansea and by Gerald Gabb of the Swansea History Project. Their work and that of Dr Fred Cowley and Bob Roberts, Fellows of the University of Wales, Swansea, must serve as the foundation for any study of Swansea's industrial remains. Valuable comments on the present project have also been made by Dr John Alban, Dr Fred Cowley, Jeremy Lowe, David McLees, Bernard Morris and Dr Peter Wakelin. The form and structure of this volume developed under the supportive guidance of the Royal Commission's Editorial Committee under the successive chairmanships of Professor Ronald Brunskill and Professor Ralph Griffiths, whose extensive knowledge of Swansea's industrial and social development has been much appreciated.

The site surveys and photographs on which the Commission's contribution to this study depend were undertaken over a period of 25 years by colleagues past and present. Those involved in survey work include Douglas Hague, Brian Malaws, David Percival and Richard Suggett. Aerial photographs were taken by Toby Driver and Chris Musson, and terrestrial mainly by Iain Wright and Stephen Hughes, with further photographic assistance from Fleur James. Graphics and drawing office support came from James Goodband, Charles Green, John Johnston, Brian Malaws, Medwyn Parry, Dylan Roberts and Geoff Ward. The database of Welsh chapels held by the Royal Commission which has further informed this project includes much valuable work undertaken by John Pritchard for a former Board of Celtic Studies project on denominational documentary sources, as well as by the Royal Commission's own staff, notably Olwen Jenkins and Penny Icke. A record of Swansea chapels has been commissioned from the Glamorgan-Gwent Archaeological Trust and is available in the National Monuments Record of Wales in Aberystwyth.

A substantial grant towards the publication of this volume has been made by Swansea City Council, to whom the Commissioners express their gratitude.

Preface

The redevelopment of the modern city of Swansea is steadily obliterating one of the key landscapes of the world's first Industrial Revolution. The pioneering reclamation of the lower Swansea Valley in the 1960s destroyed many of the then surviving industrial monuments, and the subsequent development of the Enterprise Park, the creation of areas of urban woodland, the construction of housing and roads, and the recovery of derelict land will soon complete the process. The remains of the earliest industrial landscape in Wales have now almost disappeared. It is therefore important that the topography of what was once an industrial centre of international importance should be mapped before the evidence is totally destroyed. This will allow decisions regarding future physical development to be made within an adequate historical and archaeological context. However, while few structures remain to be seen above ground, that does not mean that there are no remains: under the vast amounts of slag and other waste with which the ravines and gullies have been filled, substantially intact structures are waiting to be revealed.

Swansea was the first industrialised region in Wales. Its position on the western side of the south Wales coalfield, where the coal measures reach right down to the sea, led to the early development of the coal extraction industry. Then, as the nearest source of coal to the metal mines of south-west England, Swansea became the centre of the copper-smelting industry, not only in the United Kingdom, but throughout the world, a position which it held until the end of the nineteenth century.

The most important factor in the interpretation of this early industrial landscape is the interplay between the position, types and demands of smelters, mills and furnaces in relation both to the linear elements of water power and transport systems and the availability of minerals. Divisions dependent on the ownership of land or on the tenancy of individual plots are of secondary importance. This study, therefore, is not a plot-by-plot analysis, but focuses on the type and range of industrial installations, the locations of entrepreneurial and workers' dwellings

influenced by a variety of transport and water-power systems, and the linear patterns which determined the structure of this pioneering landscape. The dynamics of change at Landore, the district where industrial development was both longest and densest, are analysed in greater depth so that this complex landscape palimpsest can be explained and understood.

The situation with the social remains of this important metropolis is different in that much more has survived of the fabric of the workers' settlements than the industrial plant which generated the wealth making their building both possible and necessary. These employees' communities were sited near the copperworks dispersed along the length of the navigable river and were separate from the old medieval town. At the heart of each settlement were substantial educational and religious buildings. Indeed, much of the significance of these and the extent of their survival has remained unrecognised. The earliest and also the largest surviving schools built by works proprietors for the children of their workers have substantial remaining structures converted to new uses. The wealth and population of the lower Swansea Valley by the mid-nineteenth century gave rise to two of the most influential chapel architects in Wales. One of these, the Reverend Thomas Thomas, has some claim to be the first Welsh architect who practised over the whole of Wales. A core of internationally significant industrial housing and community buildings yet remain to remind us of what a town at the heart of the world's first Industrial Revolution represented.

This volume largely consists of a discussion of significant themes within the emergence of this internationally important industrial centre. Descriptive work on individual monuments and structures can be found on the Royal Commission's Internet site at www.rcahmw.org.uk; in S.R. Hughes and P.R.Reynolds, *A Guide to the Industrial Archaeology of the Swansea Region* (2nd ed. Aberystwyth, 1989) and amongst the files of the National Monuments Record of Wales in the offices of the Royal Commission in Aberystwyth.

The structure of the study

Swansea, as the centre of two internationally important industries, has already attracted a certain amount of attention.[1] The source material used hitherto has very largely been the documentary evidence; the archaeological resource of the buildings and engineering have not previously been used as a major historical source, indeed, neither have the surviving architectural and engineering plans been utilised in spite of both being a vanishing commodity.[2]

The significant threats to these resources, such as the programmes of road building over the line of the Swansea canal, have resulted in prolonged programmes of field-recording, some of which are reproduced here.[3] This recording work has already resulted in the publication of a guide to the most significant industrial remains in the Swansea region.[4] The scope of the present publication ranges well beyond that of individual sites. It has long been recognised during this programme of work that the linear systems providing both water-power and transport form the basis for understanding how the landscape of industry evolved.[5] The other essential elements within the industrial landscape are the productive units themselves and the elements of the communities of the workforce that came into being to service those manufactories. The density of these early industrial landscapes provided a training-ground for local artisan-engineers whose growing technical expertise in turn accelerated the development of such areas.

A host of religious buildings constructed to serve the needs of the industrial communities also produced a new class of artisan-architect whose significance spread to the whole of Wales.

References

1. For example see G. Grant Francis, *The Smelting of Copper in the Swansea District* (London, 1881) and N.L.Thomas, *The Story of Swansea's Districts & Villages* (Swansea, 1969). In 1990 two notable collections of essays on the development of the city were produced, edited by Professor Ralph Griffiths - *The City of Swansea: Challenges and Change*, (Stroud, 1990) and Professor Glanmor Williams *Swansea: an Illustrated History* (Swansea, 1990).

2. The Lower Swansea Valley Study was headed by a keen amateur industrial archaeologist but the recommended retention of significant historical remains of the copper industry was never a major consideration; the Building Control Records for the Swansea area were disposed of in 1973 (*v.i.* Bernard Morris, ex city land agent).

3. The full results of this programme can be consulted in the National Monuments Record of Wales in the Aberystwyth offices of the Royal Commission for the Ancient and Historical Monuments of Wales.

4. P.R. Reynolds and S.R. Hughes *A Guide to the Industrial Archaeology of the Swansea Region.*(Aberystwyth, 1988, 1989, 1992).

5. See S.R. Hughes, 'The evolution of an industrial landscape, *c*.1770-1897: a story told in maps', *Newsletter of the South West Wales Industrial Archaeology Society*, 24-25 (1980); S.R. Hughes, 'Landore, the Evolution of the Water-power' and 'Aspects of the Use of Water-power in the Industrial Revolution', *Melin: the Journal of the Welsh Mills Group*, 3 4, 43-59 & 23-37 (1988-89) respectively; S.R. Hughes, 'The Swansea Canal: Navigation and Water-power', *Industrial Archaeology Review,*. 4(1) (1979-80).

Measurements in this volume

The Industrial and other monuments recorded in this volume were designed in imperial British units of measurement. This system of measurement has been retained in the text. Conversion to metric measurements are as follows:

1 inch = 25.4 millimetres
1 foot (12 in) = 0.3048 metres
1 yard (3 ft) = 0.9144 metres
1 mile (1,760 yds) = 1.609 kilometres
1 square foot = 0.929 sq. metres
1 acre (4,840 sq. yds) = 0.405 hectares

1 cubic foot = 0.0283 cubic metres
1 gallon = 3.785 litres
1 stone = 6.35 kilograms
1 cwt = 50.80 kilograms
1 ton = 1.016 tonnes

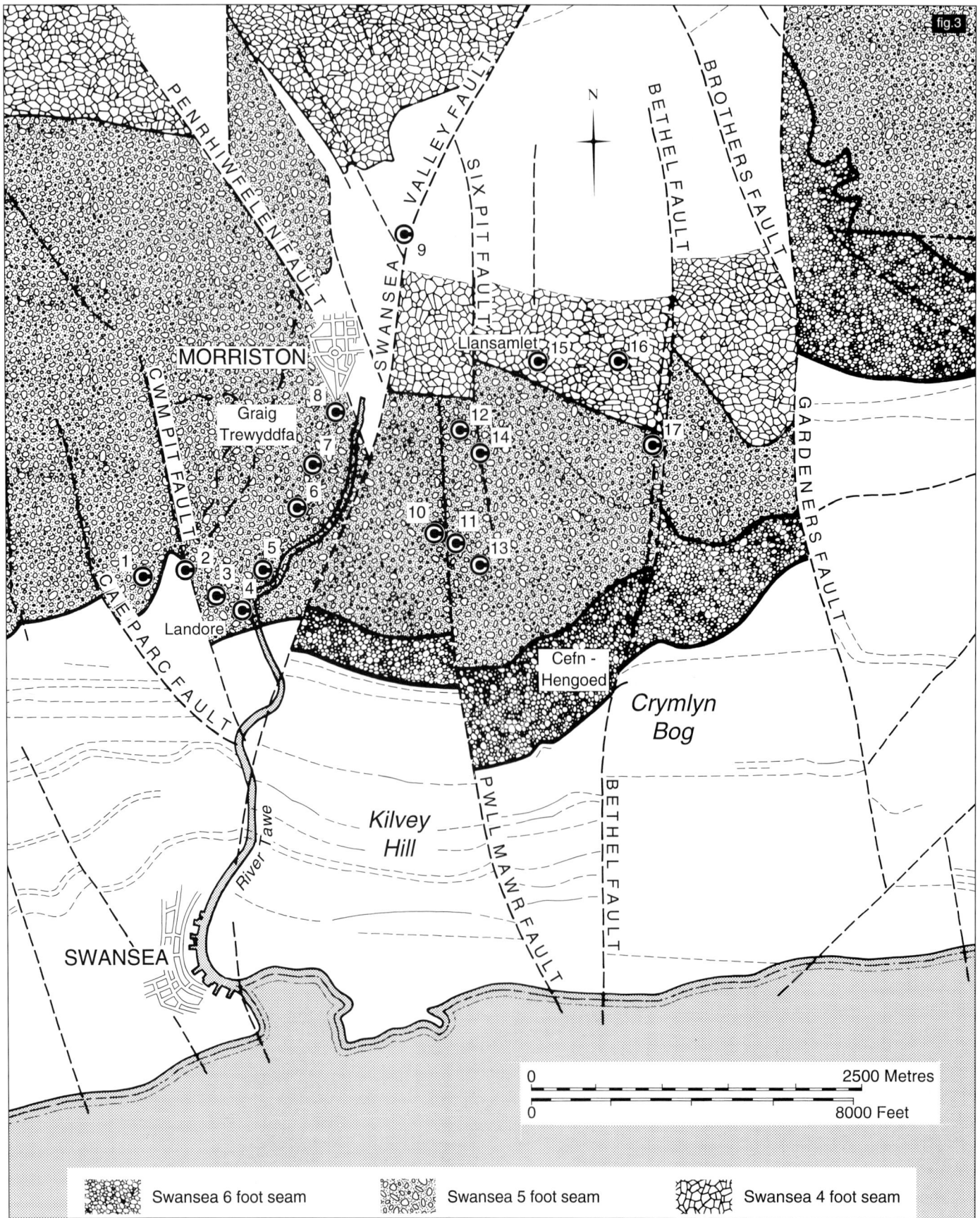

fig.3

PENRHIWFELEN FAULT

SWANSEA VALLEY FAULT

SIX PIT FAULT

BETHEL FAULT

BROTHERS FAULT

GARDENERS FAULT

CWM PIT FAULT

CAE PARC FAULT

PWLL MAWR FAULT

BETHEL FAULT

N

MORRISTON

Graig
Trewyddfa

Landore

9

Llansamlet

15

16

8

7

6

12

14

17

10

11

13

1

2

3

4

5

Cefn -
Hengoed

Crymlyn
Bog

Kilvey
Hill

River Tawe

SWANSEA

0   2500 Metres

0   8000 Feet

Swansea 6 foot seam   Swansea 5 foot seam   Swansea 4 foot seam

x

CHAPTER 1

THE INDUSTRIES

The Determinants of the Swansea Landscape

Three factors enabled the Swansea area to become the international centre for one of the key industries of the world's first Industrial Revolution. The first two factors are the balance and sources of the raw materials required for the copper-smelting process: the copper-ore was mined in Devon and Cornwall but without the availability there of the coal needed to fuel the refining process. The fact that three times as much fuel was required as ore for the refining process, meant that it was more economic for the latter to be brought to the Swansea area, rather than coal being transported for the use of large smelting works situated in Cornwall.

The other vital factor was the location of the two materials in relation to each other and to transport - Swansea was the nearest point to the Cornish metal-mining area where a coalfield was most readily accessible from the sea.

All other factors are subsidiary to these. The key transport considerations were the relatively short sea-crossing from Cornwall, the navigation of the

River Tawe which enabled sea-going vessels to enter the coalfield and the positioning of copper refineries on riverside sites easily accessible from the navigable river.

The copper-refining industry was not energy intensive and therefore the need for a significant power source was not the predominant requirement in the siting of the works. By the middle of the eighteenth century the multiplicity of furnaces used in the production of copper were reverberatory furnaces which did not require a powered mechanism to provide a blast.

Fig. 3. (opposite page) 1/10,000 scale map of the geology of the lower Swansea Valley coalfield. The rise of Swansea as the copper-smelting centre of the world was facilitated by its having the nearest easily accessible coal seams to the copper-mining areas of Cornwall and Devon. Major collieries included: 1. Pentre (Gethin) Pit 2. Pentre (Penyfilia) Pit 3. Townsend's Steam-engine (or Calland's) Pit 4. Townsend's Glandwr Pit 5. Landore Pit 6. Plas-y-marl Pit 7. Graig Pit 8. Copper Pit 9. Tir Canol Pit 10. Pwll Mawr (pumping) Pit 11. Pwll Mawr (winding) Pit 12. Marsh Pit 13. Double Pit 14. Six Pit 15. Church (or Charles) Pit 16. Round Pit 17. Park Pit.

Fig. 4. Rows of reverbatory furnaces were required in the copper industry to facilitate the repeated roastings of ore with the large quantities of coal needed to produce the refined metal. The furnaces were tended day and night by skilled artisans pictured here at the Hafod Copperworks ladling the refined copper into cast-iron moulds. (City & County of Swansea: Swansea Museum Collection)

Fig. 5. The coal magnate, Richard Calvert Jones, became one of the pioneers of photography in the 1840s. In this scene several sailing-ships, including the 'Countess of Bective' and the 'Mary Dugdale' have brought copper-ore into the old medieval port of Swansea and await loading with a return cargo of coal to fuel mine pumping-engines. (National Maritime Museum, London)

Fig. 6. Many of the early coal-mines at Swansea were mining tunnels which have usually collapsed. One mining tunnel remains open in the lower west side of the Clyne Valley on the bank of the Clyne Wood Canal. The bed of the canal is visible in the foreground. (960206/4)

Fig. 7. Coal-shaft mounds in the Clyne Valley. The mounds consist of a depression indicating the position of the collapsed shaft surrounded by a ring of spoil. (920048/7)

Fig. 8. A detail of a 1736 map showing two of the coal-pits on Tre-boeth Common noted as belonging to Mr. Lockwood (Lockwood, Morris & Co., copper-smelters), later deepened as the Pentre and Cwm Pits, and on enclosed land a third pit belonging to Mr Herbert. (National Library of Wales, Badminton Collection, Group II, 1454)

The Industrial Structure

The key to understanding the development of the industrial landscape of Swansea is an appreciation that the medieval town and its suburbs lie in a narrow valley which overlies successive seams of coal and is bisected by a river which was navigable by eighteenth- and early nineteenth-century sea-going vessels. This coal was the nearest main source of smelting fuel which was easily accessible to the non-ferrous metal mines of Cornwall and Devon, this simple fact explains the growth and intensity of the industrial landscape at Swansea.

Some idea of the intensity of the industrial production that transformed the landscape of the lower Swansea Valley is given by the following tables:

A summary of the output of the lower Swansea Valley in the mid-nineteenth century.

The following summarises the average annual turnover of raw materials and metals in the valley area in the mid-nineteenth century at a time when the water transport system and its associated railways was at its height.

Output and input of the non-ferrous metals and associated coal-producing area (about 550 calciners and furnaces in use).[1]

21,940 tons of refined copper produced
450,000 tons of coal used in copper-smelting and refining
165,000 tons of copper-ore used from Devon and Cornwall
35,000 tons of copper-ore used from Ireland
47,611 tons of copper-ore used from other countries.

Coal shipped from Swansea 507,955 tons.[2]

Britain (with production concentrated on Swansea) came to dominate the largest established producers of copper from the eighteenth century. The figures represent the annual output in tons.[3]

	1530	1560	1650	1712	1770	1800	1850	1880
Germany (Mansfeld)	2,000				400	750		
Sweden (Falun)			3,000	1,000	700+			
Sweden (Ostergotland)					100+			
Britain		100		1,000		7,000	22,000	80,000

The overall international comparisons for the copper industry are just as spectacular in emphasising the dominance of British copper production (and hence Swansea as the world centre of the industry). In the decade 1801-10 international production of copper was 91,000 tons of which the various main national producers of copper produced the following:[4]

United Kingdom:	65,000 tons	71%;
Sweden	7,000 tons	8%;
Norway	4,800 tons	5%;
Germany	3,700 tons	4%;
Others (estimated):	10,500 tons	12%.

Coal and Coal-based Industries Coal

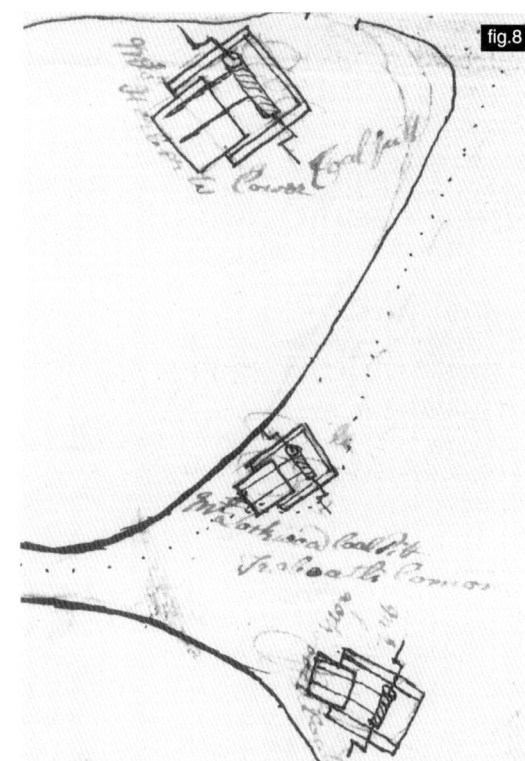

Coal was used as a domestic and light industrial fuel in the Swansea area for centuries before the establishment of the copper industry. The Romans used coal at their south Wales settlements - it has been found during the excavations of Roman Carmarthen.[5]. The earliest reference to coal-mining in the Swansea area occurs in 1306. By the sixteenth century Swansea had a flourishing export trade in coal. By the seventeenth century it was probably the third largest coal port in the country after Newcastle and Sunderland.[6]

No. 1.

A Plan of a
Small Piece of waste Ground
belonging to his Grace
the Duke of Beaufort
at Treboth
Taken Anno Dom: 1768
by Wm Bevan.

A Scale of Gunters Chains

22 Yards to each Chain

fig.9

Reference Table

a Continuation of Treboth Common as it now appears
b Private road to the Coal Pit
c Coal Pit sunk in 1758
d Engine and its Appurtenances now in elevation erected in June 1767
e Ware house, Colliers Lodge and Stable
f Yard for Coal Pit Timber
g Breaks to describe Stones carried in May 1767, towards making the farther Improvements now proposed
h Plan of Head Colliers Houses
i Gardens for Do
k Where the Dots are, shew the extent of John Franklens grant on the West Side
  Content of the whole in Statute measure is One Acre.

Fig. 9. The construction of a new deeper coal-shaft at Tirdeunaw (Tre-boeth) in 1758 necessitated a change from human to horse-powered winding. In 1767, the circuit of the horse powering the engine (or 'gin') was enclosed in a circular wall. Gearing supported by the tops of the wall transferred power to the top of the small circular shaft set in a gap in the surrounding walls. Other breaks in the wall led into a warehouse, colliers' lodge and stables. (National Library of Wales, Badminton Collection, Group II, 1282)

Fig. 10. The leat feeding John Morris II's New Mill in the Clyne Valley was extended southwards down the east side of the valley to drive the water-powered winding gear of the successive Ynys Collieries. It is a rare survivor of the many watercourses that powered collieries in the Swansea area during the eighteenth and early nineteenth centuries. (920048/12)

Fig. 11. A Pennant sandstone retaining wall carries the later Ynys Colliery watercourse around the capped shaft of the first Ynys Colliery in the Clyne Valley. The site has been buried under the fringe of a rubbish tip since being photographed. (960205/1)

fig.10

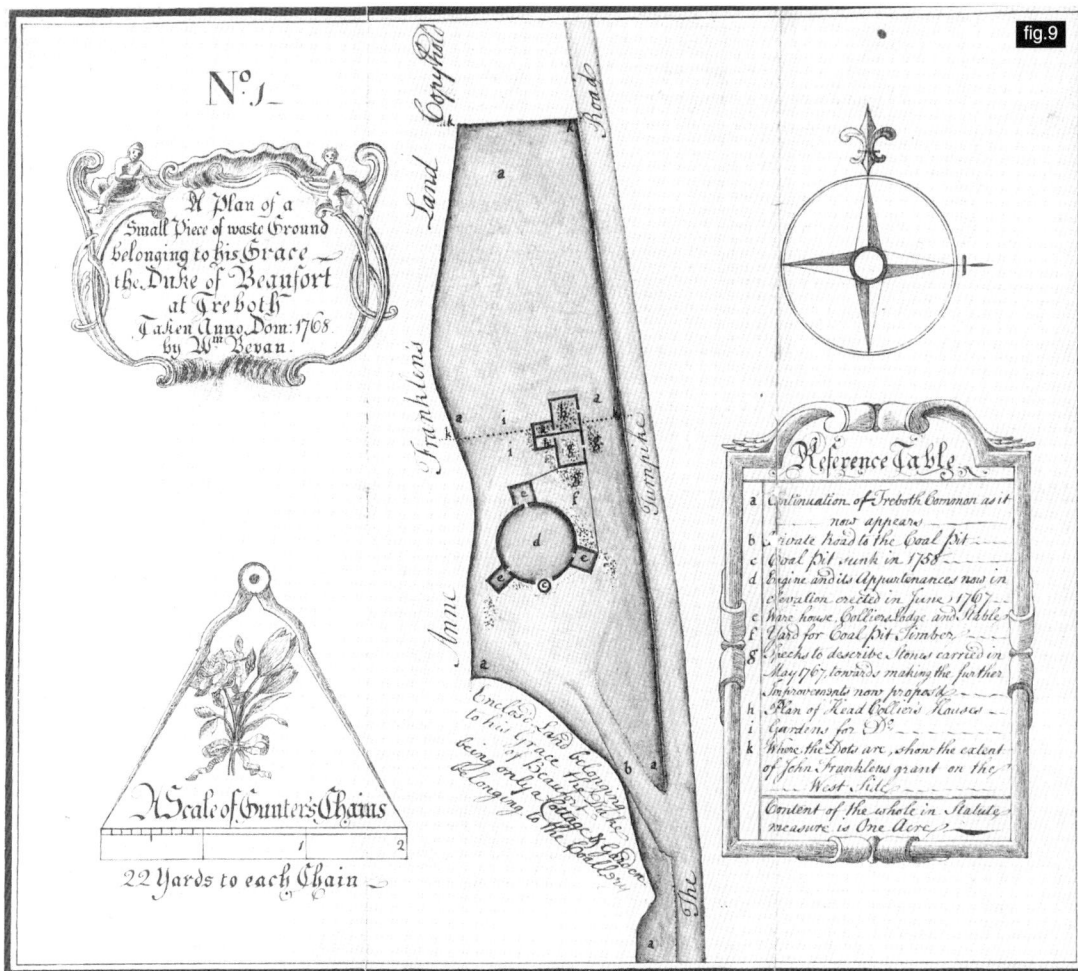
fig.11

The existence of cheap and readily accessible coal was a major factor that contributed to the location of the metal-smelting industries in the Swansea area, and in turn these industries stimulated demand for coal. Output of coal in the lower Swansea Valley probably peaked at the end of the eighteenth century. During this period the industry was dominated by two industrial dynasties, the Morris family on the western side of the valley and the Townsend/Smith family on the east. Initially, both families ran integrated smelting and mining businesses but by the end of the century they had pulled out of smelting and preferred to concentrate their efforts on supplying coal to independent smelting companies.

The geology and topography of the coalfield in the Swansea area is such that in many cases coal was accessible by mining tunnels - gently rising drainage and transport 'levels' or downward-sloping 'drifts'. Primitive methods of winding, pumping and ventilation prevented the use of deep shafts before the eighteenth century. One of the earliest ways of working coal was by multiple shallow pits (probably with short tunnels from the pit bottom)[7] and examples can be found in the Clyne Valley and in the Cwmllwyd woods to the west and north-west of Swansea. A vivid illustration of this has survived depicting multiple well-like windlasses dotting the landscape of the common land at Landore to the north of the city.

By the beginning of the eighteenth century, and probably before, it had become necessary to sink deeper shafts for the coal. The lighter task of winding coal could still be achieved using horse power. By the 1760s some had sophisticated horse-engines ('horse-gins'); one of these at Tre-boeth had the shaft perforating an adjacent circular walled horse-walk, with the wall further interrupted by the colliery stores, stable and colliers' hut. By 1767 the head colliers of this colliery owned by the copper-smelters (Lockwood, Morris & Co.) lived in purpose-built semi-detached cottages alongside. Three additional workers' houses were planned but never constructed.

These pits required to be pumped and as early as 1717 an 'engine' is known to have been at work on an unidentified site operated by Lord Mansel on the eastern side of the Tawe. There are grounds for assuming this was an atmospheric engine powered by steam: Mansel was probably the wealthiest of the local land and colliery owners with pits and adits (i.e. tunnels) producing some 80-100 tons a week in busy periods and with tunnels 600 fathoms in length. He clearly had the finances and need to introduce new steam-pumping technology (Newcomen's atmospheric-pressure engine had been used in coal-mines since 1712) and an estimated annual upkeep of £120 must mean that this was such an engine.[8] It was the eleventh recorded beam-engine built and was probably the second or third built in Wales, the first being at the Woods Mine near Hawarden in Flintshire in 1714-15,[9] while another at the Talargoch lead mine at Desert in Flintshire, may have been at work as early as 1714.[10] The most common cylinder diameter for these early engines was only 16-17in. pumping from about a depth of 150ft.[11] Water-wheel driven pumps were used until the 1840s in the lower Swansea Valley and in the Clyne Valley to the west. The water channels, or leats, feeding the water to these wheels were often very extensive: Chauncey Townsend's of 1757 on the eastern side of the lower Swansea Valley was some three miles long. Considerable remains of these channels still mark the open spaces between Swansea's suburbia. The Landore Pit water-wheel of 1754, at least, was a combined pumping and winding installation.

Deep steam-pumped shafts were constructed by the last of the local squire-coalmasters of the Swansea area. Coalbrookdale Ironworks (in Shropshire) supplied a 27cwt cylinder (bottom 5cwt, suggesting a diameter of about 30ins) to J. Griffith at Swansea in 1731 (James Griffiths had been the leading partner in the Cambrian Copperworks until c. 1730).[12] This was the ninety-second recorded atmospheric beam-engine built (from 1720-21 the engines had begun to be exported outside the British Isles) and probably the fourth recorded engine in Wales (another having been built in Flintshire in about 1719).[13] The spread of the use of steam-engines to other areas of south Wales came later. When Gryffydd Price of Penlle'r-gaer House died in 1787 he left an old 'Fire Engine' (probably the 1731 pump) as well as its succeeding 'New Machine' valued at £3,000 located at his colliery in what is now the Manselton area to the north of the present city centre.[14] It was left to the new copper-works/colliery concerns to bring in the capital to build steam-engines for the rich Swansea-Six-Foot seam by the 1750s.

This first use of steam-engines at Swansea on a large-scale is obscure. Lockwood, Morris & Co on the west side of the River Tawe had the resources necessary but probably did not invest the considerable sums required until after the building of their new copper-smelter at Forest in 1747-52. They were the first copper-smelting concern to begin coal-mining on their own account after leasing Herbert Mackworth's Trewyddfa Colliery in 1728. That the early eighteenth-century colliery infrastructure was relatively primitive is suggested by the recorded simple well-head type windlasses over multiple pits on Tre-boeth Common and around the Pentre (Penyfilia) and Cwm Pits in 1736. The larger collieries on the same Swansea Six-Foot coal seam towards the floor of the valley also had to pump the 'dip' of this seam dry.

The '£100 engine' that John Morris I was considering erecting at Plasmarl (i.e. Plas-marl or Graig) Coalpit in c. 1730[15] may have been an elaborate horse-engine ('gin') like the one erected in 1767 at the Tre-boeth Pit of 1758. Alternatively, it may have been similar to the winding and pumping water-wheel installed on the Swansea Six-Foot seam in 1754.[16] The London merchant Chauncey Townsend on the east side of the valley also had the resources necessary to construct steam-engines when developing his Llansamlet coalfield in the 1750s but he did not build his first one until the early 1760s.

Fig. 12. Photograph showing a figure standing in the sunken remains of the eighteenth-century Colt Pit; a railway ran around the far lip of the shaft. The expanding Winch Wen Industrial Estate has now obscured these remains. (960211/3)

The only reliable source for information on the building of steam-engines in Swansea in the mid-nineteenth century runs from February 1741 to January 1769.[17] In this period local engineers had to assemble their own atmospheric steam-engines with iron-castings supplied by the Coalbrookdale Iron Company of Abraham Darby in Shropshire. These were brought down the River Severn and through the hands of the Bristol merchant Thomas Goldney III, who was a shareholder in the Coalbrookdale Company and its Bristol agent: Goldney's records survive.

In November 1754 a cylinder of some 30in was supplied from Coalbrookdale to Robert Morris (managing partner of the Lockwood, Morris & Co. smelting concern) and other engine parts followed in 1755.[18] This is likely to have been for his Plas-marl Coalpit on the Swansea Six-Foot vein: his large Landore Colliery was still pumped by a big water-wheel.

The London merchant Chauncey Townsend built his large new non-ferrous metals smelters on the east side of the River Tawe in the 1750s and started developing his extensive collieries in order to feed these coal-hungry works and to use for export. Goldney's records seem to show that Townsend received the parts for three engines from Coalbrookdale in the years 1763-66,[19] which fits with the three we know about from other sources. These were distributed around the four main coalfields where he acquired mineral leases: Eaglebush, Neath, in 1748-49; on the contiguous estates of Gwernllwynchwith and Briton Ferry at Llansamlet in 1750; at Llanelli in 1752 and at Pwll-yr-Oer, Landore, in 1762.[20] Townsend's engines were larger than the early one supplied to Lockwood, Morris & Co. and one of them was very large indeed.

Townsend's Engine Pit (at the shaft later known as Calland's) at Landore was sunk in 1762.[21] This was sunk down to the same area of the Swansea Six-Foot seam reached by John Morris I's nearby water-wheel pump of 1754 at Landore and his Plasmarl steam-pump of 1754-55. A delivery worth £359 was made from Coalbrookdale to Townsend in 1763; that is a sum large enough for a substantial steam-engine with parts.[22] The next delivery made to Townsend was of a 40in cylinder (10ft long) with piston and -bottom, working barrels, pipes etc, delivered in 1764, the cost being £194.[23] In 1765 Townsend started his capital-intensive development of the Llanelli coalfield and in 1766 started sinking his 240ft deep shaft down to the nine feet thick 'Swansea Five-Foot Vein' there. The engine would have been in operation in 1768 'but for a rascall of a Mason who built the houze so ill it fell down.'[24] Subsequently Townsend concentrated (and reputedly exhausted) his resources in developing the Llansamlet Coalfield.

Townsend's largest recorded engine was delivered in 1766 with a 65in cylinder (10ft long) and cost the large sum of £432. This was followed in 1767-68 by six separate consignments of boiler-plates with pit barrels, a steam-vessel and other small items which cost £520. By 1770 he had at last located the Swansea Six-Foot coal-seam towards the southern edge of the Llansamlet Coalfield by borehole. The 'Great Pit' (Pwll Mawr) was commenced and sunk 450ft to the vein with the 65in pumping-engine erected above it. A second shaft, the winding or 'coal pit' 498ft deep, was sunk in the ground in 1772 and was probably wound by a water-wheel operated by Townsend's great water leat. The shafts operated

until 1831-32 and the tall engine-house remained for years afterwards, being sketched by the first surveyors for the Geological Survey in c. 1849. Now only a shallow depression remains in the ground at the pumping-shaft with the remains of a second beam-engine house of 1881 at the original winding-shaft up the slope of the valley scarp.

The Pwll Mawr engine was large for its time but was not the largest then built. Fifty Coalbrookdale cylinder castings were supplied through their agent (Goldney) at Bristol, who serviced Wales and the south-west of England, between 1741 and 1769. The deep Cornish tin and copper-mines (fuelled of course with Swansea coal) generally took the larger cylinders which became progressively bigger in this period. However, in this period there were only six engines of 65in cylinder diameter or above starting

Fig. 13. One of several steam-engines at Swansea that were built to a large size with cylinders 65-66ins in diameter. This atmospheric pumping-engine was necessitated by Townsend's deep driving to find the elusive Swansea Six-Foot seam at Pwll Mawr. (W.E. Logan's Survey Book No. 11, 1837-42)

Fig. 14. Aerial photograph showing the engine-bob wall of a late nineteenth-century pumping-engine at Chauncey Townsend's 'Great Pit' (Pwll Mawr) to the left of the two railway lines at the lower centre. A track on the nearside, roughly parallel to the left-hand railway, is the former course of Chauncey Townsend's Railway. (935054/55)

Fig. 15. The twin shafts of Chauncey Townsend's Great Pit. The collapsed top of the lower shaft in the foreground was where his large engine house stood. (960214/3)

ENGINE-HOUSE AT GWERNLLWYNCHWITH, LLANSAMLET

Fig. 16. The deeper coal-shafts of late eighteenth-century Swansea required greater winding-power than was available from human, animal or water-driven sources. Close to the M4 at Gwernllwynchwith (Llansamlet) is what seems to be an engine house built between 1772 and 1782 to enclose a long winding-engine. This building of international significance has partly collapsed since the 1970s.

Figs. 17 & 17a. The second-hand winding-engine at 'Plas Y Marl' (Plasmarl) colliery, converted from a pumping-engine erected a few years previously at the Cornish copper-mine of Wheal Towan. The site of the shafts lies between the old Neath road and Ffordd Cwm Tawe at Birmingham Mount, Plasmarl. (Birmingham City Archives; Boulton & Watt Collection Box 2/18/2)

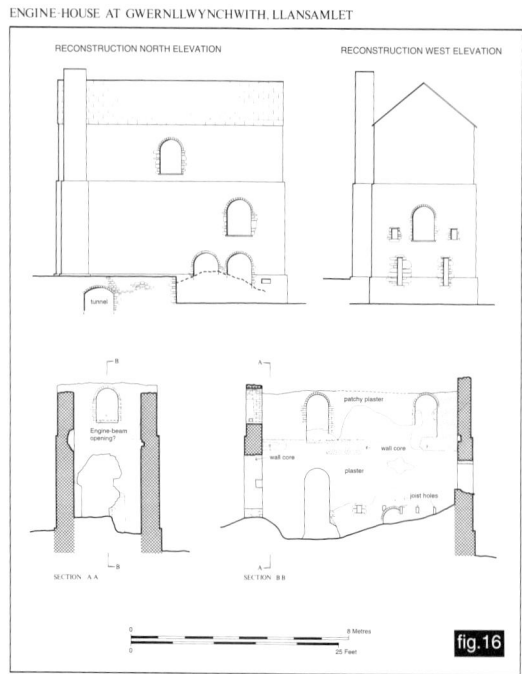

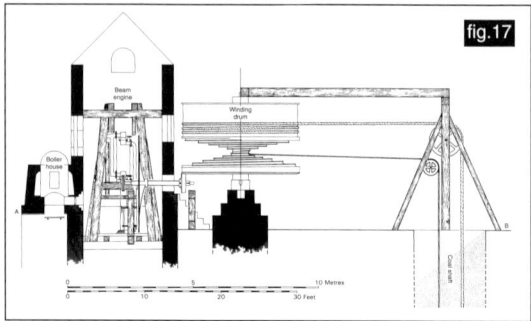

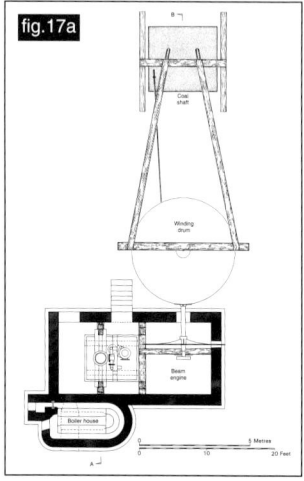

with the 70in cylinder produced for the Herland mine in Cornwall in 1753 while two large cylinders of 66 and 75ins were delivered for use in the great Newcastle Coalfield in 1765.[25] Pwll Mawr was then almost at the forefront of the development of deep steam-drained coal-mining with the largest coastal coalfield at Newcastle upon Tyne. In south Wales it had no equal.

As the shafts became deeper horse-engines were insufficient to power effectively the haulage from larger mines centred on a single pithead. It is often generally assumed that rotary steam-engines, capable of colliery-winding or of driving a factory, were not evolved before James Watt fitted his condensing-engine with a crank and flywheel and applied it to colliery winding in 1784.

Evidence exists, however, for a widespread adaptation of atmospheric engines for colliery winding[26] and for a general early application in the Swansea and adjoining Llanelli coalfields. The general historic view may have been distorted by the great (mainly civil) engineer John Smeaton, who insisted that the only effective way of making the Newcomen-type of engine effective in producing rotary motion, was to make the engine pump water back from the tail-race to the head-race of a water-wheel. Smeaton and other nationally known 'eminent engineers' were convinced that the non-rotative engine could not be made rotative by the simple addition of connecting rod and crank because its stroke was variable. The actual experience of lesser-known engineers working in each coalfield may have been quite different.

The simple addition of a crank to an atmospheric engine, in fact made the length of each stroke controllable. The theorists believed that the variable stroke of the atmospheric engine would smash any form of positive connection. For this reason, many of the efforts made to provide connections for rotative drive made use of more complex toothed racks, ratchet-wheels, pawls and gear wheels.[27]

There is evidence in the Swansea area for the use of steam-powered feeds to water-wheels as well. Within Wales there is the possibility of a much earlier pump in use to supply water to the upper floors of the Yellow Tower of Gwent at Raglan Castle but the first firmly recorded pump using steam was that patented by Thomas Savery in 1698. This was a simple atmospheric pump which created a vacuum in twin receivers. It had no complex working parts, or beam, but created vacuums in twin boilers and receivers placed on the pipes of a vertical pumping-line. Unfortunately it only had a lift of 20ft and so was unsuitable for effective mine pumping.[28] The only successful application of the Savery-type machine involved the use of a sucking stroke from below and not the forcing stroke where cold water and steam came in direct contact and rendered the machine ineffective.

In this simple type of machine the steam in the receiver was condensed by cold water injection and was thus drawn-up by the vacuum created and then discharged into a cistern where it was used to drive an overshot water-wheel. Joshua Rigley of Manchester built a number of these simple machines in Lancashire in the 1760s, in which the steam and water injection valves were actuated from the water-wheel shaft and in which the wheel also drove a small force pump to supply the injection water.[29] A number of these were used, or even built, in the nineteenth century.

What is interesting is that 'Mr. Powell', John Morris I's blacksmith-engineer in *c*1763-83, had designed and built such an engine at Landore, to the north of Swansea town. Close to the Landore Coalpit water-wheel winder and pump of 1754, and described by an agent of Boulton and Watt in 1783,[30] was: 'a small but curious Engine to throw the water when it is scarce and they are short of it - They tell me that it is of singular simple Construction his own Invention and without some customary movements but I do not know enough of the matter to describe it.' This augmented the feed to the double water-wheel which also received mine-water pumped-up from the adjacent steam-pump.

There is definite evidence of a widespread use of atmospheric steam-engines rotary-engines in the area. We should not be surprised at this when wider evidence of early mine winding-engines is emerging. George Richardson, a Cumbrian colliery engineer from Whitehaven, visited London in 1734 in an attempt to obtain a patent for raising coal using a Newcomen-type engine. He then advertised his machine in a broadsheet. Drawings exist of the mine-winding adaptation which Marten Triewald intended to make in 1730 of an engine exported to Sweden, and this may have been based on his earlier experiences at mines in Britain.

A patent ratchet system was used by Joseph Oxley at Hartley Colliery in Northumberland in 1763, another by John Stewart in a Jamaican sugar mill in 1766-68 and a third by Matthew Wasborough, a Bristol engineer, in 1779. These unsuccessful devices were superseded by the rod and crank with which James Pickard, a Birmingham button maker, made his Wasborough engine perform successfully. The engineers of this latter engine carried this engine over its centre by the use of a heavy flywheel mounted on the crankshaft itself.[31]

This provides a perspective on a reference to Chauncey Townsend's engine of 1766-68 at Genwen in the Llanelli coalfield when it was reported in April 1769 that some person had sabotaged the engine by breaking the flywheel and throwing stones into the pumps.[32] Orthodox Newcomen-type beam pumping-engines did not have flywheels. It is possible his 1763 engine at Landore may have had one too.

In 1763 'Mr.Powell' had arrived as John Morris I's blacksmith and soon after had started designing and building steam-engines. When Matthew Boulton's agent arrived in 1783, Mr Powell had just been scalded to death in a maintenance accident but the former was able to report on Powell's rotary engines which were in successful daily use:[33]

'And the Otherside the [Landore Colliery] Engines, upon one other pitt he has erected a small steam engine that an apparatus very singular of his Invention is fix'd that draws up the Coal and only one man necessary to receive and deliver the Coal and attend the apparatus and one other to the Engine for the coal being drawn up the basket strikes a small contrivance that shifts the barrel itself. lowers down the basket on the Bank - the man having only to empty the basket and return it to the mouth of the Pitt, when by his pulling down a handle the Barrel is thrown into contact with a reverse wheel and it descends again, but its movements are made by the beam - the most ingenious part being that Whether the beam ascends or descends the Barrell is made to go round in one direction.' This early version of a 'steam-whimsey': the commonest type of steam-engine in the early nineteenth century, seems to have been very successful: 'this small Engine draws 120 Tons of Coal in 24 hours'.

Mr Powell also seems to have built, or adapted, one of the large pumping-engines on the Plasmarl pits to function as a winder: 'another apparatus of the same kind is fix'd to the beam of One of the large Engines over another Pitt and draws about 130 Tons of Coal in 24 Hours.' These were 40 and 46in diameter engines pumping from depths of 180 and 240ft respectively.

Substantial remains of the Landore Pits have been revealed by recent reclamation work but have now been covered over again. On the Llansamlet coalfield there remains, at Gwernllwynchwith, the substantial remains of an engine-house that by its form must have enclosed a small rotary engine (dated from before 1783).

However, the atmospheric non-condensing engine, in all its forms, had a limited capacity. Such engines did continue in use in the Swansea area until the 1840s but the increasing depth of the pits meant that more sophisticated machinery had to be employed. By the 1770s the coppermaster John Morris I had already visited Matthew Boulton's manufactory and was sending repeated letters asking to be able to introduce separate condensers to make his existing engines more efficient, and to be allowed to do this at a reduced rate for being the first to introduce the improved steam-engine to the south Wales coalfield! Not surprisingly a protracted correspondence ensued. It became a desperate battle to de-water the Swansea Five-Foot seam at Landore Colliery, as Townsend had stopped his pumps working at his nearby engine-pit. Morris's two large steam-pumps at Landore and Plasmarl Pit were working 24 hours

Fig. 18. The low-pressure Boulton &
Watt steam pumping-engine at the
Landore Coalpit of the copper-smelters
Lockwood, Morris & Co. was the
largest yet built at Swansea. The
Pennant sandstone engine house stood
in 1948 but was subsequently
demolished and buried beneath
landscaped waste. The site is under
open ground east of the Cwm Level
Road/ Ffordd Cwm Tawe roundabout
at Landore. (Birmingham City
Archives; redrawing of Boulton & Watt
design drawings by Hayden Holloway)

Fig. 19. After 1875, new colliery
winders took the form of horizontal
engines in totally enclosed engine
houses (B), often replacing traditional
vertically-cylindered beam-engines in
partly-enclosed engine houses (C). At
Mynydd Newydd Colliery to the north
of Morriston, the traditional stone-
tower headframe was retained and
raised. No structures are visible on the
site. (Redrawn from an original
drawing; University of Wales Swansea,
Library & Information Services,
Archive, T. Bryn Richard Collection)

a day with automatic feeds to their ranges of boilers.
They could not even be stopped for minor repairs
which explains how Mr Powell was scalded to death.
Eventually the pit was drowned in 1794.

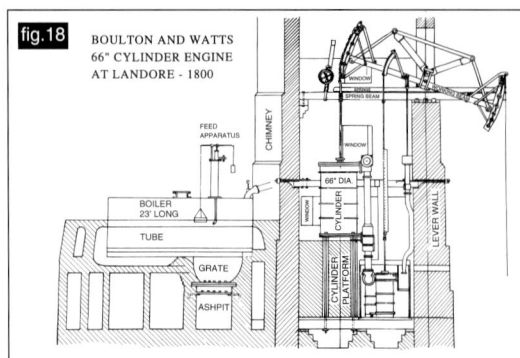

The 'stupendous engine' noted by tourists at
Landore was pumping the Landore Pit dry again
from 1800. This was a Boulton and Watt pumping-
engine with a 66in diameter cylinder, i.e. it was of
almost identical size to Townsend's Pwll Mawr
engine of 30 years before. A second-hand Boulton
and Watt pumping-engine from Cornwall had
already been re-erected as a steam-winding engine at
nearby Plasmarl in 1785, one of the first of his
engines so used.

Another means of extracting the coal, which may
have been pioneered at Swansea, was by boat level:
the earliest example known was built by Robert
Morris at his Clyndu Level, the site of which is now
covered by a large traffic island at the south end of
Morriston. This may well have been built in the
1770s although one other underground canal in south
west Wales may have pre-dated the famous Worsley
(near Manchester) navigation levels of 1759. The
best surviving example at Swansea, however, is
comparatively late, at the Rhydydefaid colliery in the
Clyne Valley, and is dated to about 1840.[34]

Other coal-based industries

Swansea may have become known as
'Copperopolis', but the same availability of large
coal reserves within easy reach of the sea which led
to the spectacular rise of that one industry also led to
the growth of others. In the words of the Swedish
traveller Eric Svedenstierna when visiting Swansea
in 1802-3:[35]

> 'The same seams supply several alum works,
> porcelain factories, and a host of other coal-
> using establishments with coal, besides what
> goes to limekilns and the household
> requirements of the continually growing
> population in the towns and in the country round
> about ...'

The Alum Industry

The availability of pyrites in the shales (aluminium
silicates) of coal seams, and in existing coal-tips, led
to an attempt in the seventeenth century to found an
alum industry in the Swansea area. An Alum Works
was built by George Owen, who leased 'mineral
waters' and 'minerals' from the Duke of Beaufort on
23 December 1735. The works were founded on the
western hillside of the Swansea Valley above the
later site of Morriston, and the area of the now-
buried remains are marked by the proximity of the
appropriately named Bath Lane (see below) in the
area of Cwm Bath.

Alum manufacture had become increasingly
important from the sixteenth century onwards as a
mordant for fixing dyes in what has been described
as Britain's 'first chemical industry.'[36] The British
industry became focused on a range of successful
works on the coast of north-east Yorkshire using
upper lias shales. Other scattered coastal sites in
Britain, such as that at Swansea, were short-lived
and unsuccessful. What is particularly interesting
about the Swansea Works is that it seems to have
been a precursor of the successful early nineteenth-
century alum works of the south Yorkshire and
Lancashire coalfields, that used the coal measure
shales to be found in colliery waste.[37] Its site, near
the top of the valley scarp at Morriston, was adjacent
to the old mining-tunnels ('levels') which penetrated
the outcrop of the thick Swansea Five-Foot Seam
that ran along the crest of the valley scarp.

The first stage of the production process involved
shale, brushwood and small coal being piled onto
(flued) clay calcination bases and ignited. The
burning mound was then often built up to some 50ft
and then sealed-in with a layer of wet clay or shale to
continue burning for some months. During
calcination, iron pyrites in the shale was oxidised to
iron sulphate and in turn reacted with the alumina to
form aluminium sulphate and red iron oxide.

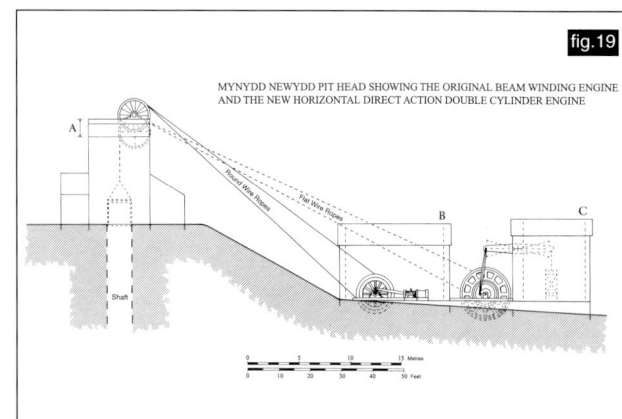

fig.20

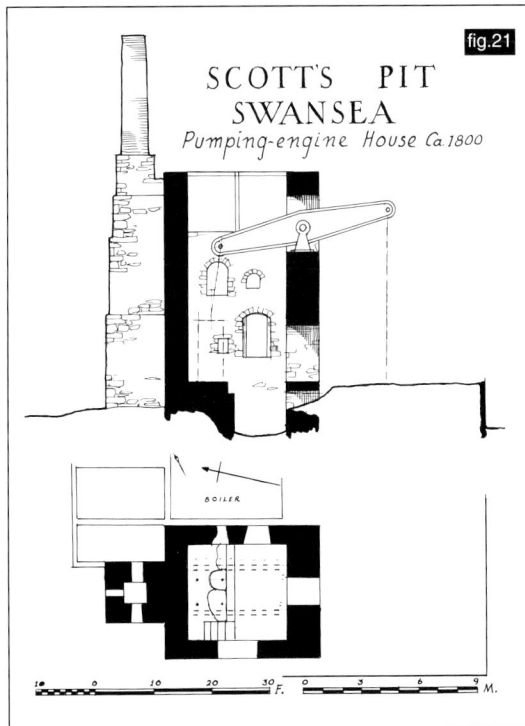

fig.21

SCOTT'S PIT
SWANSEA
Pumping-engine House Ca.1800

BOILER

The shale residue in the tanks was washed at least three times with the strengthening resultant 'liquor' and lastly with fresh water before being discarded: the process lasted up to two weeks.

The raw liquor consisted of a solution of aluminium sulphate, iron sulphate, magnesium sulphate and traces of other metallic sulphates together with any unreacted sulphuric acid.[38] It was stored in a raw liquor cistern and fed to the alum house by gravity via a wooden or stone-lined liquor trough.

The seventeenth-century alum house was normally located on the coast or at the base of an inland scarp to facilitate the transport of fuel, alum and other materials. The situation at Swansea was slightly different in that the coal fuel needed for calcination and for evaporating the liquor was available on site. The building was below the steepest section of the valley-side scarp where coal was extracted, but away from the riverside waterfront and the lower scarp sloping down to it. Access was presumably obliquely down the lower scarp on the line later used by John Morris for his Clyn-du coal road. The expense of using pack horses along this route may explain the failure of the works.

Fig. 20. Scott's Pit Engine House at Llansamlet has been conserved and survives alongside the M4. It was built in 1817-19 to house a low-pressure Boulton & Watt pumping-engine and was used with that engine until c. 1842. In 1872, a new high-pressure pumping-engine was fitted in the tall engine house and remained until 1930. The bolts of a winding-engine are in the foreground. (881162/5)

Fig. 21. The plan of Scott's Pit in 1976 shows the location of the 500ft deep shaft which is now capped in concrete. It also shows the site of the c. 1872 Cornish boiler which replaced an earlier circular-based 'haystack' boiler.

fig.22

Burnt pink shale was then placed in stone, wooden or lead-lined 'steeping-pits' so that the valuable soluble salts could be leached out of the burnt red shale. A cistern, or reservoir, often fed up to 12 steeping-pits, so arranged that the solutions could either flow by gravity or be pumped from one tank to another. It is the probable survival of these after the works closed that led to the local name 'Cwm Bath' and hence 'Bath Lane' in the area above Morriston.

At Swansea the main range of the works was aligned south-west / north-east across the slope of the valley-side. Two subsidiary wings extended downhill from either end of the main range. In such seventeenth-century works the liquid from the circular, covered, raw liquor cistern was fed into the first of the works buildings. In some works this was the 'clearing house' where the liquid was rapidly boiled in shallow lead pans above a coal fire. The

Fig. 22. An early twentieth-century steam-powered haulage-engine lying abandoned in Coed yr Ysgol at the north-west end of the Clyne Valley Country Park. This pulled trams from an inclined entrance tunnel at Clyne Wood Colliery. (920047/12)

resultant concentrated solution was then passed into a liquor settler cistern, where iron oxide and silicate impurities were precipitated out, and then on into the cleared liquor cistern.

At Swansea what may have happened was that the raw liquor immediately entered the long main range, housing a line of lead evaporating pans that were supported on iron plates held above firebrick-lined hearths which were fuelled with small coal, and the liquid boiled for between 24 and 48 hours. Urine (ammonia) or kelp lees (potassium derived from leached seaweed ashes) were then added to the solution to produce either ammonium or potash alum.

The two lower wings at Swansea, that could have had liquor fed-in by gravity, were probably the settling and cooler houses. In the 'settling house' the partially evaporated liquor was fed from the main 'boiling house' into a lead settling cistern where the remaining iron sulphate, calcium and magnesium sulphates were deposited. After two hours this solution was transferred to the 'cooler house' and left for four to five days in stone-built tanks where alum crystals formed around the edge as the crystals closed. The residual liquor was then transferred back to the evaporating pans.

In the late eighteenth and early nineteenth centuries this works would have consumed substantial amounts of the coal produced on the western side of the valley. By 1864 it was abandoned and the buildings were in ruins.

The Fire-clay Industry

The Swansea-area coal-seams also yielded low refractory fire-clay which was used to make copper-furnace linings and firebricks.[39] Higher refractory materials, capable of resisting a higher temperature, had to be obtained in the early period from Stourbridge, in the west Midlands of England, and later largely came from the silica-brick works at Dinas at the head of the Vale of Neath.

The White Rock Copperworks, the third built in the lower Swansea Valley (1737), is recorded as having a clay mill (i.e. a pug mill) that would have been used in the preparation of local fire-clay. A detailed map of the hillside adjacent to the Forest Copperworks (1747-52) shows three 'old clay pits' flanking the coal-road descending from Tirdeunaw and Tre-boeth Collieries.[40] A 'Brick Yard' was sited on the same Graig Trewyddfa Common to the north-west of the copperworks. The third operator of the Upper Bank Zincworks (1754-75) is also recorded as producing refractory bricks as a sideline. [41]

This trade increased with the number of furnaces used in the copper and other non-ferrous works. Further north by the mid-nineteenth century a fire-clay mill (pug mill), powered by the Swansea Canal, was built alongside the railway bringing coal from the southern Trebannws Levels and a wheelpit remains on site.

After 1798 the Swansea Canal traffic books show that silica-sand was being brought down to Swansea from the head of the canal and allowed the Swansea smelters to make their own high-refractory bricks and furnace bottoms. A larger brickworks, the Morriston Refractory Brickworks, producing high-refractory bricks using silica from the head of the Vale of Neath, was built by the end of the nineteenth century on the hillside of Craig Trewyddfa, in order to supply the large local market provided by the non-ferrous metals smelters.

The industry was a further substantial consumer of coal in the Swansea area.

Miscellaneous Land Uses of Coal

In the first half of the eighteenth century, before the building of copper and non-ferrous smelters induced a profound change in the markets for the coal produced, the production of coal in the Swansea area was split between land and sea sale. Thomas Price, the squire/coal proprietor of Penlle'r-gaer, in 1732-33, was sending 85% of the produce of his Trewyddfa Pit for sale via sea transport. Only 15% was going for local land-sale, i.e. domestic, smithing, lime-burning and non-smelting industrial use. Similarly, in 1751-52, in the Clyne Valley to the west of Swansea in 81% of output was being shipped from Blackpill Quay and 19% was going for local land sale. In 1753-54 the figures were 79% and 21%.[42] However, some pits were still producing for this market even at the height of the copper-smelting

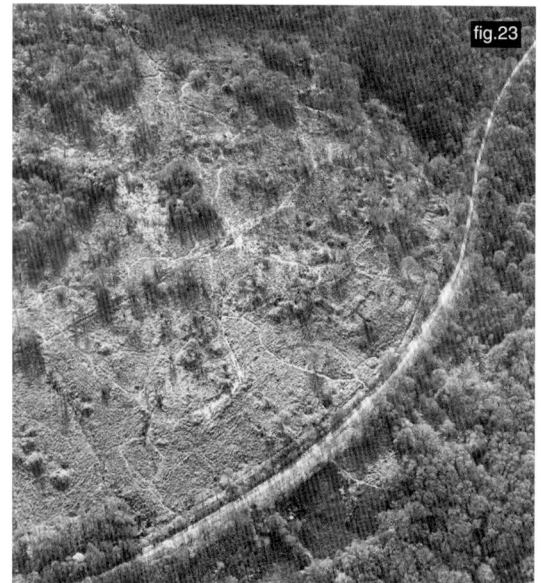

fig.23

Fig. 23. Multiple shaft-mounds of early collieries visible on the west side of the Clyne Valley. Four coal-pits on the outcrop at the south end of the coalfield are visible at the bottom left-hand corner with small spoil-tips downslope. The former locomotive railway through the valley is visible as the white line curving from bottom left to top right. (935054/42)

boom. In 1837 the engineer and agent William Kirkhouse, noted that the Double Pit at Llansamlet was producing for 'country sales.'[43]

Coal-fuelled Industries

The Salt Industry

The first of these industries was the Alum industry noted above, which not only extracted its raw material from the coal seams but also used coal as a fuel to vapourise vitriol held in large tanks. A similar use for coal but this time as a potential fuel for heating pans to evaporate sea-water led to the establishment of an eighteenth-century saltworks on the opposite, eastern, bank of the River Tawe. This had its own small dock on the eastern shore of the large expanse of Fabian's Bay at the mouth of the River Tawe. Coal was used to evaporate the large amount of brine kept in large tanks on the bay foreshore. The saltworks was reported to be out of use by 1796 upon the decease of the proprietor and was apparently not revived because of a property dispute.[44] There may be buried remains as substantial as those that have recently been exposed at Port Eynon to the west of Swansea.

The Pottery Industry

The most famous of the ancillary industries of eighteenth-century 'Copperopolis' was the pottery industry. A view of the significance of this to the local economy has been distorted by the importance attached, by modern collectors, to the fine porcelain products produced at the Swansea Pottery, for a limited period, on a totally uneconomic basis. There was a much larger and longer-lived earthenware pottery industry that involved a number of establishments and was of considerable importance to the local economy. There were also similar trade patterns to the major copper and tinplate industries, in that clay and china stone, as well as copper and tin, came from sources in Devon and Cornwall, and Swansea coal was in turn a desirable return cargo for use in West Country limekilns, pumping-engines and hearths.

The first and longest-lived of the Swansea Potteries was the Cambrian Pottery. Its foundation involved the re-use of a copper smelting building of the same name. This had been built by James Griffiths and Company in 1720. In 1764 William Coles, a partner in Messrs Coles, Miers and Lewis, took the lease. It seems likely that his primary purpose was to re-use the old copper-rolling mills for rolling wrought-iron for use in the partnership's tinplate mills and that the copperworks itself was surplus to requirements.[45]

Coles and his successors in title sub-let (by licence of the Corporation) various portions of the works, as opportunity offered, but Coles also founded the Cambrian Pottery, probably in emulation of the newly prosperous master-potters around Stoke-on-Trent (see below). The main mill building seems to have ceased being used as a metal-rolling mill and in 1776 it was leased by the pottery proprietors and was then reconditioned and adapted to the purposes of the Pottery, but was offered for sale when a new managing company was set up in 1783 after William Coles's death in 1778 or 1779. In fact, the whole Pottery complex and Greenhill Mill, further up Cwm Burlais, were offered for sale at this time, possibly as a formality in valuing the assets of the new company.

By 1790, the pottery was presumably short of capital because John Coles took into partnership the capable Warwickshire businessman George Haynes, who had spent some time trading in Philadelphia and later left the pottery (1809). He described to a Philadelphian acquaintance how he had returned to Britain:

> 'meaning to find himself in a place of retirement with some little business to employ him' but the pottery 'had been very unproductive to the former owners and that he found however on entering into it that instead of being a lot of nice easy employment he was obliged to enter into every part of it and become himself perfectly capable of conducting it.'[46]

A potter had been delegated to run the pottery for the previous 26 years and the products had been modest rough earthenware.[47] The pottery was extended beyond the 50 year old copper-smelting buildings in 1790, when a sum in excess of £1,200 was spent on new buildings within a year of George Haynes

Figs. 24 & 24a. Late eighteenth-century views of the Cambrian Pottery before the construction of the Swansea Canal. The upper view is from the Strand looking north-east from Swansea town. Three large conical kilns are visible on the far right. The lower picture is from the north-west. In the right foreground is the large mill-pool that occupied the mouth of Cwm Burlais, driving machinery in the 'Pottery Mill' on the left, which probably originated as the rolling-and battery-mill for the Cambrian Copperworks. The site is now at the junction of New Cut and Morfa Roads. (By permission of the British Library)

fig.25

fig.26

Fig. 25. Cambrian Pottery pictured from the Tawe River Navigation in 1791. At centre-right the casting-hall of Raby's iron foundry is visible to the left of four pottery kilns. To the right is the large-diameter water-wheel of the mill which may have been in use as a forge (engraving by Rothwell). (City & County of Swansea: Swansea Museum Collection)

Fig. 26. Plan of the Cambrian Pottery in 1802 with four large conical firing-kilns. Coal for firing the kilns was unloaded from the Swansea Canal into a warehouse under chambers 'd-h'. Flint and clay was imported via the river-barge dock.

Key for Figure 26

A, Counting House. *B,* Common Sale Room. *C,* Common Stores. *D,* Dish and Plate Makers. *E,* Black Cellar. *F,* Lath Room. *G,* Flag Room. *H,* Packing Room. *I,* Green-ware Room. *K,* Black Shed. *L,* Store Room. *M,* Mixing Room. *N,* Black Slip Kiln. *O,* Little Gloss Kiln. *P,* Great Gloss Kiln. *Q,* Gloss Placing Room. *R,* Sagger Room. *S,* Dipping Room. *T,* Plate and Dish Makers. *U,* Biscuit and Gloss Kiln. *V,* Biscuit Kiln. *W,* Placing and Sagger Room. *X,* Dipping and Green Room. *Y,* Biscuit and Placing Room. *Z,* Throwing Room.

a, Hardening Room. *b,* Packing Shed. *c,* Shelter Shed. *d,* Private Counting House. *e,* Private Store Room. *f,* Passage. *g,* Joiner's Shop. *h,* Crate Maker's Shop. Great coal Warehouse under the five last. *i,* Crate Shed and Rod Cistern. *j,* Straw Shed. *k,* Engine Lath Room. *l,* Stave Room. *m,* Throwers and Plate Makers. *n,* Crate Maker's Shop. *o,* Store Room. *p,* Printer's Room. *q,* Slip House. *r,* Slip Kiln. *s,* Cisterns or Flint. *t,* Benger's Room. *u,* Clay Cellars. *v,* Ash Hole. *w,* Three Necessaries. *x,* Coal Warehouses. *y,* Clay Yard. *z,* Flint Yard.

1, Flint Kilns. *2,* Flint Yards. *3,* Plaster Kiln. *4,* Modeller's Room. *5,* Sagger Makers. *6,* Great Shed. *7,* Water Closet. *8,* Enamel Dresser's Room. *9,* Muffle Kilns Room. *10,* Enamel Room. *11,* The Muffle. *12,* Passage. *13,* Painter's Shop. *14,* Laboratory. *15,* Printing Room. *16,* Waiting Room, Baths. *17,* Cold Bath, Women's. *18,* Ditto, Men's. *19,* Passage. *20, 21,* Warm Baths. *22, 23,* Dressing Rooms. *24,* Water Closet. *25,* Coach House and Drying Room over. *26,* Coach House. *27,* Stables with Hay and Straw Rooms over.

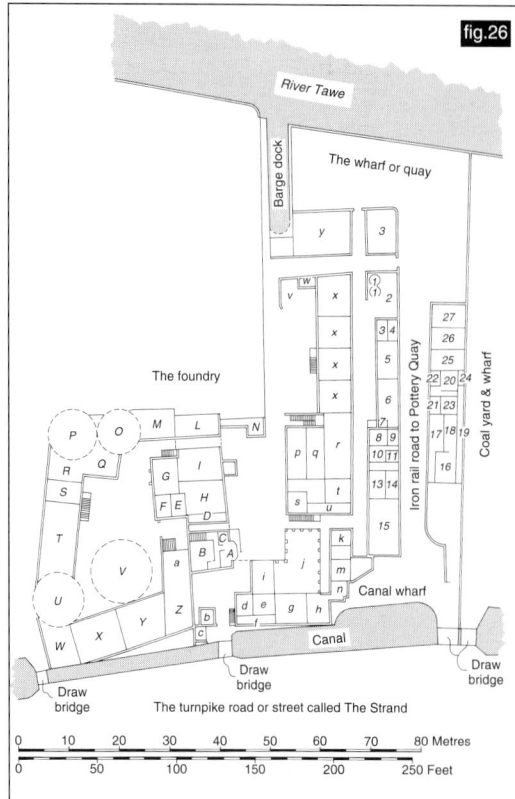

assuming direct management.[48] As well as introducing Wedgwood's ideas for the improvement of production organisation, Haynes also emulated his ideas on the blending of raw materials and in copying the body, colour and design of his creamware ('Queensware' which had received royal approval, caneware and black basalts) using moulded as well as thrown shapes.

Given the proximity of ample coal fuel and the easier access to the fine clay resources of Devon and Cornwall, it might have been thought that Swansea, rather than the inland Stoke-on-Trent area, would become the centre of pottery production in the first Industrial Revolution. It was Haynes who soon lifted production above the level of simple earthenware production and who produced his first experimental soft paste porcelain.

The development of earthenware production in the Industrial Revolution consisted of three distinct phases.[49] The initial phase of production, as in the first 20 years of the Cambrian Pottery, was in the production of earthenwares from local clays. The second period began in 1728 when calcined flint began to be added to British clays in order to strengthen and whiten the yellowish body to a cream colour (the silica so-introduced replaced the earlier use of sand). At Swansea there were at least three water-powered flintmills grinding flints for use in the two potteries, creamware being introduced after the arrival of Haynes at the Cambrian Pottery in 1790.[50]

The flint was imported from London and roasted before grinding.[51] In this second period and at about the same time (the early 1730s) in Staffordshire, the use of local marls was supplanted by Devonshire pipe clay (imported in ball form) which had considerably increased plasticity and improved vitreous properties (its 'green strength' allowed unfired pottery to be stacked high in the kiln). This in turn produced stoneware with a whiter body. These discoveries gave 50 years of great prosperity to some 80 Staffordshire master potters and may help to explain why Coles first went into the pottery business.[52]

The third, or 'industrial' phase of earthenware production dates from the period after 1775 when Cornish clay and china stone were first incorporated with local clays, or substituted for them, to give a fine white translucent appearance which was needed to compete with the newly rediscovered formula for making hard-paste porcelain of the old fine-quality Chinese type.[53]

An indication of the materials required at Swansea from c.1790 onwards was that creamware (as produced by Wedgwood) consisted of 75-85% plastic clay and 15-25% calcined flint or sand. Wedgwood made his own glaze generally available to the industry and it consisted of 30-40% flint or sand, 15-20% carbonate of soda and nitre, fritted together. To this frit was added 35-45% of red lead ground to a creamy consistency in water (red lead was made at the local lead works in the Swansea area).[54] The American industrial observer, or 'spy', Joshua Gilpin noted in 1796 that Haynes was using various mixes of mainly ground flint and south Devon ball-clay to produce his pots.[55]

What spurred on the advances in pottery production in the industrial period was the fusion of extractable resources from Cornwall and Devon with mercantile capital from Bristol, as in starting the British copper/brass industry in the Severn Estuary area and with at least one of the same families involved. In this case it was prompted by the realisation of the inferior quality of the coarse and crude earthenware of Europe compared with the delicate quality and refinement of Chinese porcelain. Haynes and Wedgewood were both involved in radical developments that helped to close the quality gap creating a mass market for their products.

Porcelain needs no glaze and is a fine white impervious translucent ware made from kaolin, a clay made from decomposed granite and fired at 1280° - 1400° (earthenware is fired at a lower 900° - 1200°). A European counterpart to the ancient Chinese porcelain was developed in Germany (Meissen, 1710) and had spread to Dresden, Venice and most of the rest of Europe by the 1750s. In Britain, the Devon man William Cookworthy discovered the kaolin deposits of Cornwall. He patented the hard-paste porcelain process in 1768 and set-up a factory in Plymouth. This was not a simple process to use because unblended kaolin consists of comparatively coarse particles that used alone form a crumbly clay that cannot easily be formed into vessels. Cookworthy's partner, Richard Champion, founded a second, more successful, porcelain manufactory in Bristol. A long and costly court case with the Staffordshire potters ensued when Champion tried to petition Parliament for a fourteen-year extension of Cookworthy's patent. The settlement encouraged an experimental and competitive spirit among Haynes and other British potters when it was announced that other potters might use china clay and stone in the manufacture of their wares, although Champion retained the exclusive right to manufacture porcelain by Cookworthy's process.

One other complementary raw material used by Haynes and the other potters was the south Devon ball clay exported via Newton Abbot.[56] It is a fine material of high plasticity and unfired strength that gains these qualities from its process of alluvial deposition during which it lost most of its original coarse quartz and granite residues. However, the large quantities of coal required for the firing of the kilns ensured that it was only economic to build potteries in the south Wales, Staffordshire, and other coalfields.

Haynes was conscious of the need to improve the constituents of the pottery sold at Swansea from the time he entered the trade in 1790. From that date into the first few decades of the nineteenth century a staple of the production at Swansea was an imitation of Wedgwood's creamware, which consisted of Devon ball-clay mixed with flint ground at the pottery and Greenhill Mills and glazed with a lead mixture. A fine white earthenware was increasingly produced, reportedly first in Swansea, and then in the rest of Britain, bearing the stamp 'Opaque China' to suggest that it compared favourably with porcelain.[57] Many and varied domestic vessels made from this clay mix were produced with fine transferred decoration and are to be widely found in both private and public collections.[58]

Lewis Weston Dillwyn (also a former resident of America) had become a partner in the pottery in 1802. George Haynes's experiments in improving the ingredients of his pottery continued and Dillwyn was sent to Cornwall to obtain supplies of 'china-stone' or 'soaprock' so that Haynes could continue his experiments on the development of porcelain at Swansea. This was to make the more commonly produced 'soft-paste' or 'frit' porcelain which was an earthenware imitation of the Chinese original in which pottery was fired at about 1100°C. Materials continued to be shipped-in from Devon and Cornwall, fired with local Swansea coal, and the business prospered.

In 1810 the 63 year-old George Haynes withdrew from the business with his assistant and son-in-law, William Baker, and left Dillwyn in sole charge, who briefly manufactured the famous Swansea porcelain in 1814-17; it was artistically fine but economically unviable. An alienated Haynes briefly persuaded business acquaintances to open a soapworks in part of the old copperworks site not in pottery use: the 'unendurable smell' produced was soon terminated in a successful lawsuit by Dillwyn. In 1818, when he extended the lease of the Pottery, an additional five kilns were added to the existing four, the expansion taking place over the now abandoned soapworks and foundry stores. Haynes was back as part-owner (three of the shares were held by George Haynes and his son) of the Cambrian Pottery in 1817-21 after Dillwyn had married the Penlle'r-gaer heiress and sub-let the business.

Competition expanded the extent of pottery manufacture in Swansea when the neighbouring 'Glamorgan Pottery' was opened in 1814 by Haynes's son-in-law William Baker, in conjunction with the local industrial/entrepreneurial dynasty of the Bevans (of Landore Ironworks etc). Another five kilns were built at this pottery, increasing the Swansea total by almost 50%. In 1812 the number of earthenware products exported from Swansea had totalled 50,993 but by 1819 this had almost trebled to 140,280 pieces.[59] Many fine transferred jugs, plates and other vessels were produced in the 'opaque china' made from Devon ball-clay and flint at the Glamorgan Pottery, particularly in the period from the early 1820s to 1838, and the two potteries copied each other's designs. Lewis Llewellyn Dillwyn bought out the neighbouring Glamorgan Pottery and it was closed in 1838. Many of its moulds and

Fig. 27. Greenhill Mill just south of the road viaduct ('Hafod or Aberdyberthi Bridge') carrying the Neath Road across Cwm Burlais. It was converted from a corn to a flint-mill with steampower added to the earlier water-wheel. In the foreground is the mill-pool of the Pottery Mill with a wagon on the horse-drawn railway through Cwm Burlais (Baxter). (City & County of Swansea: Swansea Museum Collection)

workforce went to help found the South Wales Pottery at Llanelli, the competition from which helped to close the Cambrian Pottery itself in 1868.[60]

The constituents which some of the glazes used had similar sources to those made use of in the bodies of the pots. In 1796 the American industrial observer Joshua Gilpin noted that the various 'common' yellow and blue glazes that Haynes was using at the Cambrian Pottery consisted of flint, white lead and [china] stone[61].

William Bryant, business manager of the Glamorgan Pottery, noted that his best glaze made in c.1840 consisted of 16% china clay; 34% china stone; 13% flint; 13% carbonate of lime; 14% white lead; 10% borax and 3oz blue calcite for stain.[62] Other glazes made use of some of the products of the local copper and other non-ferrous smelting industry. Josiah Wedgwood had made a green glaze as early as 1760 and this was copied at the Cambrian and Glamorgan Potteries with a mixture made up of vitrified and calcined copper, red lead, flint and white enamel.[63]

In 1796 Joshua Gilpin noted the processes and equipment being used at the Cambrian Pottery by the employees of his old aquaintance George Haynes.[64] The main stages were as follows: the Devon clay was dissolved in water and run through a fine sieve. It was then mixed with the ground London flint in the proportions desired and the resultant paste was run into a long brick-lined trough and boiled for about 12 hours using the flues running underneath; after this the solidified mass was beaten by hand and then taken to a moulder working at a potter's wheel to form cups, teapots or other round ware; these were then partially dried and taken to a turner working at a common wheel lathe who formed them to the requisite thickness and used a spring peg fixed at the centre to give them a fluted finish; handles and spouts were put on by hand by specialist workers.

Plates and dishes were not turned but were formed on Plaster of Paris moulds. Then came the first firing, with the pots being placed in large kiln-pots made of refuse flint and clay with the pots inside separated by small pieces of clay. These were stacked in the inner domed chamber of the bottle kilns which had six stoke-holes spaced around the outside of the central chamber.

After the first firing the pots were glazed, Haynes expressing the hope that the health of the workmen would be greatly improved by a recent 'invention' for omitting the white lead from the glaze (probably not then applied). The Cambrian Pottery had formerly to glaze the pottery before painting, as Wedgwood still did, but according to what Haynes told Gilpin 'more modern potters' such as himself now had the biscuit ware from the first firing painted before having a glaze applied over the pots first so that they only needed two firings as did the other pots.

The similarities between the siting, staffing and trade patterns of the Swansea pottery and copper industries are wide-ranging. Bristol, the mercantile capital of the Severn Estuary, had supplied key skilled personnel for the setting-up of the first Swansea copper-works and a similar situation pertained with the Swansea potteries. William Bryant for example, later business agent for the Glamorgan Pottery, was a Bristol-born potter who worked at both the Cambrian and Glamorganshire Potteries. The siting of the main Swansea potteries was directly analogous to the copper and other non-ferrous works, in that the Cambrian, Glamorgan the short-lived Calland's Pottery at Landore[65] and the Clay-pipe Works were all situated on sites sandwiched between the Swansea Canal on one side, and the River Tawe on the other. Clay to make the pots and glazes came by sea from Devon and Cornwall, coal for return cargoes being taken by colliery railways that crossed the pottery sites. Coal to fuel the pottery kilns was also delivered via the canal, which in addition provided the link between the Glamorgan Pottery and its flint-mill at Landore.

Gilpin noted in 1796 that:[66]

> 'The ware produced at this pottery appears in every respect equal to any [of] the Staffordshire and such as is made to imitate Wedgwood's quite equal to it. The pieces of the common kinds are equal to those of the Staffordshire but in Wedgwood's far less. So great is the demand here for the articles that Haynes tells me he often sells £600 value the month by retail from the shop for cash.'

The same factors that made the Swansea area successively the national and international centre of copper-smelting and tinplate production could also have made the town the major centre of pottery production. The copper-smelting industry had migrated from Bristol but the pottery industry did not leave the Stoke-on-Trent area to become centred on Swansea. Swansea was the nearest, and most easily accessible, coalfield to the Devon and Cornwall clay deposits, just as it was to the copper and tin deposits of the south-west; Swansea had direct sea access whilst Stoke had to wait until 1777 for the Trent and Mersey canal to provide narrow-boat access. However, the artisan skills with local clays in the Potteries area were a decisive factor, so that by 1762 there were 500 Staffordshire potteries employing 7,000 skilled and semi-skilled potters.[67] Another major factor was that the substantial population of the West Midlands of England and surrounding areas provided a ready and large market for the goods produced; certainly much more so than did Swansea's upland hinterland. However, Haynes

showed the entrepreneurial flair that might have changed this story: he reportedly developed 'white china' and in 1807 advertised 'Ware ornamented with an entirely new Golden Lustre' at the London Warehouse (62 Fleet Street) of the Cambrian Pottery Company this was well before the Staffordshire potters had the chance to develop fine quality gold lustre.[68] However, even the 14 Swansea kilns

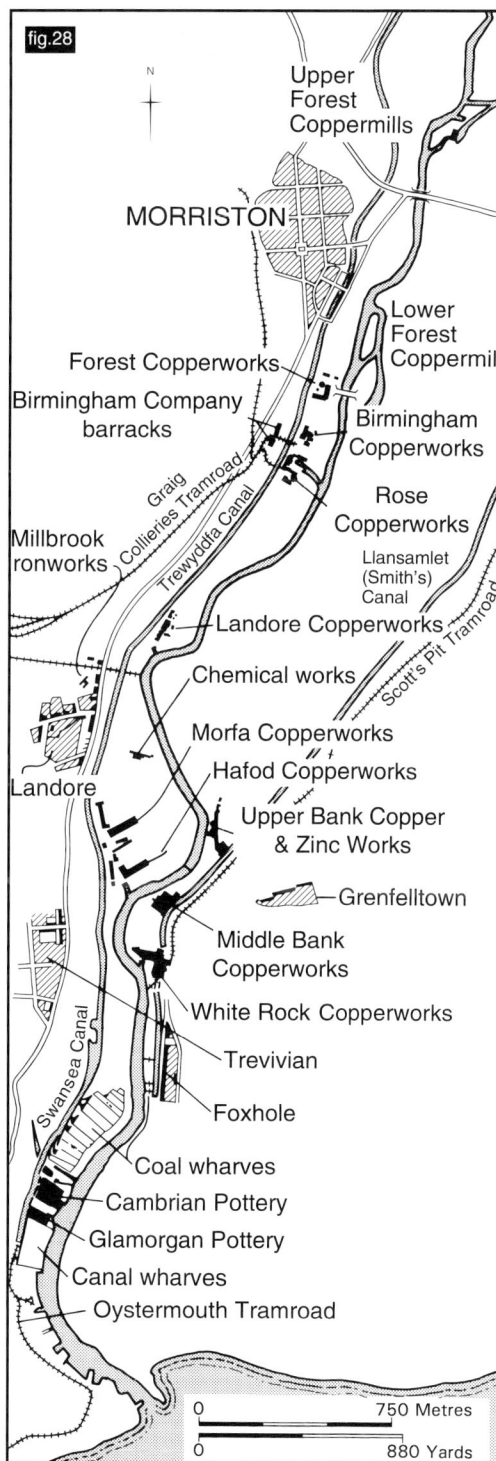

Fig. 28. Map showing copper-smelting works in use in 1830-40. Coal for the smelting-furnaces was supplied by two canal lines and a number of horse-worked tramroads. The five main workers' settlements are shown by diagonal hatching.

15

operating in the third and fourth decades of the nineteenth century were no match for the Stoke area potteries then supplying the whole of the British Empire although later potteries alongside the Swansea Canal at Morriston and Ynysmeudwy did continue to supply the local market after the Cambrian Pottery closed.

During redevelopment of the Glamorgan Pottery site in 1983, the remains of a subterranean drying-room were exposed and subsequently built over. Shards and kiln furniture from both the Cambrian and Glamorgan Potteries were recovered.[69] The remains of both potteries are now buried underground and for above-ground remains of a south Wales pottery and porcelain works one has to visit Nantgarw to the north of Cardiff.

The Copper Industry

The presence of accessible coal reserves led to the establishment in the Swansea area of one of the major industries of south Wales, non-ferrous metal smelting. In its day about 90% of Britain's copper smelting capacity was located within a 20 mile radius of Swansea and there were also smelteries processing zinc, lead, silver and other metals. This was the world centre of copper-smelting in the late eighteenth and early nineteenth centuries and it is important for the development of the 'Welsh method of copper-smelting' which consisted of repeated roastings of copper-ore in a succession of specially-designed reverberatory furnaces.

Across the Bristol Channel, Cornwall and west Devon were without coal reserves but were rich in metallic ores. With three tons of coal required to smelt each ton of copper-ore raised in the eighteenth century,[70] the coastal mining areas of south-west Wales were the most economic location for the smelting industry. The anthracite coals of Pembrokeshire were unsuitable for this purpose and the new smelters clustered further east, around the natural harbours of the Lliedi, Tawe, Neath and Afan rivers. Vessels bringing ore to south Wales could return to Devon and Cornwall loaded with coal for their home market, not least to power the engines that raised the water from the copper mines. (In 1823, as much as 80% of Swansea coal exports were carried in this way and the town had only a small fleet of its own.)[71] In this manner the coal export trade was greatly assisted.

The first non-ferrous metal smelter in the region was established in the Neath Valley by the German, Ulrich Frosse, at Aberdulais (the head of the navigable section of the River Neath), in 1584.[72] Neath remained, with Bristol, the main centre of the industry during the seventeenth century but in the eighteenth century the centre of the industry moved to Swansea, largely because of the superior harbour.

The first smelter in the Swansea Valley was founded in 1717[73] and was the fourth smelter founded on the south Wales coastal belt. The next two smelting works established in south Wales and a further four of the other nine non-ferrous smelters founded before the close of the eighteenth century

Fig. 29. Llangyfelach Copperworks, Landore, was the earliest copper-smelting works in the lower Swansea Valley. The water of Nant Rhyd-y-Filais powered the combined lead silver and copper-smelter. Three tall furnace-halls are visible, capped by square ventilators with pyramidal roofs, and flanked by nine tall chimneys needed to induce a sufficient draught to operate low, horizontal reverbatory furnaces. A stone-arched aqueduct (left) drove a large overshot water-wheel in the main 'workhouse'. The site is under St. Paul's church at Landore and the adjacent Cwm Level Road/Neath Road intersection. (Drawing in the 'King's Topographical Drawings' Album) (By permission of the British Library)

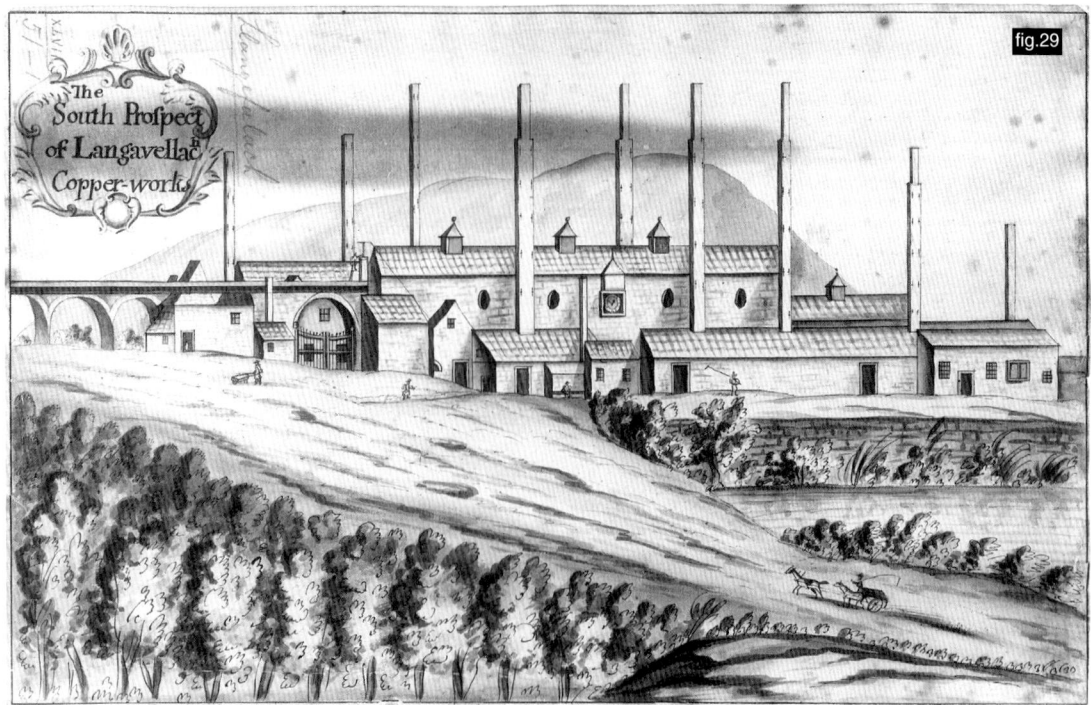

The Morris Enterprise

fig.30

In the later eighteenth century

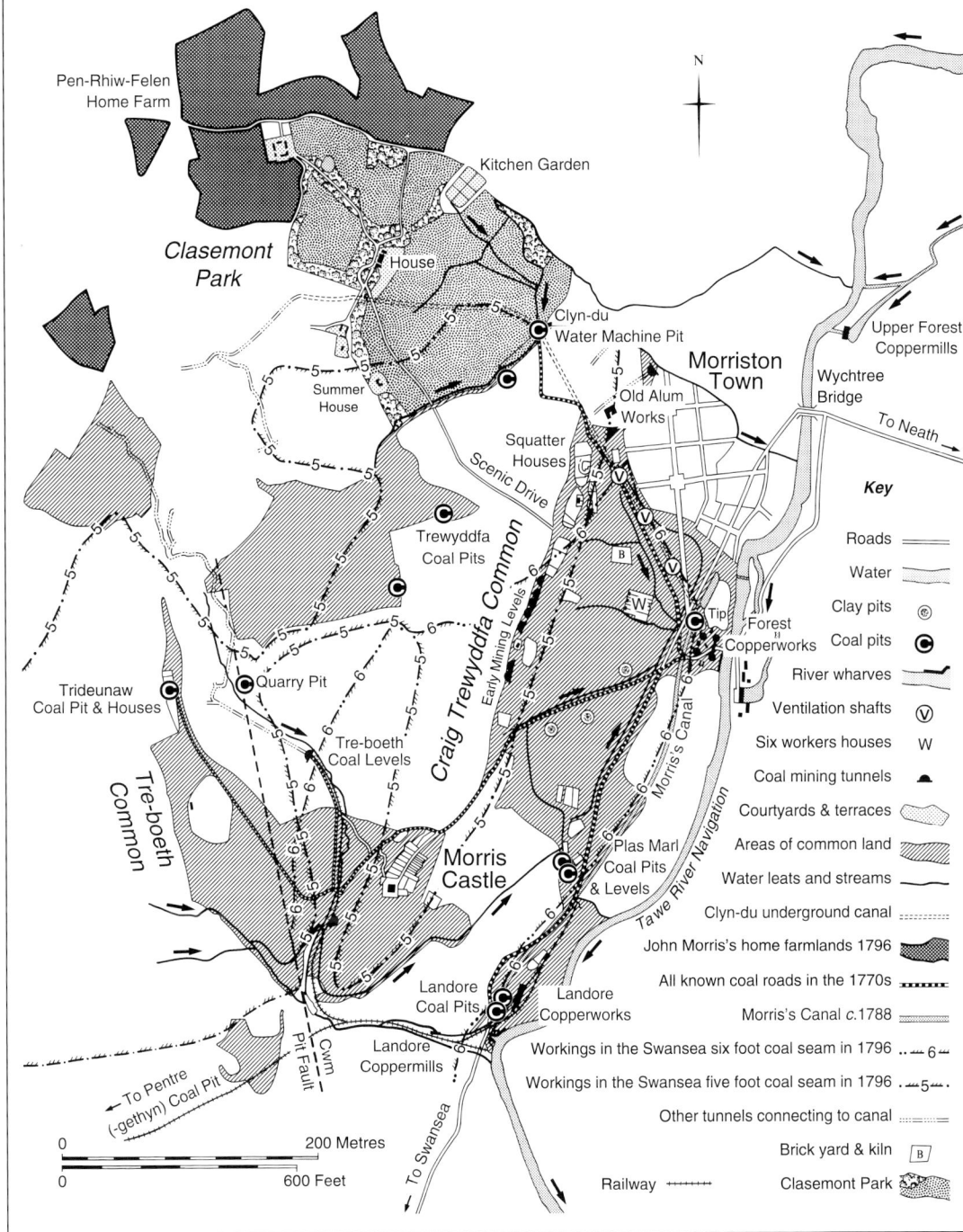

Key

Roads	
Water	
Clay pits	⊙
Coal pits	C
River wharves	
Ventilation shafts	V
Six workers houses	W
Coal mining tunnels	
Courtyards & terraces	
Areas of common land	
Water leats and streams	
Clyn-du underground canal	
John Morris's home farmlands 1796	
All known coal roads in the 1770s	
Morris's Canal c.1788	
Workings in the Swansea six foot coal seam in 1796	6
Workings in the Swansea five foot coal seam in 1796	5
Other tunnels connecting to canal	
Brick yard & kiln	B
Railway	
Clasemont Park	

Map labels: Pen-Rhiw-Felen Home Farm; Clasemont Park; Kitchen Garden; House; Summer House; Clyn-du Water Machine Pit; Morriston Town; Old Alum Works; Upper Forest Coppermills; Wychtree Bridge; To Neath; Squatter Houses; Scenic Drive; Trewyddfa Coal Pits; Craig Trewyddfa Common; Early Mining Levels; Trideunaw Coal Pit & Houses; Quarry Pit; Tre-boeth Coal Levels; Tre-boeth Common; Forest Copperworks; Tip; Morris's Canal; Tawe River Navigation; Plas Marl Coal Pits & Levels; Morris Castle; Landore Coal Pits; Landore Copperworks; Landore Coppermills; Cwm Pit Fault; To Pentre (-gethyn) Coal Pit; To Swansea; 0 200 Metres; 0 600 Feet

Fig. 30. In the later eighteenth century, Lockwood, Morris & Co. dominated much of Swansea copper-smelting. The managing partners, the Morrises, acquired large areas of land in the lower Swansea Valley with their profits; some of their interests are shown on the map. In the mid eighteenth century, many coal-mines, smelters and workers' houses were built on the common land of the area. However, the large workers' township at Morriston was built in 1779 on part of the large tract of farmland John Morris I had bought from his copper-smelting profits.

were also located in the Swansea Valley. Both the Neath and Swansea Valleys had the advantage over other parts of the coastal coalfield that they had large, navigable rivers which could accommodate numerous quays. The Neath estuary had been chosen first, probably because of the close proximity of large waterfalls which provided a ready supply of water for power generation purposes; but in 1717 it was an entrepreneur already acquainted with the area, Dr John Lane (originally from Bristol, home of the established copper-brass industry) who realised the advantages of the Swansea Valley, leaving his smelter at Neath Abbey to found the Llangyfelach Works at Landore.

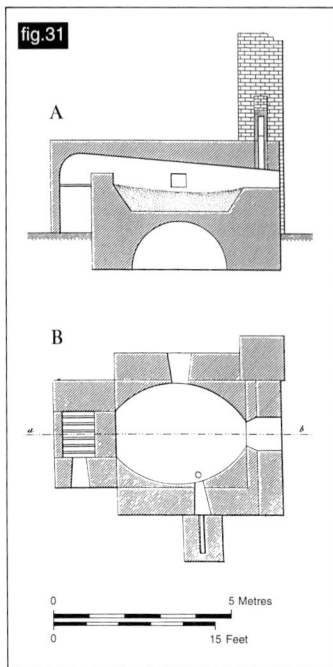

Fig. 31. European industrial 'spies' and observers of the later eighteenth and early nineteenth century studied the industrial advances at Swansea and produced drawings of what they saw for their fellow-countrymen to copy. This drawing of a copper-refining reverbatory furnace shows the bowl-like bed of the furnace in section ('A'). (Annales des Mines, 3rd Series, Tome V, p. 660)

As manager, Lane brought in a young man from the Shropshire/Montgomeryshire border, Robert Morris, who was subsequently to establish one of the major industrial dynasties of Swansea, with interests in both copper smelting and coal mining. Lane's works was the first in a succession of copperworks and over the next century a total of 13 were built on both sides of the valley between Swansea and Morriston.

The advantage which attracted Lane to Swansea, and which gave the area the lead over Neath, was the close proximity to the River Tawe of the thick Swansea Six-Foot and Swansea Five-Foot coal seams. The river is navigable and provides sheltered anchorage for three and a quarter miles between the north end of Fabian's Bay and a point just below the site of Wychtree Bridge at Morriston. At Neath these thick seams were much further from the river. But while copper smelting brought prosperity to Swansea, it also created severe environmental problems. Dense clouds of smoke, laden with sulphuric acid, billowed across the valley and blighted the sides of Kilvey Hill, and great quantities of copper-slag were produced as an unwanted by-product and deposited all over the valley to form vast, unsightly tips. It is only in recent years that the last of these have been removed.

As the nineteenth century advanced Swansea started to look further afield for supplies of ore. Cornwall was supplemented first by Anglesey and then, as the native ores started to be exhausted, by Cuba, Chile, the United States and Australia. This foreign trade is perpetuated in the names of public houses in Swansea such as the Cape Horner, the Cuba and the Mexico Fountain.

Even though the copper industry appeared to be well established at Swansea the seeds of decay had already been sown and from about 1870 the industry was in decline. The needs of mass electrification led to a surge in demand for copper which Swansea could not hope to meet, dependent as it was on imported ores. The trend was increasingly to smelt at the point of production and ship the refined copper to the point of manufacture. Swansea smelted its last copper in 1924, although the processing of copper was to continue until 1980, latterly as a very minor component of the local economy.

The Architecture of Copper Production

The development of the physical form of the industry, that it its processes, engineering and architecture, is of considerable interest.

This section is divided into three parts. The first section will deal with the form of the furnaces and how this affected the size and shape of the smelting-hall. The second addresses how the developments in the type of copper-good produced, together with the structure of the industry, affected the power requirements ie - water and later steam - of the industry. The third deals with the application of copper and copper products to architectural use both locally and nationally.

Copper Furnaces and Smelting-halls

In any consideration of this topic, Swansea cannot be studied in isolation, even if by the nineteenth century it was dominating both the copper trade of Britain and indeed that of the whole world. The introduction of the copper trade to Britain from Germany in Elizabethan times was followed by a period when the dominant focus of the industry oscillated between the two sides of the Severn Estuary, this ended decisively in the last decade of the eighteenth century when nearly all national and international copper-smelting activity concentrated in Swansea and the surrounding district.

A further complicating factor was that initially it was far from clear that works in the Swansea area would concentrate on copper rather than on the smelting of other non-ferrous metals, and as the nineteenth century progressed, Swansea once again became heavily involved in the smelting of other metals, the uses of by-products from the processes of copper-production and also in the extensive production of brass (copper-zinc) alloys.

The Copper Furnaces

In the Elizabethan period there had been a move to end the total dependence of Britain on copper and brass imported from Sweden and the rest of continental Europe. Two monopolies were established by royal charter. The one for mining was the Mines Royal Company which imported German smelting technology to the vicinity of new copper-mines in the English Lake District. It seems however, that the small water-powered blast-furnaces built at Keswick, of a type universally used in continental Europe for copper production, did not produce adequately refined copper for manufacturing use.[74]

The second Elizabethan monopoly, the Society of Mineral and Battery Works, concentrated on production and manufacturing. Calamine (Zinc Carbonate) was discovered in the Mendip Hills in Somerset, and the nearest suitable site for manufacturing, from the point of view of available water-power and transport, was deemed to be Tintern, situated on the banks of the Wye in Monmouthshire. The concern promised a weekly output of 40cwt of brass wire by the end of 1566. Instead, work was abandoned after production had been attempted for just over a year. The brass produced from the Lake District

fig.31

A

B

0   5 Metres

0   15 Feet

18

copper had proved impossible to work, lacking as it did the ductile and malleable qualities essential for battery work (i.e. water-powered hammered bowls and other goods) and wire. Some copper battery-ware was produced in the Lake District.[75]

The discovery of copper-ores in Cornwall prompted the transfer of the Keswick smelting equipment and German personnel to Aberdulais in the Vale of Neath in 1583-84, south Wales being the nearest area to Cornwall where coal was available.[76] There, records of production suggest possible trials of an early coal-fired reverbatory furnace, but the short-lived enterprise foundered from insufficient ore. This was a period when the exploitation of Cornish tin had yet to reach depths where copper was to be found in any appreciable amount.[77]

The crucial breakthrough in the technology of the eighteenth and nineteenth-century British copper industry was the introduction and sustained use of the coal-fired reverbatory furnace. Reverbatory furnaces were long, low structures that kept the coal-fuel and the metal being smelted separate and so the latter was free from any contamination by the fuel used. Heat from the coal in the sunken stoke-hole at one end of the furnace, was induced to pass over the copper-ore spread over the broad furnace bed by the up-draught created by the exceptionally tall furnace chimneys sited at the far end of the furnace structures. An early depiction of this type of furnace occurs in the fourteenth-century stained-glass windows of York Minster, which show such a furnace being used for the remelting of bronze for pouring into bell-casting moulds.[78]

Up to the late sixteenth century, the technology that had been in use for copper-smelting was derived from Roman technology current for at least 1,500 years, that is in small furnaces in which the metal ore and fuel were mixed together. Unlike medieval or post-medieval iron-furnaces of larger size, the furnaces for producing non-ferrous metals in use in continental Europe - in Keswick (1567-1583) and most of those in use at Aberdulais near Swansea, in 1584 - were adaptations of the small Roman shaft furnaces.[79] These small furnaces were square externally, only about 2ft 6in wide and 6ft 6in high. By 1520, in Sweden, this type of hearth had been raised and the stack of the furnace heightened to about 10ft. The furnaces were built in banks of six against a stone wall, behind which the bellows were driven from a single water-wheel operated shaft. Brick formed the lower part of the square shaft furnace and stone the upper, which widened somewhat at the top. The blast tuyere was placed in the rear wall, some 9in above the hearth, and there was an arched opening in the front wall for tapping. Most of the Keswick and Aberdulais plant conformed to this model. The fuel used for

repeated firings was coal, except for the final one where it was necessary to use charcoal in order to rid the charge of the sulphur contamination resulting from mixing the copper with coal.[80]

The maximum output achieved by the six small blast-furnaces at Keswick had been 60 tons per annum in the period 1567-84. From 1584, at Aberdulais where coal was easily available and cheaper, the scale of operations was large by the standards of the day, and blast-furnaces and a reverbatory furnace were used, the latter being able to smelt some 1ton 4cwt of ore per day. The smelter soon closed because of irregular supplies of ore from Cornwall and by about 1660 there was virtually no smelting of copper in Britain.[81]

The sustained move from small traditional blast-furnaces to the use of reverbatory furnaces in non-ferrous metals smelting began in the 1670s in experiments carried out by Lord Grandison and Sir Clement Clarke and his son Talbot. The partners had a dispute and the latter pair were prevented from continuing their work in smelting lead by the patent rights of Grandison.[82]

The Clarkes had moved from their original base in Ticknall (on the Derbyshire-Leicestershire border) to Stockley Vale Works on the River Avon near the later site of the Clifton Suspension Bridge at Bristol.[83] After Grandison ceased his experiments with lead-smelting, he switched to experiments in smelting copper in reverbatory furnaces, with John Coster as chief technician. The successful process was patented in January 1688.

The first new works using this process was founded alongside the River Wye navigation just as the earlier Tintern brassworks had been. Coster had returned to his home area of the Forest of Dean where his father had previously smelted iron and founded a new copperworks at Upper Redbrook (the Monmouthshire - Gloucestershire border runs through the works) which, by 1691, had been in production for a few years.[84]

Like Aberdulais a hundred years earlier, this used a technology intermediate between the German and the Welsh/English methods of smelting. The ore was calcined in ten reverbatory furnaces and then crushed. It was then roasted in two more series of reverbatory furnaces before being refined in the small traditional blast-furnace with bellows-driven blast. These were 3ft 11in high, 1ft 10in wide and 2ft deep and were stone built with a clay lining and capped by a furnace stack which was 15ft 4in high and 1ft 6in wide.[85] This hybrid process was still in use in 1724-25 (the works had 26 furnaces in total).

Coster's partner, William Dockwra, had a brass works at Esher on the River Mole in Surrey that was producing 80 tons of manufactured copper goods for the London market every year. By the end of 1692, a second concern had been founded at Redbrook within quarter of a mile of the first works. This was the works of the 'Governor and Company of Copper Miners in England' (The English Copper Company), with Gabriel Wayne as its chief technical man. This may originally have used intermediate technology; presumably its founders were drawn to Redbrook because of the availability of skilled labour, but by 1724-25 it was entirely dependent on repeated firings in reverbatory furnaces (there were 16 at this works) in what was later developed at Swansea into the more sophisticated 'Welsh Process'. The works developed at Bristol in this period also seem to have entirely used reverbatory furnaces. It may have been the differences in the market that determined the divergence of practice at this stage. The original Upper Redbrook Works required a more refined metal that was malleable enough to produce battery-ware while the relatively impure copper produced at those works with only reverbatory plant was used at this stage for the production of coins.[86]

There are some complications in understanding the process developments of the first works in the Swansea region because they were actually processing lead, copper and silver and not copper alone. This carried on a long tradition first noted by Agricola when observing German practice in continental Europe in 1556. This was the process used by German smelters at Keswick in 1567-84 and presumably at Aberdulais from 1584. During the final smelting to impure (black) copper any precious metals were separated out by adding lead to the charge. This produced a copper-lead-silver 'bullion' below the matte (a mixture of copper and iron sulphides). The bullion was later treated in liquidation furnaces in which the silver-rich lead was melted out of the solid copper. This was done at Keswick by the cupellation process (oxidation of lead).[87]

By 1697 there were other copper-working concerns using the new reverbatory technology in Derby, Cumberland and Neath Abbey. The latter was founded in 1694 and was sited at what became the Cheadle Works where the River Clydach entered the River Neath. The manager was John Champion and one partner was the Master of the Mint, who also had the Wandsworth Battery Mill on the River Wandle near London. There was an associated coppermill to the smelting works, on the River Clydach near the large waterfall that later powered the Neath Abbey Ironworks Forge. The poaching of a skilled furnaceman by the owner of the nearby Melincryddan Works contributed to the closure of the Neath Abbey Works in 1717 (although it did reopen on a small scale in 1720). Two of the partners, Dr. John Lane and his assistant, then left to found the first copperworks at Swansea, a combined lead, silver and copper

works. It may then have been the case that this immediate precursor of the first Swansea Works was also producing lead and silver as well as copper.[88]

The First Swansea Copper Furnaces at the Llangyfelach Copperworks of 1717

The nature of the furnaces at the first Swansea (Llangyfelach) Copperworks of 1717 (sited at Landore) is a matter for some debate. In the first half of the eighteenth century the use of reverbatory furnaces in copper-smelting was becoming universal. As has already been discussed these were long low furnaces that kept the coal-fuel and the metal being smelted separate and so that the latter was free from any contamination by the fuel being used. Heat from the coal in the sunken stoke-hole at one end of the furnace was induced to pass over the copper-ore spread over the broad furnace bed by the updraught created by the exceptionally tall furnace chimneys sited at the far-end of the long, low furnace structures. However, there is documentary evidence to show that the Melincryddan Copperworks at nearby Neath, still in operation in this period, was using a water-powered draught to increase the temperature in some of the lead furnaces in what was then the common usage of a general non-ferrous smelter.

The surviving picture of the Llangyfelach Copperworks suggests that the five furnaces in the main smelting-hall had water-powered bellows providing an extra blast to raise the temperature for smelting inside the furnaces. These may well have been the lead and silver, rather than the copper, furnaces. There is however the verbal evidence recorded of one of the workers at the works who said categorically that under Dr. Lane, and subsequently under Robert Morris, blast-furnaces had been used for the smelting of copper (presumably being further refined in reverbatory furnaces as at Redbrook).[89] Bristol practice was influential in the design and construction as the chief mason and the eminent innovative engineer John Padmore were from that city. A high and impressive stone-built aqueduct of at least five arches fed water to the eaves level of the main smelting-house from a 200ft long rectangular reservoir sited at the end of a leat to the west of the works. The position of a large overshot water-wheel at the west end of the northern face of the main smelting-house, and the use of the power for a process carried out within the main smelting-house is suggested by the position of a control-sluice (visible on the surviving drawing of the works) at the western gable of the main smelting-house. This had a control rope passing through the roof of a southern annexe of the main smelting-hall, obviously operated by workers situated in the interior of the main smelting-hall.

The main smelting-hall was a long two-storey high chamber with the noxious fumes inside vented by three raised open-sided square cupolas, with capping pyramids surmounted by ball finials placed at intervals along the roof ridge. Extra ventilation of the warm air and fumes was provided by four ovoloid upper openings in the sides of the smelting-hall, this being an early example of the use of the vents that became universal in the late eighteenth- and early nineteenth-century metals-smelting industries of south Wales (the 'oval' shape of the openings may have been the true circular openings found in later smelters and forges; there is evidence of considerable confusion in the perspectives used in the drawing of 'The South Prospect of Llangavellach Copper-works').

Lean-to structures alongside the ground-floor walls of the main smelting-house might suggest that the stoke-holes of the furnaces were already outside the external walls of the main smelting-hall, as was later the case with the Forest (c.1746-48) and Landore (c.1790) Copperworks used by the same Lockwood, Morris & Co. that had originated in the Llangyfelach Copperworks. However, the external stoking-places in those later works were either open-sided, or completely open, and this was not the case in the Llangyfelach Copperworks. In addition, the very tall furnace chimneys, which extended over two full storeys above the height of the smelting-hall, depended for structural support on the side-walls of the main smelting-house. This would have meant that the stoke-holes of the furnaces would have had to be

tended from their opposite ends, i.e. in the central longitudinal spine of the building. The exceptionally tall chimneys were tied together by iron tie-rods at regular intervals to resist the deterioration engendered by constant changes of heat.

Three other subsidiary buildings visible on the remaining depiction of the Llangyfelach site: one parallel to the main smelting-hall seems to have had two tall furnace chimneys. A plan of the area in 1761 also shows a small cross block between the two parallel smelting-halls at their western end and a broad square building placed between their middle sections. Two other, and lower, buildings were in line with the main smelting-hall but were downhill from it, as the works descended the scarp eastwards towards the River Tawe at Landore. The low building next to the main smelting-hall also had a venting cupola, as well as one tall chimney, which suggests that this may have housed a single specialised reverbatory furnace in the refinery, where the final valuable pure metal was produced. The other two buildings may have housed water-powered functions with the tall chimneys venting reheating furnaces and hearths needed to keep the copper malleable for working and rolling.

The Buildings and Furnaces of the Cambrian Copperworks of 1720

The exact form of the Cambrian Copperworks, founded in Swansea in 1720, is obscure but the probability is that the availability of the large hall

Fig. 32. The White Rock Copperworks in 1744 showing the 'Great Workhouse' across the background and the use of water-power on-site to drive the copper-stamps and clay-mills. The extensive tips had already started to spread across land adjacent to the navigable River Tawe. (Grant Francis, The Smelting of Copper in the Swansea District)

of the former smelting-house was convenient for conversion into a new pottery ('The Cambrian Pottery') on the Wedgwood model; this being an after-thought to new owners who were primarily interested in the re-use of the mill on the dam. The 1764 lease referred to 'all that old building ... called ... the old Copper Works'.[90] Views of the 1790s, and the pottery plan of 1802, show a strange high and broad-roofed building at the north-eastern edge of the then pottery buildings. This formed part of the original northern poolside quadrangle with the four original kilns added for the pottery, which were otherwise interspersed with apparently purpose-built workshops one room in width. This large building had three very tall stepped chimneys on its northern and eastern sides and had a roof capped by a central raised ventilation cupola with capping pyramid. By 1802 it was in use as Raby's Foundry.[91]

The copper-smelting works here was started by the local Quakers 'James Griffiths and Company' which soon ceased production. A surviving letter from Griffiths to the Duke of Beaufort's agent Gabriel Powell, when the site was being considered suggests that he thought they could smelt copper without the use of water-power.[92] That would indicate that this might have been the first works in the valley where an exclusive use of reverbatory furnace technology was attempted; however, the creation of the large mill-pool on the northern side of the works (presumably for a rolling-mill) would have allowed water-power to be used.

Fig. 33. Part of the north-eastern wall ('L' on the works plan) of 'The Great Workhouse' of 1737 at the White Rock Copperworks. The brick arch would have originally spanned one of a long row of horizontal reverbatory furnaces. The ground behind was raised in 1783-4 to accommodate the cut-and-cover tunnel carrying John Smith's Canal. (960208/3)

The White Rock Copperworks of 1737

The third copper-smelter to be sited at Swansea was the first to be located on the east bank of the river and was founded by an old-established Bristol brass and copper-making concern in 1737. A detailed and labelled early view survives. Here the main range of furnaces were arranged in a very long and comparatively narrow smelter furnace hall called 'The Great Workhouse' ('L' to 'P' on the accompanying plan of the present remains) arranged north-south between the navigable river and the hillside scarp. The walls were built of thin slabs of Pennant sandstone with dressings of locally made brick. The north end of this significant structure ('L' on the plan) remains as part of the White Rock Industrial Archaeology Park.

The power source available here was the small stream and millpond of the old manorial mill in Kilvey ('M' on the plan). This was augmented by the construction of a long watercourse, also available to power the landowner's (Lord Mansel's) collieries, curving up and round the north and east sides of Kilvey Hill. Sections of this still survive near the feeder springs on the east face of the hill overlooking Crumlyn Bog. The original stream still flows off Kilvey Hill and over an aqueduct across the southern portal of the Llansamlet Canal tunnel ('R' on the map) at White Rock. Inside the southern end of the canal tunnel is a water outlet that would also have made surplus canal water available to drive works machinery from 1783-84.

The annotated early drawing of the White Rock Works makes clear that the stream and leat drove stamps and clay mills. A rolling-mill was not initially needed on this site as the Bristol concern took the ingots back to its existing rolling-mills in that city for forming into sheet, hammering into vessels or mixing with calamine to form the ubiquitous brass.

The copper-making process at this stage already required repeated roastings in reverbatory furnaces in order to produce the refined metal, it was a coal-hungry process. Most of the furnaces were of a similar size, except for the calcining furnaces which were used initially to separate the copper-rich 'matte' from the large quantities of useless material contained in the copper-ore. The White Rock Works soon acquired a 'New Calcining House' sited alongside where the ore arrived on the expanding river-quays and the works dock (still visible on site with the latter recently re-excavated: 'O' on plan). This is readily visible on the 1790s drawings of the works which show a relatively long and narrow building ('I' on the plan) with three very broad chimneys projecting at intervals from the roof, although these are more likely to be brass annealing furnaces.

As with iron-smelting, lime needed to be added to act as flux in this initial roasting of the copper-ore (later sand was used and was considered better than lime). The enigmatic base of a blast-furnace-like kiln structure, that also acted as the support for the tips railway, remains on the White Rock site ('D' on the plan) and Wood's drawing of the works and River Tawe in c. 1813 clearly shows a number of limekilns on the river side of the works where limestone could

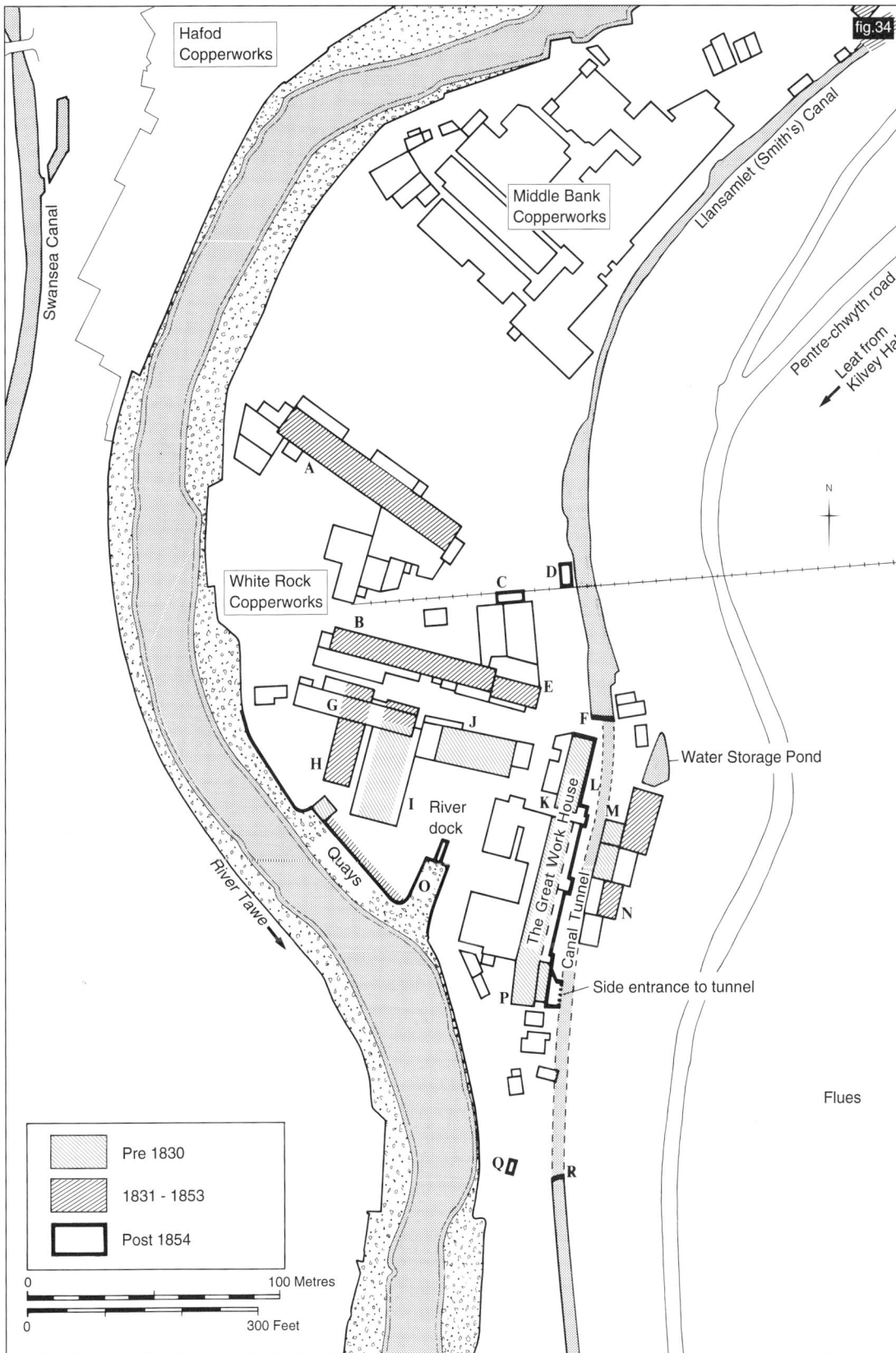

Fig. 34. Plan showing the visible remains (in thick outline) and chronological development of the Whiterock Copperworks. It is the only early eighteenth-century copperworks site in Swansea to retain substantial upstanding (and buried) remains. The site of the earlier post-medieval manorial mill which used the existing water-power resources of the site is indicated by the location of 'The Water Storage Pond' with the stream still discharging westwards into the River Tawe from an aqueduct bridging John Smith's Canal at 'R'. By the seventeenth century Bussy Mansel's 'Great Coal Road' brought Llansamlet coal down to the White Rock Quay near 'Q' to be shipped and in 1737 the 'Great Workhouse' or smelting-hall of the White Rock Works was built between 'P' and 'L'. In the 1750s Chauncey Townsend's Waggonway and from 1817-19 John Scott's Tramroad were also used for shipping Llansamlet coal at the quay and 'Q' may be a coal-shute used by Scott's Tramroad wagons. In 1783-84 John Smith's Canal was built to replace the Waggonway and to lessen disruption on the Whiterock Copperworks site was conveyed through the site in a cut and cover tunnel with side openings both to supply water to the site and later coal. One of the fourteen river docks on the river ('O') was used from an early stage and quays were extended round the whole western perimeter of the site as the eighteenth and nineteenth century progressed. Various new smelting-furnace halls ('J', 'I', 'H', 'B', 'G' and 'A'), were built around these ore-importation points including what possibly became a hall with two rows of furnaces at 'I'. The base of a later structure resembling a blast-furnace remains at 'D' and was also used to support the powered inclined-plane to the secondary waste-tips: as was a masonry arch at 'C' that also served as a stack from flues leading from the smelting-hall 'A'. The remains on this significant site have been partly uncovered and designated as the White Rock Industrial Park. (Based upon Ordnance Survey material with the permission of Ordnance Survey on behalf of the Controller of Her Majesty's Stationery Office © Crown Copyright)

be brought upriver from the Mumbles quarries. A similar limekiln can be seen on the site of the Hafod Copperworks.

Between roastings the fused mass of impure metal, or 'matte', remaining on the bed of the furnace needed to be broken into small pieces, so that the heat produced in the next roasting could easily penetrate into all the metal. Therefore, water-driven 'stamps' had to break up the impure metal mass. At White Rock, and elsewhere, the water-wheel would have had its axle extended as a heavy timber beam

fig.35

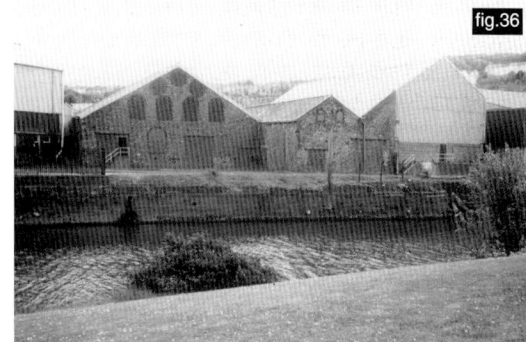

fig.36

Fig. 36. The roofed smelting-halls of the Upper Bank Copperworks are the last surviving at Swansea. Visible are the characteristic gable-end ventilators with honeycombed brick. The wider hall at left/centre probably held a double row of furnaces; the parts of the two narrower surviving halls at the right/centre may incorporate eighteenth-century masonry. (960215/4)

with projecting lugs which engaged with large timber poles set vertically in a frame alongside; the iron-shod feet of these were knocked upwards with the repeated turnings of the wheel and then smashed downwards onto the impure copper positioned in the trough below.

The reverbatory furnace beds and linings were repeatedly being fused and thinned by the heat and were in constant need of maintenance and repair. Refractory bricks were imported from Stourbridge, but the presence of a water-driven pug-mill here indicates that fire-clay from the local coal-mines was also used to form furnace bottoms with silica-sand also being brought from the deposits at the heads of the Swansea and Neath valleys by canal.

The Middle and Upper Bank Works of 1755-57

Smelting works were founded by the London entrepreneur, Chauncey Townsend, on sites on the east of the river to the north of White Rock in the 1750s. Townsend spent a large fortune in trying to maximise returns from what was an integrated industrial infrastructure. He not only developed deep mines at Llansamlet, Landore and Llanelli to both export coal and fuel his smelters but he also helped develop some of the largest non-ferrous mines in mid-Wales, notably Rhandir-mwyn (near Llandovery) and Cwmystwyth (near Aberystwyth). These latter largely produced lead and zinc ores and so it was that Townsend founded the first Swansea lead and zinc smelters at the Upper Bank site in Llansamlet, as well as the copperworks at Middle Bank, on the banks of the River Tawe, immediately north of the White Rock Works.

The use of powered processes in the latter three works is generally unspecified in the surviving

documentation and is discussed in more detail in the next section on water-power. Townsend incurred the huge cost of building a three-mile watercourse or leat to Glais, the remains can still be seen, with the avowed intention of supplying water 'necessary to the smelting' process; a leadworks like that at Upper Bank would normally have had at least one furnace with powered blast and the Middle Bank Copperworks, presumably had a stamps mill. The need for associated rolling-mills was somewhat negated with the leasing of these sites to the Williams family, entrepreneurs of the huge Parys copper-mines on Anglesey and operators of the large dams and copper rolling-mills in the Greenfield Valley in Flintshire (where the extensive remains have been excavated and conserved for display). There is, however, some evidence for the provision of rolling-mills at the Middle Bank and White Rock Copperworks in the nineteenth century, including the local copper-rollermen mentioned in the local 1841 census returns.

The Welsh Process of Copper-Smelting and the number of Furnace Types

Larger reverbatory furnaces for the initial roasting, or calcining, of copper-ore and smaller reverbatory furnaces for successive meltings in the smelting and refining process already existed in the seventeenth-century Bristol-based copper industry. By the mid-nineteenth-century the metallurgist John Percy[93] could say that basically there were still only these two types of furnaces in use at Swansea, even at the huge Hafod Copperworks where the successive ranges of reverbatory furnace built gave greater scope for detailed variation and specialisation of type in both basic calcining and melting furnaces. These were used in increasingly involved meltings, and remeltings, of various ore types to rid the refined metal of its various impurities.

One difference in the structure of the 'ore furnace': the otherwise standard small reverbatory furnace used to melt and re-melt the copper being refined was in the provision of circular water-baths sunk up to 10ft into the ground alongside or between these furnaces so that the first melting of the metal could be released into these baths to be granulated rather than solidifying into solid lumps. The baths were generally supplied with cold water by pipeline from the Morris, Swansea, Trewyddfa or Smith's Canals.

There is no doubt that ranges of fairly similar reverbatory furnaces were reserved for specific function in the 'Welsh Process' of copper-smelting that varied from works to works. At the Middlebank Copperworks (1755) two types of melting furnace were mentioned in 1796 (in addition to eight calcining furnaces, there were eight smelting and four 'roasting' furnaces); each serviced by two workers on shifts with a 'refiner'.[94]

In Jernegan's designs of *c*1793 for what was probably the Landore Copperworks there were three calcining furnaces, four ore-furnaces (with water-baths), three metal furnaces, two slag furnaces and a refinery (i.e. four different functions reserved for various of a total of 10 small melting furnaces of apparently identical size and type). One of the three large calciners was the 'metal calciner'; designed to be used after the initial granulation in the four ore furnaces.[95]

According to the idealised plan in Tomlinson's *Cyclopaedia* of 1852 the huge complement of reverbatory furnaces at the Hafod Copperworks were divided into no less than 10 different groups by function.

The first copper blast-furnace (i.e. vertical vessels containing copper-ore mixed with coal (and a flux of limestone or sand) and blown by a powered air-blast)

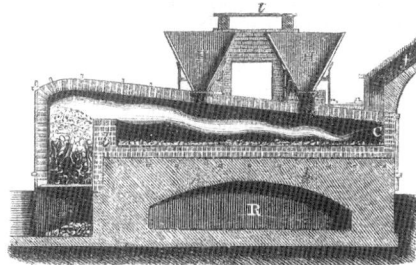

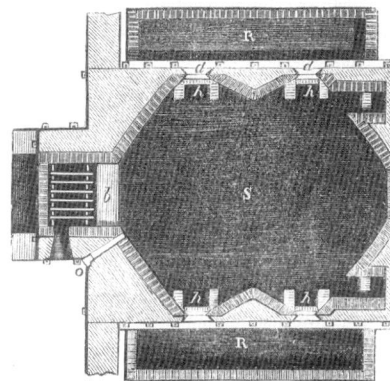

fig.37

Fig. 37. Old engraving 'Plan, Section and Elevation of a typical Calcining Furnace' in use in drawings (c. 1852) of one of the large reverbatory furnaces in the Hafod Copperworks where the initial roasting (calcining) of the copper-ore was carried-out (original furnaces of 1808-09). Copper-ore was delivered to the twin-hoppers ('H') via an elevated railway ('t') and fell on to the flat bed of the furnace ('c' or 'S'). The coal fuelling the furnace was kept separate so that it did not contaminate the ore. Heat was produced via the external fire at left which was outside the external walls of the furnace-hall: the heat passed through the furnace and was reflected down from the furnace roof on to the ore. The draught induced by the tall chimney drew the heat, smoke and poisonous fumes from the roasting off (right - 'f') via an elevated flue ('g') running alongside the central pillars of the main smelting-hall to the stack. The calcined-ore was periodically discharged via four openings ('h'- 'o') into a large vaulted-pit ('R') with an arched roof supporting the body of the furnace above. The structure of the furnace was of firebrick, able to resist the severe heating and cooling engendered by the process by having an external strapping of wrought-iron rods. (Tomlinson's Cyclopaedia, figs. 614-16)

after those used at the Llangyfelach Works (1717) was that introduced at the Hafod complex in 1859. This finally removed some of the toxic clouds of smoke that had arisen from the calcining and other reverbatory furnaces; especially when the sulphur smoke was diverted into the vast sulphuric acid manufacturing chambers which dominated the skyline at Hafod after 1865. The use of Bessemer Converters at the same works soon followed.

Layout and Planning of the Copperworks.

Furnace development in relation to the main smelting or furnace-house and the productive functions of a copperworks relative to its use of water-power, have already been discussed. What has not been examined in detail is the development of the large smelting-hall as a new building type and the overall impact of a copperworks on the landscape. Obviously the latter extended well beyond the confines of the main manufacturing site, into the coal

Fig. 38. Drawing (c. 1852) of one of the coarse-metal reverbatory furnaces at the Hafod Copperworks (original furnaces 1808-09). This was one of several smaller types of reverbatory furnaces that were used for successive roastings of the copper-ore. The hearth (again external to the main wall of the furnace-hall) was of similar dimensions to that on the calcining furnace but had a thick layer of clinker laid encouraging powerful jets of air to form that induced a much higher temperature than in the calcining furnace. Each of these smaller furnaces had an internal draught induced by individual tall chimney stacks (C - accessible via flue f) standing on an inner corner of the furnaces and extending upwards far above the furnace-hall roof. The metal was melted-down in to the bowl-like base of the furnace (h) and then released via the chute (p) into a sunken well (w) of water where it formed into small granules upon impact. Each well served two adjacent furnaces and were fed by pipes (t) leading from the Swansea Canal and away to the River Tawe. A crane (c - supported by the external wall of the furnace-hall) was then used to lift a metal basket containing the granules from the well to be re-roasted in the next stage of refining. The large amount of slag from the process was tapped at a higher level (o) on the (left) chimney end into the four moulds shown in plan: these hard metallic blocks containing much iron oxide could then be used for building purposes. (Tomlinson's Cyclopaedia, figs. 617-18)

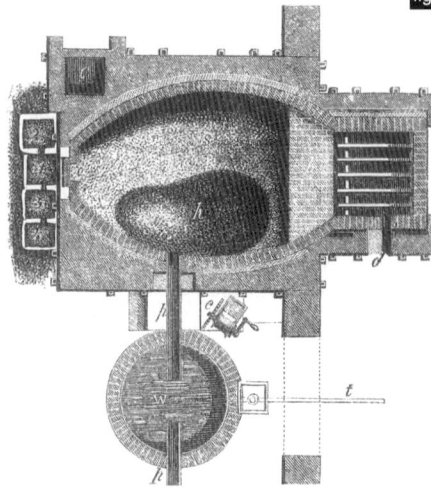

Fig. 617. PLAN OF COARSE METAL MELTING-FURNACE

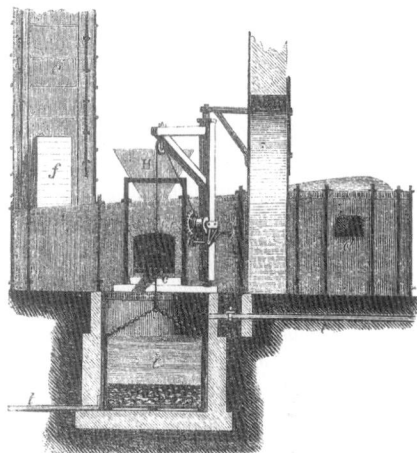

Fig. 618. COARSE-METAL FURNACE.

feeder and water-power provider systems on the surrounding hillsides, and in the surrounding settlements, houses, gardens and parks of the copperworkers and manufacturers respectively.

Many of these elements are being discussed in other chapters and this section will focus on the development of the discreet sites of actual copper-smelting and their by-products. There were 11 riverside works that can be classed as copperworks, rather than works that were planned primarily to produce arsenic or extract copper from slags. The mid nineteenth-century copper-smelters are also excluded from this detailed examination: Black Vale, Cwmbwrla and the Port Tennant Works were atypical; being hybrid tin/copper and arsenic/copper works away from the inland banks of the tidal river and with the latter being linked-in to the Vale of Neath economic zone via the Tennant Canal.

The most important determinants in the final realisation of form of the early dominant riverside copperworks were: firstly the alignment of the smelting-houses or halls (the largest and most important architectural element of a site) to the coal-feeders (later universally canals), the river, a power source, the number of furnaces; and, lastly, the overall size of the works site available in relation to the diverse needs of the development of the copper-smelting process and its by-products. It is only in the light of these factors that the contribution of the works to the landscape development of Swansea can be assessed.

The Smelting-Houses and Halls of the Early Eighteenth-century Copperworks

The first smelting site, at the Llangyfelach Copperworks at Landore (1717), was atypical to the later layout of the Swansea industrial works (potteries, as well as non-ferrous smelters) in that it was not positioned on a riverside wharfage, with its parallel landward boundary determined by a transport way feeding coal. Instead, it was positioned on, and at, right-angles to the valleyside in a position where the high-level water-power feed could drive the blast to its lead-refining furnaces and to its stamping, rolling and clay mills. Its layout, and power requirements, were doubtless complicated by its need to facilitate lead as well as copper production. However, it marked a distinct change in the scale of non-ferrous smelting-halls and their impact on the landscape.

Previous smelting sites, in the earlier copper-smelting centre of Bristol, tended to consist of irregular clusters of small smelting-houses containing no more than four furnaces apiece. This was the pattern that usually adopted for lead-smelters such as that at Gadlys in Flintshire, which resembled nothing so much as a random scattering of small dwelling-houses. The scale of the monumental smelting-hall at Llangyfelach marked a major increase in the size of these individual works buildings and their impact on the landscape, especially when this extended from the valley-side flanked by a high arcade of water-feeder arches. This was recognised by the Bristol-trained mason, David Rauleigh, who was responsible for building the Llangyfelach Copperworks, and who commented that the two large smelting-furnace halls (80ft x 40ft & 40ft x 36ft) housing most of the 20 furnaces of the works were 'larger and more useful' than any in Bristol. The 'round house' at the works may also have been the model for the four circular furnace-houses at the proprietors' succeeding works at Forest.

The wharfside site of the second, Cambrian (1720) Copperworks, with the Strand road on its landward side (with coal being fed onto this roadway via the

Cwm Burlais Coal road) was much more typical in its riverside siting but unusual in that its northern boundary consisted of a large waterpower reservoir to drive its copper rolling-mill with a low head of water. The site was subsequently re-used by the Cambrian Pottery who had a small river-dock at right-angles to the river. This allowed vessels to clear the sloping river-bed exposed by the large tidal range. The broad main building of the Cambrian Copperworks (1720) was later pictured with three main furnace chimneys and may have been a reversion to the Bristol type of small smelting-house. Even the Llangyfelach Works was only the first stage towards developing a typical Swansea smelting-hall; it had been built for a Bristol trained proprietor and its main furnace-hall was designed primarily for lead-smelting and, although larger than earlier Bristol works, it still had only five tall chimneys venting furnaces in its main furnace-hall.

White Rock, founded in 1737, by Bristol interests was the first site that established what became many of the dominant characteristics of the Swansea works, and is where the first at which surviving documentary and field evidence allows a detailed examination of the impact on the landscape of a long-lived copper-smelter; non-ferrous smelting continued here for almost two-hundred years (until 1924). The eight riverside copperworks built after White Rock followed its pattern and in addition the White Rock works remained one of the most significant copperworks for over a hundred years: from the 1750s until the 1770s its owners were one of the four largest purchasers of copper-ore in Cornwall, occasionally the biggest; in the 1820s it was the eighth largest copper producer in Britain and from the 1830s until the 1850s it was the fifth or sixth largest.[96]

Under the terms of the Whiterock lease of 1737 Bussy Mansel (4th baron Mansel) granted the proprietors six acres of riverside land for building their works, the main building was 'The Great Workhouse', probably about 340ft long and 40ft wide. This may suggest that the largest furnace-hall at Llangyfelach had set a precedent for the Swansea industry in establishing a standard width. It was the White Rock Copperworks building, however, that was to establish the great length of the main smelting-halls at each of the works, its length was no less than 8.5 times its width. This was one of the first large industrial-scale buildings both of the modern age and of the world's first Industrial Revolution. Within it, and most of its successors, were a single row of large number of reverbatory furnaces arranged along its length. According to the terms of its lease the White Rock Works was to contain at least 20 furnaces, the same number as Llangyfelach,

but unlike there most furnaces were in a single massive furnace-house. The scale of this being recognised by contemporaries calling it 'The Great Workhouse.'

Forest Copperworks (1748-52) marked a unique reversion to the cluster of small smelting-houses on an ornamental plan. Three round pavilions held the main furnace compliment of 24 reverbatory furnaces while a fourth contained four of the larger calcining furnaces. The octagonal refinery at the middle of the works probably contained four specialist refining furnaces.

The next copper-smelter founded on the east bank of the river (the Middle Bank Copperworks of 1755) followed very much the design and planning established by the early eighteenth century Llangyfelach and White Rock Copperworks. It may have started with one single long range, but by 1771[97] it had no less than five main ranges of long furnace-halls which in total extent were 1,038ft long: more than 10 times the length of the two large Llangyfelach furnace-halls built some 54 years before. As at White Rock the largest smelting-hall was ranged across the hillside, set back from the river and parallel to Townsend's wooden-tracked railway, which delivered the prodigious amounts of coal needed for copper-smelting along the open entrance

Fig. 39. The White Rock Copperworks site (right) viewed from the south-west in the 1960s. On the left is the asbestos-clad successor of the 1819 copper-rolling mills at the Hafod Copperworks. The quay walls and significant structures on both sites are probably the most important remains of the Swansea copper industry.

fig.39

Fig. 40. The Forest Copperworks in the 1790s. On the left of the bridge is the assay house and between the circular furnace-pavilions is the entrance to the works. To the left of the nearest pavilion, a cupola surmounting the central octagonal refinery is visible. The two large chimneys in the foreground vent what seem to be twin brass-annealing furnaces or copper-ore calcining furnaces. William Edwards's Beaufort Bridge and John Padmore's weir for the Lower Forest Coppermills lie beyond the works.

Figs. 41 & 41a. Two drawings by Philip de Loutherbourg showing White Rock Copperworks from the south and west in the early 1790s. Later extensions to the first smelting-hall are shown clustered around a small sailing-ship. The long building facing the river has three chimneys over brass-annealing furnaces or large calcining furnaces. (By permission of the British Library)

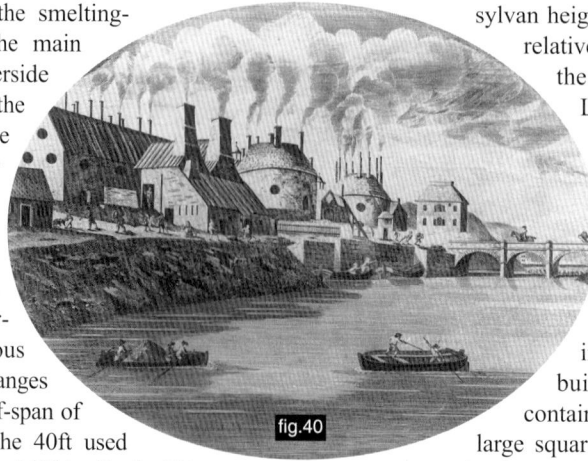

fig.40

fig.41

fig.41a

arcades at the rear of the smelting-house. This form of the main smelting-house, on a riverside site, set into the base of the valley-side parallel to the coal-road, railway or canal, was almost universally used for the next hundred years.

The main range at Middle Bank Copper-works was a prodigious 380ft long and the main ranges here may have had a roof-span of about 25ft rather than the 40ft used earlier at Llangyfelach and White Rock. This implies a massive length to width ratio for this building of 15:1; a similar parallel range on the river side of this was 263ft long. Strangely a fortnightly report on activity at the Middle Bank Works in 1796 notes only eight large calcining furnaces, eight smelting and four roasting furnaces at work in that period.[98]

By 1771 both White Rock and Middle Bank Works had lateral ranges, or buildings, extending toward the river, presumably this arrangement maximised the ventilation necessary to the intense fires burning in each range. There were four of these ranges at White Rock and three at Middle Bank which also had a third parallel wing enclosing a large courtyard in the centre of the complex: a unique arrangement for a Swansea Copperworks. The range facing the river was largely built of copper-slag blocks resplendent with Georgian half-round windows ('lunettes').

Some at least of the lateral wings shown on early plans of White Rock, and other buildings, were in fact detached smelting-houses of the earlier Bristol type. One of these at right-angles to the 'Great Workhouse' contained 'the great calciner'. Indeed, it was usual in the later eighteenth-century for calcining furnaces to be twice the size of the other reverbatory furnaces and at this stage the disparity may have been even greater, with some of the small reverbatory furnaces then in operation. Other small furnace-houses were added in a fairly ad hoc way in order to increase production.

Seven years after the foundation of the works in 1744[99] there were already 'four new smelters' in a small housing alongside 'the great calciner' and 'four new calciners' in a comparatively small building parallel to 'the Great Work House' on the side facing the river. Various artists depicted this scene of relatively small-scale industry still dwarfed by the sylvan heights of Kilvey Hill and the relatively unpolluted waters of the River Tawe. When de Loutherbourg drew this scene in the about 1790 'the great calciner' seems to have been demolished and either three substantial calciners or brass annealing furnaces inserted in the extended building that had once contained 'four new calciners'; large square chimneys then stood at the ends of the original building; piercing the apex of the gables, as did a third chimney at the end of the new extension of the building.

'The four new smelters' of 1744 were still standing in *c*.1790; like the calciner, their chimney-stacks ran along the apex of the gable at equidistant intervals. In addition, a smaller building, possibly a refinery, stood parallel to the 'Great Workhouse' on its landward side with two tall chimneys rising above the roof.

Most of the eight copper-smelters built after White Rock had a similar initial furnace-house with multiple fairly ad-hoc additions. Even the initial symmetry at the Forest Copperworks was spoilt when a long furnace-house of more conventional design was added to the south of the works before 1791.

The two copperworks built between Landore and Morriston in the 1780s and early 1790s were of this type, the Rose Copperworks had a main smelting-hall range built parallel to the hillside some 120ft long and about 38ft wide. The main block of the larger Birmingham Copperworks was built in a similar situation with mine boats from the Morris Canal able to supply the furnaces through the open arcades along the hillside of the smelting-hall as was also done at the nearby Rose, Forest and Landore Copperworks. The smelting-hall at the Birmingham Copperworks was some 210ft long and conformed to the earlier hall width of 40ft. Both these works, operated by Matthew Boulton and other Birmingham jewellery and engineering manufacturers, also had two lateral smelting wings added each on the hillside of their main buildings, which again probably gave direct access at their upper level for coal arriving via canal.

The next two works to be discussed were innovative in scale and planning and detailed partial plans of both survive (see illustrations). Both used a double width smelting-hall with central supports along which two parallel rows of reverbatory furnaces were built with their stoke-holes against openings in the outside walls.

Fig. 42. View in the 1960s of the White Rock Copperworks. To the left are the ruins of a nineteenth-century smelting-hall of the larger type with a double arcade. This hall had been widened from an earlier building which ran parallel to the Great Workhouse of 1737, and of which two semicircular arches with a doorway facing the river quays then survived. (960208/2)

Landore Copperworks of *c.*1793

This was the third smelting-works built by the Lockwood, Morris copper-smelting concern. It was built at Landore in *c.*1793 when the partnership had successfully leased the archaic multiple smelting-house, Forest Copperworks, to the Harford's Bristol Brass Wire Company. Thomas Jernegan sent copies of plans of a Swansea copperworks to the copper and brass manufacturer, Matthew Boulton, and in size and situation these conform to what is known of the Landore Copperworks.

This smelting-hall with a fairly small complement of 13 reverbatory furnaces was built across the hillside and against a retaining-wall supporting the coal-carrying Morris Canal as also were the recently completed Rose and Birmingham Copperworks. Where this works may have been unique was in siting two large reverbatory furnaces on an upper floor with all the stresses for the building structure that this implied (however it ought to be noted that two-storey reverbatory furnaces were tried at the later Hafod Copperworks). The leaving of the copper-smelting business by Sir John Morris II about a decade after the new work's completion may indicate problems with the working efficiency of the arrangement. New smelting capacity was added to the works by extending it lengthways and the two rows of furnaces were retained in the plans of the enlarged works. In the original works one arrangement for greater efficiency was the provision of a bridge directly from the canal to the upper furnaces which eased coal delivery.

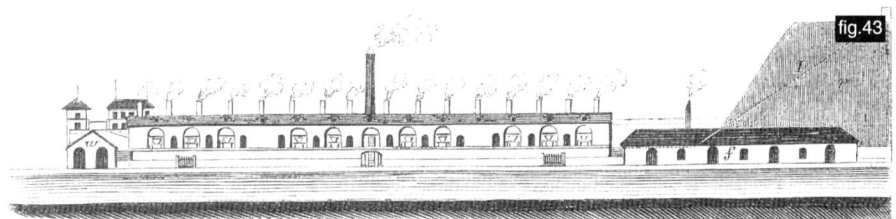

Fig. 43. Drawings of an idealised Swansea Copperworks. These emphasised the placing of each works between the river 'by which the ores are conveyed' and the canals 'by which coals, &c. are brought'. (Tomlinson's Cyclopaedia, 1852)

Fig. 619. ELEVATION OF A SWANSEA COPPER-WORK.

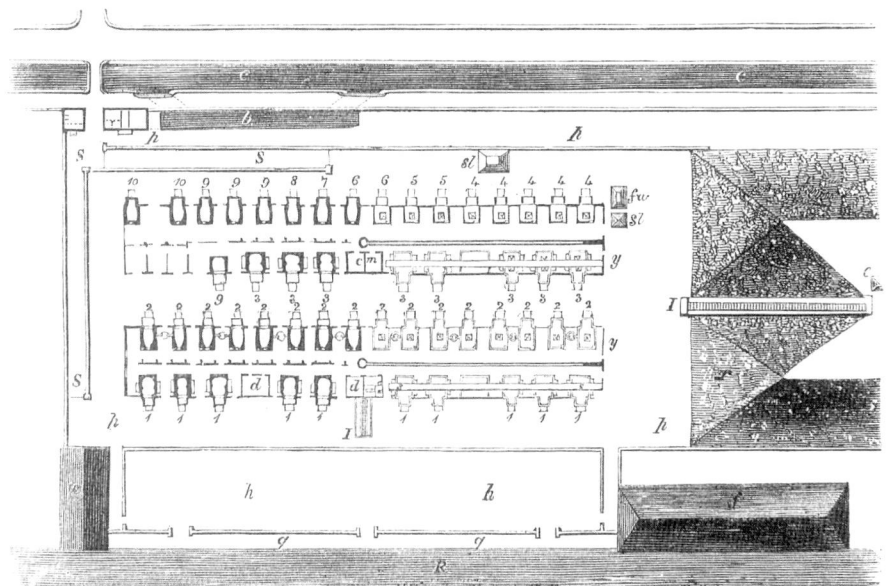

Fig. 620. GROUND-PLAN OF THE SAME.

Hafod Copperworks of 1808-09

Figure labels (fig.44):
Original slag-tip (Later crowned by sulphuric-acid chambers (1856)) · Original slag-tip railway · Elevated flue · Later copper blast furnaces (Piltz) · River Tawe · Middlebank Copperworks · Later silver blast furnaces (Piltz) · Coal bunkers · Ore yard · Ore yard · Copper wharf · Long house · Calcining furnaces · Slag house · Chill house · Crane · Ore yard · Machine shop · Coal railways · Later Manhe house (Bessemer converter) · Coal bunkers · Furnace house · Coal railways · (site of dock) · Rolling-mill No.2. (1842) · Rolling-mill No.1 · Rolling-mill (1819) · Steam-engine houses (still extant) · Coal bunkers · Hammer house · Canal dock (1808) · Offices · Towing-path · To Collieries · Swansea Canal · Canal dock (1819) · To Swansea Dock · Limekiln · 0 50 Metres · 0 200 Feet

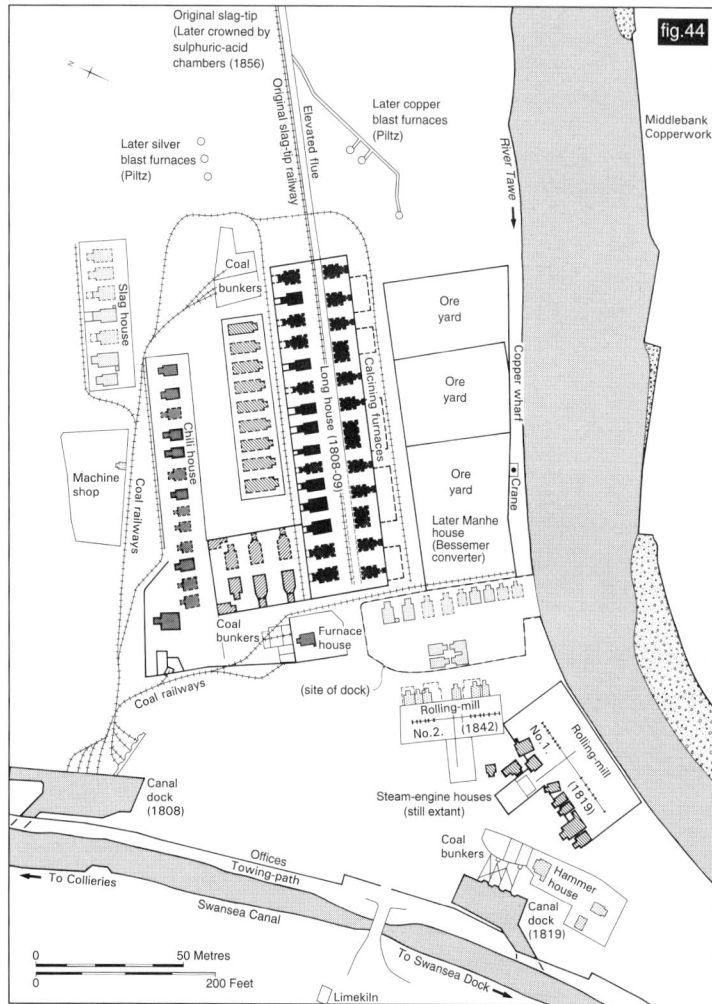

Fig. 44. Nineteenth-century reverbatory furnaces at the Hafod Copperworks. Based on a large-scale plan made towards the end of its life as a copper-smelting works which finished in 1924. The plan attempts to restore the earlier arrangement of furnaces with reference to contemporary maps and plans. (City & County of Swansea: Swansea Museum Collection)

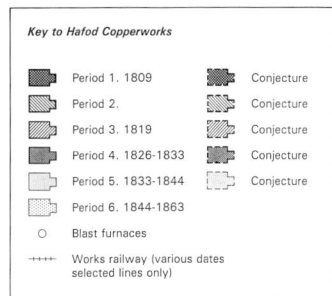

Key to Hafod Copperworks

Period 1. 1809 — Conjecture
Period 2. — Conjecture
Period 3. 1819 — Conjecture
Period 4. 1826-1833 — Conjecture
Period 5. 1833-1844 — Conjecture
Period 6. 1844-1863
○ Blast furnaces
++++ Works railway (various dates selected lines only)

The building of the Hafod Copperworks in 1808-09 marked a very significant trend in the planning and size of the Swansea Copperworks. It also marked the entry into the area of the Cornish mining interests, who had long been resentful of what had been seen as the economic dominance of the Swansea smelting concerns. The founder of the Hafod Works, John Vivian, was a Cornish copper-mining entrepreneur who had previously been heavily involved in setting up the Carn Entral (1755) and Hayle copper-smelters in Cornwall and also operated the Penclawdd copper-smelter (from 1800) to the west of Swansea.[100] He had left Penclawdd dissatisfied with its inefficient working arrangements and was determined to improve these at his new Swansea Works which his grandson Henry Hussey Vivian claimed was designed on model lines 'so as to ensure the greatest economy of working'[101] and was the 'Swansea Copperwork' whose simplified plan appeared in Tomlinson's *Cyclopaedia* of 1852.

An eminently suitable site was chosen for the new Hafod Copperworks on the southern part of a large headland enclosed by a loop of the tidal River Tawe. The Swansea Canal ran north-south on rising ground along the western boundary of the works site just south of Landore. The topography of this site allowed the new smelting-hall to be laid-out parallel to wharves constructed on the River Tawe, with an extensive copper-ore storage yard between the two. Instead of being parallel to the canal, as all previous works had been to their coal supply transport, this copperworks was built nearly at right-angles to it. In other words, it was built with expansion in mind; previous works had sited their original smelting-halls for conveyance of supply to their original long line of furnaces, at the cost of obstructing supply to any further buildings erected on their riverside sites.

An added reason for doing this was that like the Landore Copperworks (*c*.1793) the Hafod Copperworks smelting-hall had not one but two rows of furnaces, which needed feeding with enormous quantities of coal to the stoke-holes on both sides of the building and these both needed to be accessible to the canal. At Landore, this was achieved by running railways from the Nant Rhyd-y-Filais collieries down each side of the building, the upper level running alongside the Morris Canal; nearby Fenton & Co, at what later became the Rose Copperworks, also had a coal supply railway running to their smelting-hall.

In 1810 the Vivians had the first of two canal-docks built for bringing in coal from the Swansea Canal, following a precedent set earlier on local canals by the proprietors of the Upper Bank and Birmingham Copperworks. Eventually a dense network of works railways ran from this dock along the rear of each of the smelting-halls. A stretch of railway was also built in the interior of the smelting-hall, it led down the central aisle and straight out the east end of the building to slag dumps that soon extended to the riverside. By 1826 this tip had already risen high enough to necessitate a small-steam powered winding-engine to haul waste wagons to the summit: a feature that became common at all the mountainous tips of the non-ferrous metal smelting-works around Swansea. A second, and smaller, inclined-plane drew wheelbarrows loaded with copper-ore up to a landing at the central upper-level of the smelting-hall, from whence a high-level staging (Tomlinson's 'tramway') fed the hoppers charging nine large calcining kilns. Landore Copperworks also had an incline for raising copper-ore from its river-dock to its smelting-hall. Other railways around the works ended in elevated tipping-stages feeding an eventual four groups of coal bunkers.

Each furnace at Hafod, as elsewhere in Swansea, was sited by a large external arch to facilitate fuelling and ventilation. The most impressive example of this surviving until the 1960s was the continuous brick arcades of the White Rock Copperworks across the river. The only example of this to be seen today in a Swansea copperworks is in the rear (east) wall of the 'Great Workhouse' at White Rock where large intermittent brick arches pierce a mass wall of Pennant sandstone rubble (the Clyne Valley Arsenic Works also has something similar). In fact the arrangement and size of the

Fig. 45. An aerial photograph taken soon after copper-smelting at Swansea had ended in the late 1920s. The large mill buildings and smelting-houses of Morfa Copperworks are to the left middle-ground with those of the Hafod Copperworks to the right background. The slag-tips of the Hafod Copperworks are to the far left and far right middle-ground with that of White Rock Copperworks in the background. (National Museums and Galleries of Wales – Department of Industry)

openings reflects the siting and type of the reverbatory furnaces.

The Jernegan designs for a copperworks in the 1790s show an uphill elevation pierced by no less than eight semicircular headed arches rising through two storeys. Here, and at Hafod, the arches were grouped in threes with two high narrow access doorways for the furnace-keepers flanking broad arches and usually smaller and higher arches allowing maximum fuelling and ventilation access to the furnace. It is just possible that these triple grouping around a great arch gave some inspiration to influential chapel architects developing this pattern as a classic chapel facade, two of whom lived in the copper-smelting area around Landore and Morriston. The great smelting-hall at Hafod (known as 'the Long House') as illustrated in Tomlinson's *Cyclopaedia* had 11 great arches along its riverside facade arranged symmetrically in three groups with a central five flanked by blank bays with a further three arches beyond; all flanked by smaller round-headed openings that gave access to the copper-workers, the blank bays masked occasional copper-matte storage areas. Two of the giant arches designed by Jernegan actually rose through the eaves, line but this was occasioned by the exceptional circumstance of having an upper tier of furnaces.

These long, and relatively narrow, reverbatory furnace-halls of the new industrial age were not unique to the copper industry; in the early nineteenth century Glamorgan was the world centre of two metal smelting industries: copper and iron, of which both used great rows of reverbatory furnaces and it is instructive to compare the differences between the two types of building. Hafod, and later Morfa, may have been the largest copperworks in the world during most of the nineteenth century, but the world's biggest ironworks in the first two decades of the nineteenth century was the Cyfarthfa Ironworks at Merthyr Tydfil; inland and to the east. In the early nineteenth century Cyfarthfa Ironworks had three large forges housing rows of reverbatory furnaces for transforming brittle cast iron from the blast furnaces into malleable wrought-iron. Such was their heroic impact on the landscape that artists, such as Turner and Penry Williams, were recording these monumental structures just as De Loutherbourg and Rothwell were depicting industrialisation at Swansea. The nature of the metal being processed holds the key to why the two groups of structures, one by the coast at Swansea and one in the uplands around Merthyr, were different.

The Merthyr works carpenter, Watkin George, discovered that cast iron could replace timber and stone in constructional work in a way that the relatively soft copper could not. The main forge building at Cyfarthfa, some 47ft wide, was given a series of cast-iron trusses. The wall plates of the Merthyr forges were also of cast iron supported by cast-iron pillars, so that the sides were completely open to maximise ventilation and access for furnace fuelling and charging. At the lower Cyfarthfa forge, the bay system of the side arcades reflected the positioning of grouped pairs (the Cyfarthfa 'norm') of reverbatory furnaces in each alternate bay. As at Swansea, the stoke-holes of the reverbatory furnaces at Cyfarthfa were near the outside walls. Each bay in the outside wall was spanned by a cast-iron arch with circular bracing between it and the wall plate. Part of such a constructional system remains at the Crawshay's Treforest Tinplate Works Blacksmith's Shop where each bay spans 24ft 5in.

The Vivians having been part owners of the famous Copperhouse Foundry at Hayle in Cornwall and having later in the nineteenth century built the Hafod Foundry as part of their huge industrial complex at Swansea. There is little evidence of early structural ironwork before the roofing and re-roofing of much of the area of the Hafod Copperworks with steel stanchions in the early twentieth century. The adjoining Morfa Copperworks (1828 for rolling; 1835 for smelting) had a more interesting use of structural ironwork. The rolling-mills designed in the 1830s were ventilated with open arcades, with circular openings in the spandrels, between the arches. However, the great transformation of the mill in the 1840s into a huge hall-like building was undertaken with an open side to the east, with tall cast-iron columns supporting a timber wall plate and the roof, this still survives. The access to the Morfa Copperworks across the Swansea canal was also via a cast-iron bridge with the date 1840 and the name of the works cast-in. The early twentieth-century generating-house was also given dual internal metal-arched arcades that still remain (the clock-tower building).

The pattern of furnace chimneys in relation to the smelting-hall buildings was interesting; reverbatory furnaces needed tall chimneys to function in the absence of a forced draft. Structurally it might have been more convenient to put them along the outside walls of the furnace-halls as was done with the tinning-houses of the local tinplate industry with their characteristic long rows of tall chimneys above the tinning-bays set against the outside walls.

That was, in fact, the pattern adopted at Swansea's first copper-smelter at Llangyfelach (1717); apparently at the Cambrian (1720); occasionally at the White Rock (1737) but not subsequently at other works. Most of the furnace stacks in the 'Great Workhouse' smelting-hall at the White Rock Copperworks emerged at or near the apex of the roof for functional reasons. In this, and all subsequent works, the stoking-hole and grate at the opposite ends of the reverbatory furnaces were placed at or near the external walls of the smelting-hall both for

ease of charging and for ventilation. This could also help to make the exceedingly tall reverbatory furnace flues more secure by minimising the height of the stack exposed to high winds and by allowing high-level bracing from the timber roof trusses. Even better was the provision of a masonry spine wall as in Jernegan's design (at Landore) which could laterally brace the high central stacks.

The copper-furnace chimneys were originally erected singly but at the Forest Copperworks (1748-52) they were grouped in pairs (as was the case in the Cyfarthfa Ironworks Forges). The Landore Copperworks (ca. 1793) had them singly again as did the Hafod Copperworks (1808-09). However, the litigation over and general concern about the toxidity of the sulpherous copper fumes, caused the Vivians to do two things. One was to build very high chimney stacks in the hope that winds would disperse the damaging smoke; this was impractical to greatly increase the height of the multiplicity of slender individual flues, so that it was necessary to have flues from each furnace converging on a single high central stack. Secondly, the Vivians were experimenting with lime-bath and other scrubbers from the 1820s and again it was necessary to bring the furnace discharges together, so that an attempt could be made to cleanse the damaging smoke. How this was done is illustrated in the simplified plan of the main Hafod Copperworks smelting-hall as found in Tomlinson's *Cyclopaedia* of 1852. Here a central elevated flue ran down the centre of the smelting-hall where the 10 large furnaces calcining the raw ore were situated and received the particularly injurious fumes emanating from that initial processing. The scrubbers were presumably situated adjoining the high central stack which soared through the apex of the centre of the smelting-hall. The 16 individual chimneys of the smaller reverbatory furnaces on the landward side of the smelting-hall still discharged directly into the atmosphere in 1852.

By the early twentieth-century the surviving works plans of the Hafod Works[102] indicate that the furnace waste-gas flue running down the centre of the main smelting-hall (the 'Long House') of 1808-09, now extended eastwards to feed the huge sulphuric acid manufacturing chambers built on top of what had been the original waste-tip of the copperworks. These 20ft high lead-lined chambers dominated the landscape of the Lower River Tawe, but also rid the surroundings of the worst effects of sulphur contamination and which was now turned into a saleable by-product. Other copperworks followed this example set by Hafod works. A flue system running around the back of the 'Great Workhouse' smelting-hall on the White Rock site can still be seen.

The relationship between the furnaces and the outside walls of the smelting-halls has not been discussed in detail. The evidence given by Tomlinson on the Hafod Copperworks is ambiguous: on his simplified and idealised plan of the works and on his detailed furnace drawings, every one of 51 reverbatory furnaces has its fuelling-grate outside the line of the external walls. The surviving plan of the works made in *c*.1910 shows elements of no less than 6 rows of reverbatory furnaces surviving, all in separate buildings and all set within the walls of their enclosing buildings. Four rows had their stoke-holes close to the external walls but two rows were set in the middle of what were relatively narrow buildings. At this stage in the works' life the old large reverbatory calcining furnaces had been replaced by some five long rotary calcining kilns, but it seems likely that the large calcining kilns of 1808-09 did

Fig. 46. The two engine houses of the Hafod Copperworks rolling-mills in the 1990s with the copper-rolling plant visible in the open-air on the riverbank. The white-painted building at the bottom left was the later nineteenth-century Hafod works office. The former Morfa Copperworks rolling-mill is the long building at the top. (935055/62)

have fire-boxes extending beyond the line of the original external walls. Indeed there were five external pentices, or lean-to roofs, on the south side of the main smelting-hall and these may have provided shelter for the stokers of the great calcining kilns. Indeed, a map of the Swansea Copperworks made in 1822 shows the long smelting-halls at the Hafod, Landore, Rose, Birmingham and Forest Copperworks, all equipped with these intermittent canopies only on their river sides where incoming ore was taken into the large calcining kilns. The canopy was probably to deflect rain from entering the smelting-halls through the giant arches giving access to each of the calcining kilns.

The surviving detailed map of the Hafod Copperworks of *c.*1910, together with the early maps of the area, and sequential recorded totals of the numbers of reverbatory furnaces at the works,[103] enables the development of furnace numbers and smelting-halls to be reconstructed. The first smelting-hall at Hafod (later called the 'Long House'), erected in 1808-09 on model planning lines was 380ft long, equal to the length of the previous longest smelting-hall at Middle Bank erected between 1755 and 1771. In this case, however, the hall had a double row of furnaces as did Jernegan's Landore Copperworks of *c.* 1793. There were 24 furnaces in the hall: separated by a central row of roof supports; not the 26 shown in Tomlinson's simplified plan of 1854. Each aisle had a roof span of 46ft, approximating to the span of the reverbatory furnace-houses at the great Cyfarthfa Ironworks and broader than the 40ft span previously standard at the White Rock Copperworks (1737) and elsewhere. The overall width of the Long House was 92ft, producing a length to width ratio of 4:1 and so was twice the span of the earlier Cyfarthfa Forges and earlier smelting-halls.

This was soon enlarged - by 1822 the 'New Furnace House' was added as a right-angle wing to the north-west corner of the 'Long House' and seems to have contained eight more reverbatory furnaces; six of which were a slightly greater width than the 15 reverbatory furnaces arranged in a line down the north side of the 'Long House'. At the same period a steam-powered rolling-mill was added to the south-west of the Long House on the narrow neck of land between the River Tawe and the Swansea Canal on the hillside above. On this site today are the preserved rolls of a succeeding rolling-mill with the uniflow engine and engine-house of 1910 still standing on approximately the site of the original mill engine-house of 1819 (which was probably immediately to the south-east). In *c.* 1910 nine reverbatory furnaces (which worked until 1980 but are now demolished) lined the shore side of the rolling-mill to heat and reheat the copper being rolled and they, or their predecessors, were in this location for use in the original mill.

By 1826 the first of three parallel smelting-halls was added on the northern (landward) side of the Long House. It was presumably this arrangement that led to Tomlinson's simplified copperworks layout, with a parallel smelting-hall illustrated next to the Long House, in his '*Cyclopaedia*' of 1854. This third smelting-hall became known as the 'Dyfatty House' and seem to have housed a row of some nine reverbatory furnaces which were the same width as the original 15 in the Long House, but were almost twice as long (detailed drawings of these were published in John Percy's *Metallurgy* where they are revealed as being a large narrow size of calcining kiln of a design unique to the Hafod Copperworks). Their stoke-holes backed onto the same coal-supply railway that fed the long line of 15 reverbatory furnaces in the original Long House. This single row of long furnaces actually sat in the middle of a furnace-hall slightly wider than the 47ft wide aisles of the Long House and the two further smelting-halls to the north were a similar width (i.e. there was a reversion to the earlier copper and iron trade reverbatory furnace hall width).

The fourth smelting-hall (later known as the 'Chili House' from the use of Chilean copper-ores) was added in 1826-33 to the north of the New Furnace House and extended eastwards in parallel to the Dyfatty and Long Houses. It seems to have contained a single row of some 13 reverbatory furnaces, with an additional larger furnace situated in the roofed-in yard of the New Furnace House to its west. These furnaces were of the same dimensions as the original 15 in the Long House.

Thus, the total number of known reverbatory furnaces at Hafod would seem to be 65 by 1833. After this, the fourth and final parallel smelting-hall (i.e. the fifth line of furnaces) was erected towards the northern boundary of the works, it became known as the 'Slag House' and may have had a line of seven furnaces of similar dimensions to the first. All these new smelting-halls had new lines of railway laid along the facades, from which the reverbatory furnaces were constantly charged with coal day and night. The original canal dock of 1810 was extended so that two boats could be unloaded simultaneously into four coal-trucks apiece, in fans of tracks positioned at a lower level than the buttressed retaining walls holding the canal dock water and laden boats. The rolling-mill had its own canal-dock by 1826 and four lines of railways charged sunken coal bunkers which still remain in place.

Unfortunately, the layout of this functionally planned model site now plunged into chaos. The powerful Cornish concern led by Michael Williams, joined their rivals represented by Pascoe Grenfell and John Vivian in the Cornish takeover of much of the Swansea smelting industry in the first decade of the nineteenth century. The Williams started the construction of the steam-powered Morfa Rolling Mill in 1828 (supplied by their existing Rose Copperworks) on the larger half of the peninsula that the Hafod Copperworks already occupied, and in 1835 commenced the construction of a huge copperworks that equalled, and at times surpassed, in size that of the neighbouring Hafod Copperworks. The Hafod Works could not therefore expand to the north, to the east lay its mountainous waste tip, the heights of which it had eventually to ascend in order to build the sulphuric acid manufacturing chambers in 1865.

To the west lay the Swansea Canal and then (from 1845) the South Wales Railway. Eventually many of its general departments had to be moved, such as the General Office which still stands. A steeply-graded standard-gauge railway with reversing-loops replaced the vertical lift from the main line via a branch along the narrow neck of land between river and canal to the south. For half a mile south from here (past the extant engine shed constructed of Vivians' patent crushed copper-slag bricks) was the strip of riverside to the east of the workers' town of Trevivian, where the large infrastructure of the Vivians diverse expansion in the later nineteenth-century took place: the Hafod Phosphate Works; the Hafod Isha Nickel & Cobalt Works; the Hafod Foundry and the Hafod Sawmills.

After 1833 it was no longer possible to pursue a methodical development of the original copperworks site. The expansion from 65 reverbatory furnaces at that date to 95 operated by Vivian & Sons in 1844, was only achieved by building in whatever open-space existed.

With the building of the Slag House furnaces shortly after 1833, the four ranges of smelting-halls had reached northwards nearly to the original ancillary buildings that lined the northern boundary of the works. The next and final reverbatory furnace-hall was then built extending from the river bank, at right-angles to the Long House and west of the copper-ore yards. The 1826 Ordnance Survey Drawings of the site are somewhat ambiguous, but it looks as if the Hafod Works originally had a tidal river dock on this site where incoming copper-ore ships could dock and unload. The inland curving walls of the building may have been seated on the walls of the earlier dock, with the dock filled with the copious amounts of copper-slag that were available. The space above the dock was divided into two halves: that on the east was a fairly conventional smelting-hall with some nine furnaces in a row. Four of these were still in position in *c.* 1910 and were of two sizes: two were the size of the original row in the Long House and two were somewhat shorter and the smallest smelting furnaces in the works. The western hall had a unique central cluster of four furnaces. These additions to the furnace complement, together with the reverbatory furnaces at the Vivian's Lower and Upper Forest Copper Mills, would have taken their total number of furnaces to the 95 recorded as being in their works by 1844.

There is no doubt that this huge complement of furnaces allowed greater specialisation of furnace type and function. The beginnings of specialised types of reverbatory furnace had begun long before in the seventeenth-century copper-smelting centres of Redbrook and Bristol. However, the number of furnace types used in Swansea at the beginning of the eighteenth century had been quite low. At the copperworks in 1796 20 reverbatory furnaces were in use to turn 244 tons of copper-ore into 47 tons of copper within a fortnight: eight (large) calcining, eight smelting and four roasting furnaces being used.[104] As discussed John Percy considered that even at the Hafod Works there were only two basic types of furnace - the large calcining furnace and the smaller melting furnace even if parts of the large ranges of furnaces were each given specific functions to perform such as the 10 itemised in the idealised Hafod Copperworks shown in Tomlinson's *Cyclopaedia* of 1852.

The reality was that by the end of 115 years of continuous smelting in 1924, the Hafod Copperworks was a hopelessly cluttered and

Fig. 47. A nineteenth-century photograph of the Hafod Copperworks from west of the Swansea Canal.. The canal bridge on the right gave access to the works with offices and laboratories visible within. Dominating the picture is the elevated railway incline that gave access to the large secondary waste-tip on the western valley side. (City & County of Swansea: Swansea Museum Collection)

fig.47

crowded site of rolling-mills, blast furnaces (copper and silver), intermittent survivors of the great serried rows of reverberatory furnaces, rotary kilns, Bessemer converters, inclined railways, to two successive and mountainous tips that dominated the lower Swansea Valley landscape, and ancillary departments of every description.

Silver smelting had been added in the 1850s to the range of functions at the works. Blast furnace use started in 1859 and the surviving plan of the works shows three copper and three silver Piltz blast furnaces. The original riverside tipping area was full by 1865 when a large inclined viaduct was constructed to the westwards, where a mountainous tip rose immediately north of the workers' housing at Trevivian. A massive abutment of cast copper-slag blocks that survives alongside the dry bed of the Swansea Canal at Trevivian, near to the remaining limekiln.

Gaps appeared in the long rows of reverberatory furnaces by the early twentieth century as they were replaced by blast furnaces, rotary calcining furnaces and Bessemer converters. Pollution was finally greatly reduced from 1865, as long culverts or flues were constructed up the original tip to the awe-inspiring 20ft high lead-lined sulphuric acid chambers which now dominated the landscape. Some reverbatory furnaces survived in use in copper

rolling-mill no.1 until 1980 when the works finally closed. The two surviving rolling-mill engine-houses now dominate a site where only the substructures of furnaces and smelting-halls survive under a levelled area of waste-ground and new roads. These are important survivals of what was the largest copperworks in the world at the centre of the international industry.

Water to feed a large number of steam-engines and processes was fed around the works from the neighbouring Swansea Canal via a multitude of pipelines. The Pennant sandstone for building this and the majority of the works was near at hand on the adjacent hillsides.

Copper-product Manufacture

The actual extent to which the Swansea copper industry required large water-powered mills and the attendant watercourses and weirs required (before 1820s, when the first recorded steam-mills were introduced) depended on two factors: firstly the proportion of concerns with full vertical integration (i.e. those that produced finished goods for the retail market, rather than merely refined raw materials straight from the furnaces for use in manufacturing in Bristol, Birmingham, London or the Greenfield Valley in Flintshire) and, secondly, the nature of the goods produced for the market (i.e. whether they required rolling into sheet; hammering or 'battering' into shape, stretching into wire, hand-finishing into pins or alloying with other metals). These factors, and the way they affected the Swansea landscape, can only be effectively quantified by an examination of typical examples of the nature of each of the successive copper-working concerns up to the early nineteenth century.

However, before such an examination of localised detail, it is worth trying to quantify the general demands that existed in the industry. The exotic nature of slave trade goods tended to focus contemporary commentators on copper and brass (copper-zinc alloy) production during the eighteenth century. It was undoubtedly important and the 'Copper King' from Anglesey, Thomas Williams, in 1788 declared that it was this trade that induced him and his partners to invest a great sum of money in the copper industry.[105] Empty vessels leaving Bristol or Liverpool to collect African slaves needed barter goods to fill their empty holds. Textiles were the most valuable goods carried, but copper and brass goods formed the next most expensive commodity; however, this was not the dominant use of copper even in the 1770s when the trade was at its peak. The following figures graphically show the rise and fall of the trade.[106]

Fig. 50. The plate and sheet (No.1) rolling-mills at Hafod Copperworks as rebuilt in 1910. Copper slabs cast from the furnaces are in the foreground waiting to be put through the rolls and turned into the copper sheet stacked in the middle foreground. The central flywheel and wheel-drive from a steam-engine on the right are also visible. (City & County of Swansea: Swansea Museum Collection)

Copper and brass barter goods used in the African slave-trade

	Copper	Brass (copper-zinc alloy)
1721	23.7 tons	28.7 tons
1761	46.6 tons	85.3 tons
1771	52.9 tons	331 tons
1775	125 tons	130.4 tons
1780	10.6 tons	16 tons

These figures help to explain why only two large water-powered coppermills were built on the River Tawe in the eighteenth century to manufacture pure Swansea copper into goods, whereas no less than 18 water-powered works were built on the River Avon and its tributaries in the Bristol area to manufacture Swansea and Bristol copper and local zinc (from Mendip) into brass goods for export from Bristol. By 1780 trends were radically changing and this requires an examination of the changing purpose for which copper was used.

The combined export trade in copper and brass to what is now Belgium, France, Germany and the Netherlands was almost always greater (certainly after 1721) than that involved in the African slave-trade. Presumably Lockwood, Morris & Co, and others, would have run their involvement in these sales from their London warehouse.

Similarly, from 1721 the West Indian trade was greater than the African trade, with copper bottoms and plates supplied to the large sugar and rum refineries and distilleries which had developed as a result of the slave-dependent economies of the Caribbean. By 1776 this market was taking 614 tons of copper and 222 tons of brass.[107]

A major new market was opened up in 1731 when the British East India Company started to export large quantities of copper ingots to the East. In 1751 they also began the export of manufactured copper in the form of sheet or other goods. Between 1774 and 1791 an average of no less than 1,588 tons per annum were exported to the Far East.[108] For the eastern market much of the copper was 'japanned' at Swansea to give it an attractive extra lustre; this process imitated that which was used in Japan until 1873. Molten copper was cast into shallow canvas trays submerged in a bath of hot water, the red hot metal that solidified on the tray was immediately lifted out of the water into the steam and after a short period of exposure thrown back into the water. The exposure of the red hot metal to the steam produced a permanent and beautiful deep rose-coloured layer of oxide which gave the copper a very attractive appearance much valued in the Far East.[109]

The biggest change in the market, however, occurred with the huge growth in the demand for rolled copper-sheets with which to sheathe timber naval vessels; and for the production of large copper

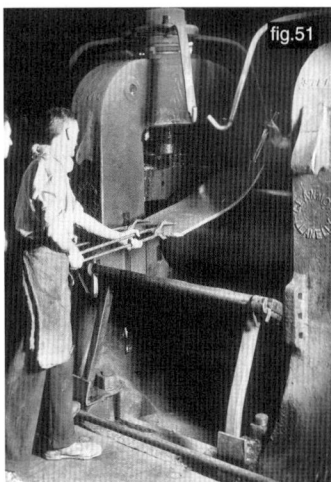

Fig. 51. The process of rolling copper cake into plate was photographed at the Hafod Copperworks at the end of copper-refining in Swansea, c.1924. (National Museums and Galleries of Wales – Department of Industry)

Fig. 52. Hafod Copperworks in c.1910: 31 locomotive boiler-ends (smoke-box ends), with holes to receive 206 longitudinal copper tubes each, readied for dispatch to a railway locomotive works. (City & County of Swansea: Swansea Museum Collection)

bolts with which to fasten the sheets to the thick timber hulls.[110] The start of this radical shift in trading patterns occurred when the 32-gun frigate 'HMS Alarm' was copper-sheathed in 1761 and it reached its zenith in 1783 when almost every ship in the British Navy had a copper-sheathed bottom.[111]

There was a third major market that was developing in the latter part of the eighteenth century. This was an ever-increasing demand for small articles such as brass buckles, buttons, ornaments and toys, the manufacture of which was centred on Birmingham. In 1762, Matthew Boulton revolutionised the size and organisation of the trade, when he founded his huge model Soho Manufactory at Handsworth in Birmingham and built workshops for over 1,000 artisans. His mint of 1783 was able to produce 30,000 coins an hour. The works expanded into the production of plated ware, of works in bronze and of composite items in 'ormolu', such as vases, candelabra and other ornamental items.

Two inventions made in 1769 facilitated mass-production in this industry and greatly escalated the demand for copper for this purpose. A London gilt toy-maker, John Pickering, patented a machine for impressing designs on sheet copper and brass by means of hammering sheet metal on to a raised die. A few months later, a Birmingham man, Richard Ford, improved on the invention by using a raised die and a corresponding sunken mould into which it fitted. The new process was particularly suited to the production of saucepans, kettles, buttons and similar articles.

In 1774, James Watt was attracted to Boulton's Soho Works, where the new foundry was soon increasing the scale of steam engineering. Many different brasses and bronzes were developed to withstand the combined effects of heat and steam pressure. Brass cylinders were required for pumps and some engines, together with a vast range of fittings, such as grease caps, gauge covers, whistles, steam valves, steam cocks and other accessories.

Matthew Boulton and his fellow manufacturers from the 'Jewellery Quarter' of Birmingham were running two of the Swansea copperworks (the Birmingham and the Rose or Ynys) by the end of the eighteenth century, in order to partially supply this huge need. A great number of special trades developed in Birmingham using copper and brass including marine compass, hinge and hat-pin makers. The extent of the export trade in these and other goods is shown by the fact that, by 1792, about a quarter of the articles produced in Birmingham were exported to France and Italy.[112]

The other great British metallurgical manufacturing centre in Sheffield also developed into a large user of Swansea copper, in a development that was to allow the growth of the Sheffield plated-goods industry. In 1742 the process by which the ancient Egyptians and Greeks had plated copper articles with silver was accidentally rediscovered. Thomas Bolsover, a knife-maker from Sheffield, noticed that copper and silver were being united by beating and rolling. At first only small articles such as buttons and snuffboxes were produced, but soon the production of 'Sheffield plate' became a major industry, the initial process being later superseded by electroplating.[113]

In 1801-5 production of copper in the United Kingdom may have averaged 6,644 tons per annum (world production was 9,100 tons in the same period),[114] of which the Swansea region produced about 90%.[115] Precise figures are hard to come by and may be distorted by the amount of copper imported and also exported, but Professor John Harris produced a workable breakdown of these figures in the varying sectors of this market c.1800:[116]

Uses and sales of British Copper c.1800

General export of copper	2,900 tons;
General export of 1,500 tons of brass	965 tons;
Exports of copper to India etc. via the East India Company	1,000 tons;
Sheathing of naval ships	1,000 tons;
Use in Birmingham jewellery, ornaments and steam engineering etc.	1,000 tons;
Coinage (a new use)	600 tons.

This demand soared throughout the nineteenth-century at a time when Swansea had a virtual monopoly of the national and world industry:

Sales of Copper in Great Britain[117]

1801-05	6,644 tons average per annum;
1821-25	11,635 tons average per annum;
1841-45	22,528 tons average per annum;
1861-65	48,118 tons average per annum;
1881-83	93,629 tons average per annum.

As can be seen, demand was doubling every 20 years, so that capacity in 1861-65 was four times what it had been in 1801-5, and in 1881-83 it was no less than eight times what it had been at the beginning of the century.

During the nineteenth century, the Birmingham jewellery, ornament and engineering markets, improved by Matthew Boulton's organisational efforts, expanded enormously. The 20,000 tons of copper used by the Birmingham industry in 1866 was equal to almost half of all the copper produced in the UK. At Birmingham, 3,000 tons were used in making brass for domestic use and 7,000 tons were used for the specialist lower copper/higher zinc (i.e. 60% copper, 40% zinc) brass alloy, termed yellow metal, commonly used for the sheathing of sailing ships. In addition, 10,000 tons were used for engineering, 6,000 tons for making copper and brass tubes and 4,000 tons for wire.[118] This level of demand explains the involvement of Birmingham firms on eventually three smelting sites from a start in the 1780s on the two riverside sites sub-leased by John Morris I between Landore and Morriston.

The almost universal use of copper-sheathing for the vast tonnage of timber sailing ships, the British merchant marine being by far the largest in the world, created a huge trade throughout the nineteenth century. From 1832, a cheap substitute was available in the form of yellow metal which was patented by G.F. Muntz in 1832, this had far-reaching implications for the form of the Swansea copper and non-ferrous works. By the 1840s the major Swansea smelting companies were producing large quantities of the alloy. In the 1860s the production abated as the tonnage of iron ships overtook that of wooden, but it carried on as an export trade into the twentieth century. The size of the trade can be gauged by the fact that in 1869 Vivian and Sons sheathed 123 ships (150,000 tons); Williams and Grenfell 169 ships; while Muntz (in Birmingham) sheaved 265 ships (using 60% Swansea copper).[119]

The Bristol brass and copper industry had dominated the early eighteenth-century copper trade. The Swansea copper trade was supreme by the last decade of the eighteenth century, but by the 1840s it is obvious that the Swansea industry had also become a combined brass and copper industry in response to market forces.

However, the early eighteenth- and later nineteenth-century brasses differed considerably from each other in both composition and uses. The brasses in use at the beginning of the eighteenth century were 'Alpha Brasses', which contained up to about 36% of zinc; these were moderately weak but very ductile and could be cold-rolled. 'Beta Brasses' such as 'Yellow or Muntz Metal' containing between 36 and 45% of zinc were much stronger, moderately ductile (they needed to be annealed after rolling) and

tough. 'Gamma Brass' containing over 45% zinc was brittle and weak and was not generally used.[120]

The great advantage of using an alloy like Yellow Metal was in reduced production costs, zinc being considerably cheaper than copper. However, the standard of purity of this alloy was very high, it was free of arsenic and almost of sulpur.[121]

The Swansea Coppermills

As has been noted, the actual production of copper required some water-power provision for the driving of fairly small-scale stamp and clay mills. In the first works, at least, water-power was also probably required for blowing the furnace. Large-scale power needs over and above these, involving the construction of substantial mills and water-leats, depended on the location of the smelter and the purpose for which the copper ingots were prepared

At Swansea there was a large early use, both of the pre-existing complex of water-powered rolling and trip-hammer ('battery') mills of Bristol and of the earlier substantial mills around London, producing pins, wire and domestic ware for what became the world's largest city and market. To an extent, mills in these two areas, and in the Greenfield Valley at Holywell in north-east Wales, continued to process Swansea copper throughout the industry's existence. Much copper production in the eighteenth century also went to Bristol to be alloyed with the zinc and then to be processed into domestic ware at the water-powered mills there and near London. Substantial remains of many of these complexes that once processed Swansea smelted copper still remain, often as attractive waterside residences.[122] The sites of the eighteenth-century, water-powered coppermills at Swansea are described later in the book and although plans survive for one of the installations, there are no visible remains to compare with the related buildings still standing and conserved in Bristol and London.

The earliest copper-smelters in south-west Wales

Fig. 53. Two pairs of iron rolls (1819) at Hafod Copperworks for 'finishing' the copper sheets as smoothly-rolled thin pieces of metal. The drive from the steam-engine fly-shaft is to the right with a gear-train reducing the speed of the rotation conveyed to the rolls. (Held by the West Glamorgan Archives Service; Neath Abbey Ironworks Collection, D/D NAI M/101/7)

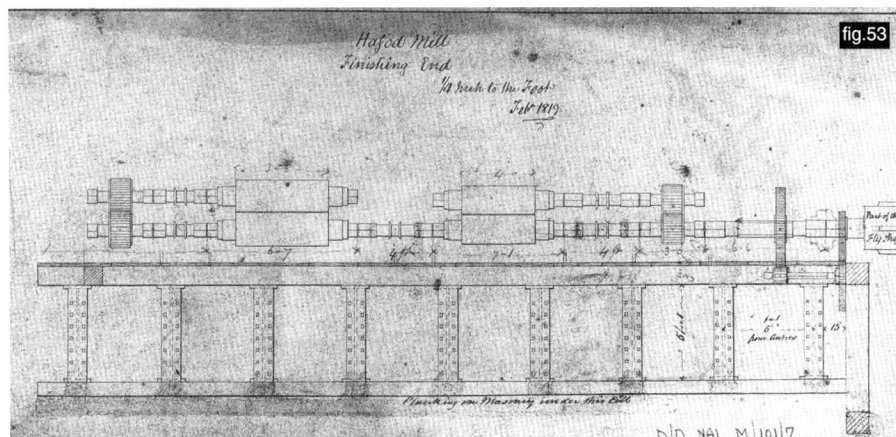

Fig. 54. Workers tending the rolls (Mill No.1) at the Hafod Copperworks in the 1920s. (City & County of Swansea: Swansea Museum Collection)

were in the Vale of Neath to the east of Swansea. The first was the short-lived, late sixteenth-century Aberdulais Works, which may well have been associated with copper battery-mills driven by the great power source of the Aberdulais Waterfalls. In 1694-95 the longer-lived Neath Abbey (Mines Royal) and Melincryddan copper and lead works were founded in the lower Vale of Neath.[123] Water-powered copper rolling-mills for their use grew up at Neath Abbey (Cwmfelin waterfall), the Gnoll (Neath) and on the main River Neath at Ynys-y-Gerwyn (nr. Aberdulais).

The growth of the Vale of Neath smelting-works had a direct effect on the development of water-powered copper-processing expertise in the lower Swansea Valley. In 1713, the large iron-forges on the River Tawe at Ynys-pen-llwch (Clydach) were taken over by Sir Humphrey Mackworth and the new 'Company of Mineral Manufacturers', operators of the Melincryddan Copperworks near Neath. By 1726 copper-rolling, slitting and wiremills were in operation there. However, the Melincryddan Copperworks were taken over by the Bristol firm of Thomas Coster & Co in 1731,[124] and the Ynys-pen-llwch Works reverted to iron-forging. The water-powered trip-hammers and rolling-mills in use there were fairly standard items in the later coppermills of the Swansea area.

The development of large-scale copper-rolling and hammer 'battery' mills at Swansea is an interesting process of adaptation rather than innovation. The two large water-power sites on the lower River Tawe, at Ynys-pen-llwch and Upper Forest (just north of the later Wychtree Bridge, Morriston), were developed as forges in the seventeenth century for the Ynysgedwyn Iron Furnace in the upper valley. In 1713, the local landlords of both these complexes realised the profits that could be made from the power harnessed at these sites and they terminated both their tenants' leases. The two water-wheels on the Ynys-pen-llwch site were then converted to coppermill use until 1731, at which time Robert Morris considered it as a possible site for rolling copper produced at the Llangyfelach Copperworks. Eventually it began a new life as the first tinplate works in Glamorgan.

Most of the experience in building large copper- and brass-rolling mills and batteries in the seventeenth- and eighteenth-centuries, was held by those employed in the complex of smelters and mills centred on Bristol. Further up the Severn Estuary in Bristol, the innovative Bristol engineer John Padmore seems to have been responsible for the high aqueduct and sophisticated water engineering of the first Swansea Valley copperworks and mills at Llangyfelach in 1717. Therefore it was natural that owners of the Llangyfelach Copperworks should seek the advice of 'Mr. Padmore, a capital mason of Bristol', when determining the site in 1728, of a new major copper-working mill on the River Tawe.

Fig. 55. Copper rolling-mills (c. 1910) stand preserved at the former No. 1. Rolling-mill (1819) of the Hafod Works. The drive from the surviving steam-engine is to the left. (960209/1)

Fig. 56. View of the Saltford Copper and Brass Mills on the Avon near Bristol. (960176/25)

powered vertical rolls with rounded gaps set on benches. These wire-rolling machines had six grooves on each 4in diameter roll, the outer strings or strips emerging from the first rolls were discarded as scrap. The oval section wire stock produced was slightly longer than the initial slitted strip. After annealing, it was usually drawn through a succession of tapering holes set in metal blocks in a series of processes designed to produce successively finer wire.

The first coppermill had come to the lower Swansea Valley in 1713 but, as noted, it was not until 1717 that the first large combined coppermill and smelter arrived in the lower Valley. Dr. Lane, founder of the first Swansea Copperworks at Llangyfelach, came to the site in 1717 from the copper rolling-mills at Cwmfelin (Neath Abbey). These latter were in existence by 1694 and John Lane was a partner there from c. 1708. Large water-power provision was made at the Llangyfelach Works which included a copper rod-mill and a battering-mill.[126] The use of water-power included the large wooden hammers which beat copper-ingots into large sheets in the 'battering-mill'; iron-made grooved-rolls which then elongated and slit these sheets in the 'copper rod-mill' and other products were fashioned in the 'smith's shop' that we know existed in this works.[127] After passing over several water-wheels in succession the water dropped onto the old wheel of the manorial cornmill at Landore.[128] Copper 'rods' were the 24-30in long pieces of wire used as barter goods in the Bristol-based African slave-trade. 'Battery or battery-ware' is a term used for copper which has been extended and shaped by beating with water-powered hammers. It was originally copper or brass sheet but by the eighteenth century its usage was confined largely to hollow-ware vessels of copper and brass. John Lane did carry-out experiments in zinc production at the Llangyfelach Works but presumably most of the output was of copper vessels. At this period two-thirds of the vessels produced were often for the West Indian trade and a third as barter items for the African slave-trade. The Swansea-produced items were mostly for Bristol-based trade but later there is much evidence for the supply of domestic plates and bowls to the London area.

Cambrian Copperworks Mills, 1720

The site originally intended for the Llangyfelach Copperworks, alongside the River Tawe on the northern edge of the medieval borough of Swansea was granted to the builders of a second copperworks ('the Cambrian Copperworks') in 1720. The site here had a lower potential head of water for power purposes than that used by the

It is unclear how many of these mills included wiremills as part of their water-powered capacity - Ynys-pen-llwch certainly did. Lower Forest Coppermills were shown on the 1820s Ordnance Survey drawings as 'wiremills', but this may be because they belonged to the Bristol Brass & Wire Company. Wire-drawing was often a preliminary stage in making pins, it required the use of wire-drawing benches. The process used at the, still extant, Avon Mills near Keynsham, Bristol, is known and is outlined below.[125]

A pair of rolls, somewhat smaller than the usual size employed in the rolling and slitting mills, was used. A coil of thin plate or strip some 4in wide was fed within guides into the slitters and separated into 17 strings, or strips, each $1/4$ in wide. A metal comb deflected alternate strips either upwards or downwards onto two levels of miniature water-

Llangyfelach Copperworks in the deep valley of the Nant Rhyd-y-Filais to the north. Therefore a large pool was created at the lower end of Cwm Burlais to provide an adequate reserve for feeding low-head breast or undershot water-wheels that discharged into the tidal River Tawe alongside. The works remained in operation between 1720 and 1745 as a copper smelter but Robert Morris remained in occupation of what may have been a copper rolling-mill until 1766. This hipped-roof building, equipped with three hearths or furnaces spread over two storeys was driven by a large diameter low-breast or undershot water-wheel on its south elevation. It remained in use for iron-rolling (for tinplate) and other uses until the late nineteenth-century. The construction of the Swansea Canal across the large mill-pool in 1794-96 meant that henceforward the mill was driven by the waste water from the canal itself.

Lower Forest Coppermills, *c.*1735

fig.57

fig.57a

The only Swansea coppermill that we have detailed drawings of was the Lower Forest mill of *c.*1735, built on the main River Tawe. However, most of the eighteenth-century mills would have been similar. Padmore, the Bristol millwright and engineer who

Fig. 57. Detail of the Lower Forest Coppermills plan showing two water-wheels driving rolling-mills and a third smaller one powering shears.

Fig. 57a. At the Lower Forest Coppermills, five water-wheels produced the power to roll copper sheets (at left) and 'batter' the sheets produced into copper bowls (at right) for Lockwood, Morris and Company. The mills later powered rollers for the Beaufort Tinplate Works. (Held by the West Glamorgan Archives Service; Neath Abbey Ironworks Collection, D/D NAI M/77/2)

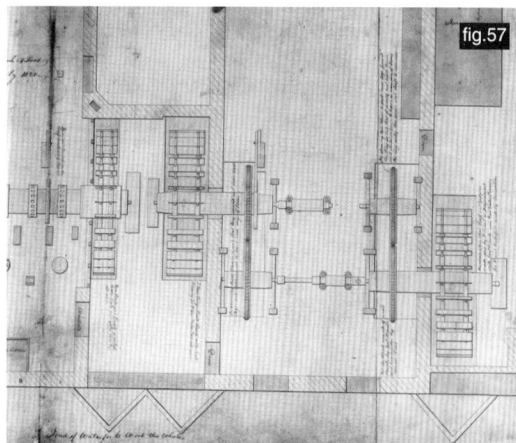

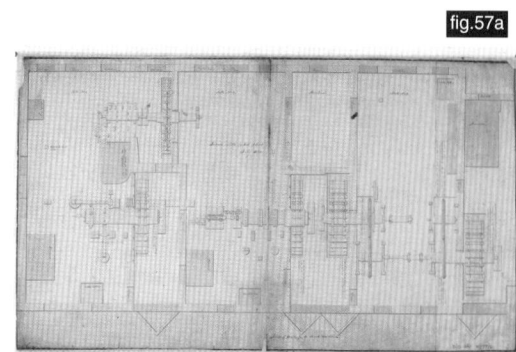

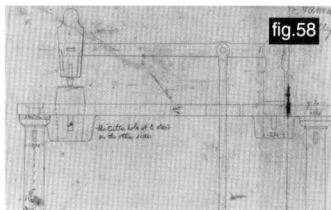

fig.58

Fig. 58. Hammer made for the Hafod Copperworks rolling-mills with the hammer-head holding a hole-punch impacting on copper products mounted on a cast-iron table supported by Doric columns. (Held by the West Glamorgan Archives Service; Neath Abbey Ironworks Collection, D/D NAI M/101/18)

designed the Lower Forest mill, designed a coppermill that resembled the common practice in the 18 or so copper and brass mills in the Bristol area. In fact the technology of immense wooden drive-shafts, wheels and large hammers was common throughout Europe in both iron and copper forges and mills.

Cogs set in the long and huge shafts running from the centre of each water-wheel drove down on the ends of huge wooden hammers that were pivoted in heavy wooden frames set close to the drive-shaft. The long beechwood shafts of the hammers terminated in a variety of cast-iron hammerheads that were held in place by wooden wedges. These bore down repeatedly on the copper bowls being shaped by being held over an anvil sunk in a large iron-bound tree trunk that was buried in the floor of the forge or 'battery-mill' (the huge wooden foundations for such equipment can still be seen at Clydach in the exposed substructure of the Upper Clydach Forge). The workers holding the bowls sat on stone edging around the sunken anvils and could stop the hammers by inserting a wooden prop under the hammer as the cogs knocked the hammer into the up stroke.

Coppermill water-wheels of this period often revolved at about 18 revolutions per minute. Each powerful and rapid tilt-hammer stroke was only carried-through some 3-3½in which was ideal for the shaping of small bowls and plates from circular pieces of plate. Many differing sections and weights of hammerheads were used for the successive stages of pan production. It was essential that the hammerheads should hit the anvils absolutely square or the brass pans being worked were liable to be damaged or to dance out of control.[129] The middle mill at the Lower Forest Coppermills had a line of three of these tilt-hammers for producing smaller domestic ware. The southern mill had one, as well as the heavier belly-helve hammer that was used in the production of heavier industrial vessels.

The heavy belly-helve hammer did not lie at right-angles to the huge timber water-wheel shafts as did the rows of lighter tilt-hammers. Instead this heavier type of hammer-shaft was constructed parallel to the shaft; from the large circumference of which projected three or four 'cams' which successively lifted the hammer-shaft at a point approaching the head. At Lower Forest, as elsewhere, there was a large rebound 'spring' beam mounted over the hammer on a large wooden frame. The use of this arrangement resulted in a far greater movement of the hammer with a much heavier, slower blow. Its purpose in the southernmost of the Lower Forest Coppermills (Mill No 1) was to produce large copper furnace-bottoms and this was how it was also used elsewhere.

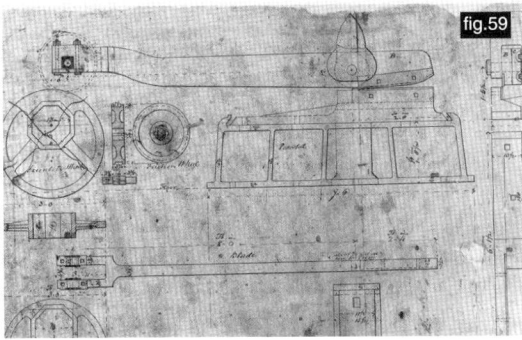

Fig. 59. Eight-feet long cast-iron shears for the Hafod Copperworks rolling-mills; a powered shaft at left raised and lowered the blade at right via an eccentric wheel. (Held by the West Glamorgan Archives Service; Neath Abbey Ironworks Collection, D/D NAI M/101/9i)

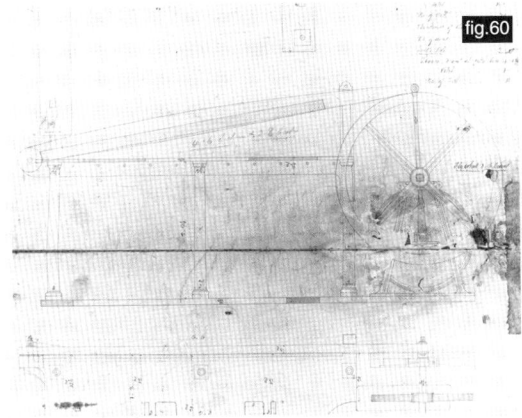

Fig. 60. The steam-powered guillotine for cutting copper sheet with a cutting-arm over seven feet long at the Hafod Copperworks rolling-mills. The cutting-table, made at the Neath Abbey Ironworks was supported on six miniature, cast-iron Doric columns. (Held by the West Glamorgan Archives Service; Neath Abbey Ironworks Collection, D/D NAI M/101/15)

At Lower Forest, as elsewhere, there were several furnaces, and a bellows-blown hearth for keeping the metal workable. The formation of a deep pan from a piece of copper was achieved by several stages of hammerings, each stage followed by a complete annealing process. A large pan would go through these annealing ovens four or five times, before reaching the final hammering stage.[130]

A flat circular sheet of copper (a 'nap' in Bristol copper/brass mill terminology) was firstly placed on the anvil after powered shears had trimmed the corners from a rectangular sheet. The resulting nap was slowly revolved and the first blows from the trip-hammers were directed to radiate from the centre to the circumference of the vessel being formed. This first hammering produced a shallow curving pan which was then annealed and doused in the water cistern. There was a water cistern in each of the three mills at the Lower Forest Coppermills (where the plan survives) but it was Mill No 2 where three tilt-hammers in line actually formed domestic ware.

After this first annealing the hammers were used to deepen the bowl with further intermittent annealings as required: more (4 or 5) were required for larger pans.[131]

The following types of vessels were the standard 'hollow-ware' products of the British battery (copper and brass) mills.[132]

1. 'Neptunes': shallow dishes of up to 2ft 6in diameter, with a depth of 3-6in were exported to Africa for salt evaporation and to the East for tea drying.

2. Milk pans: deeper bowls akin to the shape of modern (circular) washing-up bowls. These were made up to 4ft in diameter, especially for the use of the Welsh dairy trade. Smaller sizes were often called Lisbon pans and were probably used by the Bristol sugar trade in the production of 'Lisbon' sugar.

3. 'Guinea Kettles': like 'manillas' and 'rods', these bucket-like containers were originally used as barter items in the eighteenth-century slave trade. Such 'kettles' continued in British and overseas domestic use after the slave trade was abolished. They were deep and wide straight-sided vessels with a handle and no rim and were produced in various sizes. Nests of these vessels have been also found along the lines of the North American fur trading navigations.

4. 'Cheffs' or 'Compass bowls': these small rounded bowls, without rims or handles, may originally have been produced to hold ship's compasses.

Various small sizes of pans from 12in or so upwards were grouped in threes, bound one inside another by the inturned rim of the outermost pan (the 'ferrier'), for several successive hammerings.

The Lower Forest Mill was built in 1732-35. The surviving plan of this mill indicates in detail the processes that were carried-on there.

A relatively short and broad watercourse was sited immediately beyond the head of the River Tawe. This terminated in a large millpool with a dam near the bank River Tawe into which the tail water was discharged. The mills occupied a rectangular building set on the dam and sited over three main watercourses. Five large diameter but narrow undershot water-wheels drove the

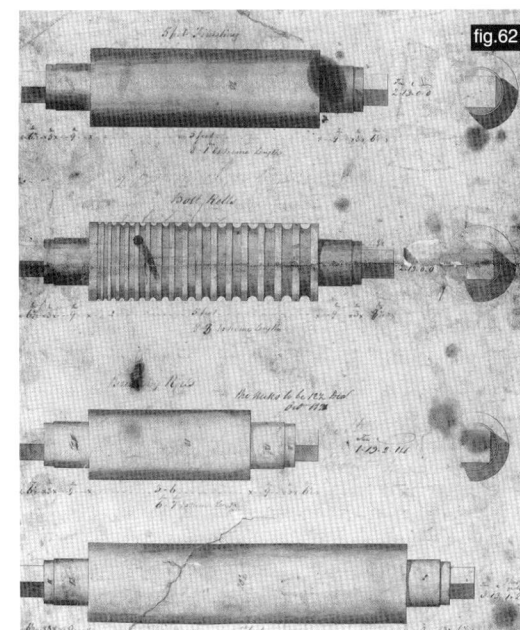

Fig. 61. Circular sheets of copper for the manufacture of hollow-ware vessels at the Hafod Copperworks in the early twentieth century. (City & County of Swansea: Swansea Museum Collection)

Fig. 62. Four types of rolls made in 1819 for the Hafod Copperworks rolling-mills: the second example was for rolling the long bolts of varying thickness that were required on a huge scale to hold timber ships and their copper cladding together. (Held by the West Glamorgan Archives Service; Neath Abbey Ironworks Collection, D/D NAI M/101/3)

Fig. 63. Four workmen shaping a large hollow-ware copper bowl whilst standing under the frame of a large steam-hammer at the Hafod Copperworks in c. 1910. (City & County of Swansea: Swansea Museum Collection)

main processes contained in the building. Two reverbatory furnaces and a bellows-blown hearth reheated the ingot and rolled-copper prior to hammering, rolling, slitting and forming the sheet metal in domestic and industrial vessels. Great wooden axles continued into the three main rooms of the building and their projecting lugs operated at least five sets of hammers: four trip hammers and one belly-helve for the forming of larger vessels. The northerly room of the mill (Mill No. 3) contained four sets of rolling-mills, arranged in two successive rows. A water-wheel on each side of the housing had its axle directly driving the lower rolls of two sets of rolls. Both wheels also had large geared wheels, engaging with second sets of geared wheels set higher and in line with the upper rolls of the mills, whose lower rolls were driven in the opposing direction by the water-wheel at the other end of the building. Smooth rolls to produce sheet-copper from ingots and grooved rolls to slice the sheets into bars were observed in action by tourists such as the American 'Industrial Spy' Joshua Gilpin and the Swede Eric Svedenstierna. Heavy industrial vessels were formed in the southern room of the building with metal kept workable by the water-wheel driven bellows of the hearth. Three water-driven shears to cut the sheet metal completed the layout.

Detailed accounts survive for the first Swansea Copperworks at Llangyfelach in the final and most developed phase of its existence in 1745, by which time it had also been using the large water-power resources of the Lower Forest Coppermills for about 10 years.[133] These show that two-thirds (15 tons) of the copper held by the firm in the main London

market was 'battered', i.e. had been worked and rolled into sheet at the Lower Forest Copper Mills. One third of the London stock in December 1745 was 'Fine Copper' (goods) in the form of bowls and plates. A considerable part of the output seems to have been worked into fine vessels or pins for the London market at the Merton Copper Mills (near Wimbledon) of William Thoyts.[134]

William Thoyts owed the company of Lockwood, Morris & Co. £6,134 1s 2d, i.e. over twice the value of the other copper that was held by them in London. The two companies shared 'copper smiths' and a 'copper warehouse' (presumably in London) worth £7,911 6s 5d to the Swansea concern alone. Lockwood Morris and Co. also had the leasehold of a house in Salter's Court in London. The firm had selling agents in Bristol (the main centre of the brass industry), Birmingham (the main centre for manufacturing copper and brass domestic ware, ornaments and toys), Dublin and Lisbon. The main cargo shipped in December 1745 included five copper teaches (worth £59 2s 5d.) en route for Lisbon. The 'teaches' were the hollow bottoms for strong copper boilers and would have been formed by the heavy water-powered hammers in the southern mill at Lower Forest.

By 1768 the accounts show that the firm was using its substantial copper-rolling capacity at Swansea to produce huge quantities of copper sheaving (worth £9,070 0s 3d) to protect the large wooden-hulled sailing fleets of the mighty East India Company from the depredations of shipworm. All the copper held in London was now in the form of 'Fine Copper' produced in the Forest copperworks (eight times the quantity that had been produced at the old Llangyfelach Works).

By 1779 Lockwood, Morris, & Co. had acquired the powerful water resources of the Upper Forest Mills and kept the tilt-hammer and rolling-mills there when they moved their smelting works from Forest to Landore c.1790; at the same time, they had a rolling and stamps mill powered by the Nant Rhyd-y-Filais built near their new works at Landore. The five water-wheels of their old Lower Forest Coppermills where then used by the new lessees of their old (adjacent) Forest Copperworks: the Bristol Brass Wire Company (alias the Harford Co.) who used the mills as a 'rolling and slitting copperworks.'

The American Joshua Gilpin described how the considerable Lower Forest Mills were still being mainly used for producing copper sheets for cladding the lower parts of ships in 1796:[135]

'Copper is cast into pigs and plates of such size as will meet each kind of work.

The sheets intended for sheathing Vessells [sic] are rolled cold and by these means the sheathing is said to last near double the time.

For rolling iron bolts [Gilpin almost certainly meant the great copper bolts that were such a significant part of the production of the Greenfield Coppermills in north-east Wales] the barrs are first beat down under the sledge then put under grooved rollers ... where they are formed round.

The hoops or thin pieces are cut or slit by rollers locking into each other...'

A much less palatable aspect of the copper trade was the large production of exchange-goods to facilitate the slave-trade which was such a considerable part of the commerce of the ports of Bristol and Liverpool. In the eighteenth century the merchant classes of these thriving commercial cities dominated the Severn Estuary and the Mersey Estuary and north Wales coast respectively. In 1745 Lockwood, Morris & Co.'s Bristol agent, James Laroche, was processing goods worth no less than £1,174 15s 8d, almost half the value of the 22$\frac{1}{2}$ tons of battered and fine copper (bowls and plates) sold in London.

The 1768 accounts of goods produced in the partnership's Forest Copperworks specify the type and quantity of goods produced for the slave-trade more closely. In Bristol, James Laroche had goods in his hands worth £617 6s 4d, these consisted of 3tons 11cwt 1quarter 10lbs. of copper 'rods'. As noted these were short pieces of copper-wire 24 to 30in long used for the purchase of slaves by the masters of Bristol ships. Laroche also held 4tons 12cwt 0quarter 13lb of bronze (copper and lead alloy) 'manillas' which were horse-shoe shaped items of barter for Africa weighing two or three lb each. The copper-wire was probably drawn at the Lower Forest Coppermills while the manillas were probably cast at the main Forest smelter site in the same way that manillas had earlier been cast on the White Rock smelter site. Laroche had already bought £511 12s 1d of copper goods in 1768, making the amount of copper which he had bought only slightly less than the amount purchased in 1745. In addition, it is evident that a second agent was buying large quantities of slave-exchange goods. Vaughan & Co. (of Bristol?) had bought over twice the amount of manillas as Laroche (4tons 12cwt 1 quarter 13lb) but fewer rods (2tons 1cwt 3quarters 24lb), the latter were valued at £618 1s 4d.

The total value of identifiable slave-trade goods produced for Lockwood, Morris & Co. in 1768 was 11 tons valued at £1746 19s 9d. This was outweighed by both the identifiable volume of goods produced for the London domestic market and the huge volume of copper-sheaving and general copper-trade goods produced for the East India Company.

In 1768 the copper supplied in the form of domestic ware for the London market comprised 12tons 15cwt 3quarters 8lb of copper bowls and 45tons 5cwt 1quarter 14lb of copper plates worth £6287 8s 2d. That is, over five times the amount of copper was used in identifiable domestic goods as in slave-trade ware and in value was over three and a half times as much. There were also 8tons 19cwt 2quarters 19lb of battered (sheet) copper supplied to London and worth £1095, and £9070 0s 3d of copper sheathing and general copper-trade goods supplied to the East India Company for its huge sailing-fleet (this may have amounted to 73tons of copper if the value was roughly equivalent to the battered [sheet] copper sold by the firm at £122 a ton). Copper for ships' sheathing and for trading purposes in India and Asia seems to have been the main market, along with goods delivered for the London domestic market. 'The Coppersmiths' Warehouse' was itemised as worth £9715 13s.9d and this must also have reflected the huge amount of copper for domestic ware passing through the London market. It looks as if a significant part of the output of Lockwood, Morris & Co. was processed through the nearby Esher Mills ('Ember Mills' in Grant-Francis's transcription of the 1768 accounts), which had £4000 worth of copper supplied and which specialised in the production of brass and copper wire.[137]

Consequently the majority of the output of Lockwood, Morris & Co. would have been either rolled or 'battered' with tilt-hammers at their Forest Coppermills.

White Rock Copperworks and the use of the Bristol Coppermills

The third, and one of the most substantial, of the Swansea Copperworks was the White Rock Copperworks founded by Bristol interests in 1736. John Coster (1647-1718) and his family had been active since the 1680s in the development of lead and copper-smelting at Bristol and Upper Redbrook, on the banks of the Wye Navigation on the Gloucestershire/Monmouthshire border, and (briefly) at Hayle in Cornwall. The partnership (taken over by John Percival in 1739) replaced their main smelting capacity at Upper Redbrook in the 1730s by the works at Melincryddan (Neath), which in turn was abandoned for the new works at White Rock in 1738-39.[138] From here large quantities of refined copper were sent to be fabricated, mainly as battery-ware, at their water-powered mills in the Bristol area: Swinford, Bye Mills and Publow.[139] Publow Mills (ST 625642 - now Church Farm), originally a 'frying-pan' works on the River Chew in north Somerset, was converted for rolling copper and later for copper-refining and, with the addition of tilt-hammers, for battery-ware production. A headrace bridge

Fig. 64. All the eighteenth-century Swansea copper-smelting concerns previously established in seventeenth-century Bristol kept coppermills there to process copper goods for market in the Bristol area. One of the many with surviving structures is the Swinford Coppermills. (960176/36A)

Fig. 65. Name-plaque of John Freeman & Co., smelters at the White Rock Copperworks at Swansea, on the Swinford Coppermills near Bristol. (960224/3)

keystone still bears the name of 'J F & Co 1799', i.e. John Freeman & Company (also notable as the later co-developers of the Kilvey Schools at Swansea). By the end of the eighteenth century, Freemen & Co developed Belton Mills (until 1860) on the opposite south bank of the River Chew from Bye Mills, both producing copper-battery vessels from the copper refined at White Rock in Swansea. The watercourses are now the most obvious survival of this site where Swansea copper was manufactured (ST 610638, between Pensford and Stanton Drew). The 1840 buildings of John Freeman & Co's rebuilding of the copper rolling-mills at Swinford Mill (ST 691689) remain with a two water-wheels still in place. In the 1790s Freemen & Co. took over a fourth mill site on the River Chew at Woollard Mill (ST 632644, now a house). This 1730s tinplate mill was used as a copper-rolling mill. Swinford, Woollard, Bye and Belton formed a complex of six mills closely grouped together on a navigable section of the River Chew five miles south of Bristol. Only a minority of the output of the White Rock works was manufactured on site at Swansea.[140]

The 1744 print of White Rock shows a 'Manilla House' on site: presumably a small foundry used for producing the bronze horseshoes (tin was a by-product of the Cornish copper-ore) used as a barter item in the Bristol-based slave-trade. In 1780 manillas were still being produced, a 1780 summary of production notes that 106cwt. of 'unsaleable [presumably defective] Manillas' were held by the company.

The 1744 drawing of the works gives no hint of any large rolling or battery-mill on the works site itself. However, the water supplies available to the works and formerly used by Kilvey Mill were used by the stamping and clay mills necessary for the maintenance of the main copper-smelting process. However, the terms of the 1736 lease of the works site by Bussy Mansel make it clear that Mansel's 'New Mill' (near Llansamlet Church) was also available for conversion as a 'battery mill or trial hammer' and that he would pay half the expense of transport between the works and the mill. There is no evidence that this conversion was ever made. Instead, it seems as if there was some development of copper manufacturing on the White Rock Works site itself. A new and larger water-supply leat was constructed to the far side of Kilvey Hill (a water source recognised in the 1805 lease). Even so, the great majority of the 4,753 tons of copper produced annually during the 1780s would have been sent for processing elsewhere.

In 1780 10,922cwt of the copper produced at the White Rock Copperworks were shipped from Swansea and 2,177cwt delivered to the brass works in Bristol. The valuation of 738cwt of furnace 'Bottoms' at White Rock suggests that there may have been a large water-powered helve-hammer at the White Rock Works for producing copper furnace-bottoms of the same type as those known to have been produced at the Lower Forest Copperworks. There must also have been a small rolling and slitting mill, for in 1780 the works produced 19cwt of 'Sheaving Nails, Nail metal, Gates etc.' The 43cwt of 'pot metal' produced must mean that a small amount of brass was produced at the works.

The 1805 lease formalised the use of the larger works water supply leat. This extended use of the water supply, presumably for copper-manufacturing purposes, necessitated extra provision in 1806 to pump water back around the water-wheel so that it could be used for a second time. A clause added to the lease in August 1806 provided that the partners erect within 12 months a 'Fire Engine' with the necessary buildings 'for supplying the stream ... by which the machinery of the said Copper and Brass Works ... have been hitherto worked, with a sufficient quantity of water.'[141]

Middle Bank Copperworks and the use of the Thames and Greenfield Valley (Flintshire) Coppermills

The fourth Swansea Copperworks was founded by the London merchant, Chauncey Townsend, in the 1750s at Middle Bank on the east side of the River Tawe, to the north of the White Rock Copperworks. Townsend also had a coppermill at Temple Mills near Marlow in Buckinghamshire to process copper for the London market.[142]

In the later eighteenth century, the Anglesey solicitor, Thomas Williams, came to dominate the national (and international) trade in copper from his base at the world's largest copper mine at Parys Mountain in Anglesey. By 1792 he also had control of the Cornish Metal Company; had two smelting works in Swansea (Middle and Upper Bank); two mills in the Thames Valley; a further group of mills and works in the Greenfield Valley at Holywell in Flintshire; two smelting works in south Lancashire; copper warehouses or offices in London, Birmingham and Liverpool; a chemical works near Liverpool and a bank in north Wales. He became MP for Buckinghamshire, and built a new mansion and park at Temple House at Marlow (he had control of the 'rotten-borough'). The latter was sited alongside his Temple Mills (these were Townsend's old coppermills that were still being used in the processing of Swansea copper for the London market.[143] Workers' housing still remains on part of the site of the Temple Mills today).

The Middle Bank Works was in Thomas Williams's hands by 1787, having been taken over from the firm of George Pencree & Co, copper and brass manufacturers of Snow Hill, London. He took control of Temple Mills on the Thames from the same company, holding it in his own right.[144] His associate, Pascoe Grenfell, later acquired the mills and succeeded Williams as the local MP whilst continuing to build up the Middle and Upper Bank Works business at Swansea.

The Upper Bank Works had been acquired by Williams' 'Parys Mine Company' in 1782, whilst the Middle Bank Works was taken over by his 'Stanley Company' in 1787. On Thomas Williams' death in 1804 both works went to the successor company of 'Williams & Grenfell', owned by his son Owen Williams and his business partner, Pascoe Grenfell, under whom both works continued until a split in the partnership in 1825 caused the Williams family to retire from the copper business.[145] Before 1813-14, it is likely that the Williams and Grenfell processed the copper smelted in Swansea at their Temple, Wraysbury and Bisham mills on the Thames, and that the Lancashire smelted (Anglesey) copper was

processed in their large mills in the Greenfield Valley in Flintshire. However, the Lancashire smelters at Ravenhead and Stanley closed in 1813-14[146] and some Swansea copper was also processed in the Flintshire mills.

Water-power use by the Forest and Landore Copperworks

By 1779 the output of the booming Forest Copperworks of the prosperous Lockwood, Morris & Co. could justify the use of a third large water-powered rolling-mill (they had been using the Cambrian Copperworks mill until at least 1765). The Upper Forest Mills were taken for a period of 65 years. Twin mills sat on their respective watercourses running between a large millpond and the nearby River Tawe. One wheel drove two stones of a cornmill and a rolling-mill whilst the second drove a tilt-hammer.

Lockwood, Morris & Co. moved their smelter from Forest to Landore in 1790 and gave up use of the Lower Forest Mills to the new lessees of their

Fig. 66. By 1811, steam-operated pumps were helping to power water-powered rolling-mills at White Rock Copperworks. Since the earlier 1790s' drawings, additional waste-tips had raised the ground between the three brass-annealing furnaces and the river; more buildings had been built on the raised ground and the present quays constructed. From the quays, steps lead up to the works and two limekilns built on the tips. (Wood, Rivers of Wales, 1811)

Fig. 67. After 1808, copper from the Landore Copperworks at Swansea was unloaded from barges by the crane shown and rolled into sheets at extant buildings at Walthamstow (north London). (960213/1)

Fig. 68. The Musgrave rolling-mill engine at the Hafod Copperworks in 1910. The Bolton-built engine had a large 29 in. cylinder that generated 600 horsepower. The engine and wheel have remained on site since 1980 in an increasingly derelict condition as a rare survival of an important industry. (City & County of Swansea: Swansea Museum Collection)

Fig. 69. Twin steam-hammers that stood in the hammer-house at the Hafod Copperworks next to the no.1 rolling-mill. In the background, a group of early twentieth-century workmen are beginning to hammer out a rolled sheet of copper. The building has Pennant sandstone walls pierced by large brick access-arches and round ventilation windows; the timber trusses support an elevated clerestory. (City & County of Swansea: Swansea Museum Collection)

Forest Copperworks, Harford & Co of Bristol. Twin water reservoirs fed from watercourses on Craig Trewyddfa had provided water to the Forest Copperworks, presumably for stamping. Alongside the Landore Works they set up a new stamping and rolling-mill driven by the Nant Rhyd-y-Filais. In the 1790s John Morris was enquiring of Matthew Boulton whether a new steam-engine he intended ordering to pump the Landore Colliery could also drive a new rolling-mill. Boulton answered in the affirmative; the new rolling-mill was not built although the huge new pumping-engine was.

The Morris family left the Lockwoods to carry on with the copper-smelting business when in 1803 they left the partnership that had dominated the Swansea copper industry since the 1720s. The Landore Copperworks were taken over in 1808 by a London based company that rolled their copper in (still extant) large water and steam-driven mills at Walthamstow in north London.

However, the scale of Swansea copper smelting and milling was transformed from 1809 with the foundation of the huge new copperworks at the Hafod, between Swansea and Landore, by the Cornishman John Vivian. In 1822 he took over the lease of both the Forest Mill complexes, with their five or more rolling-mills and extensive copper hammer batteries to add to the capacity of his recently completed steam-powered mills.

The age of the Steam-powered Coppermills (1819 onwards)

When the huge Hafod Copperworks was completed in 1809 John Vivian arranged for over half the initial output (250 tons per annum) to go to the large Bristol firm of Pitt, Anderson, Birch and Company who processed it in their existing water-powered mills in the Bristol (Netham) area for brass and copper manufacture. Pitt & Co. also used these mills to manufacture some of the Vivians' ingot copper at a fixed rate per ton of ingot and arranged for its consignment. By 1814 John Vivian had sufficient capital to build his own mills and wanted to cut out the coppermilling middlemen. Consequently he built a new steam-powered rolling-mill on the western riverside section of the Hafod Copperworks in 1819.[147] An engine-house (1910), steam-engine and rolling-mills remain on this site. This new scale to the powered activities at the copperworks was then copied on an adjoining site by Vivian's fellow west Cornishmen: the Williams family of Scorrier (and Foster) of the Rose Copperworks in 1828.

Boiler and condensing water for both mills was provided by the adjacent Swansea Canal and detailed plans of one of the first mill buildings at the Morfa Works survive in the drawings of the Neath Abbey Ironworks, which provided much of the rolling-mill and tilt-hammer equipment for the Swansea copperworks during the nineteenth century.

By 1840 two of the first square-built and hipped-roof mills on the site were linked by the long range that now survives on the site in use as a Swansea Museum store. The south-western corner of this larger building incorporates walling from one of the early steam-powered mills. It has Pennant sandstone rubble walls interrupted by large brick-arched openings to give ventilation and relief from the great heat generated by the reheating furnaces and hot metal used inside. The rear western face of the long 1840s rolling-mill was built into the high retaining bank of the Swansea Canal, but to the north and south huge brick arches punctuated and ventilated the gables and made working conditions inside more tolerable. This ventilation was

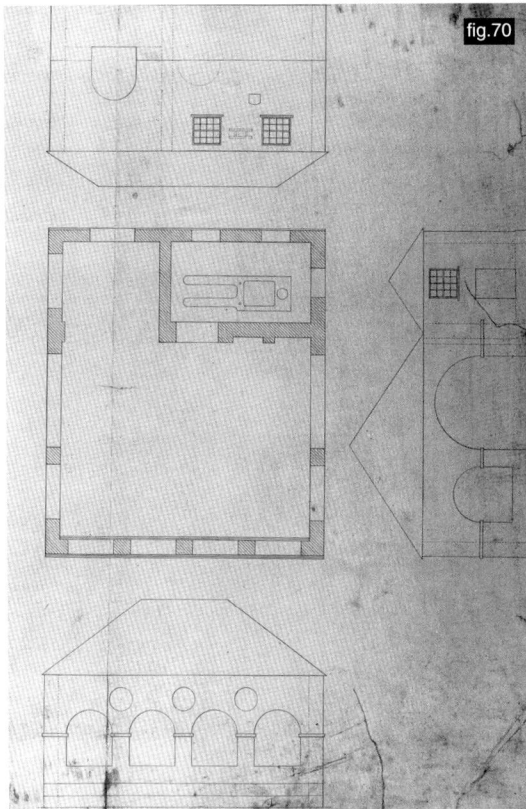

increased by the open eastern side of the long rolling-mill where tall cast-iron columns support the wooden wall-plate beams upon which a later roof now fits.

This advanced competition forced John Henry Vivian to enlarge and improve his own rolling-mills. At first he planned to take over the existing water and steam-powered coppermills at Saltford, near Bristol, in 1841 in conjunction with the mills operator Charles Ludlow Walker, but he abandoned the idea when the buildings were found to be in a very bad state of repair.[148]

In 1842 he built four new pairs of rolls on the main Hafod Works site powered by a 60 h.p. steam-engine (part of the engine-house remains on the site today incorporated into the 1863 engine-house). Consequently, in 1845, Vivian gave up his lease of the Upper Forest Mills and these instead became the iron-rolling nucleus of the Upper Forest Tinplate Works, one of three water-powered tinplate works that established the lower Swansea Valley as the world centre of the tinplate industry from the early 1840s. The Lower Forest Rolling Mills similarly became the iron-rolling plant of the newly established Beaufort Tinplate Works in or before 1860. Apart from weirs and the remains of the larger feeder watercourses, structural remains of these two significant water-power installations are now underground.

Fig. 70. The first rolling-mill built on the Morfa Copperworks site at Landore in 1828. The cast-iron bedplate for a small beam-engine can be seen in the enclosed housing in the corner with an archway accommodating the drive to the rolling-mills. Provision was made to dissipate the heat from the hearths and mills by having open arcades and circular ventilation windows in the side walls. (Held by the West Glamorgan Archives Service; Neath Abbey Ironworks Collection, D/D NAI M/139/2)

Fig. 71. In 1840, the steam-powered copper-rolling mills at Morfa burnt down and were mostly replaced by the present building which was the largest such building in the world. The spine-wall of the earlier 1828 mills, with two arched openings, seems to have been incorporated in the eastern wall of the new mill and are visible beyond the buttress at centre. (960203/3)

Fig. 72. The facade of the new Morfa rolling-mills was open on its east elevation with large Baltic timber lintels supported on cast-iron columns. A large brick arch pierced the northern elevation. The upper, north elevation has round-arched windows. The rear of the building is built against a retaining wall that carried the Swansea Canal and an unloading dock primarily for coal. (960203/4)

Fig. 73. Part of the engine house of the 1842 rolling-mill at the Hafod Copperworks survives within the northern of the two extant engine houses (north wall illustrated). It was largely rebuilt in Pennant sandstone in 1860-2, and about the end of the century the roof was raised on brick walls to accommodate tracks for an internal travelling-crane. (920054/5)

Fig. 74. Plaque on the northern copper rolling-mill engine house at the former Hafod Copperworks marking its rebuilding in 1860-2 by Vivian & Sons. (960211/2)

Copper and its Products as Architectural Materials

At the turn of the eighteenth and nineteenth centuries, Glamorgan was the world centre of two metallurgical industries: copper and iron. With so much of the material being produced, it was inevitable that attempts should be made to use these metals as a building material. Almost 2,000 years before, the Romans had used bronze (copper and tin alloy) beams in furnace lintels. Part of the reason for the early use of copper-alloys in this way was the high melting temperature needed to produce substantial beams in cast iron. Two millennia later this presented no problem in the ironworks of the Industrial Revolution. When Eric Svedenstierna visited the Penydarren Ironworks at Merthyr Tydfil in 1802-3, he noted:[149]

'the requirement for buildings ... is all the greater since everything which in our forges is usually made of wood, is here of cast iron. Besides the beams which carry the roof, the floor of the hammer mill and a long footway outside it are of cast iron, and I was assured that a pine beam of a certain strength cost more than a piece of cast iron which would be adequate for the same purpose.'

Up to the end of the eighteenth century it had only been bronze and wrought-iron that could be fabricated into substantial lengths suitable for structural purposes.[150] Cast-iron sheets were used as doors in some of the structures associated with the Glamorgan ironmasters.[151] The use of copper in this context goes back at least to *c*.1471-1449 BC in Egypt, tomb-paintings of Rekh-mi-Rë at Thebes show the casting of copper doors for the great temple at Karnak.[152] The bronze doors of the Pantheon in Rome were erected in about 124 AD and 64 sets of bronze doors older than the sixteenth century survive today.[153]

Copper glazing-bars are the earliest architectural use for which there is evidence in the Swansea area. John Johnston, the Leicester architect, used them in the house of the coppermaster, Sir Humphrey Mackworth, in the 1760s and probably also placed them in John Morris I's house at Clasemont, Swansea, in the early 1770s. Mackworth's castellated banqueting house at Aberdulais, called the Ivy Tower, also has copper glazing-bars. John Johnston designed Morris Castle for 24 of Morris's colliers and presumably the tenements there had glazing-bars made of this material. Indeed, the elaborate glazing-bars of the circular windows in the earlier smelting rotundas of the Forest Copperworks of 1748-52 may have had them too.

By the end of the eighteenth century other

coppermasters were using copper-glazing bars. Thomas Williams, of Anglesey, who owned the copper-smelters at Middle and Upper Bank in Swansea, had them in his mansions at Llanidan on Anglesey and Temple House at Marlow on the Thames. The former are still there. By the end of the eighteenth century other magnates were using copper or copper-alloy glazing-bars in their houses: Thomas Boulton's house at Soho in Birmingham still has them. The extent of their use in industrial buildings remains unclear. The surviving cast-iron lighthouse at Whitford Point has glazing-bars in its lantern which seem to be made of this material.

Copper Roofs

It was as a roofing-material that copper has made a substantial and lasting impact. The Egyptians used the metal for the capping of stone obelisks. The Romans covered the concrete dome of the Pantheon with 200 tons of copper-sheet and held this in place with four tons of copper nails (the Hafod and Middle-Bank Works at Swansea later had copper nail-making workshops). On the Pantheon there was also an outer covering of bronze tiles, and the central eye or 'occulus' in the dome has an inner rim of copper still in place.

In Britain in the early modern period copper was used as a roofing material on a substantial scale. The seventeenth-century roof of the Chapel Royal at St James's Palace in London is in this material, and Parliament proposed it for the late seventeenth century dome of St Paul's Cathedral. By the eighteenth century copper was in use as a roofing-material in Japan, and it became very widespread in Scandinavia, much of it provided from the early, large opencast copper-mine at Falum in Sweden.

The later eighteenth-century additions to John Morris's Forest Copperworks had large copper sheets roofing the smelter in a series of convex vertical ridges.[154] The great Regency architect John Nash spent the first decade of his professional life at Carmarthen, and must have been familiar with Swansea and its copper-smelters. He was an experimenter with the new architectural materials of cast iron and copper. His re-erected royal rotunda at Woolwich has a roof of copper sheets. John Rennie, the famous civil-engineer, also roofed his long blocks of warehouses in London Docks with copper (1801).

It was obviously in the interests of the Swansea coppermasters to promote this and other uses of copper and a publication of 1820 on the subject survives in the Vivian manuscripts.[155] Thomas Williams roofed his copper rolling-mills in the Greenfield Valley in Flintshire with copper: the mills

there also processed Swansea copper.[156] The copper roofing-sheets on Whitford Point Lighthouse survived until recently.

Efforts to popularise copper as a roofing material within Britain may well have borne fruit. The great dome of the British Museum Reading Room (1857) is of this metal and so are the domes of the Old Bailey and Bank of England in London. Liverpool Cathedral is roofed with 30 tons of copper-sheet and Guildford Cathedral is also covered in the material.[157]

Copper-slag Blocks

There was widespread use of one particular by-product of the copper-trade for building purposes: copper-slag blocks, cast from moulds positioned alongside the reverbatory furnaces in the copper refining process in order to receive the slag. Each of the successive copper-smelting centres in Britain used these hard durable blocks of hard glistening iron oxide as a building material. Their use may be one of the first of a manufactured alternative to brick in the construction of structures. The use of these new building blocks expanded as attempts at an alternative use by extracting the iron from this material failed; the blocks continued in use as an early type of large cast building-block.[158]

It is not certain when these blocks were first cast. The hardness of the slag must have been readily apparent from the early days of large reverbatory furnaces for copper production, and it was only a matter of time before they were formed into blocks suitable for building. The first major smelting centre after the ending of the Mines Royal monopoly of copper mining and smelting in the 1690s was Bristol. Here, between 1748 and 1769, the charismatic copper smelter and brass manufacturer William Champion built a mixed industrial and parkland landscape. A 13 acre lake was built in front of his mansion of Warmley House in order to store water to power his rolling and battery-mills. In the centre of this was (and still is) a three-storey tower of copper-slag rubble pieces with iron-ties, that had cement rendering applied to their rough surface to form a titanic figure of the god Neptune, it was visible from the Warmley mansion capping an eminence on the lake shore.

The Warmley Brook, as it enters the broad lake site, is also spanned by an arch of Pennant Sandstone carrying a three-storey crenellated tower formed completely of cast rectangular copper-slag blocks. The quoin and dressing blocks project from the wall faces and the northern face is white rendered with the projecting copper-slag blocks left bare in striking contrast. This lakeside folly-house must have offered

Fig. 75. The castellated stables of the Bristol coppermaster William Reeve, and associated servants' housing, built partly in copper-slag blocks at 'Black Castle' constructed ten years before Morris's Castle at Swansea. (960176/1)

Fig. 76. Servant's House, Black Castle, Bristol. (960176/17)

Fig. 77. Blocked doorway with cast copper-slag quoining into the former outbuilding (supposed coachhouse) at the gate of Sir John Morris's now-demolished Palladian mansion of Clasemont at Morriston. (960206/2)

a striking aspect to visitors to the Warmley Works and Park. What seems to be the intake housing over the sluices controlling the water into the works' water-wheels also has copper-slag quoins.

The Bristol Brass Company, which took over the Warmley Works from 1770, acquired the Forest Copperworks and Lower Forest Mills in Swansea in 1790. The surviving walls of Walmley Park are capped with tall triangular coping-blocks of a style that is peculiar to the Bristol area; they are found around Kelston Park and in Stoke Bishop. The copper-pin factory at Warmley, built before 1767, has rectangular building-blocks with rounded corners to stop works' traffic from damaging the building. This type of block has not been found in Swansea.

The *tour de force* of cast copper-slag block construction is without doubt the 'Black Castle' (*c.* 1760) at Arno's Vale in the suburbs of Bristol: described by Horace Walpole as 'the devil's cathedral'. The six tall towers, curtain-wall and attendant dwelling are all built of this material. Special moulds were prepared to build the circular towers at the corners of this impressive structure. It housed the stables, service buildings and some of the staff for nearby Arno's Court and its elaborate traceried windows mirrored the Strawberry Hill Gothick of other park structures executed in golden Bath Stone.[159] William Reeve, owner of Arno's Court and builder of these structures, had married into the Harford family, partners in the Bristol Brass Company.

The first datable use of copper-slag block construction in Swansea may be as the dressings (the quoins and crenellations) of Morris Castle in 1773-75. A difference from earlier Bristol constructions was in its use of projecting string-course bands of rectangular copper-slag blocks. This may suggest that, like the Warmley House lake intake house, the exterior mass walling was rendered white to provide a striking contrast to the black copper-slag dressings. Copper-slag blocks were also used in the construction of buildings in and around the parklands of John Morris I's new mansion (near what is now Morriston). The 'coach house' flanking the entrance to the Clasemont mansion drive has copper-slag quoins, and window and door dressings and presumably such structures as the banqueting house and home farm also used this building material.

There is no evidence that copper-slag blocks ever became a prestige building material in the way that brick did in areas where it was first introduced with mass walls of brick fronting houses that were otherwise of rubble-stone. The material was generally too dark and bleak for the construction of mass walls. The impression given is that the early mass use of the material for parkland structures in Bristol derived from its novelty value. Where mass-walling was used in Swansea it seems to have been as a cheap and convenient method for the copper-smelting concerns to construct structures that were not generally seen, for later extensions to smelting works and for some nonconformist chapels. In the latter context it may not have been used in Swansea but was used by Welsh workers at the Lancashire copper-smelting centre of St. Helens (where Parys Mountain ores were smelted) in order to build their own chapel and minister's house and also for the eighteenth-century Wesleyan Methodist octagonal chapel built by the proprietors (including John Vivian) of the Hayle Copperworks in Cornwall for their workers.

The large pedimented block of the 'Hostel' or 'Wychtree House' on the riverside of Morriston (founded 1779) may have been the Poorhouse (but with separate apartments) known to have been built by John Morris I in collaboration with the mason/architect/engineer William Edwards. The mass-walling of the two storeys facing Wychtree Street and the four facing the River Tawe were constructed of Pennant Sandstone rubble. As with Morris's Castle, the storeys were separated by string-courses of rectangular copper-slag blocks but not projecting, as on the castle, and the whole front facade was whitewashed. The four storeys of the rear riverside elevation that was not generally seen by passers-by had its lower storeys constructed entirely out of rectangular copper-slag blocks. The somewhat arbitrary extent of this use of blocks suggests that supply of this material was dependent on the contemporary volume of copper-smelting of which this building-block was a by-product. There are other partial uses of copper-slag block mass-walling in the Morriston area. The coal mine drainage water flowing down Water Street in Morriston was culverted at the end of the nineteenth century and the tunnel mouth at the discharge into the River Tawe has abutments of copper-slag blocks topped by walling in Pennant Sandstone rubble. On the south side of Water Street are garden walls of Pennant Sandstone rubble topped by four courses of copper-slag blocks.

Works buildings showed a substantial use of this material. Old photographs show that much of the Middle Bank Copperworks had been constructed of this material in a fairly 'polite' style. The surviving copper-slag end-gable of part of the Crown Copperworks near Neath Abbey was attached to a building with brick side walls. The most impressive surviving use of this building material in Swansea is in the huge abutment that survives on the western side of the Swansea Canal outside the Hafod Works in Swansea. This was constructed in the mid-nineteenth century to carry the large inclined railway up the side of the Hafod Works Tip. The fact that this abutment is flanked by walling in Pennant Sandstone indicates the inclination to use the more generally available Pennant Sandstone rubble whenever squared-blocks were not required. It was more worthwhile to save the trouble of working ashlar blocks of Pennant Sandstone by using copper-slag blocks mainly as quoins (cornerstones). Thus, the eastern abutment of the tips' railway bridge over the Swansea Canal at Hafod is constructed of Pennant Sandstone rubble with quoins of rectangular copper-slag blocks.

An examination of the use of this material in the settlements built for the copper-workers is instrumental. Construction of John Morris I's Morriston began in 1779, with workers building their own houses and there is no indication of the use of copper-slag in the structures that remain. The exception is the 'Hostel' or 'Wychtree House' did use fairly hidden blocks; the first church and chapel are unknown quantities and there is no use of this material in the Markethouse. The use of copper-slag blocks is not at all obvious in Pascoe Grenfell's Grenfelltown, although a redundant corner, probably a secondary addition, at the west end of the middle terrace, does have rectangular copper-slag block quoins. The surviving buildings of Pascoe Grenfell's Kilvey Hill Schools seem to be mainly of Pennant Sandstone and there is no use of copper-slag in the Grenfell's Kilvey Church.

John Henry Vivian's copper-workers' settlement of Trevivian was begun c. 1837. The surviving terraces are mainly rendered, but the unrendered early houses on Vivian Street are of Pennant Sandstone rubble but apparently the corners of the terraces are formed of copper-slag blocks. It is in the garden walls of the Neath Road houses that the commonest use of copper-slag material may be seen. The front garden walls are made up of rough copper-slag pieces, with levelling courses arranged at about one foot intervals. These are capped by semi-circular section copper-slag coping blocks.

The main blocks and teachers' houses at the Hafod Copperworks Schools have copper-slag quoins, but the windows and doors of the schools are in Pennant Sandstone ashlar. However, the church has all dressings executed in Pennant Sandstone ashlar.

Elsewhere in the area of the Swansea smelting district, the rather hard and bleak copper-slag blocks were used more widely in the homes of copper-workers at Cwmafan: here, in the mid-nineteenth century, houses were built with copper-slag blocks as edging to the doors and windows as well as for quoining. In the other eighteenth-century copper-smelting centre at Bristol, there are indications that the cast copper-slag blocks may have been widely used on the hidden side walls of town mansions in the middle-class resort of Clifton; much in the same

Fig. 78. At Morriston, whole garden walls were built from rectangular blocks of copper-slag as that at Water Street. (9500100/23)

Fig. 79. At the Hafod Copperworks, an inclined railway carried away the waste slag, the line being supported by the pier on the right and the abutment wall on the left as it crossed the Swansea Canal. The pier has copper-slag quoining, whilst the much larger abutment face to the left is completely of copper-slag blocks. (960202/4)

Fig. 80. Clifton Court, built in c. 1745 by the Bristol coppermaster Nehemiah Champion. The dark-coloured main side wall has copper-slag blocks visible, although the main elevation has a veneer of Bath stone. (990236/2)

53

Fig. 81. Cast copper-slag quoining in the former coachhouse at the residence of the general managers of the Hafod Copperworks at Aberdyberthi House, Trevivian. (960210/1)

Fig. 82. Half-round examples of coping-blocks made from cast copper-slag capping the garden walls of a worker's house in Water Street, Morriston. (9500100/13)

Fig. 83. The surviving Vivian & Sons railway locomotive shed south of the Hafod Copperworks built of copper-slag bricks. (960205/4)

way as moulded concrete-blocks are used in housing today. Clifton Court (1746) a Bath-stone fronted mansion (now the Chesterfield Nuffield Hospital) still has part of a cast copper-slag side wall visible.

Continental European visitors such as Eric Svedenstierna in 1802-3, noted the use of copper-slag paving blocks in the construction of the western harbour pier around Fabian's Bay, but he was told that Pennant Sandstone blocks were preferred as they were considered to be more durable:[160]

'At the above-described harbour wall a part of the pavement was laid with slag bricks from the nearby copper works, but this is seldom done, because they can get enough stone, which is more durable.'

Roman numerals are visible on rectangular blocks used as coping-stones for the canal bridge at Neath Abbey and this was presumably just a simple numbering of the moulds in the works.

The use of cast copper-slag blocks certainly was a precedent for, and analogous to, the use of modern concrete blocks. Mass use of the material was largely on hidden walls and its most common use was as an easily produced alternative to worked ashlar Pennant Sandstone blocks. It is not known to have been used on coppermasters' houses, chapels or churches. There was one exception to this, namely, the common and universally acceptable use as hard and durable coping-stones on park walls.

It has already been noted that triangular-shaped coping-blocks found a common use around the earlier copper-smelting centre of Bristol and on large rural gentry parks such as the Kelston estate. Likewise at Swansea, the Vivians used copper-slag coping-blocks both in their workers settlement at Trevivian and at their mansion at Singleton Abbey. The coping-stones commonly used in and around Swansea were half-round in section. These were also sometimes used in Bristol: on the Ashton Court

estate and on the Bristol side of Bitton.[161] In the Swansea area it is uncommon to find coping-blocks of a more complex section; an exception is the double ogee-section curves of the coping-stones on the tramroad bridge at the centre of the Neath Abbey Ironworks complex.

There is one other building material that was a by-product of the copper-smelting industry. This was the use of Vivian's widely-advertised 'patent' brick, made of a light-grey material based on crushed slag. Unfortunately, this has proved nowhere as durable as cast copper-slag blocks and is badly spalling on the structures on which it is has been used. The riverside 'Vivian & Sons' Beyer-Garrett locomotive shed at the Hafod Copperworks has its walling made of this material which has spayed badly. The upper part and rear of the earlier rolling-mill engine-house at the Hafod Copperworks and the whole of the later engine-house are also of this material.

The production of copper-alloys

The commercial economics of smelting the various non-ferrous metals was radically different. The relative totals of copper-ore and copper metal produced by the Lockwood, Morris & Co. partnership suggest that a small amount of refined copper could be produced after repeated meltings of the ore by enormous quantities of coal. The same was not true of the other non-ferrous metals, such as tin and lead, the smelters of which were often sited near the source of their respective ores, because of the lesser amounts of coal required to refine them. The refining of these other metals was not concentrated in Swansea in the way that copper-refining was and there was not the development of a specific refining method at Swansea for these metals as there was in the 'Welsh method' of copper-smelting there.

It seems to have been often assumed that the brass industry was associated with Bristol and copper-smelting with Swansea. The division was not as clear-cut as this and several Bristol-based concerns opened or took-over large works in Swansea to take advantage of the coal-resources available there. Bristol was the earlier smelting centre and it was here that Dr. John Lane, builder of the first Swansea copperworks at Llangyfelach, gained his experience. In *c.* 1712 the Bristol industry was producing 400-533 tons of refined copper, of which half was forged and manufactured to be marketed as copperware and half as brassware.[162]

Copper was often required as a constituent for brass (alloyed with zinc) and bronze (alloyed with tin), and there is incidental evidence that the Swansea copperworks routinely produced artefacts in these metals. This has already been touched-on with the mention of slave-trade bronze 'manillas' produced at the various early eighteenth-century 'copperworks'. The Cornish ores used in Swansea and Bristol contained appreciable quantities of tin, causing a high-tin bronze to form at the bottom of the refining furnaces.[163] This was either used to cast bronze goods at the works or sold to bellfounders. The metals from the Llangyfelach Copperworks, held in stock at the London warehouse included 'pot metal' (brass, produced in pots) and 'bell metal' (bronze) worth £345 15s.[164]

It was the alloying of copper with zinc to produce brass that seems to have continued in the later Swansea copperworks. For example, Lockwood, Morris & Co.'s Forest Copperworks' accounts reveal that in December 1768 there was 'Shruff and Old Plate' at the copperworks to the value of £345 4s. 4d. The significance of the presence of these materials at the Forest copperworks is illuminated by the following description of brass-making in 1697:

'compos'd with about two sevenths of fine copper, four sevenths of lapis calaminaris [i.e. zinc-ore], and one seventh of shruff, which is old plate brass. This put into pots containing ten or twelve pounds of metal each, and set in a furnace, where there is freedom of air at bottom, melts and coalizes, or joins in ten or twelve hours time into a new thing called brass. Then eight or ten of these small pots are poured into one larger; and when 'tis perfectly digested, and well scumm'd from the dross, 'tis then poured between two stones of a tun weight, or more each, which are elevated at one end for to make the metal fill the whole vacuity, and then 'tis set horizontal to cool, and thence comes out a plate of about seventy pounds weight.'[185]

The partners of the White Rock Copperworks did decide to produce brass there and indeed late eighteenth-century drawings do show what look like

fig.84

three brass annealing furnaces at the works. Similar drawings show two such furnaces or ovens at the Forest Copperworks. Much brass production using Swansea copper also took place in Bristol and Brimingham. Large-scale production at Swansea boomed after 1837 with the development of the brass known as 'yellow-metal' and its widespread use as a cheap substitute for more important copper-sheathing. Several Swansea copper-smelters were adapted to zinc-smelting for the purpose of producing this alloy.

Others Metals Industries

The Early Iron Industry of the Lower Swansea Valley

Some elements of the iron industry of south Wales originated in the Swansea Valley. Its metalliferous industry may have started in a forge at Swansea in the fourteenth century. There was also an early furnace at Ynysgedwyn in the upper Swansea Valley, although its origins are obscure. Ironmasters may have been attracted here by the dense woodlands and iron-ore outcrops as early as 1712,[166] and quite possibly to other valleys in south Wales for similar reasons at about the same time. The Ynys-pen-llwch forge in the lower Swansea Valley was probably opened as an adjunct to the Ynysgedwyn furnace, it was situated on the route from the furnace to the navigable water of the lower valley.

Yet it was the eastern, not the western, valleys of south Wales which provided the most attractive natural locations for the iron industry. The southern edge of the south Wales coalfield, with its iron-ore and limestone, was partly lost under Swansea Bay; further east it could be easily worked near the port of Cardiff. By 1740 a relation of the Thomas Popkin who built the Upper Forest Forge, Thomas Lewis, had opened a forge and furnace at Pentyrch to the north of Cardiff. The Thomas Price who was his partner may have been Popkin's grandson. This works was quickly followed by the founding of the Melingriffith Forge. Melingriffith attracted Francis Homfray to the Taff Valley, he was to play a large part in developing what became the largest ironworks in the world, further up the Taff Valley at Cyfarthfa, Merthyr Tydfil. Thomas Price was one of the partners in a third forge founded on the Taff, this time at Cardiff in 1751-52. A significant step was made by Thomas Lewis who was one of two men taking mineral rights on the north-eastern rim of the coalfield in 1757.[167] The first of a huge number of coke-fired furnaces in south Wales was built on part of Lewis's taking at Dowlais.[168]

The foundation of an iron industry in the Swansea Valley in the late seventeenth and early eighteenth

Fig. 84. The new steel-framed mill erected in 1920-2 over the yellow-metal rolling-mill at the Morfa Copperworks with travelling-crane overhead. Behind this is an older 500 h.p. steam-engine driving the mills, with the old masonry walls of the mill in course of demolition. (National Museums & Galleries of Wales – Department of Industry)

centuries thus led to developments of great consequence. The immense scale of the coke-fuelled ironworks developed on the north-eastern rim of the south Wales coalfield in the late eighteenth and early nineteenth centuries, was not emulated by the old charcoal-fuelled works further west.

Tinplate Manufacture

A later, but equally important, industry in the Swansea region was the manufacture of tinplate, this began locally in the middle of the nineteenth century. Similar locational factors applied to the tin industry as was the case with copper and pottery. Swansea was where the nearest coalfield to the useful mineral resources of Devon and Cornwall touched the sea and was easily accessible. It also had an iron industry which was rapidly expanding to the west with the development of anthracite-fuelled iron-smelting

Fig. 85. The statutory 22ft high waste water weir at Trebanos; colliery water supplying the canal can be seen cascading off a timber trough in the right background. Canal waterpower was important to the early tinplate industry.

after 1837. It developed so successfully, that Swansea and the surrounding district soon became the world centre of the tinplate industry. In 1913, four out of every five tinplate workers in the United Kingdom lived within 20 miles of Swansea.[169] Like the non-ferrous metals smelting industries, tinplate has contributed to the formation of Swansea's industrial landscape.

The origins of this industry go back to the eighteenth century, when the availability of water-power in the lower Swansea Valley attracted a series of iron-forges that used the product of the upper Swansea and Neath valleys. The furnaces could not participate in the late eighteenth-century transition from charcoal to coke firing because of the nature of the local anthracite coal. The iron and steel industry in the Swansea region did not therefore develop on such a scale as the iron industry in the eastern Heads of the Valleys region, until the dual development of anthracite-smelting in the upper Swansea Valley after 1837 and the later rise of the vast coastal steel works in the late nineteenth century which used imported ores.

Tinplate (iron or steel bar rolled into sheets and coated with a wash of tin) was first manufactured in Britain at Pontypool in the seventeenth century. Previous continental European practice had been to hammer sheets flat with large wooden hammers prior to coating them with tin. During the next hundred years a number of works were established in south Wales and elsewhere, but initially the concentration was around Pontypool and not further west.

At this stage, some the existing water-powered forges had ranges of tinning-bays added to them. This was still the situation in 1843, when of the 23 works active in south Wales, 15 were situated east of Cardiff; and of the eight in south-west Wales, three were in the Neath/Port Talbot area, four in the lower Swansea Valley and one was in Carmarthen. This is put in a context by the knowledge that in 1850, 71% of the tinplate works in the United Kingdom were situated in south Wales.[170]

An interesting factor in the growth, and eventual concentration, of the world tinplate industry in the Swansea area was the way in which the tinplate industry is related to the rise, change in technologies and eventual fall of the copper industry. It also inevitably depended on the change in fortunes of the local iron and steel industry. The reason for this can be appreciated if the nature and form of a tinplate works are considered. The two major items of plant required are a rolling-mill with reheating furnaces and a tinning-building with tinning-bays arranged down its sides. The first type of building was one that was common to the existing water-powered iron and copper forges and it is apparent that these existing

fig.86

100 0 100 feet 200

the lower Swansea Valley had their origins in the water-power of the Swansea Canal, a 1790s waterway originally built to take advantage of the new copper-trade harbour at Swansea. In 1843, the Primrose Forge at Pontardawe was built by William Parsons, using the water of both Swansea Canal and the Upper Clydach River to drive its three rolling-mills; this was the nucleus of a works that eventually, under the Gilbertsons, had no less that 16 tin mills and a similar number of sheet mills. In 1838, William's brother, John Parsons, had built the Pheasant Bush Tinplate Works, with its twin rolling-mills, a short distance to the south, alongside a large waste-water weir of the Swansea Canal - one of several weirs on the canal where water had to be returned to the river under the terms of its Act, so that the water supplies to the coppermills on the river could be maintained. The latter works was enlarged in 1890-99 using a water-turbine that was driven by the bypass water of the adjoining two locks on the canal.[172]

The advent of steam rolling-mills at the copperworks from the beginning of the nineteenth century meant that the earlier off-site water-powered mills eventually became surplus to requirements. This happened to the Forest Coppermills and explains why Morriston became a centre of the tinplate industry. In 1845 a single water-driven tinplate mill was put in use by Messrs. William Hallam, in what had been built as iron rolling-mills and used as coppermills for almost a century. This eventually became the nucleus of the huge Upper Forest and Worcester Tinplate Works that had 16 tin and five sheet mills.[173]

The Lower Forest Coppermills is usually said to have been transformed into the Beaufort Tinplate Works in 1859-60, by the addition of a tinning-house (which survives) to the existing rolling-mill of c.1735 (now demolished above ground). However, twin 34in cylinder Neath Abbey Ironworks engines had been added in 1836 to provide auxiliary drive for the water-power of the two initial tin-mills. The surviving annealing-house was later added to the end of the rolling-mill. Eventually the old copper rolling-mill served as the nucleus for 12 tinplate mills.[174]

At the Landore Tinplate Works the steam-engine may only have been an auxiliary-drive provided to what seems to have been the existing (water-powered?) rolling-mill of the Landore Ironworks, where experiments had taken place to roll iron-sheets from iron-rich copper-slag. There were eventually seven tinplating mills here.[175] G.L. Morris of the former copper-refining dynasty was one of the partners.

The tinplating industry was significant for the number of local, low-capital entrepreneurs entering

Fig. 86. The Pheasant Bush Tinplate Works founded in 1838. The Pheasant Bush site had a water-turbine added in 1890 that used the canal-water bypass water to locks 8-9 and returned it to the canal; the Trebanos Overflow can be seen in plan at the top of the area illustrated. (GWR Book of Swansea Canal Maps in the RCAHMW Archive)

installations helped to attract the tinplate industry to the area.

The Ynys-pen-llwch rolling-mill at Clydach in the lower Swansea Valley became the first tinplate works in Glamorgan in 1747; a tinning-house was added to a rolling-mill founded as an ironmill and forge, by Bristol merchants a hundred years before, subsequently it was a coppermill, until Rowland Pytt of Gloucester turned it into a tinworks. His son-in-law and successor, William Coles, turned the Cambrian Copperworks Rolling-mills into iron rolling-mills that may well have supplied the Ynys-pen-llwch Tinplate Works with iron sheets for tinning. A similar course of events seems to have taken place at the second Glamorgan tinplate works at Ynys-y-gerwyn (1753), in the nearby Vale of Neath.[171]

The large-scale and significant shift of the centre of the tinplate industry to the Swansea region began in the middle of the nineteenth century. Interestingly, the first of the new works at the top of

Fig. 87. The Forest Coppermills were converted into iron rolling-mills for new tinplate works. One conversion had probably taken place by 1836 when twin beam-engines were added to pump water used by water-wheels back up to the mill-pond in order to provide sufficient water for increased use. (Held by the West Glamorgan Archives Service; Neath Abbey Ironworks Collection, D/D NAI M/420/1)

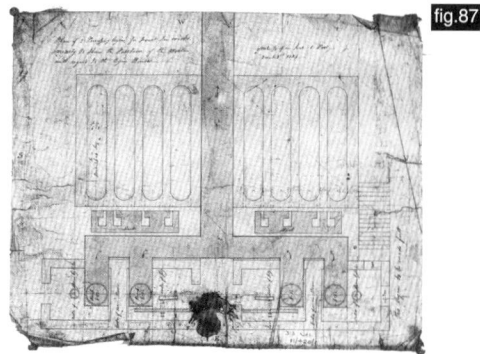

the industry. Hence, it was of great importance that these water-powered sites, buildings and rolling-mill machinery were capable of reuse and adaption. This recycling of resources, which in large measure aided a relocation of the tinplate industry to south-west Wales, went beyond metal-centred industries in the lower Swansea Valley.

The Cwmfelin Tinplate Works was on the site of the Millwood Colliery in Cwm Burlais (at what is now Manselton) and used the reservoir that had fed the colliery's three water-wheels. It was founded in 1858 by David Davies & Son and eventually had 20 tinplate mills.[176]

The first three works on the fringes of the Swansea area also reused the rolling-mills of pre-existing iron forges: Margam (Port Talbot, 1822), Copper Miners (Cwmafan, 1825), and Aberdulais (1830).[177]

By the end of the nineteenth-century, there were 11 tinplate works in the lower Swansea Valley, from Trebanos southwards; Swansea had established itself as the centre for world tinplate production. Dynamic growth followed, with the number of works in south Wales rising from 25 in 1850 to 90 in 1891; this represented a rise in the proportion of the UK total of works from 71% to 92%.[178] By the decade 1870-80, tinplating activities in South Wales were concentrated within a 15 mile radius of Swansea and by 1880, 41 of the 64 tinplate works in south Wales were located west of Port Talbot.

There were 10 works in the Swansea Valley area up to Morriston; to the west, Llanelli had seven works; Pontardulais had seven; Gorseinon/Loughor had four and to the east-Neath/Briton Ferry had eight and Port Talbot had six.[179]

The full list of the Lower Swansea Valley Works from Morriston to central Swansea is as follows: Upper Forest (1845); Landore (1851); Cwmfelin (1858); Beaufort (1836-60); Cwmbwrla (1863); Worcester (1868); Morriston (1872); Duffryn (1874); Midland (1879); Birchgrove (before 1880) and Aber (1880). In a dramatic rise, the five Lower Swansea Valley tinplate mills of 1845 had grown to 106 by 1913.[180] This period of growth coincided with the decline of the copper industry, so that a

labour force, skilled in the metallurgical industries, was available for employment in the booming tinplate trade. By this stage, supplies of sulphuric acid, used in the pickling stage of the tinplate manufacturing process, were available as a by-product of copper-smelting. The availability of coal, navigable rivers and water were also important in attracting the industry to south-west Wales.

The factor above all others, however, which made the Swansea area the centre of tinplate production, was the perfection of open-hearth steel making by William Siemens at the vast Landore Siemens Steelworks (one of the four biggest in the world) and the application of steel, rather than iron, to tinplate manufacture in collaboration with Daniel Edwards of the Dyffryn Works at Morriston. The universal spread of the new base material created huge 'integrated' tin and steel works which generated the expansion of towns such as Morriston, Pontardawe, Llanelli, Gorseinon, Neath, Briton Ferry and Port Talbot.[181] There were also many smaller works which bought their steel from outside suppliers and whose requirement working-capital led to the foundation of the 'Swansea Bank' in 1873, several of whose directors were connected with the tinplate industry.[182]

It was the American market that really boosted the tinplate trade. It proved a cheap material for all the domestic utensils that a homesteading family would need, including a roof for their house and a covering for their walls. It formed the cans in which the meat-processing factories of Chicago packed their bully beef and the drums in which the nascent oil industry transported petroleum. Until the 1870s, this huge trade was generally re-exported from Liverpool and, to a lesser extent, Bristol. With the completion of the Prince of Wales Dock in 1882, however, Swansea was transformed from a copper-trade port into a major transatlantic tinplate industry port. Swansea's share of the tinplate trade consequently grew from 35,900 tons (out of a total of 265,000 tons) in 1882 to 229,000 (out of a total of 421,800 tons) in 1890. This shift in trade patterns led to the foundation of the Swansea Royal Jubilee Metal Exchange in 1887, to house the national tinplate trade dealings that had formerly taken place in the Liverpool Metal Exchange. Almost all of the important London and Liverpool merchants established offices, or were represented by agents there, and were subsequently joined by a group of local merchants.[183]

The Swansea tinplate industry was too successful in the American market, in 1891 President KcKinley imposed a tariff on imported tinplate. Swansea was badly hit, many of the workers migrated to the United States and contributed to the success of its growing industry, while those who remained had to face a period of depression until the trade could re-establish itself with new outlets on the European mainland and in the Far East.

Tinplate, as traditionally manufactured, was a labour-intensive industry, it had a low rate of worker-productivity compared to the more highly automated American industry. After World War II, therefore, in order to ensure that Wales retained a tinplate industry, a process of rationalisation took place. This saw the replacement of the old handmills by the modern automated stripmills at Trostre (Llanelli) and Felindre (north of Swansea). Both used tinned steel-coil manufactured at Port Talbot. It was a noble attempt to maintain production in both urban centres, but only one centre of production was sustainable and the last active Swansea tinplate works at Felindre closed in the early 1990s and was demolished in 1996, so finally ending Swansea's 250 year connection with the tinplate industry.

The annealing and tinning buildings of Beaufort Tinplate Works (1860) are the sole survivors of the trade in Swansea, although the latter has a partially new replicated exterior and the converted and rebuilt copper rolling-mill of *c*.1735 has been demolished. The partial survival of the distinctive tinning building with the remains of bays and chimneys down its sides is a notable survival when compared with the tinplate works museum sites at Aberdulais and Kidwelly which have no such standing remains. The most remarkable remains of the early south-eastern phase of the industry are the great, and fairly complete, tinplate works of the Crawshays (1836) at Treforest.

The great arches in the outer walls of the surviving annealing-house at Beaufort, echo those in the Morfa copper rolling-mills and demonstrate that a major architectural imperative in the design of all of these buildings which housed metallurgical processes, was the need to make the working-conditions around the furnaces and their hot metal comfortable enough for the workforce to function effectively.

The Lead and Silver Industries

The lead and silver-smelting industries arrived in the Swansea Valley in 1717, at the same time as the copper industry for the three were long associated. However, by the mid and late eighteenth-century, specialist copperworks predominated in the area and it was not until the growth of the huge integrated works of the mid-nineteenth century that the lead-smelting industry grew in the Swansea Valley once more. The lead-smelters of Swansea never dominated the world industry as did the copper-smelters or tinplate manufacturers and neither did they control the industry nationally as did the zinc smelters at a later date.

Fig. 88. The Beaufort Tinplate Works Annealing House of 1874 with its multiple ventilation arches and openings. The Annealing House had been added onto the much rebuilt Lower Forest Copper Rolling-mill of c.1735 (background — now demolished). (849919)

Fig. 89. Tinning-bay arches of the Beaufort Tinplate Works before modern replication of the exterior facades. The surviving tinning-house had originally open arches to dissipate the heat, like many of the earlier copper-smelting and rolling buildings.

The economics of production were radically different from those of copper, with the richer lead-ore containing some 70% of lead. The smelting of the lead-ore, therefore, generally took place on, or near, the ore-field. Two of these major areas of mining and smelting in Britain were Derbyshire and Flintshire. North Ceredigion produced and exported large amounts of lead-ore from Aberystwyth and this coal-free location could send its output for smelting to either Flintshire or Swansea. Lead also existed in association with the tin and copper deposits of Cornwall and Devon; small concentrations also existed in the Vale of Neath and at the head of the Swansea Valley.[184]

The largest non-ferrous works founded in the Swansea area at the end of the seventeenth and beginning of the eighteenth centuries were of the integrated type processing several of these minerals. The works included the Melincryddan Works at Neath of Humphrey Mackworth and the Llangyfelach Works of Dr. John Lane at Landore near Swansea, both of which originally had far larger numbers of lead furnaces than of copper.

The Llangyfelach Works (founded 1717) was still operational in 1745 when there is also evidence in the accounts for silver and lead smelting although by then the relative quantities of these metals produced were much smaller. The comparative amounts of metals that had recently been produced and were still held in stock at the works in December 1745, were as follows: silver, 78oz 10cwt; lead worth £53 5s. 1d. and battered copper (11tons) worth £1,430 10s. 1d. However, the metals held in stock in London suggest

that much more lead was normally produced. At that date Lockwood, Morris & Co. held 60 tons of lead in London (worth £597 11s. 1d.) whilst only holding 7 tons 9cwt of fine copper (worth £835 12s.) and 15tons of battered copper (worth £1,960) there. There is no evidence that the new copperworks of Lockwood, Morris & Co. also smelted silver and lead, and the tons of the respective ores held by the firm suggest that the production of lead by the company was just about to end. Only 13 tons of lead ore (that presumably produced silver as well) were held by the firm in Cornwall, which also held 816 tons of copper-ore in the same county.[185]

Both the Llangyfelach and Melincryddan Works had been founded at a time of changing technology in the industry. Much of the smelting process was initially developed on the continental mainland of Europe, but it was the development of the reverbatory furnace for lead-smelting, as for copper, that led to the British dominance of the world trade in this industry in the eighteenth century. Variations on the German or 'Almain' small blast furnace (already described in the evolution of copper-smelting) had been used as the 'Slag Hearth' used for the recovery of lead from ore hearth slag. Such a furnace, there called an 'Almond Furnace', had been used at Ynys-hir smelting-works on the north Ceredigion ore-field in about 1670.

The ore hearth was the first peculiarly British type of furnace to be developed. It originated on the Somerset Mendips and used human foot-power to drive the bellows on a low lead-smelting hearth that somewhat resembled that of an ordinary blacksmith. It had been taken to the large and influential Derbyshire lead-smelting mining area by William Humphrey in about 1577. The stonework of the furnace was first replaced by iron-castings at Wanlockhead, south-west Scotland, by 1682 and became known as the 'Scotch Hearth'. Eighteenth-century ore hearths consisted of a shallow hearth basin sited on a raised base below a chimney with a water-driven blast from a tuyere blowing through a pipestone into the rear of the hearth.[186]

The Melincryddan Works at Neath were constructed at the dawn of the new processes which resulted in the large-scale expansion of the non-ferrous industries in the Swansea area which were based on the reverberatory furnace. The Clarke/Grandison experiments in the Nightingale Valley in Bristol, have already been discussed with reference to the evolution of the copper-smelting reverbatory furnace. Flintshire lead-ores and slags were used in the Bristol experiments which finally proved successful in 1683. The Clarkes were prevented from the use of this process, in lead-smelting, by Grandison's patent. When it expired in 1692, a 'cubalo' (a 'cupola' is another name for a reverbatory furnace) was erected at Flint by an associate of Grandison's which, in 1696, was sold to a London silver refiner, Daniel Peck. Peck also became involved with Sir Humphrey Mackworth at Neath where the reverbatory furnace was used on a large scale. Workers were poached by the nearby Crown Copperworks and Dr. John Lane soon left this area to found his Llangyfelach Works in the adjoining Swansea Valley.

In 1706 the newly formed London Lead Company built the Gadlys Works in Flintshire which was 'More regular and fitt, their method of working better than any in that country' and other local works were built in imitation, so that by 1730 the use of the ore hearth had been discontinued in the county. The London Lead Company installed it at Bentham in Shropshire; at Ashover in Derbyshire; at Acton, Eggleston and Nenthead in the north Pennines; at Wanlockhead in southern Scotland and in Ireland. However, the small and flexible ore hearth remained in use in Newcastle until 1960.

The early eighteenth-century lead-smelters never used the reverbatory furnace in isolation, but retained a mix of reverbatory and air-blown furnaces. This explains why Melincryddan had an air-blown refinery and Llangyfelach needed such large water-power provision with a 'blast house' operated by bellows. The Gadlys smelter of 1706 is recorded as having a mix of four smelting furnaces, two slag hearths and four refining (cupellation) furnaces. So in the first works equipped with reverbatory furnaces no less than six of the 10 furnace structures required a water-powered air blast. Water-powered stamps for crushing the products and slag for remelting were also required. Ponds are still visible among the earthworks and buildings on the Gadlys site in Flintshire.

It was the use of the cupellation furnaces for refining silver produced as a by-product in these lead works. In Europe, the normal silver content of the lead ores varied from 60 to 750 g/t of lead, and silver was always worth recovering in the larger part of this range (250-750 g/t); it was a simple and ancient process. Molten lead was oxidised to litharge by a blast of heated air blown across lead in the bowl of a modified reverbatory furnace and a small deposit of pure silver was left. The litharge could then be remelted to lead or sold; for which there was a growing market in this period.[187]

The Llangyfelach Works produced lead and silver by these processes for about 30 years but ceased production in *c.* 1748-50. Chauncey Townsend then built a new lead-smelter across the river in Llansamlet on land acquired in 1754 and with his son-in-law, John Smith, opened it in 1757. They obtained lead-ore by directly operating two of the largest mid-Wales mines: at Cwmystwyth and Rhandirmwyn. Both of these were highly capitalised and had large new access tunnels - in the latter case by an underground canal similar to that at the Forest Works in Swansea (the 'Clyn-du' Underground Canal). The Upper Bank leadworks carried on in production for 30 years, latterly as a mixed copper and lead-smelter, and from 1782 seems to have concentrated on copper alone. As with tin, lead-smelting seems largely to have remained concentrated on such ore-fields as had coal reserves such as those of Flintshire.

It might have been thought that the popularity of the copper/lead alloy, pewter, would have encouraged the continued presence of a lead-smelting capacity at the international centre of copper production, but by 1782 the widespread use of pewter tableware was already largely superseded by the products of such large-scale manufactories as the Swansea Pottery. The dominant use of lead was for the roofing of prestigious buildings and for pipework and these simple manufacturing processes could easily be carried out nearer the large domestic markets. Red (oxidised lead) and white lead were further manufactured products used in paint manufacture and the latter also for pottery and its glazes. Red lead had been manufactured at the Llangyfelach Copperworks.

When silver was again produced in the Swansea area it was as a result of the growing sophistication and size of the copper-smelting process. The massive infrastructure of the neighbouring Morfa and Hafod Copperworks encouraged diversification. Henry Hussey Vivian said 'I had seen the success of our neighbours, the Williams, who were engaged in a variety of undertakings and I resolved if possible to emulate them.'[188] He had added silver, zinc, yellow metal and coalmining to the Hafod firm's activities and by 1854 these were equal to half the firm's profits of some £56,000. His technical training in Germany allowed him, in 1851, to take out a patent for the separation of nickel and cobalt from copper-ores.[189] Some of the buildings of the Hafod Nickel and Cobalt Works survive in industrial use to the south of the copperworks.

There are the remains of some of the infrastructure required to rid the immediate area of some of the very toxic remains which resulted from the silver-refining process - the 10ft high base remains of the 200ft tall stone chimney of the Morfa Silverworks. This was demolished in the 1960s and the large diameter circular base is to the south of the Morfa Copperworks Rolling-mill (now the Swansea Museum stores).

Two inclined trenches ascend the western slopes of Kilvey Hill to the east of the White Rock Works site. These were the silverworks flues leading to large runs of condensing chambers which were built on top of the earlier White Rock Copperworks Tip and to two exhaust chimneys.

The Zinc Industry

The same locational balance of Cornwall/Devon raw materials needing huge amounts of coal from the nearest accessible coalfield, were not applicable to the smelting of calamine-ore to produce zinc. The early 'English Process' needed some 25 tons of coal to smelt every ton of ore, but the main known concentrations of calamine on the Flintshire Hills and the Somerset Mendips were near, or on, the Flintshire and Bristol Coalfields and so it was at these centres that the early smelting of zinc or spelter developed. However, calamine-ore and zinc were essential for mixing with copper to produce the ubiquitous alloy of brass, though much of the brass made from Swansea copper in the eighteenth century was produced in Bristol. The demands of this market led to a number of experiments in early zinc production in Swansea. It was the need for zinc in the production of alloys from the mid-nineteenth century onwards that eventually led to Swansea becoming the principal base for zinc production in the United Kingdom.

Spelter, or zinc, was not produced in its metallic form in Britain until about 1738, when William Champion, a Bristol copper-smelter, developed the

Fig. 90. The Morriston Spelter Works, formerly the Birmingham Copperworks. Background (right): Trewyddfa Canal with the Great Western Railway and behind the chimney to the right a former horse-worked colliery railway; foreground (right): elevated narrow-gauge railway that brought coal from the canal. By 1911, a new conveyor had been built across the roofs. The raised clerestory roofs to the east and south denote smelting-halls. The pantile-roofed building in the middle foreground may have dated from the 1790s. (City & County of Swansea: Swansea Museum Collection)

Fig. 91. The building and flues of the Loughor Zinc Works were some of the most complete to survive into the 1980s; there were probably eight furnaces in two rows above the openings in the tunnels indicated on the plan and section.

Fig. 92. The 'Belgian Process' of zinc production. Double horizontal rows of lit retorts in each furnace. The works closed in 1928 but later housed a munitions factory. Part of the buildings of the works survive as the Addis Plastics factory. (City & County of Swansea: Swansea Museum Collection)

Fig. 93. Aerial photograph (looking south) probably taken between 1924 and 1926 showing in the left foreground fumes rising from the Upper Bank Zinc Works attached to the older Upper Bank Copperworks beyond. Over the river to the right are the roofless acid chambers on top of the original tip of the Hafod Copperworks. (National Museums & Galleries of Wales – Department of Industry)

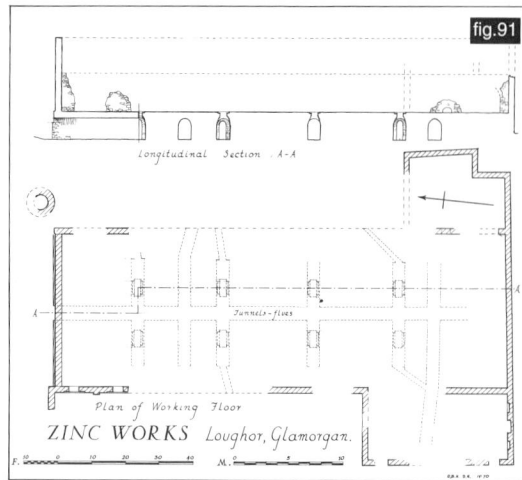

process. In 1754, Chauncey Townsend, sought to maximise the potential of his newly acquired Llansamlet Coalfield by founding smelters for

copper, zinc and lead.[190] The zinc-smelter at Upper Bank consisted of three conical furnaces and it continued in use under three operators, who seem to have brought-in specialist workers from the earlier works in Flintshire. The main activity switched to copper-smelting with the conversion of the Upper Bank lead-smelter in 1775. After the closure of the Upper Bank zinc-smelter, there was no other zinc works in the valley for some half a century.

Early in the nineteenth century, small works were established at Loughor, to the west of Swansea, and in mid-Glamorgan. In 1835 Evan John started the intermittent operation of a zinc-smelter, the Cambrian Zinc Works, with three English-type furnaces and two calciners alongside the Llansamlet Canal, at a point north of where Townsend had built his similarly sized works; but it was under-capitalised and John had to cease working in 1840. In c.1858 it was restarted by Dillwyn of the Cambrian Pottery and replaced with the more advanced Silesian and Belgian processes which had been developed in mainland continental Europe.[191]

Zinc-smelting was reintroduced to Swansea on a large scale as a result of George Muntz patenting of 'yellow metal' in 1832. This cheap copper-zinc alloy undercut copper as a meterial for the sheathing of ships. The Muntz Patent Metal Company started to smelt zinc on a large-scale in 1838-42 at the Upper Bank Works.[192] After litigation, the principal copper-smelters in the Swansea area obtained a licence to manufacture yellow metal themselves. Subsequently, Vivian & Sons purchased the disused Birmingham or Ynys Copper Works in 1841 and converted it to a zinc-smelter. The old Birmingham Copperworks Barracks were filled with the Belgian workers who were brought in to introduce the new advanced processes, in which temperatures were critical and the risk of oxidation high; in 1868 the Vivians made similar changes at the old Forest Copperworks. A shortage of raw materials caused the closure of the Ynys works in 1875 and the Forest Works was then renamed the Morriston Spelter Works; it continued in operation until 1926, when the works closed after some 180 years of non-ferrous smelting.[193]

The development of Muntz metal was followed in 1837 by the development of the process for coating iron sheets with zinc. This galvanising process completely revolutionised the demand for the metal. Galvanised sheets (of 'corrugated iron') were exported by the million to the New World, where the frontier towns of the West were springing-up and the position of the port of Swansea made it an ideal centre for the industry.

The total of 25 tons of coal required to produce one ton of zinc in the English process, fell to six tons of coal to produce the same amount of refined metal by the Belgian process. The charge of roasted ore and coal was placed in a series of horizontal retorts made of refractory clays which were then carefully heated and the zinc distilled off as a vapour that was condensed to liquid zinc.

The demand for brass increased dramatically in the latter part of the nineteenth century. In 1830 the world production of spelter had been less than 5,000 tons, but by 1837 a steady increase in production had begun and reached 17,000 tons and 29,000 tons by 1845. A huge further 12-fold leap in production had taken place by 1890, when production reached 342,000 tons.[194]

There were six zinc-smelters in operation in the valley in 1922, employing some 2,000 workers: four of them were sited along the line of the Swansea Vale Railway in Llansamlet. In the Swansea region, the expansion of zinc-smelting reflected the boom in the tinplate industry; the declining copperworks freed buildings and sites for the development of a second new metallurgical industry.

Between 1860 and World War I, 14 works in south-west Wales ceased to smelt copper-ore and no less than 11 of these started to smelt zinc. Swansea became the principal centre of zinc production in Great Britain and by 1914 was smelting 54,000 tonnes, 20% of the national total. Thereafter, large capacity plants were built on, or near, the ore-fields of America, Canada and Australia, using the new electrolytic processes. The works in Swansea closed between 1924 and 1926, at the same time as the closure of the copper-smelters in the valley, leaving only Swansea Vale Works, modernised with the aid of government funds, producing zinc, lead and sulphuric acid until its demise at the end of the 1960s.[195]

The Arsenic Industry

Arsenic is an element with many metallic properties that was often found in copper-rich ores from Devon and Cornwall. Several smelting sites there have left substantial remains; some have been conserved. The Clyne Wood Arsenic Works in Swansea has left the most substantial remains of any smelting plant in the Swansea area. From early disuse and reuse as hay sheds, it has survived into the twentieth century as impenetrable woodland in what is now the Clyne Valley Country Park in the western coastal suburbs of the city: well away from the wholesale clearances in the lower Swansea Valley to the east.

Henry Kingscole from Devon founded this works in 1844-45 but by 1852 it was primarily being used as an arsenic works by J. Jennings and Son, possibly until 1860 when the works became disused.[196] Nicholas Jennings & Co. started production of

Fig. 94. Eastern elevation of the southern calcining-furnace at the Clyne Wood Arsenic Works. (960180/1)

arsenic and copper at Dan-y-graig on the east side of the mouth of the Tawe in 1858 and moved to Llansamlet in c. 1867. Dan-y-graig then became purely a copper works under T.P. Richards & Co. from 1859, and became a copper-smelter only from a date between 1869 and 1895. Another combined copper and arsenic concern was the Landore Arsenic & Copper Co. that began smelting in the old buildings of Calland's Pottery from 1863, possibly also with workers from the defunct Clyne Wood Works. By 1880 this had been taken over by Thomas Elford, Williams & Co. (the Landore Copper Co.) which concentrated exclusively on copper-smelting.

Clyne Wood provided a site on the west bank of the Clyne Brook that was conveniently near to the sea and the Blackpill Wharf on the foreshore, from which the copper/arsenic-ores could be unloaded; they were then taken up to the works by cart. Traces of the track remain in the wood, the works was never connected to the Oystermouth Railway. The buildings were constructed of Pennant sandstone rubble quarried in the hillside to the north of the works. The works was sited on the western uphill bank of the Clyne Wood Canal which remains as a dry ditch skirting the northern tip and other tree-grown structures to the south.

The south side of the works is formed by a high, narrow gabled two-storey block which has a sunken (store?) room at its east end connecting with a basin or dock on the canal by a doorway. Coal may have arrived here by small boat from a colliery tunnel, the remains of which still survive on the western bank of the canal to the north. Coal was barrowed from the coal face to the mouth of the level. The canal's primary function was to supply water to the earlier Blackpill Mill to the south. There may have been a steam-engine in this southern block of the works in order to drive the circular revolving floor of the Brunton Calciners which were possibly situated in the north-west part of the works and connected via line-shafting to the engine.

The track from the quay branched west around the uphill side of the works and looped round to

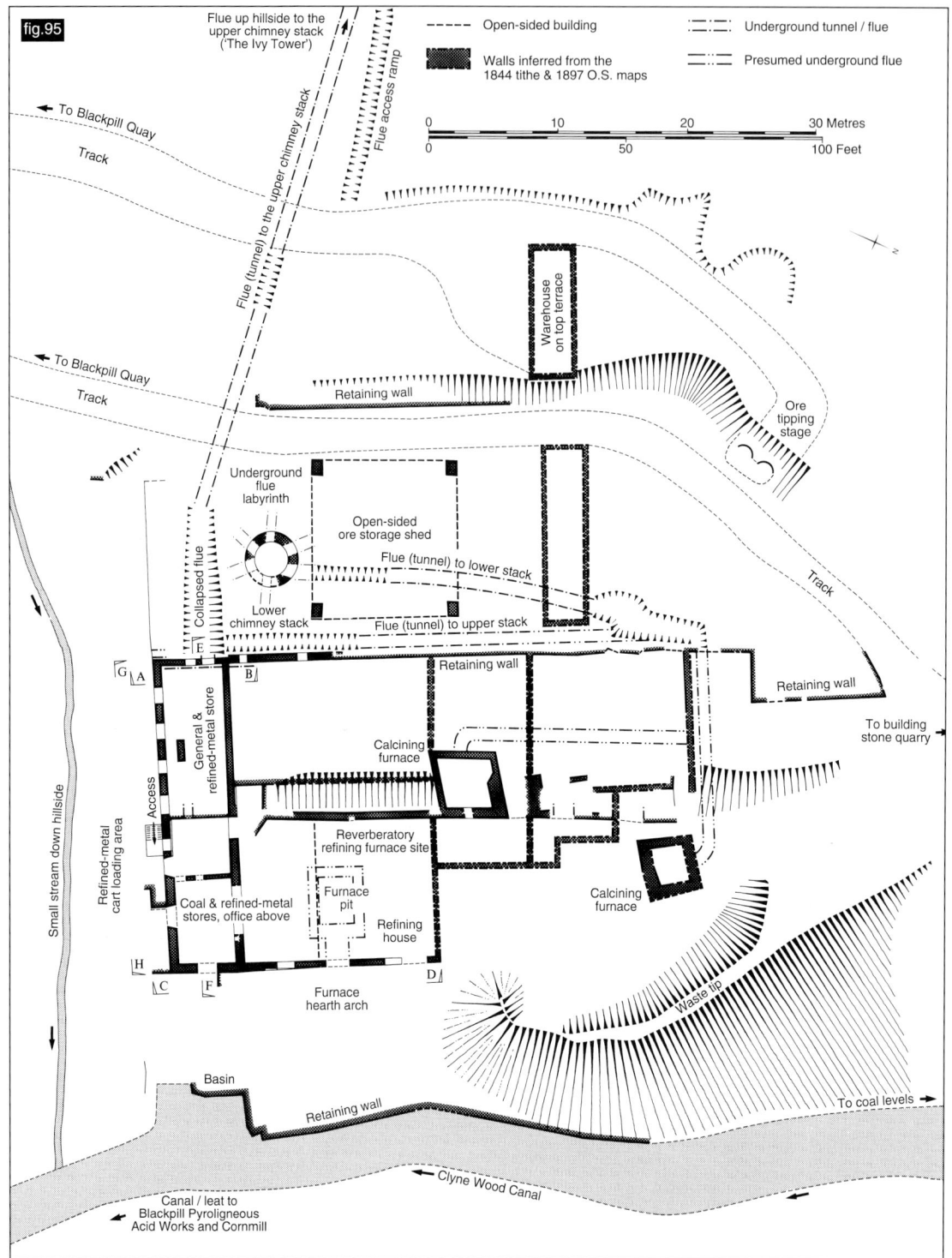

fig.95

Flue up hillside to the
upper chimney stack
('The Ivy Tower')

Flue access ramp

← To Blackpill Quay

Track

Flue (tunnel) to the upper chimney stack

← To Blackpill Quay

Track

Open-sided building

Walls inferred from the
1844 tithe & 1897 O.S. maps

Underground tunnel / flue

Presumed underground flue

0   10   20   30 Metres
0   50   100 Feet

Warehouse
on top terrace

Ore
tipping
stage

Retaining wall

Track

Underground
flue
labyrinth

Open-sided
ore storage shed

Flue (tunnel) to lower stack

Collapsed flue

Lower
chimney
stack

Flue (tunnel) to upper stack

E

G A

B

Retaining wall

Retaining wall

To building
stone quarry

General &
refined-metal store

Calcining
furnace

Access

Refined-metal
cart loading area

Reverberatory
refining furnace site

Calcining
furnace

Small stream down hillside

Coal & refined-metal
stores, office above

Furnace
pit

Refining
house

Waste tip

H

C F

D

Furnace
hearth arch

Basin

Retaining wall

To coal levels

Clyne Wood Canal

Canal / leat to
Blackpill Pyroligneous
Acid Works and Cornmill

Figs. 95, 95a, 95b. The Clyne Valley Arsenic Works. The plan shows a possible interpretation of the form of the works. Brunton calciners with a rotating cast-iron furnace-bed were developed locally, but there is not an obvious power source at this works unless the general and refined-metal stores originally housed a steam-operated beam-engine. The upper chimney-stack ('The Ivy Tower') is to the top of the main plan.

terminate at a tipping stage above the upper terrace of the works. The rubble abutment of the stage still remains. The upper (western) terrace has the remains of stone pillars that once supported an overall roof. Ore may have been stored here before being fed to the calciners below. Collapses on this platform have revealed the network of flues below, at least one can

be traced to the row of furnaces on the western (nearer) side of the platform below. On the south of the platform is the base of a circular stack with circular-arched inlets from flues all round the foot of the interior, this may originally have been the base of the main exhaust chimney from the works.

Below this platform are the remains of a north-

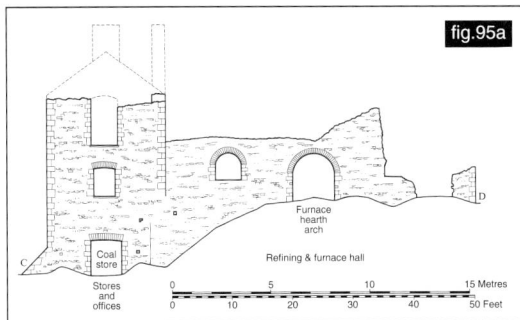

fig.95a

Furnace hearth arch

Refining & furnace hall

Coal store

Stores and offices

0   5   10   15 Metres
0   10   20   30   40   50 Feet

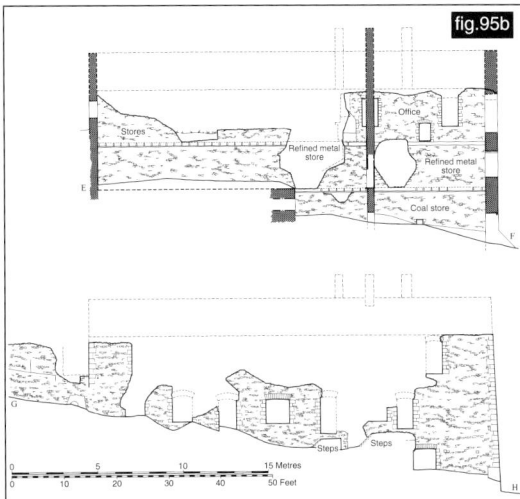

fig.95b

Stores

Office

Refined metal store

Refined metal store

Coal store

Steps   Steps

0   5   10   15 Metres
0   10   20   30   40   50 Feet

south line of furnace structures, with the upper level charging area to their west and a lower level building to their east. If they are Brunton Calciners, then a wrought-iron drive shaft would have run from the engine house to their south and along the eastern side of the furnaces with bevelled-gears leading drives for the circular furnace beds in through the small apertures still visible on the eastern side of the furnaces.

The eastern wall of the lower furnace-hall still stands overlooking the dry bed of the Clyne Wood Canal. It is built of Pennant sandstone, with a large brick opening similar to the large openings ventilating and giving access to reverbatory furnaces on the facades of the smelting-halls at Landore, Whiterock and Tomlinson's idealised Hafod copperworks. Inside (i.e. west of) the opening is a large rectangular depression as generally constructed under a copper calcining reverbatory furnace.

From the vicinity of the first circular chimneystack a large flue can be seen leading westwards up the hillside for a considerable distance to the castellated base of a second exhaust chimneystack at the top of the slope. Possibly the very poisonous arsenic fumes were not vented adequately by the first stack on the valley bottom or the works may originally have been primarily designed to produce copper rather than arsenic.

In many ways it is not surprising that the works was closed and abandoned. It is true that the Morris's themselves were opening coal-mines on the eastern side of the nineteenth century. However, the works with its high stack and flue was throwing noxious arsenic fumes eastwards onto the houses and parks of the Morris's and the Vivians: the most powerful of all the internationally important industrial dynasties of Swansea. Sketty Park House, its Belvedere prospect tower and scenic drives all overlooked Clyne Wood; and the Morris's having moved from the prospect of fumes that had developed below their mansion at Clasemont would not have wanted even more noxious fumes to destroy the ambience of their new home.

fig.96

Fig. 96. Remaining buildings on the middle terrace at Clyne Wood Arsenic Works. (812764)

fig.97

Fig. 97. Remains of the base of the lower chimney-stack at the Clyne Valley Arsenic Works with at least six openings into what seems to have been a flue labyrinth for condensing arsenic crystals. (812762)

The final reason for the move may have been the acquisition, in 1860, by William Graham Vivian (younger brother of Henry Hussey Vivian) of the mansion on the adjoining site to the west. This was the castellated Woodlands Castle built by the architect William Jernegan for the Llanelli coal-owner General George Warde in 1819-20.[197] William Graham was more interested in fine surroundings than in industrial development such as the Hafod Copperworks and by 1863 had leased Clyne Woods as extensive pleasure grounds from the Duke of Beaufort.[198] The upper parts of the high exhaust stack of the arsenic works were demolished

Fig. 98. Arsenic-rich ore was calcined in the furnaces at the Clyne Wood Arsenic Works and the gases produced were fed into a long tunnel leading-up the hillside to a tall chimney-stack. In the foreground is a collapsed section of the ascending flue with one of the substantial sections of intact tunnel. (920046/4)

Fig. 99. The collapsed top of the flue from the Clyne Wood Arsenic Works with the remaining base of the top chimney stack later adapted to be the 'Ivy Tower'. (920046/4)

and the lowest part was converted into a gothic pavilion with door, windows and spiral staircase giving onto a scenic castellated parapet. The refashioned stack, now alongside a sylvan driveway, was renamed the 'Ivy Tower'. It still looks across the woods and valley to the Morris's now derelict 'Belvedere' castellated tower on the opposite eastern edge of the Clyne Valley Country Park.

References

1. The statistics relating to the copper industry are from: C. Tomlinson (ed.), *Cyclopaedia of Useful Arts, Mechanical and Chemical, Manufactures, Mining and Engineering, Volume I, Abbattoir to hair-pencils* (London and New York, 1854), 427-28 and 442. The total output of refined copper at Swansea is calculated from the statistic that 'one foundry smelts 47,000 tons of ore every year, producing 6,250 tons of saleable copper'.

2. A.H. John, *The Industrial Development of South Wales 1750-1850* (Cardiff, 1950), 191. The export of coal from Swansea showed a large fall from 593,290 tons in 1840 to 368,449 tons in 1850, and then returned to 507,955 and 522,777 tons in 1853 and 1860 respectively. The bituminous coal exporting ports of Cardiff and Newport further east showed large and continuous increases in the shipment of coal throughout this period. The depression in coal exports at Swansea may have been caused by the sudden large use of pure anthracite coal in the new iron-furnaces of the upper valley. Therefore the shipping figure for 1853 has been used in this table as being more typical of the period.

3. R.F. Tylecote, *History of Metallurg* (London, 1976), 95.

4. J.R. Harris, *The Copper King; a biography of Thomas Williams of Llanidan* (Liverpool, 1964), 127.

5. S. Hughes, B. Malaws, M. Parry & P. Wakelin, *Collieries of Wales* (Aberystwyth, 1994), 1.

6. W. Minchinton, 'An old Gower coal mine', *Gower*, IV (1951), 56-58; Jones, *Swansea Port*, 330, 331 and 358-59; M. Williams, 'Industrial development in Clyne Valley', *Gower*, XIII (1960), 27-30.

7. The fifteenth-century coal-mines excavated in Coleorton, Leicestershire had developed galleries and were much more advanced than had been previously thought by historians, who believed that primitive shaft-mines were merely belled-out bottoms.

8. A.H. John, *The industrial development of south Wales 1750-1850* (Cardiff, 1950), 7.

9. L.T.C. Rolt and J.S. Allen, *The steam engine of Thomas Newcomen* (Hartington, 1977), 146.

10. Ibid, 151 and p.c. Christopher J. Williams, Clwyd County Archivist.

11. Ibid, 146.

12. Ibid, 152.

13. Ibid, 154; a mine at Bagillt Marsh in Flintshire.

14. J. Childs, 'Inside Penllergaer House two hundred years ago', *Gower*, XLI (1990), 22-38.

15. John Morris' Commonplace Book, *c.* 1730 (Morris Mss).

16. Ibid, 26.1.1775.

17. K.H. Rogers, *The Newcomen Engine in the West of England* (Bradford-on-Avon, 1976) quoting from MS. 473/295 in the Wiltshire Record Office.

18. Ibid, 48.

19. Ibid.

20. P.R. Reynolds, 'Townsend's Waggonway', *Morgannwg*, XXI (1977), 42-68.

21. Roberts, *Gower coalmining* (1953).

22. K.H. Rogers, *Newcomen Engine in the West of England* (Bradford-on-Avon, 1976), 48.

23. Ibid.

24. M.V. Symons, *Coal Mining in the Llanelli Area: Volume One: 16th Century to 1829* (Llanelli, 1979), 76-78.

25. Rogers, *Newcomen Engine*, 51-2.

26. L.T.C. Rolt and J.S. Allen, *The Steam Engine of Thomas Newcomen* (Hartington, 1977), 117-22.

27. Ibid, 118.

28. Ibid, 26.

29. Ibid, 30.

30. Matthew Boulton Papers, G. Watson to M. Boulton, 4 January 1783, Box W (Birmingham Reference Library).

31. Allen & Rolt, 117-22.

32. Symons, *Coal Mining in the Llanelli Area*, 78.

33. Matthew Boulton Papers, G. Watson to M. Boulton, 4 January 1783, Box W (Birmingham Reference Library).

34. For a gazetteer of colliery and other industrial archaeology sites in the area see Hughes, S. and Reynolds, P., *A Guide to the Industrial Archaeology of the Swansea Region* (Aberystwyth, 1988).

35. E.T. Svedenstierna, *Svedenstierna's Tour of Great Britain 1802-3* (Newton Abbot, 1973), 42.

36. A. Morrison, *Alum; North East Yorkshire's fascinating story of the first chemical industry,* (Solihull, 1988), 1. S. Gould, *Monuments Protection Programme: the Alum Industry; Combined Steps 1-3* Report (London, 1993), 1.

37. Ibid., Gould, 1.

38. Ibid. much of the information on this process is derived from this source.

39. E.J. Cocks and B. Walters, *A History of the Zinc Smelting Industry in Britain* (London, 1968), 9.

40. Map attached to 'Case of the Duke of Beaufort', (Swansea Reference Library), *c.* 1790.

41. G. Grant-Francis, *The Smelting of Copper in the Swansea District* (London, 1881), 120.

42. p.c. Paul Reynolds.

43. Kirkhouse, 'Letter Book', 16.1.1837.

44. Joshua Gilpin, *Tour of Wales in 1796*, transcribed by A. Woolrich (Philadelphia University Library).

45. A detailed history of the Cambrian Pottery is given in the section on the Swansea Canal and the western side of the valley placed on the Royal Commission web site.

46. Joshua Gilpin, *op.cit.*

47. H.H. Hallesy, *The Glamorgan Pottery, Swansea, 1814-38* (Llandysul, 1995), 2.

48. Ibid.

49. G.B. Hughes, *English & Scottish Earthenware* (London, *c.* 1980), 15, 122 and Hallesy, *op.cit.*, 2.

50. The Pottery, Greenhill and Landore flintmills.

51. Gilpin, *op.cit.*

52. G.B. Hughes, *op.cit.*, 48.

53. L.T.C. Rolt, *The Potters' Field: A History of the South Devon Ball Clay Industry* (Newton Abbot, 1974), 16-19.

54. G.B. Hughes, *op.cit.*, 107.

55. Gilpin, *op.cit.*

56. Rolt, *op.cit.,* 18-9.

57. G.B. Hughes, *op.cit.,* 175.

58. Hallesy, *op.cit.,* 18, 74.

59. Ibid, 6-8.

60. Ibid, 10.

61. Gilpin, *op.cit.* (1796).

62. Ibid., 13.

63. Ibid., 45.

64. Ibid., 33-40.

65. Site to the north of the Landore Viaduct adjacent to the old Glamorgan Pottery flintmill; the pottery was converted into the 'Little Landore' Copper Works by 'The Landore Arsenic and Copper Company.' (Grant-Francis, *op.cit.,* 146).

66. Gilpin, *op.cit.*

67. G.B. Hughes, *op.cit.,* 48.

68. Ibid., 195.

69. Hallesy, *op.cit.,* 17.

70. C. Tomlinson (ed.) *Cyclopaedia of Useful Arts, Mechanical and Chemical, Manufactures, Mining and Engineering. Vol. 1* (London, 1854), 424-42

71. J. Day and R.F. Tylecote (eds), *The Industrial Revolution in Metals* (London, 1991), 150.

72. R.O. Roberts, 'The development and decline of the copper and other non-ferrous metal industries in South Wales', *The Transactions of the Honourable Society of Cymmrodorion*, (1956), 78-115, 86. Recent historical research undertaken in connection with archaeological work at Aberdulais Falls has thrown some doubt as to whether the first non-ferrous metals smelting site in South Wales was there or at Neath Abbey (v.i. Richard Poole, site director).

73. The numbers of smelters used are drawn from ibid., 86-91 and M.V. Symons, *Coal Mining in the Llanelli Area, Volume One: 16th Century to 1829* (Llanelli, 1979), 47. Possible qualifications to these are firstly the obscurity concerning the Cwmfelin Works at Neath Abbey. The site is more suggestive of that suitable for a battery mill; it is very narrow and is away from navigable water and for the purposes of this study is assumed to be a battery mill only. Contemporary maps show 'Upper Bank' to be separate works: the lead smelter was the one later converted to a copperworks while the zinc smelter, situated about half-way between the Middle Bank Copperworks and Chauncey Townsend's leadworks, later fell into ruins.

74. Day and Tylecote, *op.cit.,* 135.

75. Ibid., 136-37.

76. D.E. Gibbs and R.O. Roberts, 'The copper industry of Neath and Swansea: record of a suit in the Court of Exchequer, 1723', *South Wales and Monmouth Record Society, Publications No. 4*, 125-162, 127; and Phillips, *History of the Vale of Neath,* 266.

77. Day & Tylecote, *op.cit.,* 137.

78. R.F. Tylecote, *A History of Metallurgy* (London, 1976), 71.

79. Ibid., 93.

80. Ibid., 93-94.

81. Ibid.

82. Day & Tylecote, *op.cit.,* 138.

83. Day, J., *Bristol Brass* (London, 1976), 205.

84. Day & Tylecote, *op.cit.,* 138.

85. Tylecote, R.F., *Metallurgy*, 95.

86. Day & Tylecote, *op.cit.,* 143.

87. Tylecote, *op.cit.,* 93.

88. Day & Tylecote, *op.cit.,* 150.

89. J. Percy, *Metallurgy, Copper* (London, 1861), 292-93.

90. A.F. Peplow schedule 109, 19 Sept. 1764, EA1/135.

91. See the plan reproduced.

92. J.R. Harris, *The Copper King: A biography of Thomas Williams of Llanidan* (Liverpool, 1964), 39.

93. Percy, *op.cit.*, 314-328

94. Harris, *op.cit.,* 181.

95. See chapter three for Jernegan's work.

96. Most of the information in this section on the White Rock Works is taken from R.O. Roberts, 'The White Rock Copper and Brass Works, near Swansea, 1736-1806', *Stewart Williams' Glamorgan Historian*, 12, 136-151.

97. Information given for 1771 in this section is based on B. Jones, 'Map of the River Swansey' (Swansea Museum 304/463A).

98. J.R. Harris, *op.cit.*, 181.

99. The layout for the White Rock Copperworks in 1744 is known from the engraving of the works by Thomas Lightfoot reproduced in Grant-Francis, *Swansea Copper.*

100. D.B. Barton, *A History of Copper Mining in Cornwall and Devon* (Truro, 1961), 24.

101. *The Cambrian*, 3.12.1858, speech given by Henry Hussey Vivian on 'Metals'.

102. Three mounted display copies of detailed plans of the Hafod Copperworks in the early twentieth century were located in the Swansea Museum Stores in the old Morfa Copperworks Rollingmill building. All information given on the works in the twentieth century is drawn from there, and deductions on the full complement of reverberatory furnaces in the nineteenth century are based on information contained in those plans.

103. R.R. Toomey, *Vivian and Sons 1809-1924, A Study of the Firm in the Copper and Related Industries* (New York, 1985), 187-89.

104. J.R. Harris, *op.cit.*, 181.

105. Ibid, 177.

106. Ibid, 10.

107. Ibid.

108. Copper Development Association, *Copper Through the Ages* (Radlett, Herts., 11th. ed. 1954), 39.

109. Ibid, 34.

110. Harris, *op.cit.,* 39.

111. Copper Development Association, *op.cit.,* 56.

112. Ibid, 38-39.

113. Ibid, 34-35.

114. R.R. Toomey, *op.cit.*, 47-48.

115. P.R. Reynolds 'Industrial Development', *Swansea: an Illustrated History*, ed. G. Williams (Swansea, 1990), 29-56, 32.

116. Toomey, *op.cit.,* 39.

117. Ibid., 48.

118. Ibid., 39.

119. Ibid., 49.

120. L. Aitchison and W.R. Barclay, *Engineering Non-ferrous Metals and Alloys* (London, 1923), 133.

121. Ibid., 139.

122. For details of the Bristol mills see Day, *Bristol Brass*, 204-18.

123. R.O. Roberts, 'The Smelting of Non-ferrous Metals', In: *Glamorgan County History, Vol. 5, Industrial Glamorgan* (ed. A.H. John & G. Williams, 1980), 47-85.

124. Ibid, and H. Green, 'Melincryddan Copperworks & Village', *Neath Antiquarian Society Transactions* (1978), 79-85, 79-80.

125. Day, *op.cit.,* 156-59.

126. D.E. Gibbs and R.O. Roberts,'The copper industry of Neath & Swansea', *Publications of the South Wales and Monmouth Record Society*, 4 (1957), 123-64 and B.G.Mss. II, 1496.

127. Ibid, 134.

128. See the entry on Landore Mill in the section on Landore.

129. Day, *op.cit.,* 167.

130. Ibid, 168.

131. Ibid.

132. Ibid., 169.

133. Grant-Francis, *op.cit.*, 106-07.

134. Ibid. Much brass manufacture had long been located to the immediate south-west of London. The Esher Mills had been a main manufacturing centre for brass pins in the seventeenth century.

135. Gilpin, *op.cit.,* See the section on 'Lower Forest Mills' in the Llansamlet chapter for fuller details.

136. Grant-Francis, *op.cit.,* 113.

137. Day, *Bristol Brass,* 40-41. The Esher Mills company was still in production in 1743 under the name of the 'General Joint United Stock of the Societies of Bristol and Esher for Making Brass, Battery and Brass wire.'

138. R.O. Roberts, 'The White Rock Copper and Brass Works, near Swansea, 1736-1806', *Stewart Williams' Glamorgan Historian,* 12, 136-51, 137.

139. Day, *Bristol Brass*, 68.

140. Ibid., 204-19.

141. R.O. Roberts, 'The White Rock Copper and Brass Works, near Swansea, 1736-1806', *Stewart Williams' Glamorgan Historian,* 12, 136-51, 149.

142. Day, *Bristol Brass*, 137.

143. J.R. Harris, *op.cit.*, XV.

144. Ibid, 180.

145. R.O. Roberts, 'The Smelting of Non-ferrous Metals since 1750', *Glamorgan County History, Volume V, Industrial Glamorgan* (Cardiff, 1980), 47-96.

146. J.R. Harris, *op.cit.,* 183.

147. R.R. Toomey, *Vivian and Sons 1809-1924: A Study of the Firm in the Copper and Related Industries* (New York, 1985), 361-63.

148. Day, *Bristol Brass*, 136.

149. E.T. Svedenstierna, *Svedenstierna's Tour of Great Britain 1802-3 (The Travel Diary of an Industrial Spy)* (Newton Abbot, 1973).

150. G. Delanie, 'Structural Innovation in Classical Building'; P. Needham, *History of Science and Technology in China*. It was also used in medieval Britain, e.g. in the steeple of Salisbury Cathedral.

151. Cog Farm at Sully had them in addition to cast-iron roofs: a model farm associated with the Guest family of Merthyr Tydfil (v.i. Richard Suggett, after field survey).

152. Copper Development Association, *op.cit.,* 48.

153. Ibid., 51.

154. See Rothwell's 1790 view of the works.

155. 'On the Advantages of Copper Roofs', copy in Vivian MSS (NLW).

156. *C.J. Williams, Greenfield Valley.*

157. Copper Development Association, *op.cit,* 50.

158. William Bevan tried unsuccessfully to extract iron from copper-slag at Nant Rhyd-y-Filais Works (Landore, Swansea) in 1814.

159. These include the long gothick colonnaded screen wall of the Bath Building at Arno's Court (1760) that now stands in the central garden of Portmeirion, Gwynedd.

160. E.T. Svedenstierna, *op.cit.,* 41.

161. Day, *Bristol Brass, op.cit.*, 219.

162. Ibid, 44.

163. Ibid, 55.

164. Grant-Francis, *op.cit.,* 106-07.

165. Day, *Bristol Brass, op.cit.,* 27-8.

166. Hughes and Reynolds, *op.cit.*

167. All the information quoted in the paragraph above comes from A.H. John, 'Introduction: Glamorgan, 1700-1750', *Glamorgan County History, Volume 5, Industrial Glamorgan*, ed. A.H. John and G. Williams (Cardiff, 1980), 1-46, 27.

168. Dowlais followed on from Cyfarthfa in becoming what were reputed to be 'the biggest ironworks in the world' at the end of the first quarter of the nineteenth century.

169. K.J. Hilton (ed.), *The Lower Swansea Valley Project* (Swansea, 1967), 27.

170. P. Jenkins, *Twenty by Fourteen: A History of the South Wales Tinplate Industry* (Llandysul, 1995), 28.

171. E.H. Brooke, *Chronology of the Tinplate Works of Great Britain* (Cardiff, 1944), 163-67.

172. PRO, SCCM, RAIL 876.3, entry for 4 September 1838 and E.H. Brooke, *Chronology of the Tinplate Works of Great Britain* (Cardiff, 1944), 60.

173. Brooke, *op.cit,* 119.

174. Ibid, 24-5.

175. Ibid, 149.

176. Ibid, 43-44.

177. Ibid, 176; 31, 40-1, 8-11.

178. P. Jenkins, *op.cit.*, 28.

179. Ibid, 29.

180. Hilton, *op. cit.*, 27-30.

181. P. Jenkins, *op.cit.*, 31.

182. W.E. Minchinton, *The British Tinplate Industry: A History* (Oxford, 1957), 104.

183. Ibid, 4.

184. D.J. Rowe, *Lead Manufacturing in Britain* (London, 1983), 6.

185. Grant-Francis, *op.cit.,* 105-06.

186. Much of this section is drawn from Tylecote, *A History of Metallurgy*, 134-35 and Day & Tylecote, *The Industrial Revolution in Metals*, 93-94.

187. Day and Tylecote, *op.cit,* 119 and Tylecote, *op.cit.,* 137.

188. R.R. Toomey, *op.cit.,* 210.

189. Ibid, 210-11.

190. Hilton, *op.cit.,* 21.

191. Ibid, 25 and E.J. Cocks and B. Walters, *A History of the Zinc Smelting Industry in Britain* (London, 1968).

192. Hilton, *op.cit.*, 21.

193. Cocks and Walters, *op.cit.*, 13-4.

194. Hilton, *op.cit.*, 24.

195. Ibid, 25-6.

196. The dates of working of these various works are drawn from: R.O. Roberts, 'The Smelting of Non-ferrous Metals since 1750', *Glamorgan County History, Vol. V, Industrial Glamorgan from 1700 to 1970*, 47-96, 90.

197. R.A. Griffiths, *Clyne Castle, Swansea: a history of the building and its owners* (Swansea, 1977), 21, 30.

198. v.i. David Leighton, RCAHMW, quoting the Badminton Papers, NLW.

CHAPTER 2

PATTERNS OF TRANSPORT & POWER

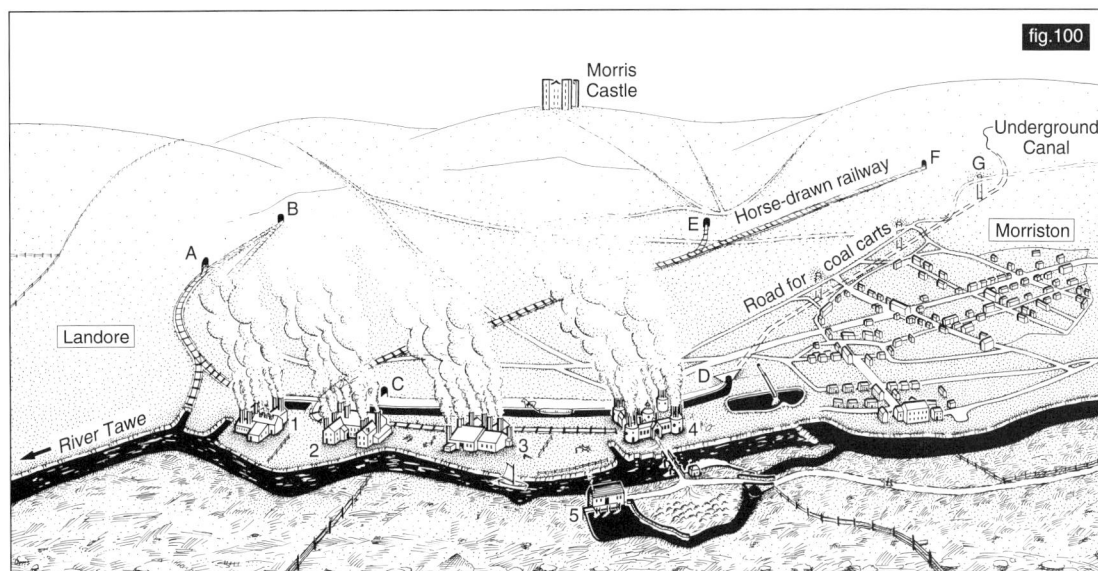

Fig. 100. Schematic drawing showing four copperworks built in the later eighteenth century between Landore and Morriston with their methods of bringing coal from nearby mines. From the 1750s (at Clyn-du Water machine Pit (G)) until the 1770s and 80s carts were used to bring coal down the adjoining hillside of Graig Trewyddfa to the copper-smelting-works. By the 1770s coal was also being brought out of the underground Clyn-du Canal (D) to fuel the Forest Copperworks (4). Rose Copperworks (2) also had its own Graig Level mine (C) going into the adjoining hillside. In the early 1780s the underground canal was extended southwards on the surface (as Morris's Canal) towards Landore to serve three new smelting-works either built by the Lockwood, Morris & Co. partnership or on land sub-leased by them. By the 1770s a wooden-tracked railway had been constructed from Cwm Level (A) to the river but by the 1790s had iron plate-rails and was supplying coal to Landore Copperworks (1). At the beginning of the nineteenth century another horse-drawn railway system was built across the Graig to supply coal from Treboeth (B), Hen Lefel y Graig or Park level (E) and another Clyn-du Level (F) to supply Rose (2), Birmingham (3) and Forest (4) Copperworks. From 1796 four ton boats had been replaced by 21 ton capacity craft on the canal which as the enlarged Trewyddfa Canal now formed part of the main Swansea Canal line with access to many more collieries. The River Tawe was used to bring copper-ore upstream and refined copper down. The copperworkers in the area largely lived in Morriston and some of the colliers in Morris Castle above. (5 = Lower Forest Coppermills).

Transport Development

Introduction

The development of any industrial complex depends on an ability to import raw materials and to export manufactured products, and both at relatively low cost. The following table gives an indication of the rate and extent of innovation in the Swansea area. The dense concentration of coal-hungry smelters generated a need for a network of successive effective transport arteries. Such was the intensity of this network that a variety of technical innovations took place. This has not generally been recognised, but examination of the historical and archaeological records suggest that several innovations which have been attributed to other industrial centres may have had their origins in Swansea. Two examples are considered in detail: the cast-iron tramplate (which formed the basis of overland and underground transport in mining and metal-working districts, especially in south Wales until it was superseded by locomotive railways) was a major contribution to the industrial development of south Wales and in its time was every as important a means of transport as the

contemporary horse-worked edge railway; and the mining-level accessed by boat was of importance too.

Sequence of Early Transport Development in the Swansea area

1650	the first known coal-road for pack-horses had been established
1730s	first use of coal-carts
1737	probable date of first river-dock
1746	probable date of first power generation obtained from a transport waterway
c.1756	first surface railway in use
1764	first Turnpike Act adopted in Glamorgan
1770	first railway tunnel into a mine; probably the first self- acting inclined-plane
1770s	probable date of first underground canal
1776	first iron railway in use underground, and probably on the surface soon afterwards
c.1776	first deep-water river quay
c.1783-4	first mainly surface canal in use
c.1790	first largely gravity working of a graded railway route
c.1790	first confirmed use of vertical transfer of loads from railway to shipping.

71

Figure 101

Average single-horse loads according to Telford & Smeaton

Type	Formation	Journey Load
Pack-horse	Path	0.125 tons
Stage waggon	Soft road	0.63 tons
Stage waggon	Macadam road	2 tons
Waggon	Waggon-way (wooden rails)	6.2 tons
Barge	River	30 tons
Barge	Canal	50 tons
Waggon	Iron-plated Waggon-way	8 tons

Fig. 101. Table of average single-horse loads according to Telford and Smeaton. Thomas Telford and John Smeaton experimented to establish what a single horse could haul on level ground. By studying early industrial areas such as Swansea, the true capabilities of haulage teams coping with the adverse effects of gradient, climate and muddy surfaces were established.

Fig. 102. Table of the varying capacities of coal-carts on coal-roads in the Swansea area showing how much more efficient they were than pack-horses but how considerably less than the smooth surfaces provided by wooden waggonways and localised canals.

Fig. 103. The lower half of the River Tawe Navigation looking southwards towards the old medieval core of Swansea. The site of the Middle Bank Copperworks lies under the roundabout in the foreground with the remains of White Rock Copperworks with the last of the fourteen river docks visible above it. The Hafod Copperworks quays and rolling-mill engine houses are visible in the right foreground. (983515/19).

Figure 102

Eighteenth century transport container capacities recorded in the Swansea area

Type	Date	Formation	Owner	Dimensions of each container	Journey load	Propulsion
Pack-horse	Pre-1702	Path	–	–	0.05 tons	1 horse
Cart	1728 (1748)	Public-road	Lockwood, Morris & Co.	4' 0½" x 1' 10½" x 1' 1"	0.41 tons	4 bullocks
Cart	1733 (1752)	Coal-road	Lord Mansel	4' 0" x 2' 1" x 1' 3"	0.45 tons	?
Waggon	1750 (1769)	Surface Waggon-way	Chauncy Townsend	5' 6" x 3' 4" x 4' 9"	3.1 tons	1 horse
Cart	1702 (1770)	Coal-road	Thomas and Gruffyd Price	4' 0½" x 1' 10½" x 1' 1"	0.36 tons	?
Cart	(1776)	Public-road	Thomas Popkin	4' 0" x 2' 0" x 1' 1"	0.38 tons	?
Cart	c.1764 (c.1771)	Turnpike & Coal-road	Thomas Price	?	0.48 tons	3 horses
Cart	c.1771 (c.1771)	Turnpike and Coal-road	Gruffydd Price	?	0.33 tons	2 horses
Boat	1746 (1841)	Underground Canal / Surface canal after c.1787-91	Lockwood, Morris & Co.	20' x 3' x ?	4 tons	Water flow + man on return
Waggon	1788-1790 (1802-3)	Underground + surface waggonway	Lockwood, Morris & Co.	?	1.3 tons in 2 waggons	Gravity + horse on return

(Bracketed dates are those of sources held in the National Monuments Record).

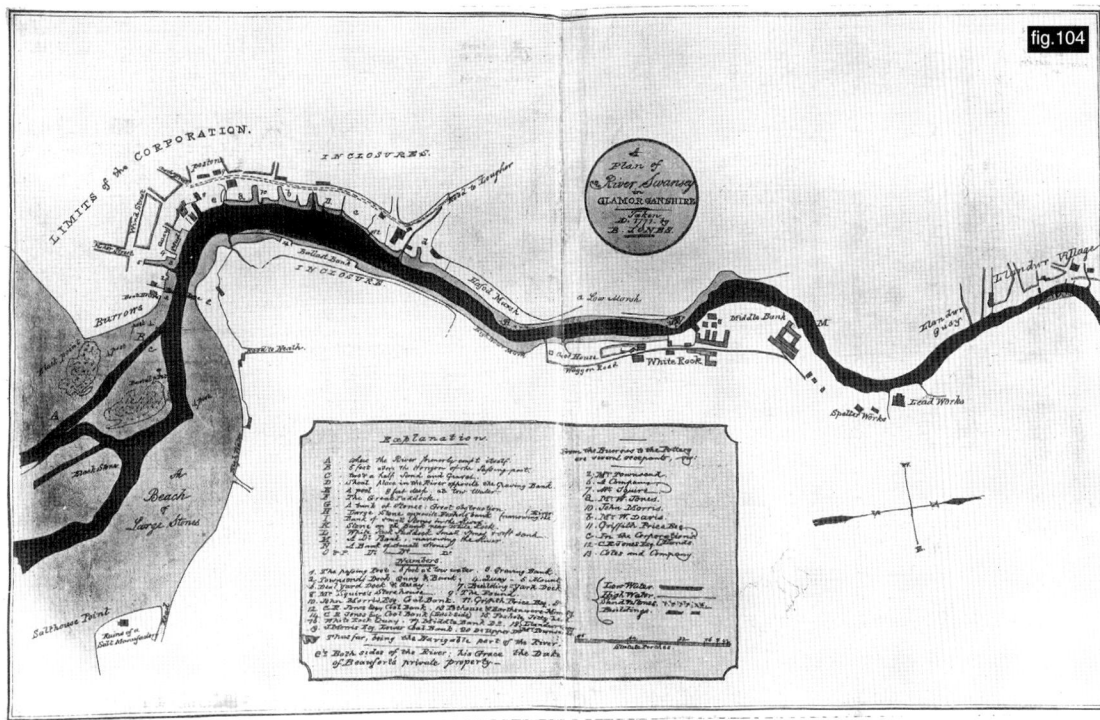

fig.104

Fig. 104. Plan of the three-mile long navigable section of the River Tawe in 1771 showing its relatively unimproved eighteenth-century condition with shallows and no towing-path. (City & County of Swansea: Swansea Museum Collection)

The River Tawe

The navigability of the unimproved River Tawe as it bisected the coastal rim of the south Wales coalfield is the explanation for Swansea's development as a major metallurgical centre.

Navigation

The River Tawe is one of a number of rivers which rise in the Brecon Beacons and flow south to form the main corridors of settlement and commerce in upland Glamorgan. None of them is navigable very far inland and no ambitious programmes of work had been undertaken to extend navigation by means of cuts and locks.

The widely meandering course of the River Tawe was subject to large variations in flow which made its upper and middles reaches unsuitable for either navigation or power generation. However, its lower course, whilst subject to a wide tidal range, was navigable for three miles inland, to the site of later Morriston, by quite substantial vessels.[1] The penetration of the coalfield by a tidal river was therefore a crucial factor in the development of industry at Swansea.

The export of coal through Swansea was well established in the sixteenth century. Early techniques for handling these shipments were primitive: the vessels simply tied up alongside the river edge or sailed up a narrow pill at high water and settled on their bottoms as the tide went out. Once loaded, they waited for the next tide to float them off. The first improvements came in the sixteenth century, when

quays were built along the waterfront in Swansea, and on the sands of Swansea Bay between high and low water. In the seventeenth century the first docks - tidal, of course - were constructed. Attempts were also made to extend river navigation inland, for beyond the limits of navigable water coal was virtually worthless. The agent of the Duke of Beaufort, who had considerable mineral interests on both sides of the Swansea Valley, stated *c.*1750 that '... coals [i.e. potential coal mines] at a distance from navigation were considered of no value', while in about 1795 the promoters of the Swansea Canal said of the valley that '... limestone, iron ore, and coal, in almost inexhaustible quantities ... have been but little worked (the lower part adjoining Swansea excepted, there being but one Iron Furnace and a Tin Work higher up ...)'.

Following the establishment of non-ferrous smelters in the Neath Valley in the seventeenth century and, from 1717, in the Swansea Valley, the export of coal started to be supplemented by the

Fig. 105. View looking northwards up the River Tawe Navigation c.1809 showing several sailing-ships passing to and from the Landore Quays. Smoke rises on the left from the Landore Copperworks, in the centre from the Rose, Birmingham and Forest Copperworks, and on the right from the Upper Bank Copperworks. (Wood, Rivers of Wales, 1811)

fig.105

fig.106

Fig. 106. Between 1790 and 1810, Fabian's Bay at the mouth of the River Tawe was enclosed with two large piers of copper-slag and Pennant sandstone blocks to form a large tidal harbour. The surviving western pier can be seen as the quay running up the lower left-hand side of the photo. The end of the former eastern pier is the square masonry block at the right of centre. (935054/47)

fig.107

Fig. 107. Redundant quays and high tip-retaining walls alongside the west bank of the River Tawe, south of Landore Viaduct before reclamation. The wharf in the foreground succeeded one of 1750 that had the world's first recorded intermediate tunnel on a surface railway. (81/33/2)

import of copper-ore and lead-ore. This trade tended to be handled at riverside wharves along the Neath and Tawe rivers. The import of copper-ore and the export of coal grew together, the one encouraging the other. The growth of the coal trade produced many major schemes for harbour improvement. The first floating dock in south Wales was constructed by Humphrey Mackworth at Neath in the early eighteenth century. The first public floating dock was opened in 1834 at Llanelli, but it was not until

1852 that Swansea, where the harbour had been managed by a public Harbour Trust since 1791, had its first floating dock. Until that date, the industrial landscape of the district was entirely centred on the use and development of the tidal river navigation.

The first recorded development of the Tawe navigation is early in the eighteenth century. At that date navigation was not attempted above the lower three miles of the river, with Forest Mills the farthest point which vessels reached. However, before the construction of the Lower Forest (Fforest Isaf) Mills weir, just south of the present Wychtree Bridge at Morriston, navigation as far as Ynys-pen-llwch Tinplate Mills at Clydach may have been possible, although there is no evidence that vessels ever reached this point. Construction of the weir between 1728 [2] and 1735 [3] is said to have curtailed navigation, although other sources state that the Upper Forest (Fforest Uchaf) Mills were just above the head of possible navigation.

At Upper Forest, Thomas Popkin built a bridge in the early eighteenth century to supersede an earlier ferry. In c.1746-48 Lockwood, Morris & Co. had William Edwards build the three-arched Beaufort Bridge just south of their Lower Forest weir.[4] This was the only fixed bridge, and it had insufficient clearance for sailing vessels to stand on the tidal river until the present river bridge at Swansea was built in the 1960s. Immediately downstream of the Beaufort Bridge was the quay for the Forest copperworks. The river navigation was intensively used throughout the nineteenth century and only declined with the non-ferrous metals' trade that it primarily served. Seventeenth-century coastal

fig.108

Fig. 108. Birmingham Copperworks Dock in the 1860s. A river-barge and two other craft lie in the dock with the sluice controls for using Swansea Canal water to flush out mud visible in the right foreground. On the left is what seems to be the broad chimney of a brass-annealing furnace. The early nineteenth-century steam-powered stamping and rolling-mills of the Rose Copperworks are on the further bank of the River Tawe. ('Le Tour du Monde') (Reproduced by permission of Birmingham Library Services)

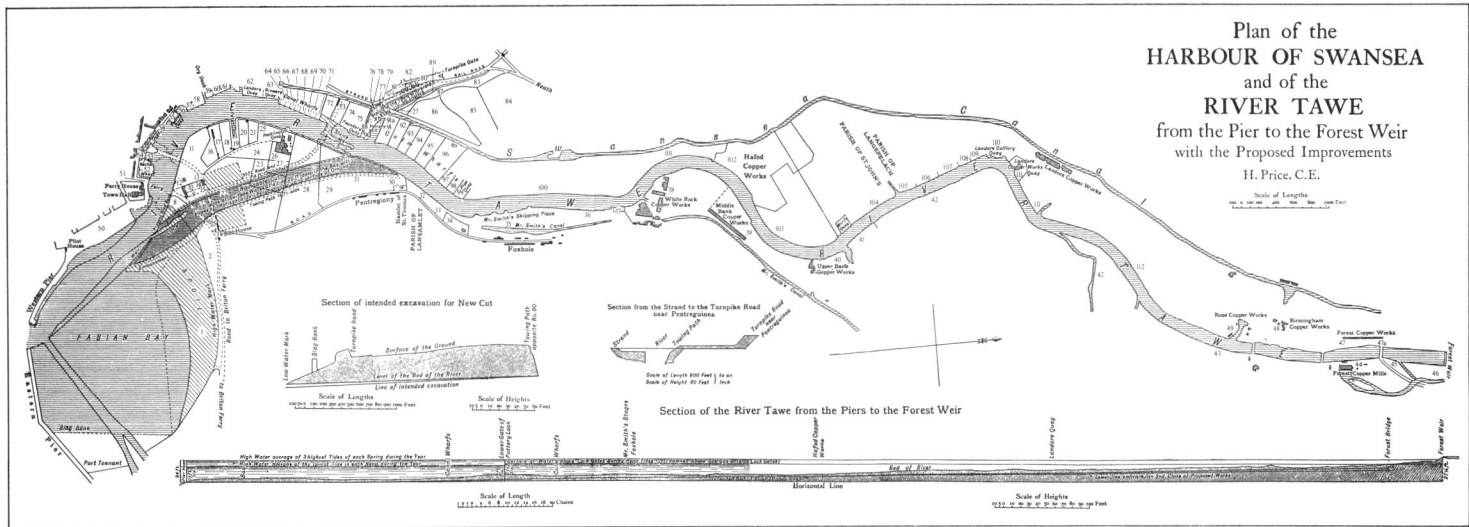

Fig. 109. Map showing the River Tawe in 1836, drawn to accompany one of the bids to construct the New Cut Canal. It also shows the ownership of the copperworks and wharves alongside the river. The substantial Eastern and Western Piers enclosing Fabian's Bay are clearly depicted. (Jones, 'History of the Port of Swansea', 1922)

Key for Figure 109

1. Beach and Mud overflow by the tide (Unoccupied).
2. Uninclosed land between high water mark and the road to Briton Ferry (Unoccupied).
3. Cottage (Elizabeth Williams).
4. Cottage (Gwenllian Morgan).
5. House and garden (Thomas Rosser).
6. Masons Arms Public House (Rachael Rees).
7. Uninclosed land (Unoccupied).
8. (Mrs. Martha Martin).
9. House and garden (Wm. Phipps).
10. Red Lion Public House (Robert Davies).
11. Old Ballast Bank (Unoccupied).
12. Cottage and garden (William Morgan).
13. Cottage and garden (William Clement).
14. Cottage and garden (Mary Payne).
15. Cottage and garden (Elizth. Philips).
16. Part of Tyrlandwr Farm (Edward Griffith).
17. Ship Builders yard (Willm. Meager).
18. House garden & yard (Representatives of the late Willm Meager).
19. Patent Slip (Patent Slip Company).
20. Copper Ore yard (Henry Bath).
21. Copper Ore yard (Henry bath).
22-29. Part of Tirlandwr farm (Edward Griffith).
30. Cottage and garden (William Williams).
31. Towing path (Edward Griffith).
32. Stone Quarry (Trustees of Swansea Harbour).
33. Stone Quarry &c &c (Freeman & Co.).
34. House and garden (Robert Mills).
35. Coal Bank & Shipping Place &c (Charles Henry Smith Esq).
36. Uninclosed land adjoining the River Quay (Freeman & Co.).
37. (Freeman & Co.).
38. White Rock Copper Works (Freeman & Co.).
39. Middle Bank Copper Works (Pascoe Grenfell & Sons).
40. Upper Bank Copper Works (Pascoe Grenfell & Sons).

41. Old Brick yard (Vacant).
42. Marsh (Vacant).
43. Stamping Mills Marsh &c (Benson Logan & Co).
44. Forest Copper Works (Vivian & Sons).
45. Slag Bank &c (Benson Logan & Co).
46. Waste land by River side (Benson Logan & Co).
47. Forest Copper Works (Benson Logan & Co).
47a. Forest Copper Works (Benson Logan & Co).
48. Copper Works (Pennyvilla Vein & Co).
49. Marsh (William Foster & Co).
50. Burrows (Corporation of Swansea).
51. Quay &c (Swansea & Liverpool Steam Packet Co).
52. Copper Ore Wharf (Henry Bath).
53. Graving Bank (Corporation of Swansea).
54. Public Quay (Corporation of Swansea).
55. Quay warehouse &c (Silvanus Padley).
56. Uninclosed ground (Grove & Co).
57. Bonded Timber Yard (Grove & Co).
58. Coal Wharf (Sir John Morris Bt).
59. Dry Dock Ship Builders yard &c (Edmond Richards).
60. Timber yard (Francis Richardson & Co).
61. Ship Builders Yard Dock &c (Charles Llewhelling).
62. Landore Coal Quay (Sir John Morris Bt).
63. Brewery Wharf (Henry Huxham).
64. Coal Wharf (John Strick).
65. Coal Wharf (Perkins & Heiniekin).
66. Coal Wharf (Pennyvilla Vein Co).
67. Coal Wharf (Yniscedwin Iron Co).
68. Coal Wharf (John Strick).
69. Coal Wharf (Williams & Rowland).
70. Coal Wharf (Cathelid Culm Co).
71. Coal Wharf (Perkins & Heinekin).
72. Glamorgan pottery (Bevan & Co).
73. Coal wharf & wharehouses &c (CH Smith Esq).
74. Cambrian Pottery (LW Dillwyn Esq).
75. Foundry Quay (Sir John Morris Bt).
76. Pond (LW Dillwyn Esq).

77. Canal (Swansea Canal Co).
78. Smiths Forge &c (Evan Hopkins).
79. Shed &c (David Jones).
80. Rail Road (Sir John Morris Bt).
81. (Unoccupied).
82. House and garden (Evan Hopkins).
83-87. Part of Hafod farm (David Jon Thomas).
88. Mill Pond &c (Morgan Hussey).
89. Pond &c (Messrs Williams).
90. Coal wharf (Sami Jenkins & Ed Brown).
91. Wharf (Benson Logan & Co and John Parsons).
91a. Smiths forge (Evan Hopkins).
91b. Limekiln (George Huxham).
92. Wharf (Graigola Coal Co).
93-95. Wharf (Treacher & James).
96. Wharf (British Iron Compy).
97. Wharf (Richard Morgan & Co and Sam Jenkins).
98. Wharf (James & Aubrey).
99. Wharf & pipe works (Ed & Nicholas Kermicott).
100. Copper Ore yard and part of hafod farm (Williams & Co).
101. Site for depositing Ashes (Vivian & Sons).
102. Hafod Copper works (Vivian & Sons).
103. Slag Bank & Marsh (Vivian & Sons).
104. Morfa Quay copper works marshes &c (William Foster & Co).
105. Old Quay &c (Christopher James).
106. Garden and uninclosed land near the river side (Henry Thomas).
107. Uninclosed land adjoining the River (Unoccupied).
108. Uninclosed land adjoining the River (Sir John Morris Bt).
109. Uninclosed land adjoining the River (Pennyvilla Vein Co).
110. Coal wharf (Sir John Morris Bt).
110a. Coal wharf (Sir John Morris Bt).
111. Landore Copper Works Quay Marsh &c (Nevill Sims Druce & Co).
112. Marsh (Pennyvilla Vein Co).

Fig. 110. The New Cut Canal at Swansea extends from the river bend at bottom-left to just beyond the further road bridge. Construction of the canal allowed the loop of the medieval port to be transformed into an enclosed and permanently flooded port in 1852 (now filled in). (915021/3)

sailing vessels were usually only 6-10 tons in burden. The average size of such boats increased throughout the seventeenth century and after 1700 a burden of 40-60 tons was not unusual.[5]

The increased size of boats necessitated the construction of both quays and docks (open to the river) for easy loading. Up-river, docks were probably built on the lines of the 'Tonne [i.e. Town] Dock' of 1624.[6] A 'New Quay' had been built on the town reach of the river in 1616 and may also have had imitators.[7] The quays built before William Edwards'[8] New Quay of 1772 at Landore were probably banks of earth heaped inside timber palisades. The form of these is clearly shown on a lease of the site of at least one quay on the nearby River Loughor.[9] Thomas, 2nd Baron Mansel already had a dock for the shipping of his coal at White Rock when the copperworks was built there in 1737.[10] The copperworks' proprietors took over the dock and

by the end of the nineteenth century at least 17 docks had been built upstream of the town reach and the harbour proper. Each dock was scoured of its silt by a natural stream or leat from one of the adjacent canals. The elaborate dock of the Birmingham Copper Company was the only one to have a proper scouring reservoir.[11]

A towing-path for ships and, perhaps, barges existed on many sections of the river by the early nineteenth century. Plans for bridges over the navigation had to be submitted for approval to the Harbour Trustees and had to include adequate clearance over both the towing-path and the navigation channel. The decreasing depth of the navigation from the 15ft 3in available just above the old town wharves to the 4ft 6in just below the head of navigation is illustrated by the following table.[12] The plan numbers refer to the map of the river navigation that accompanies this section.

fig.111

Fig. 111. In the eighteenth and nineteenth centuries, the lower three miles of the River Tawe became increasingly lined with shipping quays. This 1985 photograph shows the thickness and construction of the mid nineteenth-century quay, with mooring bollard, on the west side of the New Cut at Swansea. (851016)

Fig. 112. A later nineteenth-century coal-shipping wharf for the Swansea Vale Railway on the east bank of the New Cut. The facing of ashlar Pennant-sandstone blocks has collapsed revealing the rubble core beneath. (851017)

Fig. 113. The nineteenth-century bridges over the lower River Tawe had to allow for the passage of tall sailing-ships. The first drawbridges were built at the mouth of the New Cut in the mid nineteenth-century for both road and rail transport (Jones, 'History of the Port of Swansea', 1922)

Fig. 114. Isambard Kingdom Brunel's South Wales Railway crossed the Swansea Valley on a high and long viaduct, built in 1847-50. It had composite decking with pine for compressive members and wrought-iron rods for elements in tension. A tertiary steel viaduct-deck (1978-9) sits on four original masonry piers (drawing from contemporary published plans).
114a. Details of jointing on Brunel's Landore Viaduct showing the composite nature of the wooden and wrought-iron elements.

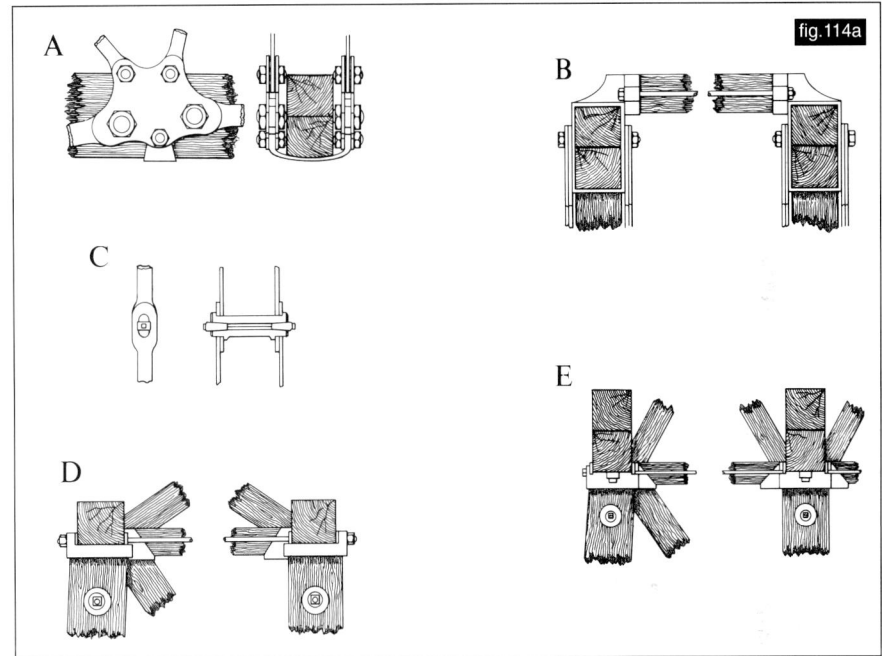

fig.114a

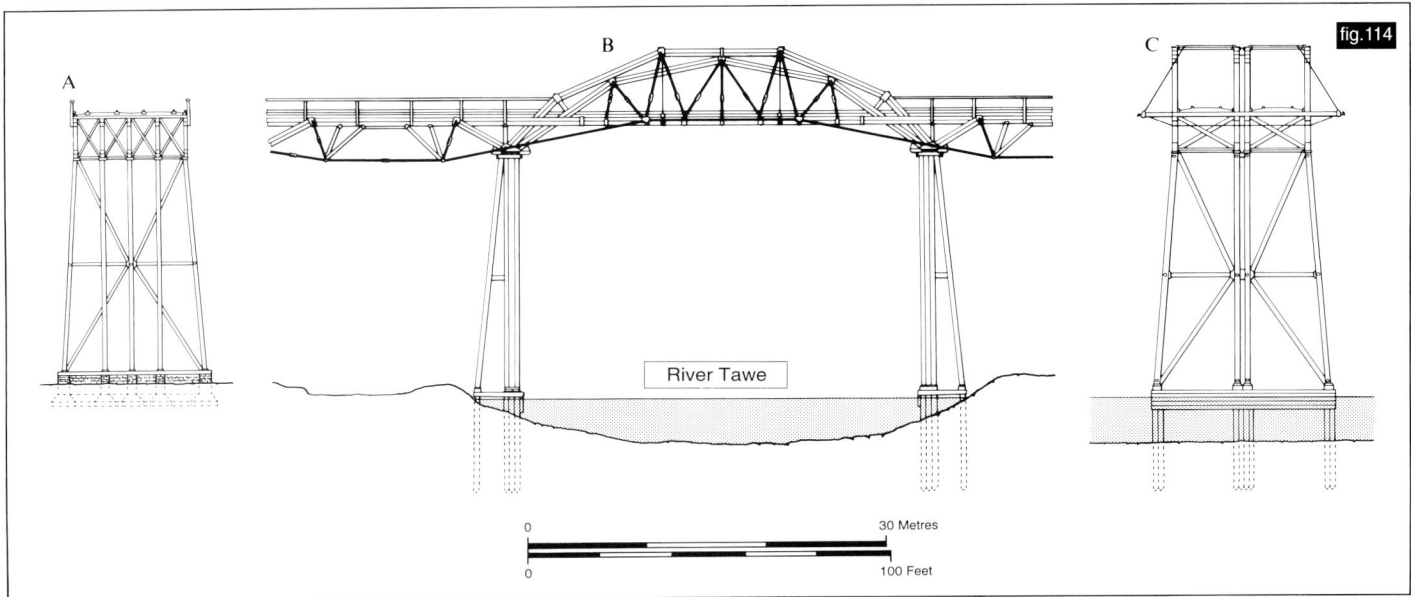

River Tawe

0   30 Metres

0   100 Feet

fig.114

Figs. 115, 115a, 115b. The Morfa Lifting Bridge (photographed during construction in 1910) was one of several large bridge structures needed to carry the copper-slag waste when the original riverside tips had been filled. The heavy lifting-deck carried railway-trucks of waste across from the Morfa Works furnaces on the west bank to tips north of Upper Bank Works on the east. (University of Wales Swansea, Library & Information Services, Archives)

	Neap Tides	Spring Tides	Number on plan
At Forest works	4ft 6in	9ft 3in	47
Birmingham works	5 6	10 3	45
Rose works	6 0	10 9	49
Landore works	9 3	14 0	110
Upper Bank works	10 6	15 3	40
Middle Bank works and Hafod works	11 3	16 0	39/102
Mr Smith's shipping stages at Foxhole	14 3	19 0	35
Pipe Quay	14 11	19 8	99
Parson's culm wharf	15 3	20 0	91
Within the float at all the wharves	17 3	24 0	52-72

Periodic campaigns were made to clear the 'paddocks', or shallows in the river, allegedly the result of scouring operations in the upper valley. Sophisticated staithes, developed in the mid-nineteenth century in Northumberland and Durham to lower coal unbroken onto ships, never made their appearance on the coal quays of the River Tawe. Timber spouts for loading were used from the early eighteenth century [13] until at least the building of the Swansea Vale Railway in the 1850s.[14]

A canal or 'cut' across the marshy meander to the south of the Landore quays was proposed as early as the 1770s. The Hafod and Morfa Works were later built there and their tips built up the level of the marsh. This cut was never built, but a 'New Cut' to straighten the lowest meander was built in the 1840s in order to facilitate the construction of the Town (or North) Dock on this loop of the river.

The New Cut is 840yds long and 55-70yds wide. It was initially intended to be the 'waste-water cut' for the new dock. Work had begun by May 1840 and *The Cambrian* noted that Mr Burrows, the contractor, had 200 navigators at work. In January 1843 the Harbour Trustees decided to make the canal into a navigable tidal cut, and by the end of 1843 preparations were being made for its use as a waterway. In February 1845 the dam across the old original course of the river was almost complete and soon all ships travelling up the river had to use the New Cut.[15]

On 11 March 1845 the barque *Charles Clarke* was the first vessel to sail through the New Cut. 'She passed through amidst the firing of guns, etc., in most gallant style without touching either sides or bottom, or meeting any obstruction to her free navigation.'[16] In May the harbour master was provided with £450 with which to purchase a dredger to deepen the river, 'particularly above the Pottery' (i.e. north of Swansea town wharves), and

he was dispatched to Totnes where it lay in the river, in order to pay for it and to bring it back to Swansea.[17]

It is interesting to note the number of workers required for a civil engineering project such as the construction of the New Cut. The 1841 census notes 366 excavators' and three 'navigators' resident in Swansea at this time. Of these, 16 had English names and Timothy Flanagan was the only Irish navvy present. Only four of the people with Welsh names were born in Glamorgan. This compares with the 200 local monoglot Welshmen who had been employed at the top of the Swansea Valley in the 1830s on the extension of the Brecon Forest Tramroad; 32 of the 37 men involved in the construction were resident in Kilvey Terrace, St Thomas, while one unfortunate was in Swansea Infirmary; 10 of the English contingent of 16 were lodged in one house in the terrace! Another six, three of whom had English names, lodged in a nearby house in the same terrace.

The only similar cut attempted on a river in south Wales was that on the River Neath at Neath Abbey in the 1870s, which was destroyed by flood waters and never completed. The other rivers of south Wales were navigable to varying degrees, although the lowest part of the Taff to Cardiff was considered unsuitable for late eighteenth and early nineteenth-century sailing vessels, which were diverted into the one mile basin of the Glamorganshire Canal and the subsequent 'Cardiff Ship Canal' (the later Bute West Dock). The estuaries of the Cleddau and Tywi were navigable as far as Haverfordwest and Carmarthen respectively, while a view of a river-boat near Dryslwyn Castle, and the existence of Quay Street at Llandeilo, suggest at least one unimproved navigation in rural mid Wales. The Neath Navigation was the only south Wales river to have locks (at Aberdulais) and it is even in limited use today. However, with the exception of Neath, none of these other navigations had the underlying coal reserves in their lowermost reaches which prompted the intensive industrialisation and transport use at Swansea.

Swansea was also the western terminus of the Tennant Canal. This had at least three locks to the River Neath, built at various times, and it was originally envisaged as a barge-canal between the Neath and Tawe Rivers. The indications are that this scheme was never effectively implemented; the Swansea and Neath Valleys remained separate zones of economic development and hence the detailed history of the Tennant Canal is not considered in this volume. In the Swansea Valley the canal locked into the old Salt House Dock on the eastern fringe of a fairly exposed reach of Fabian's Bay at the mouth of the River Tawe. There is evidence for at least limited consignments of sand, extracted from the Burrows at Jersey Marine, being locked into the enclosed bay and barged up to the White Rock Dock (sand was used to form the copper-furnace beds and was also preferred as a flux to lime).[18] The Tennant Canal later had two successive locks to the various floating docks which formed part of the development of Port Tennant. The river navigations could have developed together as part of a considerable barge system, but the Neath and Swansea Valley systems were duplicates rather than complementary, with the exception of the minor trade in sand. The Swansea Canal was not connected directly to the river but after 1853 had a lock to the North Dock.

The accompanying illustrations show some of the types of craft used on the river navigation. Besides the sailing-ships,[19] an enlarged version of a rowing-boat seems to have been used, presumably mainly to take copper-ore up river from the larger vessels moored in the town reach. The Swansea Canal had taken over some of this traffic by 1832.[20] This transfer of trade to the canal became more common with the advent of larger ships and the possibility of loading canal boats alongside incoming ships in the North Dock. Some coal was probably taken downstream for loading onto larger vessels, as was copper-slag for the reinforcement of harbour works. Versions of small barges with sail, resembling pilots' cutters, were used to tow sailing ships along the river. By the 1860s, illustrations show not only the smaller barges but also vessels almost twice as large

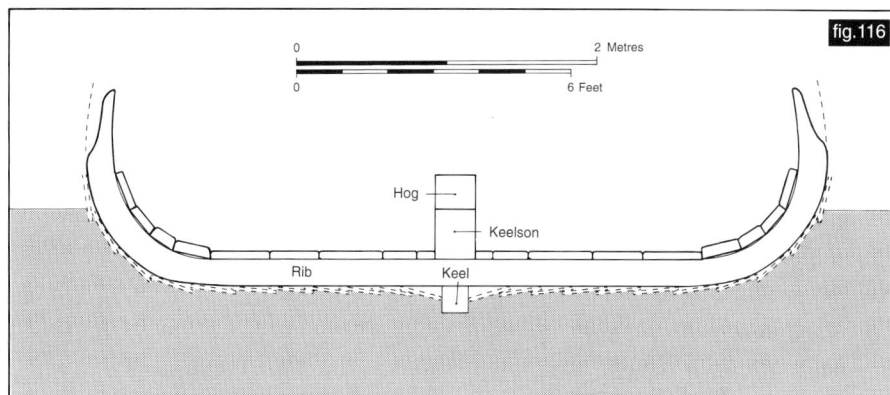

Fig. 116. The remains of a flat-bottomed vessel on the river bank south of Morfa Quay. River barges were used to take ore to the shallow tidal waters three miles inland to the Forest Copperworks, but were later increasingly used in the nineteenth century to bring ore upriver from the larger sea-going sailing vessels.

with a partly covered deck and central hold. The use of the latter may have been facilitated by steam dredging of the river after 1845. The remains of what is probably one of these larger barges have been located at the south end of the Morfa quay, and the remains of the keelson and lower ribs have been measured. Early photographs of Swansea North Dock show trains of rectangular barges that must have been pulled along the river by steam tug. A specialised barge dock had been built at the Cambrian Pottery by 1802 and measured 15ft wide by 60ft long.[21] This may give some idea of the size of contemporary river craft. The cost of a new coal barge in 1775 had been £105 19s.[22]

An idea of the minimum number of people employed on the river barges, and hence of the possible number of such craft, can be deduced from the 1841 census.[23] Nineteen entries are identified for individuals who probably worked on barges on the river. A further six bargemen and boatmen living further north along the line of the Llansamlet Canal probably worked on the boats in that canal. All except one were born in Glamorgan. Two of the 'bargemen' listed were under 20 years of age, one a boy of 16, the other of 12. Ann Samuel (60 years old and married to an agricultural labourer) was the only female 'bargeman'. Her sons (25 and 20 years old) also worked on the river barges, which might suggest that the barges had a crew of three. This seems reasonable, since the much smaller canal craft had crews of two. The areas of residence of the bargemen were as follows:

'Graig Trewyddfa'	4 bargemen
Morriston	3 bargemen
Plasmarl	1 bargeman
'Traveller's Rest', Llansamlet	1 bargeman
Lower Forest	1 bargeman
Llysnewydd	1 bargeman
'Nant y Gwindy'	1 bargeman
'Ty Pistill'	1 bargeman
'Gwindy Bach'	1 bargeman
Foxhole	3 bargemen

Fig. 117. The barge remains survived into the 1980s on the west bank of the River Tawe south of the Morfa Copperworks; the quay of the Upper Bank Copperworks is in the background. (803747)

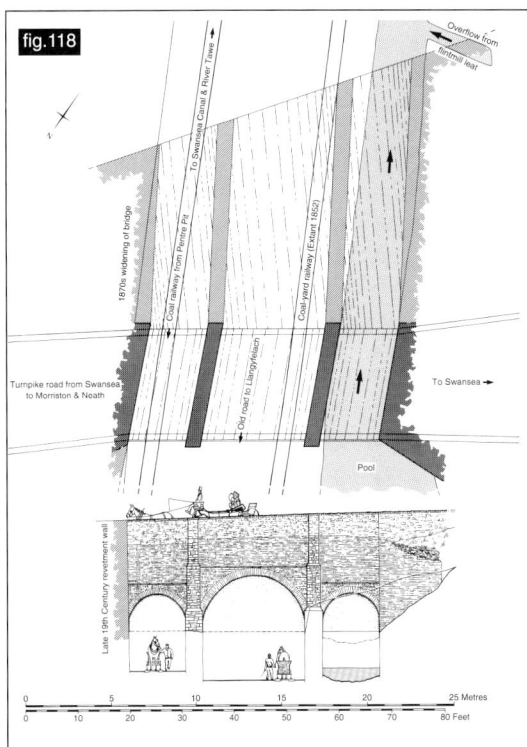

Fig. 118. Hafod (Aberdyberthi) Bridge spans the deep and narrow valley of Cwm Burlais that formed a funnel for coal transported from the north-west of Swansea. In the 1790s, the new turnpike road to Neath was carried across the valley on the three-arched viaduct illustrated.

Fig. 119. 'The Great Coal Road' of the Mansels was constructed in the later seventeenth century from their wharf at White Rock to their collieries in Llansamlet. This view shows the 'zig-zag' end of one branch of the road that led from the Glan-y-wern Colliery, the water reservoirs of which are visible in the central upper part of the photograph. (935057/66)

Fig. 120. Table showing the constraints imposed on the economic viability of collieries by their distance from navigable water, and the comparative costs of different means of transport.

Figure 120

Limits of possible working at 26% a ton profit of eighteenth-century collieries
(average profit of four Swansea pits under two owners recorded in 1772)

1773 LIMITS (WITH MINING COSTS AT 18.4d. A TON, SELLING PRICE AT 46.8d. A TON)
- With transport by pack-horse at 15d. a ton per mile: 1.3 miles
- With transport by cart at 5d. a ton per mile: 3.8 miles
  (greatest distance for which evidence exists of an actual mine so served: 3.6 miles)

1772 LIMITS (WITH MINING COSTS AT 27.4d. A TON * AND THE SELLING PRICE AT 47.7d. A TON)
- With transport by cart at 7.5d. a ton per mile; 1.5 miles
  (greatest distance for which evidence exists of an actual mine so served: 1.8 miles).
- With transport by wooden waggonway at 3.7d. a ton per mile: 2.9 miles
  (greatest distance for which evidence exists of an actual mine so served: 3.4 miles).
- With transport by canal (Swansea Canal costs were later to be 1.6d. a ton per mile): 6.6 miles**

* *All costs drawn from P.R. Reynolds, 'Townsend's Waggonway' but calibrated to 1733 market values.*
** *It was very cheap to open a lucrative mine in the unexploited upper valley so that it was economic for coal to travel over 16 miles down the valley.*

Concentrations at Graig Trewyddfa, Morriston and Foxhole are readily explicable. Foxhole was the main wharf for the Llansamlet Coalfield; the southern end of Graig Trewyddfa at Landore had large coal wharves serving the area from Landore to Lower Forest; and Morriston was at the northern end of the coal-mining and copper-smelting complex at the foot of Graig Trewyddfa.

The names of only three other workers along the river navigation are noted in the 1841 census: two quaymen living at St Thomas on the east bank of the river, and a 'coalheaver' next to the Llansamlet Coalfield quays at Foxhole who possibly heaved coal from the Llansamlet Canal boats. The myriad 'coppermen' noted as living near the non-ferrous smelters along the river bank may also have laboured at the copper-ore quays.

Generally, it is very difficult to differentiate between the Swansea tidal harbour prior to 1852 and the river navigation. Indeed, the river navigation was administered by the Swansea Harbour Trust from the formation of that body.[24] The main port from medieval times had been the curve of the river passing the town wharves at Swansea and the continued importance of this area is shown by its conversion, into Swansea's first floating harbour, or non-tidal dock, in 1852.[25] This area received the deep-sea ships that could not go up the river.

Detailed information on the river navigation is not as easy to obtain as for the adjoining Swansea Canal but it should be noted that the navigation was much longer-lived than the canal, and that it had been essential in the first phase of industrial growth in the valley and continued to play a central role in the economic life of the lower valley during the nineteenth century.

Use of Water-power

The River Tawe was an important source of water-power at and above the head of navigation. Three large industrial complexes, dependent on the availablity of water-power, grew up at Ynys-pen-llwch, Upper Forest, and Lower Forest in the late seventeenth and early eighteenth centuries. By their nature, such sites were viewed as major fixed assets for power generation which could be adapted to a wide variety of uses, sometimes simultaneously. Each of these sites had substantial weirs requiring constant maintenance, long headrace channels or canals and multiple release channels which were needed both to accommodate the large seasonal surges of water down the main river and to provide enough sites for the water-wheels to drive the industrial processes; each site has some remains of its former water-power channels. One building remained on the Upper Forest site until the 1970s and a partly rebuilt tinning building of the Beaufort Tinplate Works remains on the Lower Forest site, complete with reconstructed water-wheel. Substantial remains of all three sites doubtless remain underground and opportunities for excavation may occur.

Traffic Feeders

Even alongside the navigable river the valley needed efficient artificial feeders to transport coal to the river itself and to the new works on its banks. The adits and shallow shafts nearest the river navigation had been worked from medieval times and the Swansea coal market, for works'

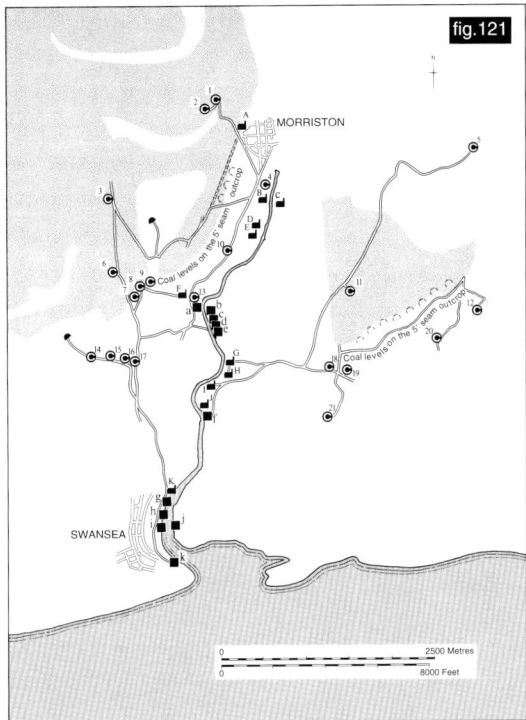

consumption (largely in the copper industry) and export, was growing at an increasing speed during the eighteenth century. The large coal reserves to the north east of the river in Llansamlet were owned and worked until 1750 by the Mansels. The earliest form of coal-carrying transport used by them, which required no specialised engineering works, was the pack-horse. The Mansels' nearest colliery, however, was 1.33 miles from the navigable water, slightly more than the maximum distance of economic working for this means of transport (1.31 miles, see table 'B'). Thus, an early local specialist transport system was essential for the development of the lower Swansea Valley.

From 1702 until the opening, in 1796, of the lower part of the Swansea Canal, and the Trewyddfa Canal which formed an integral part of it, many relatively sophisticated transport lines were built. Evidence for many miles of this early transport network still exist today. As this network mostly preceded the introduction of joint-stock communal transport systems, the resulting system was a disparate combination of privately constructed coal-roads, railways and tub-boat canals, above and below ground. The underground ways should definitely be regarded as transport tunnels and not merely as mines. A noted contemporary coal viewer from northern England was reported as stating that 'the depth of the valleys and heights of the hills worked by tunnels .... and the coals .... are let down into the boats on the canal tunnels, or as loading for tram-waggons in the tunnels below; through which they are conveyed to open day, and thence to the .... place of shipping'. The terrain presented gradients that were difficult for coal-roads and railways to negotiate and, of course, surface canals were confined to the valley floor. It is not surprising therefore that this intensively developed area possessed the world's first known tunnel on a surface railway. It was built 34 years before the more famous ones elsewhere. Early mine access railway tunnels were also plentiful in this area years before Kitty's Drift was built (in Co. Durham). One of only four tunnels on the main lines of surface canals in south Wales existed in this area. For these reasons it is worth examining in some detail the largely unpublished and forgotten transport features which ante-date the trunk canal.

Coal Roads

No details survive of the tracks used by pack-horses to convey coal down to the navigable river, possibly from Roman times but certainly from the beginning of local recorded coal production in the thirteenth century. The routes may have been those later used by coal-carts and waggons from the seventeenth century onwards. How sophisticated these routeways were must be open to question. For the most part they were probably unsurfaced tracks on rudimentary shelves cut into the steep valley sides. The earliest must have been down to the shipping stages on either side of the navigable river.

On the west side, the old-established coal-shipping

Fig. 121. Map of 'coal roads' in the lower Swansea Valley. A dense system of pack-horse tracks and cart roads was developed by the late seventeenth and early eighteenth centuries to convey coal from collieries to the copper-smelting works and to shipping wharves on the River Tawe. (Based upon Ordnance Survey material with the permission of Ordnance Survey on behalf of the Controller of Her Majesty's Stationery Office © Crown Copyright)

Fig. 122. The mid eighteenth-century coal road curving (top left to bottom right) across Graig Trewyddfa Common from the Tirdeunaw and Treboeth Collieries to Forest Copperworks. An earlier track for conveying coal from the mining tunnels on the outcrop of the five-foot seam is visible horizontally across the upper middle of this view. (935055/69)

Fig. 123. Woodfield Street, the main thoroughfare of Morriston, on the floor of the Swansea Valley (upper right); top right: a former coal road (Clyndu Street) runs up the western valleyside to the middle right; at the summit it bends west towards Clydu Water Machine Pit. (983515/10)

wharves in Swansea town were accessible to the collieries to the north-west via trackways down the Cwm Burlais ravine, running along the northern boundary of the medieval borough. The various shaft-heads of the Prices of Penlle'r-gaer were linked to both the town and Landore Wharves. 'The Great Coal Road' of the Mansels led from Glan-y-wern and Bon-y-maen to the White Rock Quay. Further extensive systems were built by Robert and John Morris and their partners from the collieries on the western valleyside to the Forest Copperworks and other copper smelters on Graig Trewyddfa.

Railways

Note on Railway Terminology in this volume.
The term railway is used in this book as a generic term for any railed way. Contemporary usage was not consistent but generally in this study a 'waggonway' refers to a wooden-tracked railway of the eighteenth century; a 'tramroad' or 'plateway' is generally a railway with 'L' shaped rails or 'plates' and an 'edge-railway' resembles modern locomotive track.

The large capital resources of the Society of Mines Royal allowed Sir Humphrey Mackworth to construct an extensive system of horse-worked wooden-tracked railways, which connected his collieries to the Melincryddan Copperworks and to his shipping canal in 1699-1701. At this time, the less well-funded Swansea copper-smelting concerns were still generally using pack-horses for localised transport. The first highly capitalised Swansea smelter was the White Rock Copperworks built by the wealthy Bristol copper-smelting concern of Joseph Percival & Company. However as Bussy Mansel was supplying all their fuel needs at a fixed rate they had no incentive to improve the local transport infrastructure.

This situation changed dramatically with the arrival of the rich London merchant Chauncey Townsend, who leased the large Llansamlet and Gwernllwynchwth coalfields on the east bank of the River Tawe. He built railways and canals on every one of the coalfields he developed at Swansea and Llanelli. He brought in his engineer from Tyneside, George Kirkhouse, a man used to working amongst what had become the largest web of surface railways in the world around Newcastle. After 1756 he developed at Llansamlet a fairly large mileage of wooden railways to multiple shaftheads laid-out on the Tyneside model. These, constructed under the terms of his leases from the ground and mineral landowners, led to his coal-shipping place at White Rock, the adjacent copper-smelter and to his own newly-built copper-smelter at Middle Bank and his zinc and lead smelters at Upper Bank. Technologically significant was the first intermediate railway tunnel on a surface railway on Townsend's Landore Engine-pit Railway of 1762.

Figure 124

Fig. 124. Costs of building the wooden-railed railway (waggonway) from Plas-y-Marl Colliery Level to the Landore shipping wharf in 1769-71.

'New Waggon Way from Plas-y-Marl to Landore'
(contemporary costs of construction recorded in John Morris's Commonplace Book).

1769	December 31	To Forrest Copperworks, supplied to it (sic)	£84. 1s 0d.
1770	March 2	To James Thomas for Oak Timber & waggon rails	£19. 15s. 9d.
	November 7	To Thos. Maddock for 12 tons of Oak Timber	£36. 0s. 0d.
	December 31	To Forrest Copperworks for sundries	£73. 7s. 10d.
1771	–	For cash paid Thos. Morris for laying it down	£30. 13s. 0d.
		Total	£243. 17s. 7d.

Townsend also built a second short double-track wharfside railway at Landore in 1768-70.

The lease of Graig Trewyddfa, on the west side of the River Tawe, to the Lockwood, Morris & Co. partnership specifically mentioned the right to build railways. However, this partnership had started by taking over the bankrupt Llangyfelach Works of Dr. John Lane and was not notable for the capitalisation of more advanced technology until the 1760s. It had coalpits using the most elementary pithead gear in 1737 and built a network of basic coalroads, rather than railways, to its new smelter at Forest in 1747-52. The company also fostered the employment of gifted local artisan-engineers such as William Edward(s) and 'Mr. Powell' who had little, if any, experience of engineering practice in coalfields outside south Wales. This may explain why their first recorded railway was not built until 1769-71 – it ran from a coal-level at Plasmarl, between Landore and Morriston, to a shipping wharf at Landore. The building accounts of this oak-railed track, laid by a Thomas Morris, survive.[26] As with Shropshire railways and Sir Humphrey Mackworth's Neath railways, these railways continued into the mining tunnels in a manner that was more appropriate to the local Swansea topography.

Meanwhile, local landowning families who were still directly involved in coalmining were also building railways: Calvert Richard Jones from his Cwm Level (Tunnel) to his shipping wharf at Landore (in 1768-70); and the Prices of Penlle'r-gaer underground in their pits which are under modern day Manselton (by 1787).

The landscape on each side of the river now began to be significantly altered by the construction of significant sections of line cut into gently sloping terraces along the steep hillsides on both sides of the valley. From the 1770s Swansea was also involved in significant developments in railway track technology. Lockwood, Morris & Co. ordered all their iron castings from the Coalbrookdale Company at Ironbridge in Shropshire. In 1776 John Morris ordered substantial iron plates, some of the first iron rail laid anywhere in the world.

The Landscape Impact of the Early Railway System

The period from 1756-83 was an era of wooden-tracked railways carrying coal, largely for export, via the navigable rivermouth. It was a period in which the earlier coal-roads of local entrepreneurs were replaced by the more highly-engineered railways made possible by the greater resources of incoming capitalists. In the 27 years from 1756 to 1783, just over 13 miles of largely wooden-tracked surface railway (or 'waggonway') were built in the Swansea

Fig. 125. The route of Chauncey Townsend's collieries railway main line which paralleled, at a lower level, the course used by the present eastern valley road which runs from one-third of the way up the right side of the photograph. (983515/13)

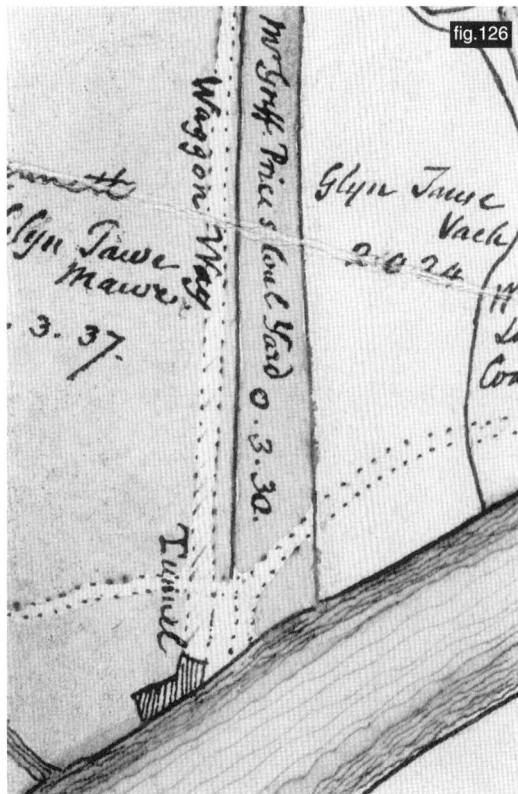

Fig. 126. Detail from a 1768 manuscript map, showing the earliest known intermediate tunnel on a surface railway ('Waggonway') anywhere in the world. (The National Library of Wales)

Fig. 127. Clyndu Street at Morriston, to the left, is a former coal-road (c.1747) to Forest Copperworks. The formation of Pentremalwod Street to the right is that of a former early nineteenth-century colliery railway that ran down Graig Trewyddfa to the Birmingham, Rose and Forest Copperworks. (9500357/3)

area (many more miles were built underground), of which the majority were considered by the London merchant, Chauncey Townsend, than by any other entrepreneur.

However, his substantial investments were

fig.128

Fig. 128. Map of Railways 1756-83. The effectiveness of the river as a means of cheap transport in enabling collieries to develop away from the river bank, depended on means being available for cheap bulk transport to the river bank and the copper-smelting works. By the seventeenth century coal roads were being built (see below) and from 1755 wooden-tracked railways ('waggonways') provided a more evenly-graded and consistently smooth and hard running-surface, and so a cheaper form of transport. By 1783 the most extensive system was serving the multiple shafts on the Llansamlet Coalfield (G) on the east bank of the river and the three smelting-works at White Rock, Middle and Upper Bank. In the 1750s short railways to pits at Landore on the west bank were built (A & B), and by the 1770s railways extended underground at Cwm Level (C) and Plas-y-Marl (E). Lines led to Landore Wharf from Plas-y-Marl (F) and Landore Coalpits (D). The Forest Copperworks lay at the mouth of an underground canal (the Clyn-du Level) that supplied it with coal from at least the 1770s (A-Townsend's Steam-engine Pit; B-'Tir Landore'(Glandwr) Pit; C - Cwm Level; D-Landore Pit; E-Plas-y-Marl Level and F-Plas-y-Marl Pit).

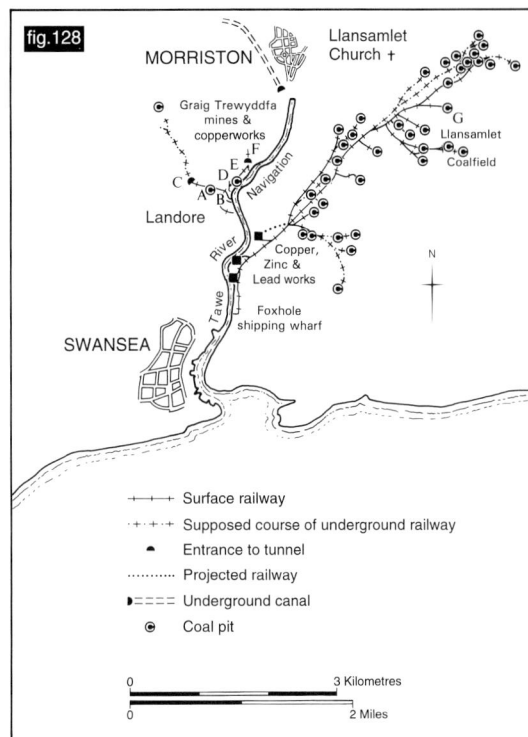

fig.128

MORRISTON · Llansamlet Church †

Graig Trewyddfa mines & copperworks

F / E / D / C / A / B

Navigation

Landore

River Tawe

Copper, Zinc & Lead works

Foxhole shipping wharf

SWANSEA

G Llansamlet Coalfield

N

———— Surface railway
·+·+·+ Supposed course of underground railway
▲ Entrance to tunnel
.... Projected railway
▶==== Underground canal
⊙ Coal pit

0   3 Kilometres
0   2 Miles

Fig. 129. The curving track running from Llanerch Place at the bottom right-hand side to the colliery spoil-heaps at the upper right-hand corner of this aerial photograph is one of the last surviving formations of the extensive wooden-tracked railway system constructed by Chauncey Townsend in the later eighteenth-century. (935054/49)

Fig. 130. At Gwernllwynchwith a terrace remains that was once part of the formation of the wooden-tracked railway system built by Chauncey Townsend. (960213/2)

unrewarded because of a lack of knowledge of local mineral resources. This resulted in large sums being spent on inappropriately placed collieries, quarries and railways. The large network built by Townsend and his successors on the east side of the valley in contrast to the network of short lines on the west reflects the differences between two systems of mining. That on the west was the so-called 'Shropshire' system of surface lines running directly to the coal-face underground, via tunnels that also drained water from the collieries by gravity. On the east Townsend had his mines built by a Newcastle engineer who imported the principle of an extensive surface railway system serving a multitude of shafts; each shaft there had to have its own winding and pumping facilities.

A development of a different type was the Clyn-du Underground Canal at Morriston which itself had

underground feeder railways. The first all iron rails for the valley were designed in 1779 and a variety of flanged platerail and edgerail were probably in use by 1783.

The period from 1784-93 was marked by a growing predominance of the use of the mining tunnel rather than of vertical shafts and the huge growth of the non-ferrous smelting market for coal which characterised Swansea's growth as the national and international centre of the copper industry.

There was a consequent development of a different rationale for both mining and railway practice - not over hill and down dale after a multitude of shafts, instead water levels and tunnels were now seen to be adequate.

In the nine years from the building of the first of the local surface canals in 1784 some five and a half miles of railway were built. Both the Lockwood, Morris & Co. Canal (c.1787-91) on the west side of the valley and Smith's or the Llansamlet Canal (1784) on the east side, were short private canals carrying coal (and incidentally supplying water to the works), both for delivery to the rear of the riverside smelters and to shipping on the river. The latter canal was built in order to redeem the fortunes of Townsend's successors as part of a transport and mining system suitable to the area. Five largely underground railways replaced the previous expensive and extensive surface system. On the west side of the valley Morris' canal penetrated a mile underground and was fed by a canal side railway tunnel. Lockwood, Morris & Co., using funds generated by over 60 years spent in the copper-smelting trade, took over the old Pentre Colliery to the north-west of Swansea town. Local topography dictated that longer lengths of surface railway should replace an intended canal to carry coal for shipment to the port of Swansea, whilst a second railway led from the colliery to the southern end of the Lockwood, Morris & Co. Canal.

At the end of the eighteenth and in the first two-thirds of the nineteenth century a complex pattern evolved of a fairly fixed pattern of coal-consuming

fig.129

fig.130

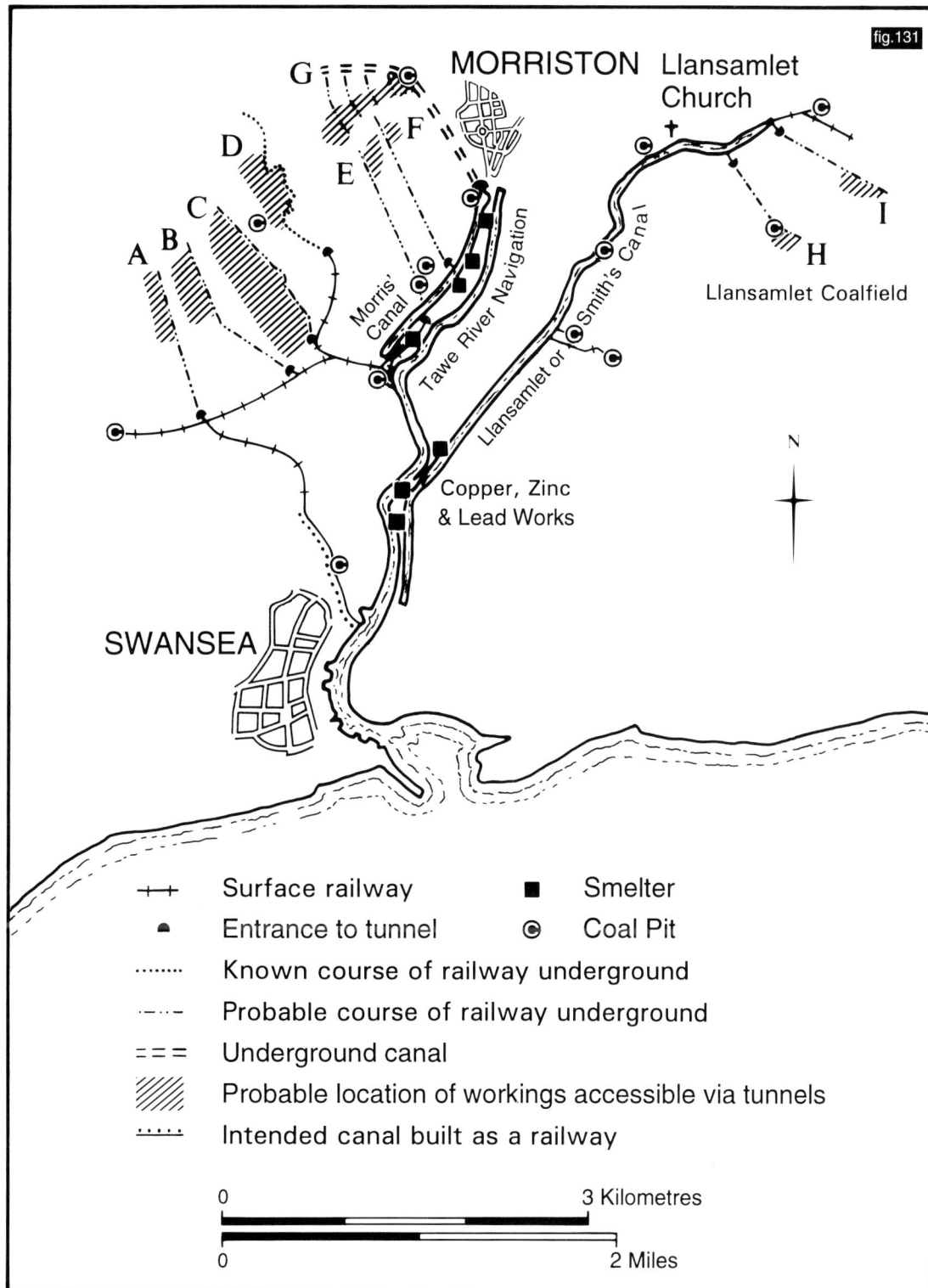

fig.131

G
D
C
E
F
MORRISTON
Llansamlet Church
A B
Morris' Canal
Tawe River Navigation
Llansamlet or Smith's Canal
Llansamlet Coalfield
I
H
N
Copper, Zinc & Lead Works
SWANSEA

+—+—+   Surface railway
●   Entrance to tunnel
········ Known course of railway underground
—··—·· Probable course of railway underground
≡≡≡   Underground canal
▨▨▨   Probable location of workings accessible via tunnels
·····   Intended canal built as a railway

■   Smelter
◉   Coal Pit

0   3 Kilometres

0   2 Miles

Fig. 131. Map of canals and railways at Swansea in 1784-93. By the 1780s, the operators of coal-mines on both sides of the River Tawe had constructed short canals which both delivered coal to copperworks on the riverbank and, by short iron railways, to ships on the river. One canal extended into mines underground (G) but more commonly iron railways extended from the banks of the canals and river into mining-tunnels (A- Pentre-gethin Level; B - Brynhyfryd Level; C - Cwm Level; D - Treboeth Level; E - Plas-y-Marl Level; F - Graig Level; H - Cwm Level; I is the Level near Gwernllwynchwith. The copperworks from north to south are the Forest (c.1752); Birmingham (1793); Rose (c.1780); Landore (c. 1793); Upper Bank (c.1777); Middle Bank (c.1765) and White Rock Works (1737).

non-ferrous smelters, these consumed coal supplied from a number of pits and levels which had a shorter life-span than these capital intensive works and which were served by a frequently changing web of short transport links.

After 1796 railways built to the newly-opened Swansea Canal, and constructed under the terms of its Act, were public railways open to other users upon payment of a toll equivalent to that paid to the Canal Company. However, the evidence is that for

Figure 132

Swansea Valley Tramroad Plates (listed in possible chronological order)

Tramroad or Plateway (i.e. Railway using flanged-rails)	Length of running bed	Thickness of running bed	Whether ribbed	Is rib under wheel path	Height of flange above running bed	Is flange fish-bellied?	Top width of flange	Weight per yard	Has the plate a sleeper seating?	Lobe at junction?	Bevel of whole flange
1. Cribarth Tramroad	?	1/2"	No	–	1 1/8"	No	1/4"	a.22.5 lb.	No	No	1/8"
2. Cribarth Tramroad	?	1/2"	No	–	1 1/4"	No	1/4"	–	No	No	1/8"
3. Cribarth Tramroad	?	5/8"	No	–	1 1/2"	No	3/8"	–	?	?	1/8"
4. Cribarth Tramroad	?	5/8"	No	–	1 3/4"	No	1/2"	–	?	?	1/8"
5. Cribarth Tramroad	?	5/8"	No	–	1 1/2"	No	3/8"	–	?	?	1/8"
6. Cribarth Tramroad	?	5/8"	No	–	2 1/4"	No	1/2"	–	?	?	1/8"
7. Oystermouth Tramroad	3' 0"	1/2"									
8. Lefel Fawr Tramroad	?	1/2"	No	–	2 1/4"	No	3/8"	–	No	Yes	1/8"
9. Lefel Fawr Tramroad	?	5/8"	No	–	1 1/2"	No	3/8"	a.22 lb.	No	?	1/8"
10. Cwm Lefel/Penyfilia Pit Tramroad	4' 0"							42 lb.			
11. Northern Graigola Levels Tramroad	?	3/8"	No	–	2"	No	5/16"	a.25 lb.	Yes	Yes	No
12. Southern Graigola Levels Tramroad	?	5/8"	No	–		No	3/8"	a.26 lb.	?	?	1/8"
13. Southern Graigola Levels Tramroad	?	3/4"	No	–	1 3/4"	No	1/2"	a.30 lb.	No	Yes	1/4"
14. Southern Graigola Levels Tramroad	?	5/8"	No	–	2 1/8"	No	3/8"	–	No	Yes	1/4"
15. Scott's Pit Tramroad	3' 5 1/2"	5/8"	No	–	2"	No	5/8"	30 lb.	Yes	No	1/8"
16. Scott's Pit Tramroad	?										
17. Brecon Forest Tramroad	3' 2"		?	?	1 1/4"–1 1/2"	Yes	3/8"	a.33 lb.	?	Yes	?
18. Brecon Forest Tramroad	3' 2"	3/4"	Yes	Yes	2 3/8"	Yes	1/8"	–	Yes	Yes	5/8"
19. Brecon Forest Tramroad	3' 2"	3/4"	Yes	Yes	2"	Yes	3/8"	–	Yes	Yes	1/8"
20. Brecon Forest Tramroad	?		?	?	a.1 1/4"	?	?	–	?	Yes	?
21. Landore Colliery Tramroad	3' 8 7/8"	5/8"	?	?	1 3/4"	?	1/4"	36 lb.	?	?	1/4"
22. Oystermouth Tramroad	4' 0"	1/2"	Yes	No	2"	No	1/2"	–	Yes	Yes	No
23. Garth Pit Tramroad	?	5/8"	No	–	1 3/4"	No	1/4"	–	Yes	No	1/8"
24. Northern Graigola Levels Tramroad	3' 0"	3/4"	Yes	No	1 1/2"	Yes	3/8"	–	No	No	No
25. Northern Graigola Levels Tramroad	?							30 lb.			
26. Northern Graigola Levels Tramroad	?							33 lb.			
27. Northern Graigola Levels Tramroad	4' 0" ?							37.5 lb.			
28. Northern Graigola Levels Tramroad	4' 0" ?							41.25 lb.			
29. Gwauncaegurwen Colliery Tramroad	3' 0"	1 1/8"	Yes	Yes	2 1/8"	No	3/4"	45 lb.	Yes	Yes	No
30. Ystalyfera Ironworks to Gurnos Wharf Tramroad	?	5/8"	No	–	1 1/2"	No	3/8"	a.25 lb.	No	No	1/4"
31. Ystalyfera Ironworks to Gurnos Wharf Tramroad	3' 0"	5/8"	No	–	2 1/8"	No	5/8"	35 lb.	Yes	Yes	1/4"
32. Wernddu Coal Levels Tramroad	3' 10 1/4" ?	3/4"	No	–	1 3/4"	No	5/8"	–	No	Yes	No
33. Hen Neuadd Tramroad	3' 0"	3/4"	No	–	2 3/8"	No	3/8"	a.40 lb.	Yes	Yes	1/8"
34. Outram's Recommended Plate: 1799 (Non S.V.)	3' 0"	–	–	–	2 1/2"	Yes					
35. Penydarren Tramroad (Non S.V.)	?	1/2"	Yes	No	2 1/2"	No	1/2"	–	Yes	Yes	No
36. Penydarren Tramroad (Non S.V.)	?	3/8" or 3/4"	Yes	No	2 1/2"	No	3/4"	–	No/Chair	No	3/4"
37. Penydarren Tramroad (Non S.V.)	3' 0"	1"	Yes	Yes	2 3/8"	Yes	5/8"	a.60 lb.	Yes	Yes	No

Swansea Valley Tramroad Plates (listed in possible chronological order)

Bevel at flange base	Distance from wear to bevel	Distance from bevel to greatest wear	Width of running bed	Width of wear	Type of joint	Possible date of plates	Original Engineer(s) of Tramroad	Original Entrepreneur(s) of Tramroad	Find spot of plate	Present location of plate	
No	–	–	3"	No wear	Projection/Recess	c.1798-1844	Thomas Sheasby ?	Daniel Harper ?	SN 8222 1340	Swansea M.&I. Museum	1.
No	1"	1"	3"		Projection/Recess	c.1798-1844	Thomas Sheasby ?	Daniel Harper ?	Not known	Rattenbury Collection (Ironbridge Museum)	2.
No	–	–	3"	No wear	?	c.1798-1844	Thomas Sheasby ?	Daniel Harper ?	SN 8213 1326	Swansea M.&I. Museum	3.
No	3/16"	3/4"	3"	1 1/4"	?	c.1798-1844	Thomas Sheasby ?	Daniel Harper ?	SN 8213 1326	Swansea M.&I. Museum	4.
No	–	–	3"	No wear	?	c.1798-1844	Thomas Sheasby ?	Daniel Harper ?	SN 8213 1326	Swansea M.&I. Museum	5.
No	0"	3/4"	3"	1 7/8"	?	c.1798-1844	Thomas Sheasby ?	Daniel Harper ?	SN 8222 1340	Swansea M.&I. Museum	6.
			3 1/2"			1804-1828	Edward Martin; Evan, Roger & David Hopkin	John Morris I & others	Documentary ref. Lefel Fawr Entrance		7.
1/4"	0"	0"	2 3/4"	5/8"	Butt	1805-af.1855	William Watkins ?	Daniel Harper & William Shaxby		Nat. Museum of Wales	8.
1/4"	0"	0"	2 3/4"	1"	Butt	1805-af.1855	William Watkins ?	Daniel Harper & William Shaxby	SN 8152 1244	Swansea M.&I. Museum	9.
					Cone/Socket	c.1805		John Morris I & others	Penyfilia Pit Site	J. Lerwell	10.
1/2"	–	–	?	No wear	Projection/Recess	c.1799-c.1858	Edward Martin ? & William Bevan ?	Richard Parsons	SN 7158 0200	R.C.A.H.M. Wales, Aberystwyth	11.
1/4"	–	–	3"	No wear	?	c.1799-1852	Edward Martin ? & William Bevan ?	Richard Parsons	SN 7066 0143	Swansea M.&I. Museum	12.
No	0-1/2"	0-1/2"	3"	1 1/4"	Projection/Recess	c.1799-1852	Edward Martin ? & William Bevan ?	Richard Parsons	SN 7066 0143	Swansea M.&I. Museum	13.
No	0"	0"	?	1 1/4"	Butt	c.1799-1852	Edward Martin ? & William Bevan ?	Richard Parsons	SN 7066 0143	Swansea M.&I. Museum	14.
1/8"	–	–	3 1/4"	No wear	Cone/Socket	1817-1845	–	John Scott, James Cox & T.H. Ewbank	Scott's Pit Site	Paul Reynolds	15.
					Butt			John Scott, James Cox & T.H. Ewbank	Scott's Pit Site		16.
?	?	?	?	?	Butt	1821-1862	David Jeffreys & Joseph Jones	John Christie	Near Crai	Paul Reynolds	17.
–	0"	0"	3 1/4"	1 1/4"	Butt	1821-1862	David Jeffreys & Joseph Jones	John Christie	SN 8570 1225	R.C.A.H.M. Wales Aberystwyth	18.
3/4"	3/8"	1 1/8"	3 1/8"	1 1/8"	Butt	1821-1862	David Jeffreys & Joseph Jones	John Christie	Not known	Rattenbury Collection (Ironbridge Museum)	19.
?	?	?	?	?	Butt	1821-1862	David Jeffreys & Joseph Jones	John Christie	Christie's Quarry Penwyllt	Paul Reynolds	20.
?	?	?	3"	?	Cone/Socket	bef.1826-27	?	John Morris I & others			21.
No	–	–	3 3/8"	No wear	Butt	1804-1855	Edward Martin; Evan, Roger & David Hopkin	John Morris I & others	Not known	Swansea Museum	22.
No	–	–	3 1/2"	No wear	Cone/Socket	c.1830-b.1845	?	?	SN 6866 0017 Old coal tips	Rattenbury Collection (Ironbridge Museum)	23.
No	–	–	3 1/4"	No wear	Projection/Recess	c.1799-c.1858	Edward Martin ? & William Bevan ?	Richard Parsons	Ynys-y-Mond Fm	Nat. Museum of Wales	24.
						c.1799-c.1831	Edward Martin ? & William Bevan ?	Richard Parsons	Documentary ref.		25.
						c.1799-c.1831	Edward Martin ? & William Bevan ?	Richard Parsons	Documentary ref.		26.
						c.1831	Edward Martin ? & William Bevan ?	Richard Parsons	Documentary ref.		27.
						c.1831	Edward Martin ? & William Bevan ?		Documentary ref.		28.
1/2"	–	–	3"	No wear	Projection/Recess	1832	Roger & David Hopkin	Roger & David Hopkin	SN 7077 1132	Swansea M.&I. Museum	29.
No	–	–	3 1/4"	No wear	Projection/Recess	1838-1852	Joseph Martin ?	Treacher, James & Joseph Martin	SN 7721 0923	Swansea M.&I. Museum	30.
1/4" ?	–	–	3 1/2"	No wear	Butt	1838-1852	Joseph Martin ?	Treacher, James & Joseph Martin	SN 7721 0923	Swansea M.&I. Museum	31.
No	–	–	3"	No wear	Projection/Recess	af.1830-b.1897	?	John Parsons	SN 7340 0241	Wern-ddu Isaf	32.
3/8"	–	–	3 1/2"	No wear	Butt	c.1797-1844	Thomas Sheasby	Swansea Canal Company	SN 8095 1259	Craig-y-Nos Visitor Centre, B.B.N.P.	33.
						1799	Benjamin Outram				34.
1/4"	–	–	3 7/8"	No wear	Butt	1800-c.1880	George Overton, Roger Hopkin	Richard Hill & others	Not known	Nat. Museum of Wales	35.
No	–	–	3 3/8"	2 1/4" ?	Butt on chair	1800-c.1880	George Overton, Roger Hopkin	Richard Hill & others	Not known	Nat. Museum of Wales	36.
3/8"	–	–	3 3/8"	No wear	Butt	1800-c.1880	George Overton, Roger Hopkin	Richard Hill & others	Not known	Nat. Museum of Wales	37.

Fig. 133, 133a, 133b. The railway and canal infrastructure of early nineteenth-century Swansea (excluding arterial railways). The copperworks shown on the plans are 1. Whiterock (1737); 2. Middle Bank (c.1765); 3. Upper Bank (c.1777); 4. Hafod (1809); 5. Nantrhydyfiliast (c.1814); 6. Landore (c.1793); 7. Rose (c.1780); 8. Birmingham (1793); 9. Forest (c.1752); 10. Morfa (1835); 11. Black Vale, Cwmbwrla (1852); 12. Port Tennant (1852); 13. Dan-y-graig (1858).

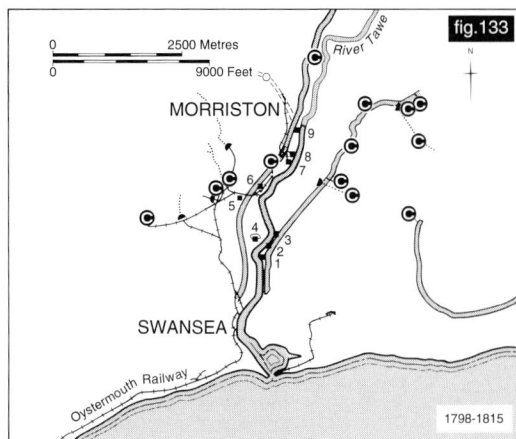

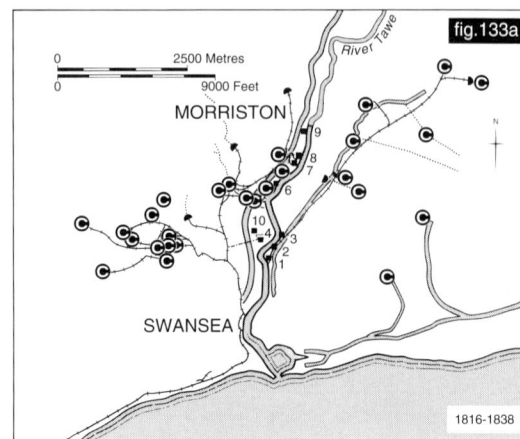

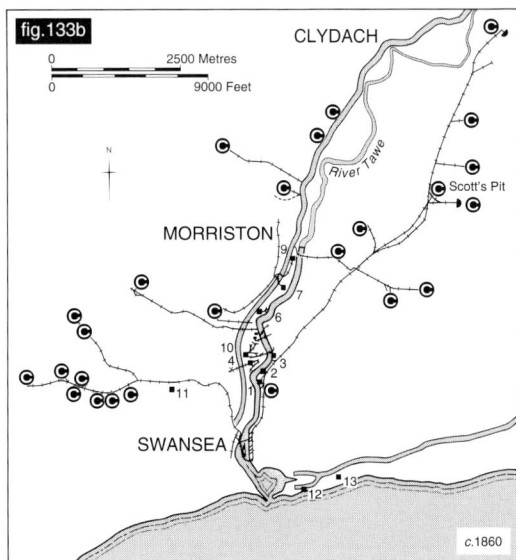

Fig. 134. Table showing the great lengths of lower valley transport ways that existed below ground.

distances of under two miles each concern preferred to build its own lines, which were sometimes parallel to those of its competitors. Public locomotive lines were not used for localised transport in the valley

until the construction of the Swansea Vale Railway in the 1860s.

In the period from 1806 to 1814 railway development was centred around what might be termed the Swansea Bay speculation. The coal traffic along the Swansea Canal almost trebled in this period. The first large-scale Ordnance Survey plans produced at the end of this period reveal that the railway mileage increased by 12.3 miles in order to carry this traffic. A substantial part of this increase in mileage was due to the construction of the Oystermouth 'Rail-way or Tramroad.' This scheme had first been proposed as a canal connected to the Swansea Canal. The copper-smelting company of Morris and Lockwood with their landlord, the Duke of Beaufort, already owned one such waterway - the Trewyddfa Canal between Landore and Morriston. These partners were keen to supply the copper-smelters alongside their Trewyddfa Canal with the bituminous coal that they thought would become cheaply available by extending the Swansea Canal southwards. The copper-smelters preferred their coal to be as bituminous as possible and that found

Figure 134

Table showing comparative above and below ground extent of track in 1784-95

These had railways unless stated otherwise	Letter on plan	Extent of railway or canal above ground	Contemporary Estimate of railway or canal above ground	Extent of track or canal Underground	Contemporary estimate of track or canal Underground	Extent of tunnel to main coal workings	Extent of tunnel inside coal workings
Pentre Level	A	2,600 yards	3,520 yards	1,320 yards	5,280 yards	660 yards	660 yards
Penyfilia Level	B	840 yards shared plus 400 yards	–	1,760 yards	–	840 yards	960 yards
Cwm Level (landore)	C	840 yards	–	1,650 yards	–	NONE	1,650 yards
Treboeth Level	D	840 yards shared plus 900 yards	3,520 yards	1,320 yards	1,760 yards	440 yards	880 yards
Plas-y-marl Level	E	570 yards	–	1,540 yards	–	1,320 yards	220 yards
Graig Level	F	60 yards	–	1,380 yards	–	1,000 yards	380 yards
Cwm Level (llansamlet)	H	140 yards	–	920 yards	–	700 yards	220 yards
Gwernllwynchwith Vei Day Level	I	80 yards	–	1,040 yards	–	640 yards	400 yards
Total of railway Lengths		**5,590 yards**	–	**10,930 yards**	–	**5,800 yards**	**5,130 yards**
Clyn-du Canal Level	G	As built 110 yards	–	1,880 yards	1,100 yards	1,000 yards	880 yards

around Swansea Bay fully met their requirements. The proprietors of the Swansea Canal did not view this transport and mining scheme as being complementary to their own. Boats ascending the four or five locks from the proposed Oystermouth Canal to the copper-smelters, would have taken the water supplies required to sustain the lucrative long distance trade of the Swansea Canal from the upper Swansea Valley to the port of Swansea. The Oystermouth limestone, available from the southern edge of the great mineral basin of south Wales, would also have competed for a market with the Cribarth limestone from the northern edge of the basin. Hence the Swansea Canal Company had the Oystermouth Company's Bill changed into one giving powers specifically for a railway rather than for a canal, and they themselves began building a further railway line to by-pass their own four lower locks. The Oystermouth Company's line was a financial disaster. The broken and almost vertical coal-seams outcropping on the southern limit of the coal-basin were either not present in the areas predicted or virtually unworkable when found. The Oystermouth Act gave no compulsory powers for new branch railways and no permanent link to the majority of the Oystermouth Quarries could be established. One of the available mineral waggons on the under-used line provided the basis for the well-known introduction of passenger services. That notable historical event has masked the abysmal failure from which that event largely sprang.

In the period from 1831 to 1838 there was a general consolidation of the system. During this period Sir John Morris (II) and Charles Henry Smith were extending their systems of lines west and east of Swansea respectively, in order to supply the copper-smelters with more bituminous and semi-bituminous coal. The disadvantages of building railways without the compulsory powers available under the Swansea Canal Act were shown by the example of Sir John Morris. He had to abandon his lines west of Swansea at the end of this period when his lease of the right of way through Cwm Burlais in the town of Swansea expired. Similar disadvantages had earlier forced Morris to abandon his lines to the Oystermouth limestone quarries and contributed to the financial failure of that line.

In the period from 1839-60 there was a continued growth of the non-ferrous works at Swansea in a period when the town was at its zenith as the copper-smelting capital of the world. A series of new collieries and railways was built outside the lower valley area. This was in response to a crisis in the supply of bituminous coal available for use in the expanding copper-smelting and other non-ferrous works. Due to this crisis semi-anthracite coals had to

be extensively used for the first time in copper-smelting. By 1858 Scott's Tramroad on the eastern side of the lower valley had been transformed and extended as the Swansea Vale Railway. This was operated by locomotives and although at first largely complementary to the Swansea Canal it did replace the function of one canal railway at its northern end (at Graigola). By 1860, with the extension northwards to Pontardawe, the Swansea Vale Railway had become a fully fledged modern line carrying passengers and using its own rolling-stock. The process by which it largely replaced the Swansea Canal and its attendant waggonways as the only heavy transport system in the Swansea Valley then proceeded apace.

Landore Railways

At Landore a narrow side valley of the River Tawe channelled through it a dense network of successive and contemporary early railways serving a centre of the coal-mining, copper-smelting and coal-shipping activities of the lower valley. It constantly changed in response to varying sources of raw materials, sites of manufactories and varying patterns of entrepreneurial partnerships. It has fortunately been possible to reconstruct successive phases in this palimpsest of a landscape, thanks mainly to the scrupulous paperkeeping and extensive legal activities of the Dukes of Beaufort and their agents. That system is examined in chapter six.

The Early Railway System Underground

The concept of an early, horse-worked railway is often interpreted in terms of the modern railway. Hence the extent of the railway network has largely been assessed in terms of the layout above ground. However, the system did not stop above ground but often had a homogeneous form above and below ground. The first indication of a separation in function between surface and mining railways in the lower Swansea Valley was along the first Swansea Vale Railway extension in 1855-6. Elsewhere in the valley, unitary above and below-ground systems continued in use until the 1880s.

Early horse-worked railways were not independent transport concerns but merely part of the necessary fixed capital of a working mine. It is impossible to consider them without discussion of contemporary geological knowledge and mining practice. A consideration of that part of the early railway system that was built underground is also important in enabling us to see in which areas of the country large lengths of railway existed. Such concentrations, below and above ground, created pools of experience in railway technology at a formative stage of development. The extent to which a mining railway network lay above ground was largely governed by the type of mining practice. On the so-called Shropshire railways, small waggons with flanged wheels ran on narrow-gauge, wooden edge-rails directly from the coal-face through a tunnel or adit to the surface, and then by means of an over-ground line to the nearest shipping place or ironworks.[27] On the so-called Tyneside wooden railways, large waggons with flanged wheels ran on wooden edge-rails from a multiplicity of coal-shaft heads to the nearest navigable water. The gauge of a Tyneside waggonway was broader than Shropshire practice, often approximating to the later standard railway gauge of 4ft 8½in. Small trolleys with flangeless wheels were used underground in side passages leading to the coal face. These barrow-ways sometimes consisted simply of planks laid longitudinally to form a simple track. More frequently provision was made for guiding the trolleys by means of a central raised row of planks, or more often by nailing vertical strips of wood on each side of the running planks. [28] This presumably was the origin of the flanged rail or tramplate. In the main passages of these north-eastern mines wooden edge-railways were found (often double-track), carrying waggons of an intermediate size running on flanged wheels.[29] The underground trolleys, or 'rolleys', carried varying numbers of 'corves' or wicker baskets containing coal which transferred from smaller to larger underground trolleys and then hauled up the shaft and tipped into larger surface waggons.

The different types of railway and mining methods can be seen by reference to the map of railways in the period 1756-83. Line G to the east of the River Tawe served the coal-shafts of the Llansamlet coalfield, an area where Tyneside practice was followed. This resulted in a number of separate shafts, all rising to ground-level and each requiring its own branch from the waggonway spine. On the western bank, Shropshire practice was preferred, and this is reflected in the number of short lines, each running from a level or mining tunnel to the river. Within the levels, it was not uncommon to find shafts descending to a lower seam

The map of railways and canals in the period 1784-93 shows the probable extent of all the valley lines then existing, both above and below ground. Attempts at such comprehensive coverage for other periods are impossible because of the lack of plans. The waggonway tunnels around Swansea were, in the main, not just small mining tunnels entering a hillside at an outcrop of the coal and then following it back into the hill; such shallow levels or drifts had existed around Swansea since the fourteenth century and by 1717 in Llansamlet, for example, there were brick-

lined colliery tunnels nearly a mile long.[30] Of the nine levels shown on an old mining map of the area in the Badminton Collection, one level goes directly into the outcropping of the Swansea Six-Foot seam; this is the first level in which a waggonway is known to have been laid (between 1768 and 1770).[31] The other eight combined drainage and transport levels were needed to penetrate deep into the hillside beyond earlier workings. The function and type of these tunnels were described by a number of contemporary observers.

According to Edward Martin, the mining engineer based at Morriston:

'Through a large district of South Wales, their highly valuable veins of coals, ... are gained at comparatively trifling expences, compared with most of the Newcastle pits; the depth of the vallies and heights of the hills in that part of the country, allowing several successive and thick vains (sic) of coals to be worked by tunnels into the hills above the level of the rivers, or springs of water, in the vallies; and the coals, and the valuable iron ore which also abounds, are let down into the boats on the canal tunnels, or as loading for tramwaggons in the tunnels below; through which they are conveyed to open day, and thence to the iron-works [i.e. 'copper-works'] or places of shipping'.[32]

The reports of casual visits to the Pentre-gethin Colliery[33] of Lockwood, Morris & Co. ('A' on the 1784-93 map) emphasise the point that these levels were not merely minimal-bore passages leading to single coal-faces. In 1791 E.D. Clark noted that:

'The entrance is vaulted, and perfectly level, and continues so for about one hundred yards, when our guides made us turn off to the right, to a sort of staircase, which they call the horse-road. By this we descended to the depth of eighty fathoms, and came to a spacious area, where the miners were sending up the coal in baskets, through a shaft, to the vaulted level we had just quitted. It is there put into carts, with friction wheels, and drawn by oxen to the mouth of the mine ... We did not proceed more than half a mile under ground; but were sufficiently fatigued with the excursion. If a person has spirit and strength sufficient to explore the whole of this mine, he would have above three miles to walk in these gloomy abodes ...'[34]

In 1804 this picture of this tunnelling system was augmented by the Revd J. Evans:

'The Whole hill is full of coal, and is obtained by what miners term open audits, i.e. horizontal shafts driven into the hill, which form levels for draining the work, as well as ways for the delivery of the coal. There are within some vertical shafts, beneath these levels, and whimsies have given way to a more philosophical and expeditious machine, the improved steam-engine of Bolton [sic] and Watt. One of these adits, which we traced about a mile in length, admits low waggons, holding a chaldron each, which running on an iron railway, one horse with ease delivers at the quay. [35]

Such contemporary descriptions reinforce the point that these were transport tunnels and not simple passages within the thickness of one particular seam (i.e. akin to 'modern' horizon mining methods).[36] Both Martin and Evans make it clear that coal could either be lowered or raised to the tunnels from workings at some distance. The coal raised through the Pentre Level would have come from workings in the Swansea Six-Foot seam, the thick seam below the Swansea Five-Foot seam which the level was intended to strike. The Clyn-du Level ('G' on the 1784-93 map) was actually sometimes referred to as a 'tunnel'. Contemporary descriptions of that level make it clear that the 'vaulted' level at Pentre may only have been 'arched with Stone part of the way in' until hard self-supporting rock was reached. Others were vaulted for most of their length.[37] The relationship between mining and transport functions would have been similar in the Clyn-du and other levels even though the former was a canal (rather than a waggonway) tunnel. Contemporary and later mining plans[38] make it clear that coal along 3,000ft of the Swansea Five-Foot seam was being mined at a higher level than the Clyn-du and other tunnels and was being lowered to the tunnels along the dip of the seam, or through other ground. Thus, even in the area of the main seam, a transport level could be divorced from the actual mining process.

Rather less than half the length of the tunnels was within the area of active coal extraction and just over half of the contemporary waggonway mileage was underground in comparatively large tunnels accessible from the surface.[39] Even more track was underground, in tunnels accessible only from shafts. (The levels west of the River Tawe could reach the Swansea Five-Foot seam but deep shafts from the surface were required to penetrate the Swansea Six-Foot, Three-Foot and Two-Foot seams.) In Landore Colliery alone there were 3.5 miles of track by 1788.[40] The surface railway networks that are known represent a fraction of the total length of line built.

The testimony of contemporary observers may be combined with archaeological, cartographical, documentary and geological sources to produce a reliable estimate of the extent of wooden railways in the lower Swansea Valley. The existence of such an extensive transport system indicates its importance to local industry, and makes it likely that innovations in railway practice took place in the area.

The evolution of iron track

There has been a tendency, encouraged by Michael Lewis's seminal work on *Early Wooden Railways*, to see railway development in the eighteenth and early nineteenth centuries purely in terms of two schools of development, based on the Newcastle and Severn Gorge areas respectively. This is apparent in Barrie Trinder's hypothesis that '... in the eighteenth [century] almost all the major innovations in railway technology - the iron wheel, the iron rail and the self-acting inclined-plane - were first employed in Shropshire'.[41] The same thinking lies behind his statement that:

'After 1800 most of the principal innovations in the workings of railways took place in the North-East. In the years 1800 and 1810 the first inclined-planes worked by steam-engines were employed in the area.'[42]

However, Trinder notes that the north-east was anachronistic in at least one respect - it was slow to adopt iron rails and still had lines using wooden rails in 1807.[43] In fact, such wooden rails may have been commonplace in the north-east, at that date, for some wooden waggonways were still in use in their original state in the 1840s.[44] Further, in the example which Trinder cites of powered inclined-planes, the north-east was not in advance of at least one other area, for in south Wales there was a considerable inclined-plane worked by a steam-engine, planned in 1802 and put into construction in 1803 by Evan Hopkin, an engineer based in the lower Swansea Valley.[45]

There may have been other areas of innovation in railway practice, independent of the Tyneside and Severn Gorge areas, before the spread of the canal system produced a degree of cross-fertilisation between the two recognised areas of early railway development.[46] Hence the sudden increase in the construction of feeder railways which accompanied the 'canal mania' of the 1790s, is given the credit for giving impetus to the invention of the two types of all iron rail - the tramplate and the edge-rail.

'Canal mania' affected two main industrial centres, south Wales and the east Midlands. The first all-iron edge-rails are recorded at Merthyr Tydfil in 1791 (however, every coalfield was a centre encouraging artisan-engineer innovation and all iron rails may have been used at Whitehaven in Cumbria in 1738)[47] and all iron 'L' shaped plate-rails for use underground at Sheffield in 1787.[48] In 1788 these iron plate-rails appeared above ground in two locations, and their use was subsequently championed by Benjamin Outram of Butterley Ironworks in Derbyshire. When consulted about the Monmouthshire Canal railways in 1799, he recommended a wholesale changeover from edge-rail to plate-rail. He largely determined the practice of industrial railways for a full 30 years outside the north-east of England. It has been said that practically every railway owner in south Wales followed his advice.[49]

The existence of only two schools of innovation in railway technology, before the 'canal mania' of the 1790s disseminated this pioneering work, has hitherto been generally accepted. In part, this assumption has been based on an underestimate of the extent of early railways, and hence of the fund of railway engineering experience available in Britain. It seems to have been overlooked that Outram stated in his *Observations* of 1799, made to the Brecknock and Abergavenny Canal Company, that '... the railways ... I have constructed (and others copied therefrom) ... are made nearly on the Principle of those ... Tram or Dram Roads as they are called in this Country' [Wales]. Despite the reference to Welsh 'Dram Roads', contemporaries and later historians appear to have overlooked other, earlier references to iron plateways. Lewis seems to imply that Outram was the inventor of the tramroad when he writes that 'practically every railway owner in south Wales followed his advice'.[50] Yet it is clear from his *Observations* that Outram was simply following existing practice in Wales.[51]

Rails having an L-shaped, or flanged, cross-section and made of wood had been in use on the Continent since the medieval period.[52] The first convincing evidence of wooden flanged track being used in a coal-mine is from Alsace in 1543.[53] Such railways were in use in some British mines before 1756, when John Smeaton copied the type for use above ground. At this date they were also in use in the lower Swansea Valley: a letter to *The Cambrian* (22 March 1878) mentions that 'There are aged colliers now living at the Cwm, who, more than fifty years ago, came across old workings underground that were entirely unknown to anyone then alive. In these workings … Little cars have also been found, which were used in hauling the coal from the headings to the bottom of the pit. The frames and wheels of these were of wood, the boxes of plaited willows, and they ran along wooden rails formed of three pieces of wood nailed together like a water-trough. It appears that these cars were pushed or pulled from place to place by men and boys, but in one old working a windlass was found, which had been used to draw the cars from the headings.' As Paul Reynolds has pointed out, if these abandoned workings were broken into in the 1820s, and nobody was alive at the time who remembered them, that indicates that at the latest they must date to the third quarter of the eighteenth century, which is the period when the coal in this area was being worked by Chauncey Townsend (1750-70).

The Welsh examples which immediately predated Outram's *Observations* of 1799 were of iron, for he stated in that work that 'the rails ... are cast ...'. Curr's rails were first used in south Wales in 1796, but the iron plateways stated by Outram to have been called 'Tram or Dram roads... in this Country [Wales]' had a longer history than that. There are schemes recorded for 'dram' and 'tram' roads in 1790 and 1791 from Carns Mill to Cardiff and from Sirhowy and Tredegar Ironworks to the Monmouthshire Canal. The latter scheme was for 'an intended *Rail Way*, or *Tramroad* ...'. [54] Such alternative terms show that a clear distinction was made between plateways and edge-railways at this date, with the term 'tramroad' referring to a plateway. Such a distinction between terms, if it can be sustained, might make a re-examination of the following passage worthwhile. It was written in 1776 by Sir John Morris I with reference to his collieries in the Landore-Morriston district of the lower Swansea Valley:

> 'Cast-iron Tram Plates - In Nov'r 1776, I wrote to Messrs. Darby & Co. at Coal Brook Dale that I had sent them a pattern in wood about 4 ft. long and 5 in. wide, to cast-iron plates for wheeling Coal in my Collieries, each plate to weight ab't 56 lbs: that if they cou'd be supply'd at £8 p ton I shou'd want immediately 100 Tons, and that the introduction of them wou'd occasion a vast consumption of metal never before us'd for such purpose.'[55]

The order was fulfilled, for Sir John Morris later wrote in his commonplace book that:

> 'In 1788 there was about 240 tons of Cast-iron Tram Plates underground in Landore Coll'y, or three and a half miles of track.' [56]

Michael Lewis explains the 1776 passage as an order for flat plates for flanged wheels, on the pattern pioneered by Coalbrookdale.[57] This is based on the assumption that only the Severn Gorge and Tyneside areas had extensive enough early railway systems to produce the technical experience necessary to generate innovations in type. Lewis therefore assumes that innovations happened elsewhere only with the great increase of track to supplement the waterways constructed in the 'canal mania' of the 1790s. The possibility that cast-iron tramplates were an earlier and independent development in south Wales was either regarded as most unlikely or rejected. A more recent authoritative work has also stated that 'no one doubts that Curr's flanged iron plates were the earliest of their kind'. [58]

Outram's statement that the term 'tram' or 'dram',

as used in Wales, specifically referred to a type of waggonway having flanged rails, was made some 23 years after Morris's statement. The scheme mooted in 1791, 15 years after Morris's use of the term, for a 'Tram Road' from Sirhowy and Tredegar Ironworks, implied the proposed use of flanged rails on a line in south Wales. It is of course possible that railway terminology in south Wales underwent a change in that period, but we have no evidence on that.

The term 'tram', as used in 'tramroad', assumed currency in the rest of Britain as a term meaning a railway with flanged track, after that usage had become common in Wales. The term was already widespread throughout Britain but had a variety of meanings. 'Its original meaning was the shaft of a barrow or the frame of a sledge, and many early examples probably mean a sledge or even a barrow ...'.[59] This being so, an element of uncertainty must exist as to the exact meaning of the term 'Tram Plate' as used by Morris in 1776.

The origin of the iron plate-rail in the lower Swansea Valley is made more likely as a result of archaeological evidence. The accompanying table summarises the data. It collates information regarding 37 different plate-rails. Some of these were located in complete or fragmentary form in private or public possession; in other cases information has been derived from documentary or published sources. Rails numbered 34-37 are from lines outside the valley where the recommendations of Benjamin Outram or George Overton were adopted. (Overton was the engineer of the Penydarren Tramroad, among other lines.) The other examples are from the lower Swansea Valley, and it is immediately clear that this group is distinguished from other tramplates by two features.

These features were very uncommon outside the Swansea Valley. The two distinguishing characteristics are a length about one to two feet greater than the 3ft usually used within and outside the valley, and an interlocking cone-and-socket jointing between plates.[60] This latter feature may be seen on a plate found at Scott's Pit (1819). Tramroads on which this type of cone-and-socket jointing were built by the same engineers or owned by the same entrepreneurs as lines on which the unusually long type of tramplate was employed.[61] On lines outside the valley (and on many of those within it) adjoining tramplates simply abutted one another, or else semi-circular projections on one plate interlocked with a recess on its neighbour. These unusual rails with cone-and-socket joints suggest an independent origin. All known examples are probably no later than the first two decades of the nineteenth century; they were restricted to the lowermost part of the Swansea Valley and two of the four provenances for this type of plate are

collieries run at some stage by Sir John Morris I. The dimensions show a close relationship to those plates ordered by Sir John in 1776. The two plates recovered from the Pentre Pit are the best for comparative study, being complete and independent of the ambiguities of the 1826 account of plates at Landore;[62] they are also early, being from a line replaced by an edge-railway in 1806.[63] Like the plates described in 1776, these were 4 ft long and like those described in 1788 at Landore (where the Pentre Pit is sited) were 56lb in weight. Mid-eighteenth-century (wooden) plateways had track widths of up to 9in,[64] so a width of 5in is not exceptionally wide for the period. The tramplates from Pentre Pit have a width of 3.5in, but this is in line with the width of other tramplates from the late eighteenth century, when 3in was quite normal.[65] The consistency in weight, despite the narrowing of the width of plate, could be due to a thickening of the running surface. The lack of tensile strength of cast iron was a problem endemic to the comparatively thin running surface of plates of a normal length (914mm), let alone the 4ft plates used at Landore. The 0.75in thick plates found at Pentre are among the thicker of the plate-rails recorded from the valley. It may be that plate-rails made entirely of iron were in use some 11 years earlier than previously thought and were used for the first time in the lower Swansea Valley.

Further archaeological and documentary evidence disproves the supposition that Outram was the first and general introducer of the iron plate-rail to south Wales. When it was opened in 1796-8, the Swansea Canal had three lengthy railway feeders. However, two of these are labelled on the large-scale canal survey of 1797-9 as 'Rail roads' (i.e. edge-railways)[66] and at least one of them was engineered by the north-country engineer Edward Martin.[67] The third line was at the head of the canal and was partly owned by the canal company and partly by 'the Cribbarth Limestone Rock Company'; the part of the railway owned by the canal company had been planned in 1794.[68] It was built as a plateway by Thomas Sheasby, the canal engineer.[69] The extension up Cribarth mountain was, of course, of a similar track type. The main promoter of the

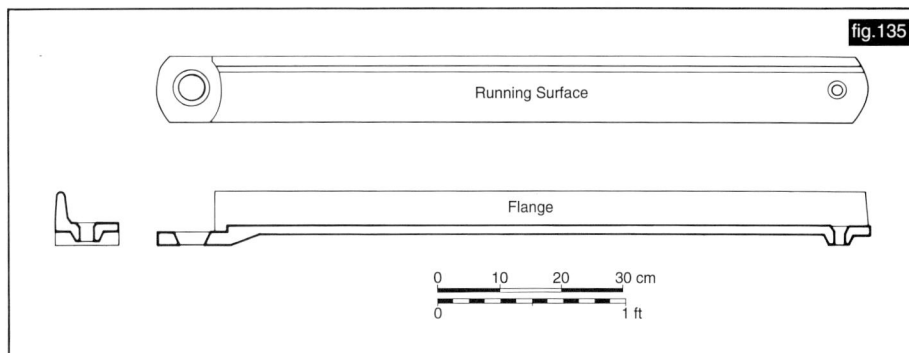

Fig. 135. Drawing of a plate-rail (from Scott's Pit) with cone and socket jointing, uncommon outside the lower Swansea Valley.

Cribbarth Limestone Rock Company (formed in 1797) was Daniel Harper from Tamworth[70] who had business connections with his former fellow townsman, Thomas Sheasby.[71] The railways laid out by Sheasby on the eight canal wharves constructed at Swansea probably had a similar type of track. Here again, therefore, are iron plateways in the valley area pre-dating Outram's 'observations' of 1799.

The plate-rails recovered from Cribarth are of an entirely different type from those specified by Morris. Plates 1 to 6 and 33 on the adjoining table are from Cribarth. The substantial plates represented by item 33 formed part of a section of plateway excavated *in situ* and showing no signs of wear, they are therefore likely to represent the last phase of the use of the railway. The similar plate 6 may well be contemporary. The tops of the flanges of this type of plate are parallel with the tread of the rails and are as small as 1.625in compared to the convex longitudinal profile of between 2.5in to 3.5in recommended by Outram.[72] The weight of the Cribarth plates is as little as 22.5lb compared with Outram's recommended weight of 40lb.

If the form of these plate-rails has an independent origin, then the question arises of what the origin was. John Curr's track came to south Wales in 1796.[73] The weight of the lightest and possibly earliest of the Cribarth track at 22.5lb per 3ft length matches almost exactly that of the 23.5-25lb of Curr's track. Curr's track was originally made in 6ft lengths but it assumed a varied length during the widespread early copying of his idea referred to by Outram and Curr himself. Outram said in 1799 that 'In mines and other works underground where very small carriages only can be used, very light rails are used, on a system introduced by Mr Curr, and these sort of light rails have been much used above ground in Shropshire, and other Counties where Coals and other Minerals ar gotten'.[74] Curr said in 1801 that 'I need say no more to recommend Rail Roads, than that about 16 Collieries out of 20 have introduced this mode of conveying coals in the Countys of York, Lancaster, Salop, Derby, Stafford, Warwick, and great parts of Wales ....'.[75]

The Sheasbys may have seen, or heard of, one of the first uses of Curr's plates while working as contractors on the Cromford Canal in Derbyshire in 1789-90.[76] If this was the case, it is surprising that their fellow contractors, the Dadfords, went on to build the first iron edge-railways rather than follow Curr's example.[77] The great promoter of the use of the iron flanged rail, Benjamin Outram, was assistant engineer on the Cromford Canal,[78] but as far as is known his advocacy of the use of iron flanged rails came later.[79] There are 22 plates (excluding those of the cone-and-socket type) recorded from the Swansea Valley area where evidence for the form of

the vertical guiding-flange exists. Only four of these had flanges cast with a convex longitudinal section ('fish-bellied') to provide additional strength, on the model recommended by Outram.

It may be that the two major influences on the spread and design of iron rails in the Swansea Valley, if not in a greater part of south Wales, were indigenous. The appearance of iron track as an aspect of the evolution of experimental or 'hybrid' railways did not occur until about the 1790s or 1800s on Tyneside, but probably dates back to 1776 in the Swansea Valley. Outram's references to earlier uses of such track (and other contemporary schemes known as iron 'tramroads' throughout south Wales in the early 1790s) suggest that the idea had already spread over a considerable area in this region. Hybrid railways as a group were built using both flanged plate-rails and edge-rails. It was only George Stephenson's opposition to the strong body of opinion advocating plate-rails that prevented the Stockton and Darlington Railway[80] (and so the modern railway) from using that sort of rail. Contemporaries viewed each type of iron track as alternatives and flanged iron rails were not seen as a dead-end aberration that some judgements have claimed.

Early iron edge-rails

It is possible that one of the earliest examples of an all-iron edge-rail appeared in the Swansea area. The first use of iron as a track material seems to have been as a more durable replacement for the upper part of the double wooden rails long used in the Coalbrookdale area of Shropshire. This type of composite rail is first recorded as having been used in 1767.[81] These cast-iron upper rails were 6ft in length, 3.75in broad and 1.25in thick. Three lugs, about 3in by 3.75in, connected the upper rails to the lower rails and to the timber cross-sleepers, which must have been positioned at distances of 3ft. The cast-iron upper rails weighed about 102.5lb or 51.25lb per 3ft. The new railway feeders built during the 'canal mania' of the 1790s, especially those constructed in the largest iron-making areas, offered an opportunity for the innovatory leap to the use of all iron rails. It is generally accepted that the first use of all iron rails was at Dowlais in 1791. William Taitt, a partner in the works wrote that:[82]

'We are now making Rails for our own Waggon Way, which weigh 44 li or 45 li per yard. The Rails are 6 ft long, 3 pin holes in them, mitred at the ends, 3 inches broad at Bottom, 2.5 Inches Top, and near 2 Inches thick.'

It is possible that John Morris's 'Cast-iron Tram Plates' of 1776 were really edge-rails, and that his claim that iron was 'never before us'd for such purpose' simply indicates his ignorance of contemporary practice at Coalbrookdale, where the earlier double rail, made entirely of wood, was in the process of being superseded by the combination rail comprising a wooden under-rail and an iron upper-rail. Indeed, it is possible that by 1776 rails made simply of iron may have been in general use.

Evidence also survives of an independent use of iron rails in west Swansea by 1787 and if John Morris's innovation was one of the first uses of all iron edge-rails, it is likely that this was copying him, which much also strengthen the case for the likelihood that Morris's plate was an edge-rail, rather than the early type of flanged plate-rail found in his collieries. Some 1,000yds south-west of Morris's Landore Pit were the mines of his neighbour, Griffith Price (1718-87) of Penlle'r-gaer House. Price was one of the last local landowners to work his own coal, following the practice of his ancestors back to the seventeenth century.[83] He was prepared to make some investment in modern practices, though in many ways his collieries (which were at the head of Cwm Burlais, under what is now Manselton) were operated on traditional lines. This is brought out in the inventory of his possessions made by two assessors on 20 August 1787, three weeks after his death.[84] Fifteen horses were listed among Price's colliery assets, along with 16 sets of harnesses for the use of horses operating the winding apparatus ('whims') at Carrpound, Fountain and Hill Pits and the 14 coal waggons (including five 'large' waggons) which took the colliery output down the 'coal road' in Cwm Burlais to the shipping wharves on the River Tawe. Alongside these traditional aspects of a colliery that had been in production since at least 1745,[85] there is evidence of more modern practice. Gruffydd Price's father, Thomas Price, c.1687-1763 (High Sheriff of Glamorgan in 1739), was operating the family colliery in the Penllanmarch/Pwll-y-domen area at the head of Cwm Burlais by 1745.[86] By the time the colliery had passed to Thomas's son, Gruffydd, it had reached such a depth as to be 'unworkable without a fire or steam-engine' and an atmospheric-engine was erected. Nineteen years later, in 1787, the 'Fire Engine and Materials' were worth £1,000 and the engine may have reached the end of its effective working life as the 'New Machine and Materials' alongside were given a valuation three times greater. Among the items listed were 'Iron on Waggon Ways - ab^t. 25 Ct. at 18; worth £22 10s. Also above ground at the Fire Engine Pit were '?H^d. of Deals together with Rails and other stuff' worth £20'. The extent of any surface waggonways is unclear, but underground there were extensive tracked ways to carry the 'six small waggons under ground' which received the 45 coal 'basketts' from the side-passages on five sledges. The underground track was itemised as follows:[87]

Hill Pitt	340 yds of Iron Planks at 40 lb. per yy @ 8 per Ct	£48 11s.5d.
Hill Pitt	208 yds of Wooden do. at 4d p. yard	£3 9s.4d.
Do. in the 3ft. (?)	304 yds. of Do. @ 4d p.yard	£5 1s.4d.
New Pit	60 yds of Do. @ 4d p.yard	£1 0s.0d.
Ca Pound Pit	347 yds. of Do. @ 4d p.yard	£5 15s.8d.

There is no correlation between the total lengths of iron and wooden planks, making it most unlikely that rails were made up of a lower wooden plank with a cast-iron 'plank' on top. Single barrowing planks are not intended because no wheelbarrows are itemised in the inventory of the mine. The weight (40lb) of these iron rails, which were apparently 3ft in length, is comparable to the weight of the 1776 rails at Landore (42lb) 6ft-long rails used at Dowlais in 1791 (45lb).

In January 1788 John Morris I took over the running of the mine on behalf of Lockwood, Morris & Co.. By 1787 the annual production of the colliery was averaging 10,457 tons per annum. The extension of the existing railway system was one way by which Morris increased the annual average production over the next seven years to 16,394 tons per annum.[88] The costs of improvements made between January 1788 and November 1795 included £1,583 17s. 1d. spent on waggonways and a wharf. Price's existing waggonway was extended down the Cwm Burlais ravine, facilitated by a lease granted in February 1790,[89] to a wharf on the River Tawe, granted to Morris in 1793.[90] Morris's improvements were centred on the sharp hill-scarp to the north-west of the existing colliery, where Price's Hill and New Pits were situated. Two of the pits were sunk to a much lower level and Morris and Lockwood were forced 'to make two new tunnels or stone drifts for the

Fig. 136. The open ground is part of the western side of the Swansea Valley, at Manselton, which at the end of the eighteenth century had the large mining tunnel of Pentre-gethin Colliery at its base (to the left of the institutional building). (935055/65)

fig.136

95

Key for Figure 137

1. Landore Copperworks
2. Stamping and Rolling Mill
3. Coal Waggon way from Penyvilla & Treboth
4. Landore Corn Mill
5. D°. Engines and Pits
6. Coal wharfs
7. Canal
8. Waggon way from Place ŷ Marl to Landore
9. Mouth of Place ŷ Marl level
10. Place ŷ Marl Coal pits
11. D°. Engine
12. Old Engine House
13. Timber yard
14. Engine pond
15. Waggon way from Coal pits to Canal
16. Coal pit
17. Mouth of Graig level
18. Messⁿ. Fenton &Co⁵ Copperworks
19. The Birmingham Company's Copperworks
20. Forest Copperworks
21. Coal pit
22. D°. And fire Engine
23. Mouth of Subterraneous Canal (to left of '24')
24. Ponds
25. Air pits
26. Fire Engine and pit
27. Pond
28. Morriston
29. Brick yard
30. Loom pits
31. Water Courses
32. Old Clay pits
33. Old Coal levels
34. Coal roads
35. Proposed Canal
36. River Tawe
37. Forest Copper mills
38. Weir
39. Bridge
40. Trewyddfa Forest [i.e. Common]

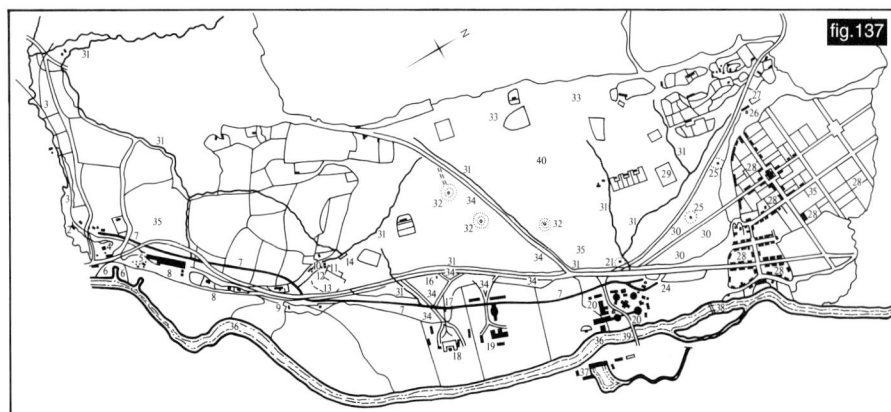

purpose of getting at the remaining coal on or under the said lands.'[91]

Traces of the approach cutting to the main level are still visible on the open hillside to the the north-west of St. John's Road at Manselton. The upper part of Brondeg Road follows the formation of the railway which originally emerged from the level. The enlarged colliery became known as the Pentre(-gethin) Colliery. In 1791 as already noted, a tourist, E.D. Clarke, visited the newly functioning system:[92]

> 'The entrance is vaulted, and perfectly level ... coal in baskets [is sent up] ... to the vaulted level ... [and] there put into carts, with friction wheels, and drawn by oxen to the mouth of the mine ... the whole of this mine ... [is] ... three miles to walk in these gloomy abodes.'

Another traveller (Revd. J. Evans), the account of whose tour was published in 1804, noted that 'One of these audits, which we traced about a mile in length, admits low waggons, holding a chaldron each, which running on an iron railway, one horse with ease delivers at the quay.'

John Morris II was ordering flanged edge-railway wheels in 1810 which would not have been for the Landore Colliery plateways and may have been for the system of lines routed through Cwm Burlais.[93] By 1805 observers from the Cumbrian coalfield were able to note:[94]

> 'Iron Roads are universally used at Swansea, both above and below Ground. - These are of Two Sorts.-
>
> 1st. The rail road where the Wheels are Flanged and where the waggon carrys about 2 Tons.
>
> 2d. The Tramroad where the plates are Flanged, and where a waggon carrys about Half a Ton.'

So once iron had been introduced as a revolutionary material for one type of trackwork by a single concern it seem quite likely that it spread throughout what was a landscape of intense activity and was adapted by each owner to suit the vagaries both of his own way of working and of the type of railway that had evolved in his mines.

Canals

In the lower Swansea Valley area one, possibly even three, underground canals were built: 11 local canals were planned, eight built and two arterial canals entered the area. Several of these waterways also provided a water-supply, water-power and a coalmine drainage function, and it is this combination of uses that makes it impossible to divorce the study of navigable watercourses from that of the wider landscape. The local canals, with feeder railways, did not have a significant effect on the development of the landscape until the period 1784-93. Large-scale canal building did not start until 1794.

The tidal River Tawe must always have been used for water-transport and the first recorded improved and artificial navigations in the Swansea area were constructed in 1699-1701. A tidal 'pill' or inlet was improved to give access to the shipping quay of the Melincryddan Copperworks near Neath, and at its upper end this had a long dock-canal with high-tide levels accessible by an upper gate.

In the nine years from the building of the first local canal at Swansea itself in 1784, some five and a half miles of railway were built. Both the Lockwood, Morris & Co. Canal (*c.* 1787-91) on the west side of the valley and Smith's (or Llansamlet Canal) (1784) on the east side, were short private canals carrying coal for delivery to the rear of the riverside smelters and to shipping on the river (and, incidentally, supplying water to the works). The latter canal was built in order to redeem the fortunes of Townsend's successors as part of a transport and mining system suitable to the area. Five largely underground railways replaced the previous expensive and extensive surface system. On the west side of the valley, Morris's canal penetrated a mile underground and was also fed by a canal-side railway tunnel. Lockwood, Morris & Co., using funds generated by over 60 years spent in the copper-smelting trade, took over the old Pentre Colliery to the north-west of Swansea town. Local topography dictated that longer lengths of surface railway should replace an intended canal to carry coal to the port of Swansea for shipment, while a second railway led from the colliery to the southern end of the Lockwood, Morris & Co. Canal.

The radical change in scale in the Swansea area came with the construction of the 17-mile long Swansea Canal in 1794-98. Prior to the building of the trunk canal up the valley, the lower Swansea Valley already had two short canals and the river

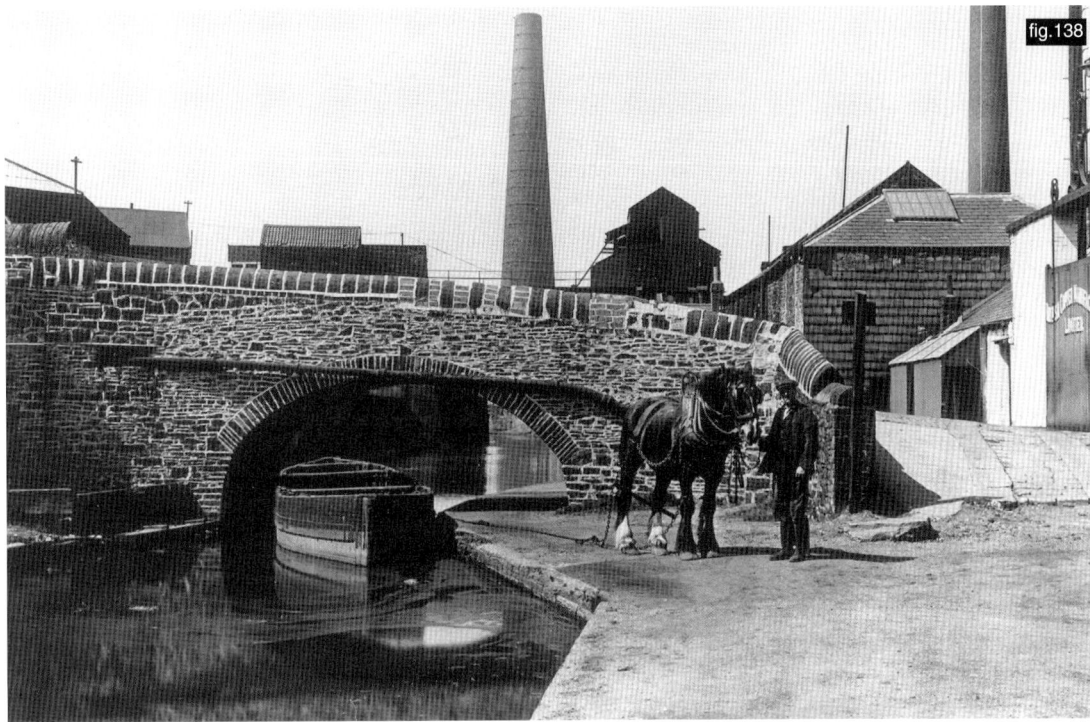

Fig. 138. The last working Swansea Canal Boat with horse and haulier at the main entrance to the Hafod Copperworks in 1931; note the pointed copper-slag blocks capping the bridge parapet. The large stack of the Morfa Silverworks dominates the background. (City & County of Swansea: Swansea Museum Collection)

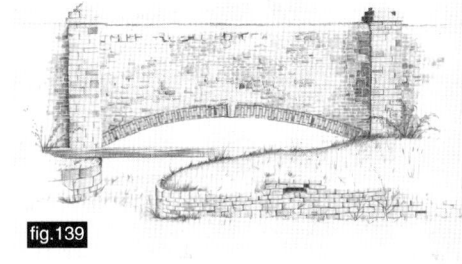

Fig. 139. Northern Swansea Canal dock entrance at the Morfa Copperworks. The remains of the timbers of the towing-path bridge can be seen in the foreground and the gable of the remaining former Morfa Copperworks rolling-mill was beyond the arch.

navigation. One of these canals, built by Lockwood, Morris & Co., was enlarged and incorporated into the Swansea Canal. Its local transport function continued as part of the main artery; it was retained under separate control, renamed the Trewyddfa Canal, and given a separate outlet to shipping via an incline at Landore. As previously noted, short-haul traffic on the main line of the Swansea Canal north of this area was virtually unknown. However, local traffic was common on the Trewyddfa Canal and later on in the extension of the copper-smelting zone alongside the Swansea Canal to the south of Landore. Any truly anthracitic coal from the upper valley, or semi-anthracitic coal from the middle valley, would have reached the non-ferrous smelters via the Swansea and Trewyddfa Canals. Unfortunately, the traffic returns from these canals are extremely fragmentary. The only surviving records for the Swansea Canal are for the upper third of the canal north of Ystalyfera. They survive for a period including most of the second and early third decades of the nineteenth century. The part of the canal which they cover coincides exactly with the zone where truly anthracitic coal was available. In all, 2,126 boat-loads, or 46,770 tons, of anthracitic coal were carried exclusively for shipping at Swansea over a 12-month period in 1812-13; by 1821-2 that figure had risen to 2,900 boat-loads, or 63,790 tons, of anthracite. None of that coal was used for the smelting of non-ferrous metals.

Fig. 140. Swansea Canal passing the south end of the Hafod Copperworks. The arch in the works wall on the right allowed boats to enter a basin supplying coal and water to the rolling-mills beyond. Wall, arch and mill engine houses still survive. (Francis, 'The Smelting of Copper in the Swansea District', 1881)

The Trewyddfa Canal returns survive for four years at the end of the second decade and the beginning of the third decade of the nineteenth century. The points of origin and the destinations of the great majority of this trade did not lie within the 1.4 miles of the Trewyddfa Canal itself and are not given. However, it is possible to determine with some accuracy the origin and type of coal carried.

By 1832, then, it appears that the tonnage of anthracite carried along the canal had reached 98,180 tons and none of this was unloaded at any of the four copper-smelting works along the route of the canal. On the other hand, some 83,280 tons of semi-anthracitic coal were carried along the canal and rather less than a quarter of this may have been used in the Forest Works. Local trade in fact continued until the demise of the canal in the 1930s and far outlasted the long-distance coal traffic on the canal which had virtually ceased by the 1890s. These differing patterns of transport reflected the very different properties of coal available in the old mining area around the non-ferrous metals' smelting works and made available via the relatively cheap transport of the new trunk canal.

Fig. 141. The Swansea Canal passing between the middle tips (left) and the works buildings at Hafod Copperworks (1860s). The tall tower included a wagon lift for lowering loaded trucks crossing the high canal bridge from the standard-gauge locomotive railway to the left. ('Le Tour du Monde') (Reproduced by permission of Birmingham Library Services)

The Use of Canals Underground

It is generally accepted that the first use of canals for transport underground in mines was in 1759 at Worsley, at the original terminus of the Duke of Bridgwater's Canal near Manchester. However, Charles Hadfield noted in 1960 that one underground canal pre-dated Worsley: that at Gwauncaegurwen on the north-west edge of the Tawe river basin. It has since been possible to locate four other canal levels in the area centred on Swansea. Two possible examples have also been located in this area. Sources suggest that two of these local examples, and a possible third canal made by Swansea-based entrepreneurs at Rhandir-mwyn, near Llandovery, predate the more famous Worsley levels. Canal transport in mines apread throughout Britain and was then copied in Europe. The possible international importance, therefore, of the Swansea mining canals requires detailed study.

Clyn-du mining canal

The Clyn-du mining canal was situated[95] to the north of Swansea and was driven into the western side of the Swansea Valley between Landore and Morriston. It may have been the first use made of a canal for mine transport.[96] In the Swansea valley context, it is important in that it determined the level, and so the course, that was subsequently chosen for the canal of Lockwood, Morris & Co., and the Trewyddfa and Swansea Canals.

No documentary or cartographic sources have been located to confirm the early date for the canal as given in nineteenth-century published sources. This mining canal was one of three levels (or mining tunnels), all called 'the Clyn-du Level', and it is probably the middle in date of these. What suggests that early date is that this and two other mining canals in south Wales,[97] allegedly nearly contemporary, are dated in local secondary sources, all of which are independent of one another. All three sources seem to be unaware of the oft-repeated claims of Worsley[98] or indeed of the use of underground canals anywhere else.

The Worsley system began in 1759 and Clyn-du allegedly started in 1747, Gwauncaegurwen in 1757 and Rhandir-mwyn, near Llandovery, in 1755-7. The existence in south Wales of at least nine similar underground, or mining canals is known,[99] including a planned example at the head of the Swansea Canal [100] and another example at Rhydydefaid Colliery, just to the west of Swansea.[101] The was latter built by the grandson of the constructor of Clyn-du. Such examples were sufficiently common in south Wales by 1806 to be named as a type of mining level that was in everyday use in the area. The knowledgeable mining engineer Edward Martin saw them as particularly suitable for the local terrain.[102]

The date of construction of the Clyn-du Level is taken from a passage in the second edition of George Grant Francis' book on the Swansea copper industry of 1881, first printed in 1867:

> '... Mr. W. Edmond was good enough to hand me the following memorandum, given him by a friend some years since, at Clase [i.e. the area where the level was built], but whence the authority for the dates was obtained by that gentleman he was unable to say:
> 'Copper Works were erected at Bank-y-Gockus, at Swansea, in 1719; were removed soon afterwards to Landore; they passed into the hands of Lockwood, Morris & Co., in 1727, and later in the same year were removed up to the then New works at 'Forest'. In 1747 an assay office was built there, and is still in existence, as well as an underground canal (also still open) through which the coal was brought for use into the Works'.[103]

The evidence in the above passage has an element of truth and is fairly accurate when it can be cross-checked. Dr Lane intended to open his copperworks at 'Bank-y-Gockus' in 1717 but was persuaded to build instead on a site at Landore. The Cambrian Copperworks was built on 'Bank-y-Gockus' in 1719, opening in 1720.[104] Lockwood, Morris & Co. did buy the Llangyfelach Copperworks at Landore (formerly Dr Lane's) in 1727,[105] but they leased the land on which to build their new smelter in the early 1740s.[106] The new smelter at Forest was *not* in production until after their old works at Landore closed down in 1749.[107]

Some circumstantial evidence also points to a date of around 1747-8 for the construction of Clyndu Level. Construction of the Forest Copperworks began at this date, and it would seem reasonable to suppose that, as part of the project, a coal-level was driven into the hillside, as was done at the other four smelting works established between 1737 and 1781.[108] It is true that none of these other levels are known to have been boat-levels, but in view of the pool of talented engineers whom Lockwood and Morris employed, there is nothing implausible in Clyndu Level having been built as such. This team, which included William Edwards, the bridge-builder, was probably responsible for another innovation in mine transport, the cast-iron plateway (see above) and they were responsible for their own design of steam and hydraulic engines; they made early use of hydraulic lime; they built multi-storey workers' flats and large-span bridges and they developed new types of reverberatory furnaces.

However, there are reasons to question the construction of Clyn-du Level at a date prior to 1761. A large-scale map of the Landore-Morriston area produced in that year[109] shows many collieries in the area, including 'old' and presumably disused levels at Morriston. It does not show the mouth of the coal level on the works site or any of the ventilation shafts down to it, although it does show the pumping shaft at Clyn-du which went down to the tunnel. The omission of the two intermediate ventilation shafts is not conclusive, for other known feaures are missing: for instance, the three 'old clay pits' ('32' on the 1793 map of Graig Trewyddfa) are not shown on the 1761 map even though the adjacent brick-kiln and yard actually shown on the 1761 map ('29' on the 1793 map) would have required them. Similarly, the position of the Clyn-du water-engine pit is not indicated on the plan, and its existence is only vaguely hinted at by a letter 'A' placed off-site. There is little doubt that the purpose of the map, like those of nearly all maps produced at the time, was to assist in the solution of a legal problem, in this case to decide the boundaries between the rights of Thomas Popkin and

those of Lockwood, Morris & Co.. The ventilation of Morris's original Clyn-du Level (a tunnel at a much higher elevation than the canal tunnel) had been stopped off and several of his miners had suffocated, and then Popkin had built a wall across a nearby brook and flooded the entrance to the level. The compiler of such a map, intended to resolve this dispute, may have regarded the new and deeper Clyn-du Level as an irrelevance. The higher Clyn-du Level was then in production and a cart-road to convey their output down the slope of the Graig to the works is shown; that route would not have been made unnecessary by the existence of the lower level or canal.

Another point for consideration is the rate at which the canal-level was driven across the Swansea Five-Foot seam. The evidence for this shows that a length of only 240yds across the seam had been worked by 1796, and a further 1,400yds had been worked out by the early nineteenth century. This is understandable, since the original 240yds needed to feed coal to only one copperworks, and that with alternative sources of supply. In the 1770s, Copper Pit was actually sunk on the smelter site itself in order to have direct access to the Swansea Six -Foot seam below. After 1800, by contrast, the level was feeding coal to four smelters along the Trewyddfa Canal: the owners of the level no longer had their primary interest in copper smelting but instead were concerned to maximise their coal output. Even if the date of 1747 in Grant Francis' document is disregarded, the Clyn-du boat-level can still be assigned to the 1760s or 1770s.

Another source does support the likelihood that Morris's canal was not built until the later 1770s by John Morris I. In May 1774 John's brother Robert walked five miles, from Manchester, along James Brindley and John Gilbert's pioneering Bridgewater Canal and, in a detailed diary entry, made no mention of any such feature already existing at his father's and brother's mines:[110]

> 'At Worsley is ye entrance into the hill from wch ye chief quantity of Coal is got: this I entered along ye continuation of ye canal in a long narrow boat; and here ye work appears to be a most useful undertaking… One is however rather hurt to see the entrances into ye hill made only through two very small apperbures; through which a man cannot I believe stand upright in passing. For utility, to be sure, nothing higher is necessary.'

We are on firmer ground in other respects. Graig Trewyddfa, the hill under which the Clyn-du Level was constructed, was notorious for its drainage problems and the partly drowned level may have made the introduction of boats inevitable.[111] The level intercepted the old water-engine shaft at Clyn-du, and the Clyn-du steam-engine shaft ('26' on the

Fig. 142. Transcript of a recollection of the Clyn-du Underground Canal in 1846. (The National Library of Wales)

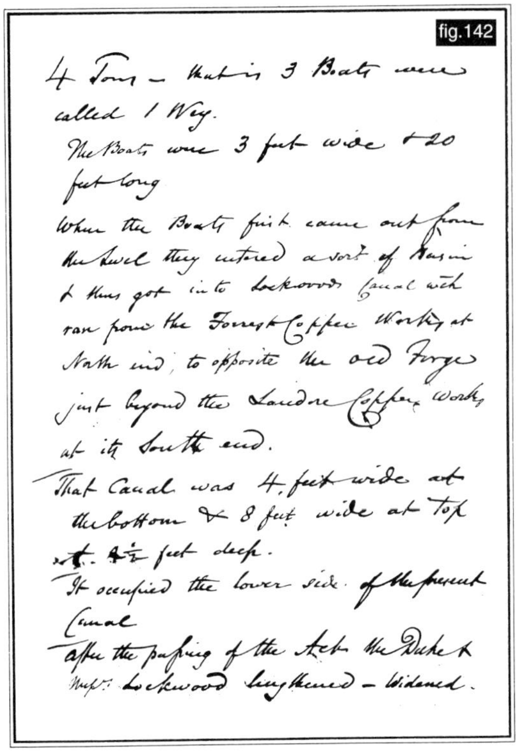

1793 map) was sunk on its line. Both probably pumped water from below the Swansea Five-Foot Seam, perhaps as deep as the Swansea Six-Foot Seam. This water would most likely have been diverted to flow along the Clyn-du Level, probably to augment the supply of water needed to drive ore-stamps or a clay-mill at the Forest Copperworks. The area of working to the canal level is shown on at least one old map. That the level predated the construction, between 1778 and 1791, of Lockwood, Morris & Co.'s surface canal is shown by the following statement made by a 73-year old man in 1846, who had worked for Lockwood, Morris & Co. before the passing of the Canal Act: [112]

'the [Lockwood, Morris & Co.] Canal ... was a continuation of the Underground Level called the Clundee Level ...'.

His description of the level continued:

' ... the Clyndee Level ... extended abt. 1100 yards. under the Mountain to the face of the Coal in the 5 feet or Penyvillia Vein. That level was 4ft. 7?inches wide and 8 feet high arched with Stone part of the way in - it still exists as a Water Way from the Coal Works as perfect as ever so far as it is arched, but further in, it is [sic] Crept together in places so as not to be of the full size.

Boats went up that level to the face of the Coal where the Waggons were tipped into them - each boat held 2 Waggons full - or 4 tons - that is 3 Boats were called 1 Wey.

The Boats were 3 feet wide & 20 feet long. When the Boats first came out from the Level they entered a sort of Basin [the original terminus near 'the eaves of the copperworks'? & thus got into Lockwoods Canal wich [sic] ran from the Forrest [sic] Copper Works at North end, to opposite the Old Forge just beyond the Landore Copper Works at its South end'.

The Swedish observer, Eric Svedenstierna, was shown round the Swansea industrial area by John Morris II in 1802-3, [113] and was given 'all the necessary information' by him and by 'a Mr Lockwood, who owned coalmines and copper works ... we also had unrestricted admission to the latter's works, and permission to inquire about the methods of working.

... the Swansea canal ... boats, one of which often only carries 2 to 3 tons [a loading of three and not four tons would have given the *correct* total loading for three boats of '1 Wey' as given in the earlier description] either go over branch canals constructed for the purpose, directly to some landing place [an allusion to the Trewyddfa Canal Incline?] or the coal is again transferred into the above-described waggons. Occasionally also these little boats go a whole English mile or more underground, and take on their loads in the mine itself.

[Other small canals] are conducted to some copper works or to the places on the River Tavey [Tawe] where there are larger or smaller collecting places, where several boats at once can be unloaded comfortably. At one of the copper works [almost certainly the Forest Works] such a collecting place or basin was laid out at the same level as the eaves of the roof of the building, so that the boats, which went to the coalmine a Swedish mile underground on one of the small canals, could on their return empty their loads directly into the smelting house'.

The exact form and length of the level can be plotted from old mining maps,[114] together with the total area of coal which was eventuallly worked from the canal level and that worked by 1796. The total length of the canal underground was about one mile. The approach tunnel to the dip of the Swansea Five-Foot seam was a straight bore heading north-west for 1,170yds from the Forest Works to the probably pre-existing Clyn-du Water Machine Pit. Two air-pits and the Clyn-du steam-engine shaft were placed at intervals on the initial half of this tunnel. [115]

Beyond the water-engine shaft, the canal curved fairly sharply to the west along the dip of the Swansea Five-Foot seam as it dropped steeply almost from the crest of Graig Trewyddfa. By 1796 workings had proceeded 260yds beyond the shaft and along the width of the seam, the tunnel already ran westwards some 820yds across the seam. The main passage continued another 660yds as a dry heading until halted by a fault almost under Llangyfelach Road.

The underground canal was still much in use in 1814, as is shown by the mention of mine boats at the basin of the Birmingham Works. By 1846, however, it is obvious that the canal had been out of use a long time and was then used only to rid the Clyn-du Collieries of water. [116] In about 1925 it was said that 'The opening of the old canal is to be seen at present on the railway near the Copper Pit siding'. [117] In the later 1970s the disused cutting of the Swansea to Morriston Railway was investigated and only a man-hole cover on the western crest of the cutting at the approximate site of the level could be located. Welsh Water record this as a surface water drain. The railway cutting has now been filled in and the road roundabout at the south end of Morriston covers the site of this historic canal.

Graig Level

A contemporary source speaks of 'underground canals' in this area. A further example may have been the Graig Level, which was located alongside Morris's Canal to the south of the Clyn-du Level, with its entrance set in the scarp above the Rose Copperworks.[118] It is marked as 17 on the map illustrated of Graig Trewyddfa in 1793. A line about 30yds in length is shown on that plan, which may have been either a surface waggonway or a branch off the canal which led to the mouth of the level.[119]

Svedenstierna may have been referring to this feature in the following passage (1802-3):

'*The Canals and aqueducts*. Besides the often-mentioned, five-feet deep and ten-feet-wide Swansea Canal, which stretches a few English miles up into the country beside the River Tavey [Tawe], there are several small canals and aqueducts, which partly carry the water to the surrounding works, and partly are laid out for smaller boats. Among the latter *some* come out of the mines, and either cross the large canals on aqueducts, or fall into them', and '*Occasionally* also these little boats go a whole English mile or more underground, and take on their loads in the mine itself'.

Svedenstierna's guide was John Morris II and most of his time would have been spent in the Landore and Morriston area. The only reason for suggesting that Graig Level may have been a mining canal is that he referred to such 'canals' in the plural but was almost certainly confused by the plethora of leats in the area. He is known not to be an absolutely reliable source[120] and hurried observation of the underground Clyn-du Canal and the adjoining aqueduct, which passed over the Trewyddfa Canal and supplied water to the Forest Copperworks, are probably the basis of his comments. Several maps distinguish between the 'Mouth of subterraneous Canal [at Clyn-du]' and the 'Mouth of Graig Level', as does the map illustrated of Graig Trewyddfa. The level mouth was probably buried beneath the G.W.R. Morriston line in 1881 and lies under the western edge of the new Plasmarl bypass.

Rhyd-y-Defaid Mining Canal

The existence of a third possible mining canal in the Swansea area is indicated by the remains of its approach cutting. This mining canal is in the Clyne Valley to the west of Swansea, a valley developed initially to supply coal to the Swansea Valley copper-smelters. A mile from the sea is a water-filled cutting at the Rhydydefaid Colliery site. The cutting is some 12ft deep to the present surface level of the water standing in the excavation at its deepest, or northern, end. Water trickling out of the northern end of the cutting indicates that this is a collapsed level or adit mouth. The water and silt in the cutting are at least 3ft deep, with no indication of the existence of any railway track, such as a raised platform along one side of the floor of the cutting.

The waterway swings northwards from the mouth of the cutting to a course alongside the formation of the upper section of the Clyne branch of the Oystermouth plateway which was built by Sir John Morris II (grandson of the probable builder of the Clyn-du Canal level) in 1840. The railway formation is perhaps 6ft above that of the water level and there is no indication of any branch having descended into the cutting. A mining plan of 1920, which shows the surface layout of the Rhydydefaid Colliery at this date, labels the wide waterway in its cutting simply as a 'canal'.[121]

Most mining canals were built in the later eighteenth century, yet the Rhydydefaid Colliery canal would not have been a late oddity if the date suggested above is correct. A few such canals were newly built in the nineteenth century, for example: the construction of a navigation level was considered at Moat Farm near Kidwelly in 1813-15;[122] Mosley Common Level, part of the Worsley complex was built in 1822;[123] and the Magpie Sough in Derbyshire was constructed between 1873 and 1881.[124]

Other Mining Canals

There were plans to build a mining canal at the top of the Swansea Canal. Daniel Harper, William Shaxby and the Swansea Canal engineer, Thomas Sheasby, came from the trades or professional classes of the small town of Tamworth in Staffordshire.[125] It is likely that Sheasby alerted his

family friends at home to the new opportunities for investment in the Swansea Valley.[126] Harper was sufficiently established in the area by 1801 to build a house and to drive 'Lefel Fawr' or 'Tymawr Level' nearby.[127] In order to give access to the level, which was on the south bank of the river, a timber bridge seems to have been built.

It is likely that the Clyn-du underground canal, in the lower reaches, gave Sheasby the idea of urging Harper to construct a similar undertaking at Abercrâf. The canal feeder weir on the River Tawe was built by Sheasby to supply a height equal to that of the level floor, even though the canal basin itself was eventually constructed at a lower altitude, with a difference equal to the depth of two deep locks. The level was driven for a mile using slaked lime[128] and was probably opened for production in 1805.[129] The lowering in level of the uppermost pound of the canal, as built, reduced it to a height well below that of the level and made a canal arm impracticable.

Economic aspects of the mining canal

Smaller mining canal schemes backed by minimal capital might have boats carrying as little as 2cwt (Tir-y-lluest Coal Level, Glamorgan, 1793) and tunnel widths of 4ft 5in. Worsley boats carried from five tons to 12 tons.

Could a small canal mining level carry economical payloads? A still-water canal-boat could carry at least 8.25 times the load of a waggon on the wooden rails, usual underground in the eighteenth century, with the expenditure of the same amount of effort. With the current in a drainage-cum-navigation mining level this useful ratio would be increased. How much advantage was taken of this in the smaller levels can be examined with reference to some of John Morris's levels in the Landore-Morriston area. The boats on the Clyn-du level carried a load of four tons through a level 4ft 8in wide by 8ft high, arched only near its entrance. The main haulage way in the adjacent Landore Colliery in 1826 had waggons 6ft long x 2ft wide with a 2ft high container carrying a load of 18cwt. Therefore, the rail haulage way needed to be only marginally smaller than the canal level, and the equivalent of a load on the haulage way requiring the effort of four and a half horses could probably be controlled by one man on a canal level. Waggons used at Clyn-du coal-face held two tons - the larger capacity may indicate that they worked in headings intended for canal expansion. In any case, the cost of constructing a rail level for such waggons would certainly approximate to that of a canal level, but the labour of two haulage horses would be required instead of one man assisted by the water flow.

An ideal solution to overcome high transhipment costs was to have the destination for the mineral positioned at the level mouth. This was the arrangement at Clyn-du where the Forest Copperworks was sited by the level entrance.

Power Systems.

Systems of water-power and supply

The prime factors determining the location of industry in the lower Swansea Valley were not to do with the provision of power. The coal-mining industry was here because of the presence both of visible outcrops of coal and underlying seams. The coal trade grew because export was possible from the navigable mouth of the river. The copper trade was, in its turn, attracted here both because this was the nearest coalfield accessible by sea to the copper-mining fields and because of the prodigious amounts of coal needed to smelt and refine copper-ore. The power required for these, and other, industrial processes had to be provided by the engineering of substantial watercourses and storage ponds in a partly urban landscape where all the local watercourses were already partly in use to drive cornmills.

The Cornmills

Every area, predominantly agricultural or industrial, urban or rural, required mills to grind wheat to produce flour for the local population. By 1786 there were 27 cornmills in the lower Swansea Valley area.[130] Cornmills supplying the Medieval town of Swansea were powered by the Burlais Brook that formed the northern boundary of the borough. A further two mills lay on the Brynmill stream that formed the western Borough boundary. On the crest of Kilvey Hill, to the east of the river, was a tower windmill with a second located on the Cefn-hengoed ridge extending towards Llansamlet on its north.

In the lower Swansea Valley there were also other mills serving the scattered rural population. Many of these manorial mills may have had Medieval foundations. The second side valley north from Swansea: the Nant Rhyd-y-Filais at Landore had two earlier cornmills on it, one of these fell out of use with the growing industrial use of the stream and the supply to the second had to be protected by litigation. Another cornmill on the east of the river at the north-western edge of Kilvey Hill was already in ruins when its water-supply was leased out to the new proprietors of the Whiterock Copperworks.

The new industrial infrastructure of the seventeenth, eighteenth and nineteenth-century affected the corn-milling provision in other ways. The large water-power installation at Upper Forest (near what is now Morriston) included a cornmill driven by massive water-power provision from the main River Tawe for a large forge. This powerful new installation of the local landowner/entrepreneur, Thomas Popkins, drew trade away from the manorial (Landore) mill of the absentee lord of the manor on the west side of the lower valley, the Duke of Beaufort. At this period in the early eighteenth-century his agent instigated inconclusive litigation against Popkins for drawing the trade of the growing industrial community away from him. This rivalry between the traditional magnates and the industrialists may also explain the density of mill watercourses whose remains form a prominent part of the fossilised landscape of the Clyne Valley which opens onto Swansea Bay in what are now the western suburbs of the city.

The Morris family, the most prominent of the contemporary coppermasters, acquired the manorial

Fig. 145. The dry course of the Clyne Wood Canal is visible on the left as its earthworks run through the western side of the Clyne Valley Country Park. The pool is caused by a collapsed culvert on the adjacent railway embankment. (812757)

rights of the east side of the Clyne Valley and in 1800-01 their 'New Mill' stood alongside a new 1,980yds watercourse adjacent to their home farm. At its southern end the watercourse also drove the machinery of the Ynys Colliery. The watercourse ran just above the 75ft contour with a minimum width of 8ft 2in. The Dukes of Beaufort were the long established proprietors of the opposite, west, side of the valley and had their manorial mill at Blackpill, near the mouth of the valley to the south. It may have been an awareness of the previous unequal competition for trade between the Duke's old Landore Mill and the new Upper Forest Mill built by Popkins on a large well-engineered and ample watercourse that prompted the Duke's agent to re-equip his Blackpill Mill with a very substantial new watercourse at the same time as Morris's New Mill was constructed. This mile long 'Clyne Wood Canal' ran just below the 50ft contour for the whole of its length and had a minimum width of 6ft. Its construction would have facilitated an increased fall at the Blackpill Mill water-wheel from a 3ft to 15ft and the provision of several sets of adequately driven mill-stones that would have allowed the miller to compete with Morris's New Mill.

Other cornmills underwent, or resulted from, a change of function in response to the competing water needs of the new industrial installations. The lowest of the older Cwm Burlais mills for example: Greenhill or Aberdyberthi Mill, became a mill grinding flints for the Cambrian pottery. Conversely, the Cambrian Copperworks rolling-mill on the Burlais Brook downstream became an iron forge and foundry and then was known as the 'Pottery Flour Mill' by the mid nineteenth-century: from 1796 this installation was driven by water from the Swansea Canal which had been constructed across its previous water-feed. In 1810 the engineer/ entrepreneur/landowner Edward Martin built an entirely new cornmill at Ynys-Tawe also driven by the waste water of the Swansea Canal.[131] In 1736 the Lord of the Manor, Bussy Mansel, 4th. Baron

Mansel, offered the use of his cornmill at Llansamlet ('New Mill') to the new proprietors of White Rock Copperworks if they should need it and could replace the mill-stones with a battery mill or trial hammer. Later in the eighteenth century the cornmill further north on the Lower Clydach River at Clydach was also replaced by the 'Forge Fach',or the Lower Iron Forge, to process iron from the upper valley furnace at Ynysgedwyn.

Generally cornmills did inevitably become firmly interlinked into the highly-developed water-economy of the industrial period. With the growth in size of the industrial population there was a tendency for the only successful mills to be large multi-stone installations adequately powered by large industrial size leats (which might also drive substantial industrial uses) which provided a considerable water-flow with a length of watercourse that ensured a large terminal drop capable of driving a substantial water-wheel. Some did also take advantage of the new industrial transport systems to distribute their flour to the growing urban and industrial population. Ynys-tawe and Pottery Mills were built into the bank of the Swansea Canal and the millpond of the latter may also have served as a loading dock. The lower Brynmill had its own branch from the Oystermouth Tramroad and indeed the miller there was one of the men who instigated the world's first passenger carrying service on the railway.

Early Eighteenth-century Copperworks Use of Water-power

In contrast to Swansea, there were other areas where the industries of the world's first Industrial Revolution were located primarily because of the water-power available. Among these were the cotton mills of Lancashire, the woollen mills of west Yorkshire or the metal-working and cutlery workshops of Sheffield. The latter is a particularly appropriate comparison for the Swansea area for the main river sites in the two areas were developed in the same period and there are also close parallels in the type of water-powered installation: among the diverse metals mills in the south Yorkshire town were those for rolling copper-plate for example. The key period for the development of the Sheffield mills was 1680-1780 and in particular the middle two quarters of the eighteenth century. On the 30 miles of the River Don and four other main streams and their tributaries in the area of the modern city of Sheffield upwards of 115 mill-sites existed. The average density of water-power sites on a watercourse was about four per mile and by 1780 all available water-power sites were in use with many watercourses serving several mills in succession.[132]

The water-power potential at Swansea was altogether different. The 12 main copper-smelting sites were dispersed along three miles of navigable river for reasons of transport and coal-supply and obviously could not easily use the tidal river there as a power source. The largest power requirement was for the beating of ingots into flatter sections and then rolling them into sheets and this was where the four and a half miles of river above the tidal limit to the north was utilised.

Even here, at a point north of what is now Morriston, a large density of mills could not be achieved because of the relatively low gradient of the river at this point: each mill-site had a relatively low head that could only be provided with the construction of long watercourses across two of the three large meanders on this part of the river. Both these installations with considerable watercourses cutting across large river-bends: at Ynys-pen-llwch (Clydach - leat 1,040yds long) and Upper Forest (Morriston - 1,310yds long), originated as seventeenth-century iron forges but both were adapted for use in processing copper: Ynys-pen-llwch became a copper rolling-mill from 1713 to 1731 and Upper Forest from 1779 to 1845. Both of these large main valley river installations, with two water-wheels each, were of further great significance in powering some of the earliest tinplate works in what became the world centre of the industry.

The third coppermill site on the River Tawe, at Lower Forest, was of a different degree of sophistication to the earlier two river mills that had been adapted from other uses. Built in the mid-eighteenth century (1732-35) by a specialist Bristol engineer with experience of the many large-river copper and brass mills built on the River Avon and its tributaries. This installation had a shorter feeder canal (490yds long) driving no less than five undershot water-wheels and water was discharged directly into the tidal river.

The copper-smelting industry itself was concentrated on integrated sites to the south where necessary water (for powering slag-stamps, test hammers, clay-mills and for filling granulation vessels) could either be provided from tributaries to the main River Tawe or by canal.

It was acceptable, if a little inconvenient, to have these large hammer and rolling-mills some distance up-river from the main copper-smelters. However the rolling-mills of the first two Swansea Copperworks used smaller volumes of water available from the side streams, with a larger vertical drop, to drive more efficient overshot, breast or undershot water-wheels. These were the mills of the Llangyfelach (1717-49) and Cambrian Copperworks

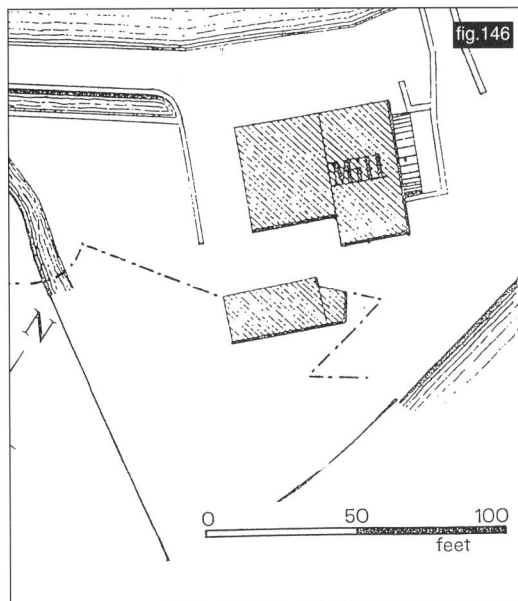

Fig. 146. In the lower four-mile pound of the Swansea Canal the only waste-water weir allowed was at Ynys Tawe, just upstream of the upper intake to the Forest Coppermills. In 1810, Edward Martin built the first of many new water-power installations on the canal at this point.

Fig. 147. A rolling-mill stand made for the Morfa Copperworks in 1831 of a type similar to those driven by water-power on each of the three large water-power sites on the lower River Tawe in the seventeenth, eighteenth and nineteenth centuries. (Held by the West Glamorgan Archives Service; Neath Abbey Ironworks Collection, D/D NAI M/571/62)

Fig. 148. One of the last water-wheels on the Ynyspenllwch Forge site, probably pictured in the 1940s. Today this area forms part of the eastern section of the Mond Nickel Works at Clydach. (University of Wales Swansea, Library & Information Services, Archives)

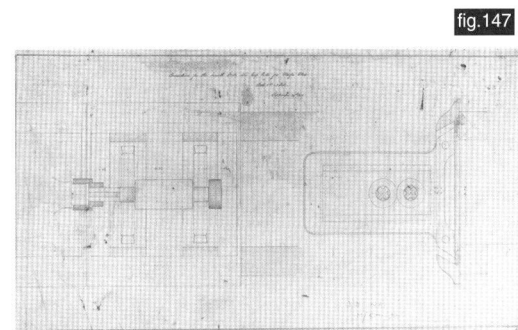

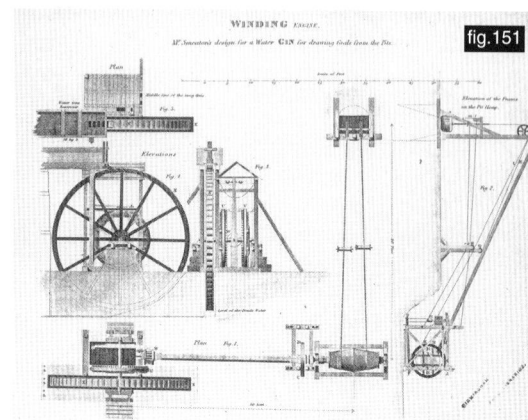

(1720-66): the former with a 550yds leat including a considerable stone-arched aqueduct giving it a high fall of water. This had a feed directly from the pre-existing Kergynidd Cornmill but by 1732 the proprietors were forced, by litigation from the Lord of the Manor (the absentee Duke of Beaufort), to return all the water used into the head pond of the manorial cornmill at Landore. By contrast, the third Swansea Copperworks at White Rock was founded very much with the encouragement of the local Lord of the Manor/landlord/coalmaster (the fourth Baron Mansel) who controlled the water-rights and pre-existing cornmills. So it was that the proprietors assumed direct control of the neighbouring, and

ruinous, 'Knaploth' Cornmill and were also offered the use of the tailwater of Melin-y-Frân Cornmill and if the Kilvey Hill water was insufficient for 'a Battery Mill or Trial Hammer for proving of copper' with half the costs of transport via a man and horse to and from Mansel's 'New Mill' which they were given permission to convert to a battery mill for these purposes.[133]

The Bristol proprietors of the White Rock Copperworks were a well-funded concern who rapidly developed the site and almost certainly constructed the elaborate leat system that served the works. The original feed to Knaploth Cornmill was from a small stream that flowed down the western face of Kilvey Hill. This was augmented by a 3,280yd long watercourse that went northwards, then east along the broad northern face of the hill and then south to a spring on the far south-eastern side where the earthworks of this considerable channel can still be seen. However, as with many of these large watercourses, more than one industry was intended to use the available power potential; in this case Swansea's original large industry of coal-mining.

Waterpower Use in Collieries

In Europe the most advanced practice in Medieval mining had been in the German-speaking areas of middle Europe where elaborate watercourses powered deep mining-shafts via the use of large water-wheels. Britain's largest early coalfield by far was the Great Northern Coalfield of Northumberland and Durham, which used water-wheel-winding ('coalmills') on a large scale by the seventeenth century. The pattern and intensity of mining in the pre-copper-smelting era at Swansea was different, with the use of multiple shallow shafts and colliery tunnels (levels) into the steep hillsides that existed here but not on Tyneside. Mining practice at Swansea was typified by the illustrated map of Tre-boeth Common that showed a multiplicity of pitheads with very primitive windlass-gearing.

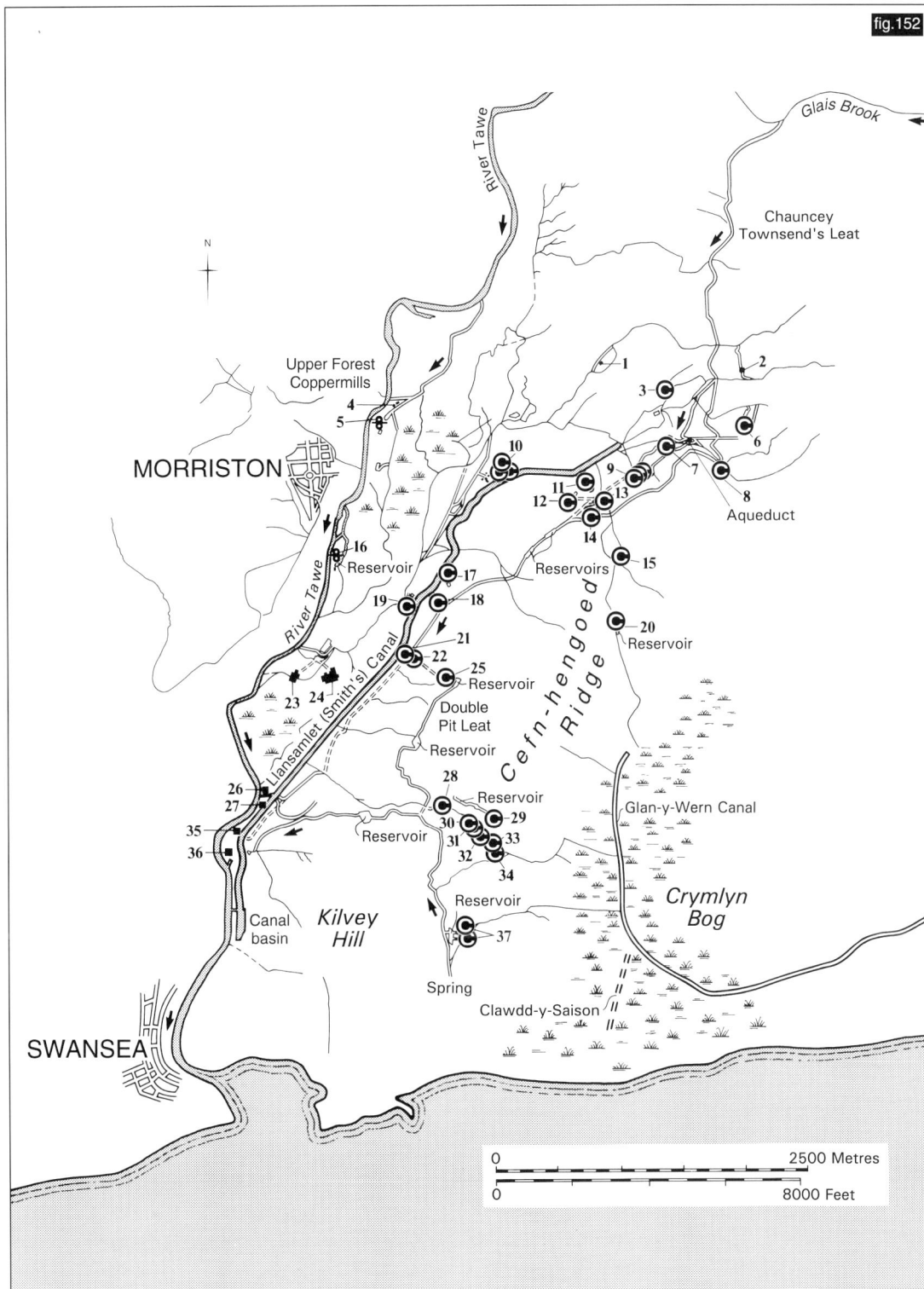

Fig. 152. Map of the basic water-economy of the eastern side of the lower Swansea Valley in the eighteenth and nineteenth centuries showing the multiplicity of large leats to provide water-power.

Key

1 Felin-Fran (corn to wool);

2 Birchgrove House (prob. churn/threshing);

3 Scott's Pit (steam-engine);

4 Upper Forest Mills (corn & iron);

5 - Upper Forest Mills (copper rolling & battery);

6 Emily Pit (steam-engine); 'Aqueduct' - over stream next to Gwernllwynchwith Pit (steam-engine);

7 Watch Pit (pumping);

8 Oak Pit; 9 - Gwern pit (probably pumping and winding);

10 Church/Charles Pits (steam-engines);

11 Round Pit;

12 Cwm Pit;

13 probably Parc Pit;

14 Parc Pit;

15 Park Pit;

16 Lower Forest Mills (copper battery, rolling & slitting);

17 Six Pit;

18 Garden Pit;

19 Middle Pit;

20 New Pit;

21 Pwll Mawr (steam pumping);

22 Pwll Mawr (winding);

23 Landore Siemens Steelworks (steam-engines);

24 Landore Tubeworks (steam-engines);

25 Double Pit;

26 Upper Bank Lead Works;

27 Upper Bank Spelter Works;

28 Unnamed Pit;

29 Ty-Draw Pit;

30-34 Fowler & other Pits;

35 Middle Bank Copperworks;

36 Whiterock Copperworks;

37 Tir-isaf Pit.

The first indication of a consideration of the large-scale use of water-power engineering to power collieries in the Swansea area is given in the lease of 1736, granting water rights to the proprietors of the White Rock Copperworks by the 4th baron Mansel. Bussy Mansel reserved the right to use water to drain his Llansamlet coalfield collieries with the provision that the tailrace water be returned to the copperworks for further use. Such an installation would have had to be in the vicinity of Bon-y-maen near the western summit of the Cefn-hengoed ridge to the north of Kilvey Hill.

Fig. 153. Much of the heavy work of colliery pumping was done by atmospheric engines at Swansea from the early eighteenth century, but at least one of the large colliery water-wheels along Townsend's Great Leat drove heavy beams suggesting it was also a pumping-engine similar to that illustrated. (960224)

Fig. 154. Earthworks at Bonymaen of a mile-long colliery watercourse that ran north from Kilvey Hill, via two reservoirs, and drove the winding and pumping water-wheels at the eighteenth-century Double Pit. (960212/4)

Fig. 155. The curving field boundary to the middle left follows part of the five-mile long watercourse built for Chauncey Townsend; with ploughed-out continuations visible to the South Wales Railway at upper left and the M4 motorway at bottom left. Lower leats are also visible. (935056/53)

A number of shafts are known to have existed in this area.

However, there are three major problems in defining exactly the extent of colliery water-power use. The first is that watercourses and reservoirs were also provided at pit-heads in order to provide both condensing and boiler-feed water to pumping and winding steam-engines, although these were generally smaller in scale than those used for water-power purposes. Secondly, eighteenth and sometimes nineteenth-century coal-mining used a large number of fairly shallow coal-shafts which were advanced relatively rapidly across the main seam being worked-on. This meant that the siting of a particular water-powered pump and its leat could be a comparatively short-lived use and the small size of the installation also meant that it was often not recorded on contemporary maps.

By 1813 a second leat, of which there are remains, extended 820yds north-eastwards along the eastern side of the Swansea Valley on the ridge from Bon-y-maen to Winsh-wen.[134] This may have been connected to the White Rock Copperworks leat but there is no definite evidence that water would have been returned to power the copperworks machinery from the far northern end of the leat. There were two small reservoirs on the watercourse, at the northern end of which was Mansel's 'Little Pit', later known as the 'Double Pit' of Chauncey Townsend.

Chauncey Townsend, a London Merchant, acquired the mineral rights of the Llansamlet Coalfield in 1750. Armed with large capital resources he brought-in the Newcastle engineer George Kirkhouse to reorganise the operation of the coalfield. He also produced new markets for the large-scale operation, with three new smelters built next to his shipping place near the mouth of the river (Middle Bank Copperworks, 1755; Upper Bank Lead and Zinc Works, c.1757). By 1756 a large railway system to multiple coal-producing shafts was operating on the Newcastle model. Steam-engines pumped the new collieries and water-wheels drove the winding-machinery required.

These water-wheels were driven by a five and a quarter mile long watercourse: by far the longest built in the Swansea area. It was designed to terminate at Townsend's new smelters at its southern end but in fact its water was probably entirely used in winding from various coalpits along the eastern side of the lower Swansea Valley. It certainly wound from Mansel's former 'Pydew Colliery' some one and a quarter miles north-east of the Middle Bank Copperworks and the water-wheel at this shaft, called the 'Piddn or Pidden Engine' is shown on contemporary maps.[135] One of the last acts of Chauncey Townsend in 1770 was to sink the Pwll Mawr ('Great Pit') to the Swansea Six Foot Seam at a depth of 163 yards. By c.1840 the southern terminus of the watercourse was at the upper shaft of the Pwll Mawr Pits (today marked by the remains of a later steam pumping-house) where a powerful water-wheel wound from the pit.[136] Downhill was a second shaft where the very large steam-powered pumping-engine stood. In the one and a quarter miles north-east from here there were several other collieries that may have been powered by the watercourse such as the contemporary operating shafts of Garden, Six, Park and Cwm Pits.

There is evidence of other subsidiary systems. The Little or Double Pit leat ran parallel to Townsend's great watercourse but at a much higher level, and flowing in the opposite direction. Its tailrace could have discharged directly down the hillside to an early 'Pidden Engine' or the later Pwll Mawr winder.

To the east of the Llansamlet Church, around the remains of Gwernllwynchwith Enginehouse, at a point two miles from the source of Townsend's watercourse are the earthwork channels of two parallel leats at a lower level. Surviving maps show how parts of these functioned and one of two feeder channels extended up through Townsend's large watercourse and may have pre- or succeeded it in operation but may also have been in use at a contemporary period.

The largest of these watercourses and the next below it, leading to Gwern Colliery, ran from the next east-west tributary south of the Nant Brân: hence not leaving the Nant Brân depleted before it flowed west and then south west to drive the long-established cornmills of Felin-Brân and then Felin-Newydd (the 'New Mill' offered for conversion to a copper battery-mill in 1736). This colliery leat seem to have had two head feeds: one on a level from this stream and the other coming down the hillside obliquely from the point where Townsend's watercourse crossed the stream. The Gwern Leat then ran south 540yds to what, by the 1840s, was a header pond for a separate watercouse to Gwernllwynchwith intersecting the Gwern Leat at right-angles. The Gwern Leat then crossed a small valley on an aqueduct and its channel is still visible as it goes west towards the site of Watch Pit

Fig. 156. On the east side of the lower Swansea Valley at Gwernllwynchwith are the remains of three parallel watercourses on the hillside which once drove colliery pumping and winding water-wheels. This view is of the lower leat to Gwern Pit. (960202/2)

which Townsend may have taken over from the Morgans. In 1772, the Oak Pit/Watch Pit working level across the Swansea Four-Foot Coal Seam was superseded by an 'Engine Pit'.[137] This was almost certainly the water-wheel 'Engine Pit' shown on a mining plan of 1808 and lying between Oak Pit and the steam-powered winding-house later built further down the seam at Gwernllwynchwith.[138] So in the time of Townsend's successor, John Smith, water would have been drawn from the Gwern Leat, powered the 'Engine Pit' and its tailrace would have fed into the head of the Cwm Pit Leat below. From Watch Pit the Gwern Leat ran south-west along the east side of the Swansea Valley for a further 440yds to the far south-western end of the rectangular Gwern Pit pond. In total, this colliery watercourse was some 1,260yds long.

Below it on the hillside was yet a third leat. This started at a point 770yds south west from the source of the Gwern leat and ran from a stream just south-west of Gwernllwynchwith at a point where the tailraces of Watch Pit, Engine Pit and Gwernllwynchwith Pit would have run into it. Part of this leat is visible on the hillside at this point adjacent to the last part of the visible formation of Chauncey Townsend's Railway. Its initial section ran south west for 440yds to a point where the tailrace from the Gwern Pit would have fed into it. Another 275yds and it projected line crossed a brook downstream of a pond that probably powered a small Parc Pit. The later Park Pit and its pond discharged water into the watercourse at a much higher level. A 110yds further south-west along the hillside and the watercourse was just above the later Round Pit pond and 160yds further brought it to its probable objective at the Cwm pit, 985yds from its start.

Both these shorter leats intersected a third at right-angles which either provided condensing or boiler water for the Gwernllwynchwith steam-powered winding-engine (c.1780) or possibly powered later threshing-machinery at the later Gwernllwynchwith Mansion barn. The line of this started on the hillside 160yds above Townsend's watercourse and cut across its line to a header

pond at a point where it was intersected by the Gwern Pit leat. It then seems to have run a further 220yds west to Gwernllwynchwith Pit enginehouse and the ruins of the barn; a line of about 550yds in all.

There are indications of a similar density of early pits on the Llansamlet coalfield to the south-west, but how many of these would have originally been powered by water-wheels is now impossible to say. The existence of earlier installations of the Mansels, and the Morgans at Gwernllwynchwith, may explain why Chauncey Townsend (admittedly endowed with considerable capital) had to build such a long five and a quarter mile long leat to reach the Glais Brook at a point east of Clydach.

To summarise there is extensive evidence for the use of waterpower to operate collieries on the eastern side of the River Tawe in the eighteenth century. The idea was familiar to at least one of the local landowner/coalmasters: Bussy Mansel, in 1736 and in 1750-56 Chauncey Townsend was carrying out the necessary engineering on a huge scale. The latter seems to have had a significant effect on the lesser capitalised coalmasters on the opposite, western, side of the river.

There is no indication that the active landowner/coalmasters on the western bank of the River Tawe (Thomas and Gruffydd Price of Penlle'r-gaer, Thomas Popkins of Upper Forest or Calvert Richard Jones of New Place) had considered the use of water-powered engines before the 1750s, although the former family did have the capital and expertise to build at least two early steam-engines. Even the Lockwood, Morris partnership (owners of the Llangyfelach Lead, Silver and Copperworks and colliery operators), and others, were using primitive hand-windlasses on and around Treboeth Common in the 1730s (including a shaft on or near the significant Pentre-Penfilia Pit near Landore). Then, in 1754 their Landore Coalpit was sunk and then operated by a water-wheel, fed from the millpond of the manorial 'Little Mill' at Landore, which is known to have both pumped and wound (some of the earlier Llansamlet water-wheels may also have been dual purpose) until the installation of an atmospheric pumping-engine in 1776.[139]

Lockwood, Morris & Co.. may have already built a second water-wheel winder some years previously. Their second smelter at Forest, built in 1746-48, had a specially made cart-road that ran diagonally up the hillside to the 'Clyn-du Water-Machine Pit' on the western valleyside above the site of what is now Morriston.

In the 1760s Townsend also expanded his coal-mining activities onto the western side of the river at Landore. His Engine Pit was sunk in 1762 and, by

1768, was being kept dry by a steam-powered pumping-engine. A 275yd long leat was built along the south side of the Nant Rhyd-y-Filais Valley from a tributary to the pit site and almost certainly powered a winding water-wheel as well as providing boiler and condensing-water for the pumping-engine. The pit was abandoned in c.1780 but reopened soon after 1800. Townsend also opened a second connected pit on the quayside at Landore by 1770 (the Tir Landore Pit) - both pits had railway connections and hence required winders. It seems likely that the Engine Pit leat was continued a further 330yds on the south side of the Nant Rhyd-y-Filais ravine to drive a water-wheel winder at Tir Landore Pit.

The 550yd lower length of the Nant Rhyd-y-Filais valley at Landore had no fewer than 14 water-power sites of which no fewer than six were colliery sites with leats. These included one elaborate mile long watercourse, constructed in about 1775 by Lockwood, Morris & Co., which left the headwaters of the Nant Rhyd-y-Filais, crossed the Nant Gelli ravine on a timber aqueduct, drove machinery at Cwm Pit, and then followed the 175ft contour north-eastwards along the face of Graig Trewyddfa to the twin reservoirs at the long established Plasmarl coal shafts. Tail-race water was then taken by a further short leat to drive machinery at the Rose Copperworks.

The long-established pit of Lockwood, Morris & Co., at the head of the valley: the Penfilia or Pentre (Landore) Pit was deepened in 1770 and a steam-winder may have been installed in 1806 when the pit was modernised.

There is then large-scale evidence for water-power use in driving collieries: much of it to drive the rotary motion, and lighter duty, of winding. The straightforward reciprocal motion of pumping was carried-out by powerful steam-powered pumping-engines.

However, even in this period of the later eighteenth century, coal-winding was not universally achieved by water-power. Lockwood, Morris & Co.. built a very elaborate walled horse-gin at Tirdeunaw Coal Pit, half a mile north of Pentre (Landore) Pit between 1760 and 1768.

Shafts were getting deeper and coal outputs were getting higher with the results that more powerful winding engines were required. Lockwood and Morris's innovative engineer 'Mr. Powell' had developed and built one atmosperic-engine winder at Landore by 1783 and had made at least two other large pumping-engines dual purpose winding and pumping engines: probably Plasmarl and the Copper Pit at Forest Copperworks. The competing coalfield of John Smith on the opposite, eastern, side of the River Tawe could not be allowed to lose its commercial effective

by failing to utilise this new innovation and the very significant remains of the rotary engine house at Gwernllwynchwith indicate similar innovation at new deeper pits; even those that were sited alongside the already existing leat complexes.

Lockwood, Morris & Co. hoped to acquire the first Boulton and Watt steam-engines in the area cheaply, as an advertisement for the Birmingham firm but Matthew Boulton's commercial instincts thwarted this plan. In 1785 Morris purchased a second-hand Cornish mine-pump that Boulton and Watt supplied as a steam-winder. The long, drawn-out negotiations over whether the large new Landore engine should be a dual-purpose machine ended in 1800 with the building of a large new pumping steam-engine. In the early years of the nineteenth-century non-condensing Trevithick winders were also used like that built at Landore Pit in 1803. As elsewhere in south Wales, dual-purpose winding and pumping steam-engines dominated local winding requirements. Water-balance winding, so common around the Merthyr area, seem to have had only isolated examples in the lower Swansea Valley area such as that at Graigola (to the east of Clydach) built by c.1860.

Water-powered winding may have fallen out of fairly universal use in the lower Swansea Valley after 1800 but it was obviously still deemed suitable by the Swansea Valley entrepreneurs for shallow new mines where appropriate. The Morris family had turned to coalmining rather than copper-smelting by c.1800 and the promotion of the Oystermouth Railway was partly to exploit the perceived coal reserves at the new Morris lands in the Clyne Valley. The huge new 1,980yd leat down the eastern side of the valley, may have powered Morris's New Mill enroute but it was designed primarily to power Sir John Morris's new Ynys Colliery at its southern end. This failed because of the tenuous nature of the coal seams on the edge of the coalfield. Howevers as late as c.1840 John Morris II was opening new water-powered shafts at two collieries on the watercourse. One at the new Rhydydefaid Colliery (superseded by steam by 1854) and the other with two wheelpits: one for winding, and a second for probably pumping at the Ynys Colliery (closed by 1877-78).

In the lower Swansea Valley area itself the Millwood Colliery in Cwm Burlais was equipped with a large reservoir with three water take-offs by c.1838: presumably to drive both pumping and winding functions (replaced by the 1870s with a tinplate works that used its water supply).

The use of water-wheel winding continued for longer in the upper Swansea Valley where many Swansea interests had opened mines. A coal-mining boom in the upper valley followed the opening-up of the area by the Swansea Canal in the 1790s. At first these mines were merely tunnels into the newly discovered outcrops. The Tymawr Level at the head of the canal in Abercrave was one where a side stream to the Tawe powered a water-wheel driving a haulage rope out of the mile-long tunnel. To the west of this the 'Feeder Jams' was a considerable watercourse taking water from the upper Afon Llynfell to the upper Cwm Twrch where it drove colliery machinery. Nearby large water-wheels providing collieries with powered winding survived in Cwm Llynfell to the end of the nineteenth century.

From the 1780s both sides of the lower Swansea Valley came to have localised canals and in 1794-96 that on the west was absorbed into the much larger Swansea and Trewyddfa Canals line. Morris Canal water may have provided a partial feed to the Landore Coal Pit from the 1780s. Otherwise it is not obvious that any of the 48 or so sites on the Swansea Canal line water system was a water-powered colliery although the two types of feature did interact in other ways.[140] Firstly mine drainage water was discharged into the canals and went to power other industry and secondly the colliery owners preferred to take clean condensing and boiler-feed water from the canals than use the dirty and potentially dirty water from the mines.

At least two levels under Graig Trewyddfa (the Graig and Clyn-du Canal Level) discharged water into Morris's Canal: enough to supply three copperworks with water. Three levels penetrated the ridge between Winsh-wen and Birchgrove on the opposite, eastern, side of the valley had entrances on the bank of John Smith's Canal and at least two of these must have discharged water into the canal. At Trebanos the mine water from one level in the hillside can still be seen discharging water into the Swansea Canal via a wooden trough opposite the large waste-water weir that once fed water to power the Pheasant Bush Tinplate Works.

On Smith's Canal a large part of the water from Chauncey Townsend's; Gwern Pit; Cwm Pit and the Gwernllwynchwith leats must also have been discharged from the colliery water-wheel tailraces along with the colliery water itself.

Morris's Canal, and later the Trewyddfa Canal along the same line, must also have received, and returned clean water to, Copper and Plasmarl Pits. More evidence survives after 1796 from the completed Swansea/Trewyddfa Canal line through the area when Tir Canol, north of Morriston, was one Pit that in 1803 received permission to use clean canal water as long as mine water was returned to the canal. Five pits in the Swansea area are known to have done this. Charles Pit, on Smith's Canal bank at Llansamlet was similar.

Copperworks Water-supply Systems in the late Eighteenth and Early Nineteenth Centuries

By 1731 the Ynys-pen-llwch Coppermills at Clydach, the most remote from the copper-smelting district, had reverted to being an iron forge (with the very significant addition of the first tinning-house in Glamorgan). The centre of the copper-milling area powered by the main River Tawe moved southwards of the tidal limit to the Lower Forest Mills built in 1732-35 by Lockwood, Morris & Co.. Their old Llangyfelach Lead, Copper and Silver Works was over 3/4 mile to the south and transport from this hydrid smelter must have been both time-consuming and expensive. Consequently their new Forest Copperworks (the fourth copperworks constructed in the lower Swansea Valley) was built opposite Lower Forest Mills in 1746-48. A complex system of watercourses ran downwards and across the northern face of Graig Trewyddfa to supply two ponds to the immediate north of the works, presumably to power slag-stamps and to provide water for the copper-granulation tanks.

The next three smelters built in the lower Swansea Valley were those constructed by Chauncey Townsend in the years 1755-57 at Middle and Upper Bank. The water-rights agreement Townsend signed with the Duke of Beaufort allowed for the use of Glais Brook water via his great leat but in fact Townsend did not pay the Duke for this use on the grounds that his metals works were not using the water. The topographical evidence shows that by the 1840s Townsend's great leat had it southern terminus at Pwll Mawr and it may be that this largest of the watercourses was never used by the metals smelters. Map evidence shows that the Upper Bank lead smelter did have a feeder watercourse on the line of

Fig. 157. The lease of water on the Glais Brook (right) allowed its diversion into the long watercourse or leat surviving as the silted terrace on the left which was designed to lead to Chauncey Townsend's three non-ferrous metals smelters at his Middle and Upper Bank Works. (960212/7)

111

the great leat but apparently terminating at the Cwm y Danas watercourse coming down from Bon-y-maen where it may already have driven Mansel's old collieries. The Middle Bank Copperworks may have tapped a similar source: it later had a rolling-mill on site but this is likely to have been steam-powered.

The next generation of copper-smelters in the 1780s were built in the era of local canals. The new Forest Copperworks was a precursor of a more general trend as it reportedly had the Clyn du Canal Level driven into the hillside of Graig Trewyddfa from its site in 1748. The level must have been at least 8ft below that of the works water storage ponds but even so the Swedish 'industrial spy' Svedenstierna reported how the canal emerged at eaves level and its drainage water in such a situation could have driven a water-wheel just as John Smith's Rhandir-mwyn Leadmine underground canals released water to drive a crushing-mill. The big change came in 1783-85 when John Smith built a three mile long surface canal from north-east of Llansamlet Church to the Foxhole shipping-wharves at its southern end. This produced what became the classic Swansea siting for its copper smelters as shown on the idealised illustration in *Tomlinson's Cyclopaedia'*. What were now the Upper Bank, Middle Bank and White Rock Copperworks had works with a coal-supplying canal running at a higher-level on their landward site and ore-receiving and refined-copper exporting quays on their riverside. As the Swedish industrial reporter Eric Svedenstierna noted in his visit to Swansea in 1802-03 'there are several small canals and aqueducts, which partly carry the water to the surrounding works, and partly are laid out for smaller boats.'

The construction of the canal seems to have removed the watercourses already supplying water to the Upper and Middle Bank Works. Smith's

Canal, however, had ample supplies of water running-in from the various colliery leats discharging into its northern end. There are still remains of the sluice in the canal bank above the reused buildings of the Upper Bank Copperworks and at the White Rock Works canal tunnel is a low level water-outlet into the works. Over the south portal of the tunnel is still an aqueduct that carries water from the west side of Kilvey Hill to discharge into the River Tawe. In the late eighteenth and nineteenth centuries water would have run from the works leats over the top of the canal tunnel at right-angles to drive the works machinery.

This type of siting for a further four copperworks built between a higher canal and the riverside was next carried-out on the opposite western side of the river in a possible imitation of what had already been achieved on the east. Here the mile-long canal of Lockwood, Morris & Co. ran from the mouth of the existing Clyn-du Canal south along the foot of Graig Trewyddfa to the northern side of the Graig Trewyddfa at Landore. The building of Morris's Canal was being considered in 1787 presumably primarily to take coal from the Clyn-du Canal Level south to Lockwood, Morris & Co.'s new smelter at Landore. It was built by 1791 and after the mouth of the Clyn-du Level first passed along the landward side of the Forest Copperworks at eaves level. The two copper-works built in the 1780s on land sub-leased by Lockwood, Morris & Co. (and mostly operated by Birmingham companies) were built immediately prior to the construction of the canal alongside the river midway between Morriston and Landore. Of these the Rose Copperworks had both a natural watercourse and a tailrace leat from the Plasmarl coal-pits which themselves were supplied by a long leat leading from the Nant Rhyd-y-Filais valley to the south.

The northerly of these copperworks on Ynys Howell was only developed on a substantial scale by the 'Birmingham Mining and Copper Company' from 1791. Sir John Morris's agent wrote to him on 9 May 1791 that:[141]

> 'Three of the Birmingham Co. were here this morning. They intend making a Pond or Reservoir to hold water by their works, and that if you please to let them have water for the use of the works they will make the cut from the Canal to their reservoir.' Morris gave permission and the Birmingham works proprietors 'made a reservoir or dock and cut the communication from it to the Canal.'

The use of canal water carried-on after 1794-96 when Morris's Canal was widened to some four times its original width to form part of the Trewyddfa Canal length of the Swansea Canal line. It was noted

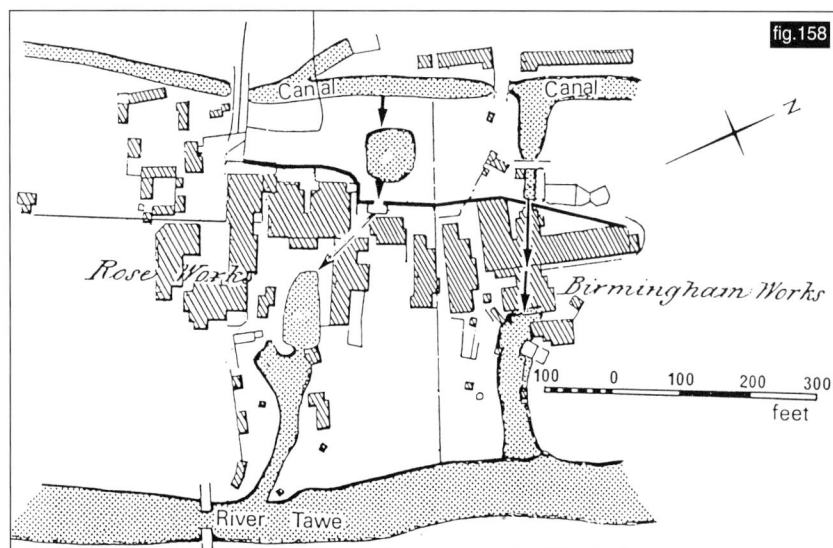

Fig. 158. Canals ran along the landward side of the copperworks and supplied coal and water. Their water was also used to scour out the river-docks alongside the River-navigation (water-flow indicated by the arrows). These processes were aided by canals leading from the valley-side coal-mines being supported above the riverside copperworks by high retaining walls (thick lines). (Redrawn from part of the plan of the Swansea Vale Railway, 1840) (The National Library of Wales)

that 'The work of Smelting Copper cannot be carried on without a certain quantity of water which the Birmingham Company have been allowed to take out of the Trewyddfa Canal.'

Initially the Birmingham Works reservoir also served as a dock for the small 3-4 ton coal boats delivering coal from the Clyn-du Canal Level but by 1814 this source of fuel may have diminished or ceased for in that year the entrance was widened to allow 21 ton coal boats from the company's Tir Canol Colliery to enter and discharge coal directly through the upper wall of the Birmingham Copperworks furnace-hall.

By 1840[142] the Rose Copperworks also had a pond between the canal and works and may have a direct feed from the Trewyddfa Canal to supplement or replace the earlier water-supply arrangements. Both works were built into a hill slope from the canal down to the river, probably similar to those illustrated on Jernegan's design for a copperworks in *c.*1790 (almost certainly the Landore copperworks) with direct access from the canal at first-floor level with a large retaining-wall parallel to the canal which is clearly shown on the plan of 1840. Both works would have needed water to drive stamps for crushing copper-slag and matte between successive roastings; possibly to drive a test-hammer for the copper produced; to fill the water-tubs alongside the reverbatory-furnaces so that copper produced into them could be in the more easily refined and alloyable form. After this use at the Birmingham Works the water went to scour-out the 200ft long dock built between the works and the navigable river. The Rose Copperworks immediately to the south had a second pond enclosed by ranges of building which was a scouring-reservoir impounding water used in the works so that it could be released in a rush to remove silt from its 200ft dock leading to the river.

The fourth copperworks, on the western bank of the River Tawe, built in a similar situation between the canal and the river was the Landore Copperworks itself. This was built at the south end of the Lockwood, Morris & Co. Canal and was the first to be considered for the provision of a steam-powered rolling-mill via an attachment to the powerful Landore pumping-engine supplied by Boulton and Watt. Instead a rolling and stamps mill, powered by the Nant Rhyd-y-Filais, was built a short way to the south-west on the site of the Llangyfelach Copperworks with a tailrace discharging into the Little Mill (Landore Cornmill). Presumably some process water was also fed into the Landore Copperworks via the adjacent and higher Morris Canal.

The next two copperworks founded were in a similar relation between two roughly parallel watercourses but by this time the Swansea-Trewyddfa Canal-line had been completed in 1794-98, absorbing the Lockwood, Morris & Co. Canal en-route. These two smelters, the Hafod and Morfa Works, were constructed on an entirely different scale to the previous water-powered works that employed hundreds rather than thousands. The representatives of Cornish mining interests were responsible for these large-scale investments and both had considerable lengths of works walls alongside the Swansea Canal that were punctuated by the openings into each of the works of watercourses that served as the entrance to pools of water held above the mains works by considerable retaining walls: these served both as coal unloading docks and reservoirs. Leading-off these pools were networks of water-pipes that fed boiler and condensing water to the works steam-engines and supplied water necessary for copper production such as that used in the granulation tubs.

Originally, as first built in 1810, the Hafod Copperworks was much like any of the previous works except that it had a rather larger rectangular furnace-house at right-angles to the canal. In 1822 the proprietors (the Vivians) also took over both the existing water-powered copper-rolling mills on the River Tawe upstream at Upper and Lower Forest. At this stage (1828) the other large Cornish concern, Williams & Foster, operating the Rose Copperworks alongside the Trewyddfa Canal to the north, built a steam-powered mill on the site next door to Hafod. This revolutionised the type of landscape feature required to drive the copper-rolling mills. The linear watercourses forming considerable features across the landscape were no longer required. Instead there arose the forests of chimneys so obvious from later views of the huge Hafod and Morfa Works and of which the solitary Hafod Rolling-mills stack is the sole survivor.

To compete with the very large steam-power rolling-mill erected at Morfa by the 1840s (which is still extant and in use as a Swansea Museums store) Vivian at the Hafod Works considered expanding his water-powered rolling-mill capacity by renting one of the largest mill-complexes on the River Avon near Bristol. This mill at Saltford was in a bad state so instead Vivian built a large new 60 h.p. steam-powered mill at Hafod in 1842 that proved so successful that he relinquished the Upper Forest Rolling-mills in 1845. That then (reputedly) became the first tinplate works in Swansea with the Lower Forest mills becoming a tinplate works by 1860.

It was the third decade of the nineteenth century, then, that there was a large-scale switch in the copper industry from a use of steam rather than water-power

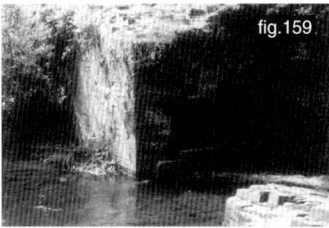

Fig. 159. A substantial dam, formerly holding back a considerable lake, at the site of the Upper Clydach Forge (1784) on the Lower Clydach River, Clydach. (960214/4)

for new powerful large-scale installations. Some of the substantial water-power installations might remain in use but by the mid-nineteenth century all the main water-power mills in the lower Swansea Valley were providing a water-power resource that helped attract expanding tinplate manufacture to what became the world centre of the industry. Significant parts of the large watercourses of all three main mills on the River Tawe remain along with the weir of the Lower Forest Forge.

The Water-power uses of the Tinplate Industry

The earliest tinplate works in south-west Wales was the water-powered Kidwelly Tinplate Works of 1719 and the second seems to have been the Ynys-pen-llwch Tinplate Works built in *c.*1750 (the first in Glamorgan) attached to the former iron forge and copper rolling-mill at this site in the lower Swansea Valley. By *c.*1753 the next valley east had the Ynys-y-gerwn tinning house which had been added to the earlier forge and rolling-mill, powered by the main River Neath, near Aberdulais. These were the beginning of what became the world centre of the tinplate industry. However, it was only in the early nineteenth century that south-west Wales rather than south-east Wales became the centre of the trade. In fact about half the new tinplate works founded in south-west Wales from 1800 to 1850 (six out of 13 works), including all those in the mid and lower Swansea Valley, were built on valley water-power sites.

The Primrose Forge and Tinplate Works of 1843, the nucleus of what became the major tinplating centre of Pontardawe, was designed to be partly powered by the water coming down the Swansea Canal and flowing along the waste-water bypasses on the two locks alongside the works. The linked Pheasant Bush Tinplate Works at Trebanos of 1833 was built to harness the large fall of waste canal water that had to be returned to the adjacent river here to safeguard the water supplies of the river mills at Ynys-pen-llwch and Forest. Later in the nineteenth century it too also harnessed the additional waste water flowing down the canal around an additional two locks.

In 1845 the former copper rolling-mill at Upper Forest had a tinning-house added and became the Forest Tinplate Works and reputedly in 1860 a similar process took place at the Lower Forest coppermills which became known as the Beaufort Tinplate Works. Elsewhere the two other lower valley works that seem to have been all, or partly, steam-powered from the beginning were the Landore Tinplate Works of 1851 that partly reused an existing iron forge or cast-house building in part of a water-powered complex and the tinplate works built on the Millbrook Colliery site in 1858 that reused its water storage pond.

Other Industrial Use of Water-power.

The early water-powered cornmills on the periphery of west Swansea, at Blackpill and Brynmill, carried-on in use into the later nineteenth century and the pottery mill at the south end of the Swansea Canal was actually converted from industrial use to the production of flour early in the nineteenth century and was still in production in c.1855. The first new mill to use the waste water that had to be returned from the Swansea Canal to drive the Forest Coppermills on the River Tawe was Edward Martin's Felin Ynys Dawe (cornmill) of 1810 which continued in production until *c.*1852. The principal users of water-power besides copper and nonferrous metals smelting, coalmining, tinplate manufacture and cornmilling were the iron and pottery industries.

An important early developer of the large water-power sites in the lower valley was the independent iron industry that both preceded the tinplate industry and carried-on an independent existence alongside it. It was the iron industry that developed the potentially most powerful water-power sites on the lower main river of the Swansea Valley. No other industry, certainly not the proprietors of the manorial cornmills, could afford the large capital works entailed in building the river weirs and the long and wide headraces required for the Ynys-pen-llwch and

Fig. 160. Substructure timbers and wrought-iron fixings to water-powered hammers and to anvils exposed in the bed of the Lower Clydach River just downstream of the dam of the Lower Clydach Forge. (960207/2)

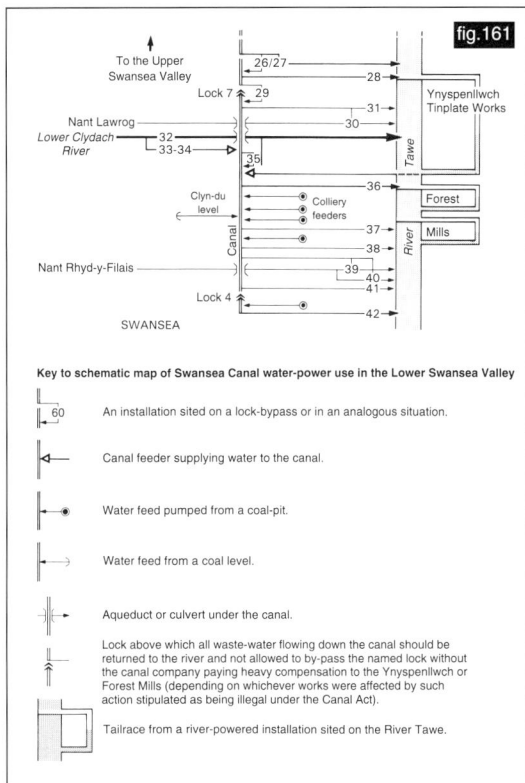

Key to schematic map of Swansea Canal water-power use in the Lower Swansea Valley

60	An installation sited on a lock-bypass or in an analogous situation.	

Canal feeder supplying water to the canal.

Water feed pumped from a coal-pit.

Water feed from a coal level.

Aqueduct or culvert under the canal.

Lock above which all waste-water flowing down the canal should be returned to the river and not allowed to by-pass the named lock without the canal company paying heavy compensation to the Ynyspenllwch or Forest Mills (depending on whichever works were affected by such action stipulated as being illegal under the Canal Act).

Tailrace from a river-powered installation sited on the River Tawe.

Fig. 161. Schematic map showing how Swansea Canal and River Tawe water-use related to each other in the lower Swansea Valley and to local water-powered installations.

Upper Forest Forges before the mid eighteenth-century growth of the copper industry. By the end of the eighteenth century the presence of the upper valley iron-furnace at Ynysgedwyn had also spawned the two iron forges driven by the Lower Clydach River (at what became the village of Clydach) in order to succeed the earlier Tawe River installations in processing cast into wrought-iron (the remains are some of the most interesting in south Wales).[143] The two largest water-power pools on the tributaries of the lower River Tawe were both used by the eighteenth-century iron industry. The Upper Clydach Forge built by Richard Parsons in 1784 was one and the other was the reuse of the Cambrian Copperworks pool at the mouth of the Cwm Burlais in Swansea. This may have been reused as an iron rolling-mill by the new propritors of the adjacent Cambrian Pottery but by the end of the eighteenth century was processing iron for the local market for the owner of the two Cwmddyche Furnaces and Foundry at Llanelli, Alexander Raby, certainly as a foundry but possibly also as a forge and rolling-mill. Raby, a London coal and iron merchant and entrepreneur (originating from Worcestershire) had built the Ynyshowell Copperworks in 1787 (sold to the Birmingham Mining and Copper Company in 1791) in conjunction with his partner, and brother-in-law, Thomas Hill Cox.[144]

In about 1790 the water resources of the Nant Rhyd-y-Filais at Landore on the site of the Llangyfelach Copperworks were reused for the Landore Copperworks stamping and rolling-mill. After an attempt at recycling the iron from copper-slag this eventually became the Landore or Millbrook Ironworks, forge and foundry with at least two water-wheels on site, one powering blowing-cylinders for a blast-furnace used in the production of specialist heavy castings for the Swansea area by the mid-nineteenth century. A specialist water-powered boring mill was later added upstream.[145]

The increasing density of water-powered installations driven by the ample supplies feeding the Swansea Canal encouraged entrepreneurs entering the valley's iron industry to view it as a ready-made water-power leat. In 1824 a large water-wheel 35ft in diameter and 8ft wide was built on, and driven by the top feeder of the Swansea Canal in order to provide the blast for the newly built Abercrave Ironworks. This was rapidly followed by a proposed 1825 scheme whereby one or two blast-furnaces to be built below Pontardawe Bridge with a blast driven by the waste-water of the canal flowing down the by-pass channels of locks 10 & 11. It was only in 1829 that the first successful scheme was built on the canal to utilise a lock's by-pass water for the generation of power. This was to provide the blast for the small cupola remelting furnace at John Strick's Clydach Foundry at Lock 6 which used iron brought down the canal from Ynysgedwyn Ironworks to cast products for the lower valley. Part of the wheel-house and one of the two succeeding water-turbines still remain.[146] The iron and brass foundry sited at the southern end of the Clyne Wood canal may also have used a water-wheel to provide the blast for a small cupola furnace.

Later iron and steel works in the lower Swansea Valley were powered by steam. The many engines at the second site of the Landore Siemens Steelworks located to the north of the Landore Viaduct had large reservoirs alongside a canalised Nant-y-Ffendrod to store condensing and boiler-feed water and a pipeline crossing a swing-bridge over the Tawe River Navigation to provide water from the Trewyddfa-Swansea Canal line.

Another noticeable use of water was in powering the various clay industries of the lower valley. The White Rock Copperworks of 1736 (one of the thirteen major copper-smelters of the valley) was noted as having a 'clay-mill' presumably a pug-mill for preparing clay for firing in reverbatory work. The later nineteenth-century fireclay mill at Graigola Basin (Clydach), powered by the Swansea Canal, was another similar installation (the wheelpit still remains). The Swansea potteries were sited on the coalfield because of similar supply-lines to the copper industry but the works never developed to the

Figure 162

Swansea Canal Water Power Installations in the Lower Swansea Valley – a summary of information

Installation	Date Built	Date of First User of Canal Water²	Date of Disuse	Type of Site	Grid Reference	Whether Canal was sole water supply	Water rent	Remains	Comments
26 **Pheasant Bush Tinplate Works**	1838	1838	be.1890-9959	Waste	SN7124 0265	Yes	£5 p.a.	None	Used compulsory waste specified in Act.
27 **Pheasant Bush Works Turbine**	1890	be.1890-9961	1890	Bypass	SN7123 0268	Yes	£5 p.a.	None	On locks 8-9.
28 **Graigola Basin Fire-clay Mill**	b.1875	b.1875	b.1897	Waste	SN7041 0171	Yes	–	W.P., H.	Used compulsory waste specified in Act
29 **Clydach 'Electricity' Turbine**	a.1899	a.1899	a.1960	Bypass	SN6972 0158	Yes	5s. p.a.	None	On Lock 7.
30 **Nant Lawrog Mill**	b.1875	b.1875	a.1896-97	Waste	SN6957 0144	No	£5 p.a.	None	–
31 **Nant Lawrog 'Electricity' Turbine**	–	a.1896-97	a.1896-97	Waste	SN6954 0143	Yes	£5 p.a.	H, I.	–
32 **Clydach Lower Forge**	1791	1796	pr.1817	Feeder	SN6885 0129	Canal Feeder	None	H.	On pre-canal leat.
33 **Clydach Upper Feeder 'Wheel'**	1891	1891	–	Feeder	SN6881 0130	Canal Feeder	£5 p.a.	H.	On pre-canal leat.
34 **Clydach Saw Mill**	1889	1889	–	Feeder	SN6885 0109	Canal Feeder	10s. p.a.	H.I.	On pre-canal leat.
35 **Clydach Foundry**	1829	1829	a.1940	Bypass	SN6871 0086	Yes	5s. p.a.	T,H,I.	On Lock 5.
36 **Felin Ynys Dawe**	1810	1810	ca.1852	Waste	SN6834 0025	Yes	£5 p.a.	None	Used compulsory waste specified in Act.
37 **Fforest Copperworks**	ca.1735	ca.1747	a.1851	Waste	SS6695 9708	Probably main supply	10s. p.a.	None	Probably used Clyn Du Level and then Morris's Canal Water.
38 **Birmingham Copperworks**	1787	1791	be.1833	Waste	SS6692 9682	Yes	10s. p.a.	None	Used Morris's Canal Water.
39 **Landore Flint Mill**	–	be.1814-17	a.1818	Waste	SS6625 9598	Probably not	–	None	Probably on pre-canal leat: Canal water-use stopped by Co.
40 **Landore Corn Mill**	b.1730	1796	b.1844	Waste	SS6620 9590	No	–	None	On pre-canal leat. Use of canal water not definite.
41 **Landore Pottery**	1848	ca.1848	1856	Waste	Not identified	Probably	–	None	Canal water-use stopped in 1854.⁹⁴
42 **Swansea Foundry Mill**	b.1793	1796	ca.1855	Waste	SS6594 9375	Yes	None	None	Feed compulsory under Canal Act.

Abbreviations used in the column headed *Remains:* There are some remains left on site of the respective features when a letter is given:
H. = Headrace; I. = Installation (no machinery survives except at Clydach Foundry; T. = Turbine remains in situ and W.P. = Wheel Pit.
Abbreviations in front of dates; *a.* = after; *b.* = before; *be.* = between; *ca.* = circa; and *pr.* = probably.

Fig. 162. Table giving details of water-power installations in the lower Swansea Valley.

same degree. However, the technical finesse achieved by the industry necessitated the plentiful addition of ground flint to the body of the clay. The later eighteenth-century Cambrian Pottery reused the large water-supply pool of the Cambrian Copperworks at the mouth of Cwm Burlais to drive one of its flintmills. The pottery proprietors also converted the lowest cornmill in Cwm Burlais to serve as a large flint-mill and this continued in this use until the mid nineteenth century with the supplementary addition of steam-power. The proprietors of the rival Glamorganshire Pottery had an early nineteenth century flint-mill driven by the old water-wheel of the Landore Coalpit with a dual feed from the old Landore Cornmill pond (on the Nant Rhyd-y-Filais) and an illegal feed from the nearby Trewyddfa Canal that functioned between 1814 and 1817.

Any other minor use requiring power in the valley was likely to use one of the many natural or artificial water-channels existing in the eighteenth and nineteenth centuries. One minor industrial use was for the water-powered 'marble mills' built alongside the Oystermouth Railway on the watercouse at West Cross on the western part of Swansea Bay. Here suitably textured blocks of limestone could be prepared for architectural use in Swansea. By the time of the 1842 Public Health Maps of Swansea this craft had been switched to a steam-powered mill in the western part of the town.

Along Swansea Bay the availability of timber products from Clyne Wood gave rise to the mid nineteenth-century pyrolignious acid factory sited alongside the southern end of the Clyne Wood Canal. The extent of a water-powered fulling and woollen industry in the lower Swansea Valley is difficult to quantify. There was a fulling, or tuck-mill, (powered by a side stream) near the later site of the Graigola Fireclay Mill to the north of Clydach. There may have been other late Medieval or post-Medieval

fig.164

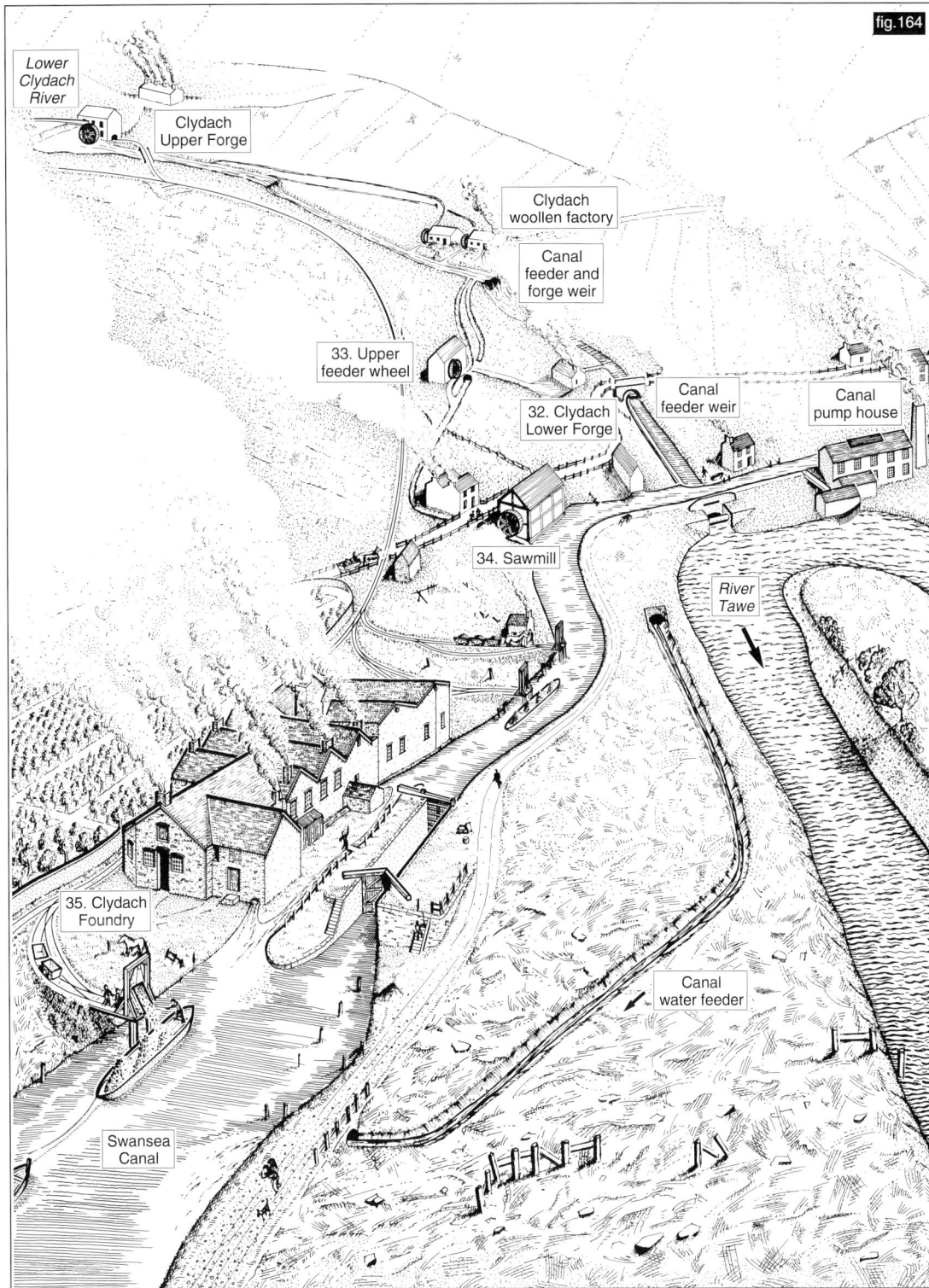

Fig. 164. *The complexity of the water economy of the lower Swansea Valley is indicated by this multi-period schematic reconstruction of water-power installations at Clydach around Lock 5. No lock by-pass water was allowed past Lock 7 at Clydach to ensure that the earlier Ynyspenllwch Mills on the River Tawe never ran short of water.*

Figure 163

	1793	1798	1798-1860	1861-1894	1894
29	–	–	–	–	B
30	–	–	W?	W	
31	–	–	–	–	W
32	L	F	F	–	
33	–	–	–	F	–
34	–	–	–	F	–
35	–	B	B	B	–
36	L	W	W	–	–
37	W	W	W	–	–
38	W	W	W	–	–
39	–	–	W	–	–
40	L	W	W	–	–
41	–	–	W	–	–
42	L	W	W	–	–

A letter indicates that the water-power installation that came to be, or was canal operated, was functioning in the period specified. The type of feed is shown by the letter used:
B = Bypass;
F = Feeder;
L = Fed by a pre-canal leat and
W = waste-water feed.

Fig. 163. *Table showing that there was so much water flowing down the Swansea Canal that the maximum number of water-power installations driven could coincide with the height of canal traffic, represented by the middle column.*

fulling-mills on the Nant Burlais along the northern borough boundary. The huge expansion of the industrial population in the nineteenth-century created a large demand for woollen products that new local water-powered factories attempted to fill.

In Cwm Burlais there was a late nineteenth-century woollen-mill to the immediate north-west of the Hafod Bridge called the Aberdyberthi Woollen Mill that was almost certainly water-powered. To the north-east of Swansea at Llansamlet another woollen

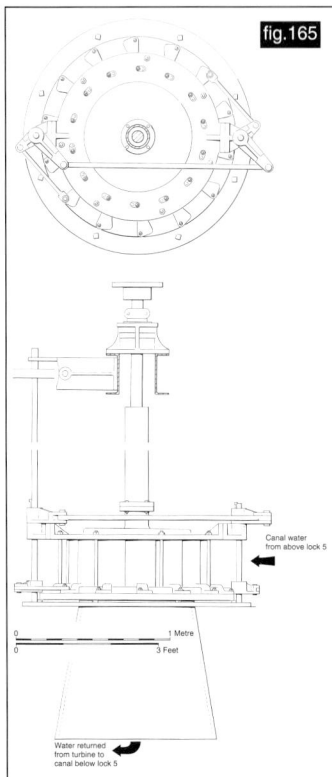

Fig. 165. The water-turbine alongside Lock 5 on the Swansea Canal at Clydach. This succeeded an earlier canal-powered water-wheel in providing power for the foundry, and probably also provided some power for the adjacent Player's Tinplateworks. The turbine is at Kidwelly Tinplateworks Museum.

Fig. 166. Comparative cross-sections of eighteenth-century leats in the Swansea area. Big coppermills on the River Tawe had larger canal-sized examples, although the Clyne Wood Canal, probably built primarily to drive a canal-age corn-mill, was also of a fairly large cross-section.

factory was built alongside the old Nant Brân Cornmill. To the north the woollen mills began to fill spaces near the mouths of the tributaries of the River Tawe that were not already taken-up by water-power installations.

On the Lower Clydach River (at Clydach) this entailed double weirs and leats sited below the tailrace of the Upper Clydach Forge. The woollen factory was sited immediately above the Lower Forge Weir at Clydach with its twin tailraces flowing into the pool above. From here a canal feeder also flowed driving another two water-wheels on its way down to the canal with a second, and lower, canal feeder designed to utilise the tailrace water from the Lower Forge. Another woollen mill on the Upper Clydach River at Pontardawe also fitted into a complex pattern of interdependent water-supply. It was situated downstream of the canal feeder weir but upstream of the canal aqueduct. To the south the Primrose iron forge of 1843 (later the Pontardawe Tinplate Works) had a dual feed from further-up the Upper Clydach River and the Swansea Canal above locks 10-11. The tinplate works operators bought the water rights of the woollen mill and conveyed its tailrace water in an aqueduct southwards over the river to the water storage ponds of the tinplate works from where a second aqueduct took the water over the Swansea Canal to drive the large water-wheel of the works and be discharged into a canal branch behind that ran back into the canal below the locks.

Other water-wheels were used for agricultural purposes. Examples of the former included the 'furze' or gorse-mill (for preparing cattle fodder) on the Nant Rhyd-y-Filais and probably the water-wheel added to the upper feeder of the Swansea Canal at Clydach in 1891. The water-wheel added to the southern end of the old Beaufort Tinplate Works in the 1980s is from the model farm of the colliery operator John Jones of Brynamman. Domestic installations included that which survived in the grounds of the Vivians' Singleton Abbey until the 1970s. This used a small stream in the north-east corner of the park to pump water up to the old walled-garden of 'Verandah' and remnants remain near the north-east lodge of Singleton Park: Vivian may have used steam-power exclusively at his great Hafod Works but a working-water-wheel was by the end of the nineteenth-century viewed as a suitable picturesque addition to the park of his domestic domain.

Some of the leats in the Nant Rhyd-y-Filais valley may have been designed primarily to supply boiler and condensing water to steam-engines; certainly such specialised watercourses are known to have been designed from such sources as the Boulton and Watt Collection of late eighteenth and early

nineteenth-century steam-engine designs.[147] What also certainly happened was that watercourses designed to provide water-power were later used to provide water for steam-engines. The archaeology of one instance of this still survives on the south-eastern side of Kilvey Hill where the later engine-ponds of the Tir-Isaf Colliery were connected to the water-power leat constructed for the use of the White Rock Copperworks in c.1736.

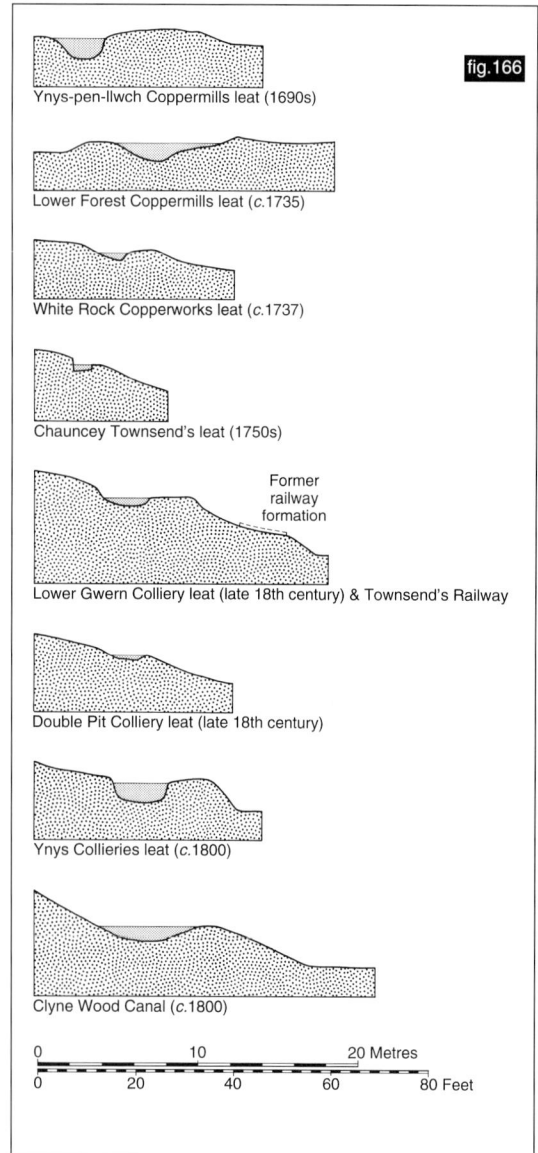

Ynys-pen-llwch Coppermills leat (1690s)

Lower Forest Coppermills leat (c.1735)

White Rock Copperworks leat (c.1737)

Chauncey Townsend's leat (1750s)

Former railway formation

Lower Gwern Colliery leat (late 18th century) & Townsend's Railway

Double Pit Colliery leat (late 18th century)

Ynys Collieries leat (c.1800)

Clyne Wood Canal (c.1800)

The Engineering of the Water Channels

Most of the water channels built were simple earthworks; the extent to which they would have been clay-lined is difficult to quantify without excavation. Their cross-section varied widely depending on the flow of water available at the point of use. The power available generally depended on the quantity of water that could be dropped through the height available in the buckets of a water-wheel except for the more simplistic undershot wheels on the main river which depended instead on the force of current of the main river. More advanced structural engineering was firstly required in enabling a limited watercourse to attain the height necessary to drive a large-diameter and powerful water-wheel, or water-wheels.

Most spectacular was the large stone arched aqueduct of at least five spans built in 1717 leading to the Llangyvelach copper, lead and silver works. Dr. John Lane, the entrepreneur, may have been trained in the Bristol copper-smelting industry but this type of installation was without known precedent in that area whose power sources were the substantial low-head flows of the Rivers Avon, Chew or Frome. However, in the upland landscape of Wales high-heads of water were available by this expedient and the Llangyfelach Aqueduct had a number of identifiable successors.

A 'tour' records that the Taibach Copperworks near Aberafan, opened in 1774, also had such a structure. However, metal works which smelted solely copper with a multitude of reverbatory furnaces (which needed no blast) had a lesser power-requirement than a lead-smelter with blown furnaces. Large ironworks blast furnaces had the largest power requirement of all and it is appropriate that the largest stone-arched water-power aqueduct in south Wales was that built by the ironmaster John Reynolds in 1823-24 at Pontrhydyfen where four elliptical arches still stride majestically across the river valley north of Port Talbot. Elsewhere in Wales a single-arched stone aqueduct can be seen conveying the headrace to the late nineteenth-century estate mill at Doldowlod in the Wye Valley in Radnorshire. Other large stone-arched water-power aqueducts in Europe include the mid to late nineteenth-century examples at Luxarnam in Cornwall, the Laxey Leadmine on the Isle of Man and at the Colonia Sedo textile factory (an 'English-style factory colony') near Barcelona.

In the lower Swansea Valley the only other masonry aqueducts, besides that supplying water to the Llangyfelach Works, were those built along the line of the Swansea-Trewyddfa Canals in 1794-96. These canals were primarily navigations rather than water-power supply channels but they did also power water-wheels from the time of their building. At the lower-end of the canal mainline the Swansea Canal cut across the large pool of the Pottery Mill and the canal instead of the Nant Burlais was used to power the mill. An aqueduct then both carried the Swnsea Canal over the Nant Burlais where it ran into the tidal River Tawe at a lower level and also spanned the early railway that ran down Cwm Burlais to the banks of the navigable river. Similarly two large masonry arches conveyed the Trewyddfa Canal across the Nant Rhyd-y-Filais ravine. One bridged the stream itself, and the second crossed an early railway incline pre-dating the canal. These aqueducts looked like the ordinary masonry aqueducts on any other navigable canal, but were in fact the first canal aqueducts in Britain to use hydraulic lime (impervious to, and capable of setting under water) rather than a bulky and heavy infilling of puddling clay. The limestone came from Aberafan in 1794-6 and was from a shallow deposit on the site of the present Port Talbot Docks.

Alongside the two Nant Rhyd-y-Filais aqueducts was an inverted syphon under the canal which fed Landore Cornmill, Landore Water-wheel Coal Pit and the Glamorganshire Pottery flint mill (with water at a reasonably high level). The inverted syphon was a well tried device much used by the Greeks and Romans to carry water across declivities without expensive masonry aqueducts or significant loss of height of water in the channel. It had been used in that prodigy of early modern canal engineering, the seventeenth-century Canal du Midi, so it was hardly an innovation when used at Landore. There were many such features built under Welsh canals to safeguard the supplies of pre-existing mills whenever the new canals and the old leats intersected on the level.

The aqueduct carrying the Swansea Canal over the lower River Clydach at Clydach is one of the larger single-arch aqueducts that was built along the line of the canal. On its southern side is an arched tunnel carrying the lower canal feeder from the river, at a point where the tailrace of the Lower Clydach Forge ran into the river, to a point on the canal below Lock Four.

Similarly, leats at a lower level to the canal could be accommodated in an aqueduct under the canal (such as the leat to Abertwrch woollen mill at Ystalyfera, in the upper Swansea Valley, from the combined tailrace of Melin Palleg and Gurnos woollen and tucking mills), or the culverts carrying the scouring tailrace channels from the large iron-ore opencast working at Gurnos, Ystradgynlais.

Timber troughs and aqueducts must once have been a very common feature: one of the few in south Wales still survives in Merthyr Tydfil carrying a leat over a later railway as part of the extensive water-power system of the Dowlais Ironworks and its mines. The earliest such trough found in Wales was part of a drainage system for the Bronze Age workings on the early opencast at the Copa Hill Mines in Cwmystwyth. The observations of the Swedish industrial reporter Eric Svedenstierna in 1803-04 noting the large numbers of aqueducts in the

Fig. 167. Section of Chauncey Townsend's Great Leat that survives as a field boundary above Gwernllwynchwith. The cross-section is noticeably larger at this point, having been augmented by several streams over its great length. (960214/2)

Swansea area have already been noted. Timber troughs are first mentioned in 1736 when the White Rock Copperworks proprietors were given permission to take the timber necessary to lead water from the hillside onto the adjoining water-wheel. In about 1775 the Plas-y-Marl Coal Pit Contour Leat was built from the headwaters of the Nant Rhyd-y-Filais at Landore, crossed the Nant-y-Gwm-Gelli on what was presumably a timber aqueduct, powered Cwm Pit and then ran around the southern flank of Graig Trewyddfa.[148] Between 1852 and 1876[149] when a small section of watercourse was still functioning to supply Cwm Pit it was shortened again, probably by the collapse of a timber aqueduct over the Nant-y-Gwm-Gelli, and a new supply channel was cut from that stream. This fell into disuse between 1876 and 1897.[150] Other aqueducts of this type took the compex of leats over the small natural watercourses and around the hillside above Gwernllwynchwith. Some of these may have predated Chauncey Townsend's great leat of the 1760s.

The cutting of the Trewyddfa Canal across Graig Trewyddfa in 1794-96 disrupted the existing pattern of watercourses and an aqueduct was constructed to carry the old leat across the Swansea Canal to the Forest Copperworks ponds. On the 16 miles of the rest of the Swansea Canal line there were two other such aqueducts over the canal; from the Pontardawe woollen-mill, and works reservoirs, to the Primrose Forge and Tinplate Works; and from the Ynysmeudwy cornmill to the Ynysmeudwy furze-mill, pottery and brickworks and subsequently to the two water-wheels of the Bryn tinplate works. The intensity of traffic that developed on the lower canal in the late nineteenth century required a greater water supply that was provided by a riverside steam-powered pump erected on the north-east side of the Clydach Aqueduct. The water was then fed across the Lower Clydach River in a timber trough supported by iron pipes.

In areas of south Wales in close proximity to ironworks some of these types of aqueducts were cast, or replaced in iron in the late eighteenth or early nineteenth centuries. This was the case in the Cyfarthfa Works Pont-y-Cafnau of 1792 and in two other water-power aqueducts supplying the

Cyfarthfa and Plymouth Ironworks at Merthyr Tydfil. In the 1820s, the Tennant Canal in the Vale of Neath was given a cast-iron trough to take it over the earlier navigable water-power channel at Aberdulais. In the 1830s the Neath Abbey Ironworks supplied cast-iron troughs to replace timber ones that carried three natural watercourses over the line of the Neath Canal. The first metal aqueduct in the Swansea area may have been the late nineteenth-century wrought-iron or steel aqueduct that replaced an earlier structure taking water over the Swansea Canal to the Forest Copperworks ponds. At neighbouring Llanelli, Alexander Raby was using large structural cast-ironwork from at least 1800 with his construction of iron railway bridges but there is no evidence that he, or any other Swansea industrialist applied this material innovation to Swansea area water-power engineering.

There is also little evidence of masonry wall construction being applied to Swansea area water-power engineering on anything like the scale of, for example, the large water-power dam at nearby Neath Abbey Ironworks or the large masonry retaining wall that once held the Glamorganshire Canal on the hillside of the Taff Valley to the south of Aberdare. However, there were some masonry dam and retaining structures associated with this engineering network. In the Clyne Valley, for example, the former overflow for John Morris II's New Mill is still visible and the retaining wall carrying his watercourse in a curve around his 1840s Ynys Pit was still visible until the 1970s when the extension of the Swansea rubbish-tip encroached on the area.

Recycling of water

With so many calls on water for industrial water-power use in the lower Swansea Valley it is not surprising that ways were sought to augment this supply. On the simplest level long watercourses were built that tapped water sources outside the valley or tapped sources available north of the intensively developed lower Swansea Valley. This was the case with the White Rock Copperworks Leat of *c*.1736 and Chauncey Townsend's great leat to Glais (east of Clydach) built in *c*.1752 respectively. The other artificial water-source, from outside the lower valley area, that was tapped for water-power purposes was the 17 mile length of the Swansea-Trewyddfa Canal which extracted water from the main River Tawe and its tributaries in the mountains of Breconshire from its construction in 1794-98. This watercourse itself was augmented by the water emerging from colliery adits and being pumped-up from various coalmines by the colliery steam-pumping engines. Consideration was also given to provide the canal with water raised from the River Tawe by steam-pump. The first of these to be considered was at Landore in 1803 in response to the possible additional water requirements of a proposed extension of the canal line around Swansea Bay to Mumbles: a proposal that never materialised. In the 1890s a steam-pump was in fact built at Clydach, downstream of the Ynys-pen-llwch Copper Mills, in response to the lack of boats then bringing lockfuls of water from the upper canal, alongside the continuation of intensive traffic amongst the works and mines at the Swansea end of the canal.

A more direct form of augmentation of supply was provided by the water lifted by the colliery steam-pumping-engines which could be used to wholly or partly drive the colliery winding water-wheels.

It is known that there were at least two steam-engines built that were especially designed to pump water back up to a water-wheel headrace from the tailrace below the wheels. One of these was probably next to the Landore Coalpit and was recorded in that vicinity in 1783. The second was built at White Rock Copperworks in 1805 to provide an adequate water-supply for the coppermills there.

The Landscape of Water-power

How typical was the water-powered industrial landscape of Swansea? This can only be commented on with reference to other landscapes of the coastal coalfield and the copper-smelting areas of the Severn Estuary. The date when waterpower was first used for colliery pumping in the Swansea area is impossible to determine, as is an exact quantification of its extent because of the destruction in the 1940s of the Jersey Estate Papers which covered most of the important Llansamlet coalfield to the east of the River Tawe. For an indication of what there might have been it is worth looking at the Llanelli area to the west where the archaeology of the early coalfield has already been the subject of a superlative study of the large surviving documentation and maps.[151] At Swansea the first surviving mention of an intended water-powered colliery is in the lease for the White Rock Copperworks by the local landowner/colliery owner Sir Bussy Mansel in 1736. At Llanelli there is a definite mention of 'John Allen of Llanelli' having two-three water-powered pits in the Wern area in 1740-49. These were probably fed by a fairly short lateral watercourse running along the east side of the River Lliedi in its last mile before entering the sea. If there were definitely water-powered coalpits at the smaller Llanelli coalfield in the early eighteenth century then it can be assumed that such power techology was applied to the pits of the much larger and active Swansea coalfield (at the beginning of

Fig. 168. Mine-water, coloured brown with iron oxide, still issues onto the riverbank at Morriston from Water Street from an old opening with cast copper-slag abutments. (9500100/22)

the eighteenth century the Swansea coalfield was producing ten times the amount of coal as Llanelli and at the end of the century about 30 times as much. In 1701 2,503 tons of coal were exported from Llanelli compared to 26,085 tons from Swansea and the figures for 1799 had increased to 12,005 and 245,000 tons respectively with in 1784 some 120,000 tons of coal being consumed by the Swansea copper smelters).[152]

In any comparison between the two areas it is worth also worth remembering the differences in topography. The Swansea coalfield lay along both sides of three miles of tidal river while the Llanelli coalfield occupied six miles of coastline. Coal outcropped at Llanelli on the edge of a low-lying estuarine marsh where extensive working by self-draining tunnel was impossible and where a very extensive use of pits dropping below the marsh had to be resorted to. Much of Swansea was different with considerable uplands dropping below the outcropping of the thick Swansea Five Foot seam and facilitating gravity drainage. However where the topography of the Swansea coalfield did resemble that of Llanelli was on the east bank of the Tawe north of the coastal Kilvey Hill where the valley scarp sloped away from the river in a north-westerly direction towards Llansamlet Church. At the more rural northern end of this area around Gwernllwynchwith there are the archaeological traces of several leat systems and some surviving map evidence also. The problem is with the intensively worked ridge between the evocatively named Winsh-wen (White winch) and Bon-y-maen where a combination of early suburban development and the 1940s destruction of the

Jersey Estate manuscripts and maps have removed detailed evidence. The other side of the ridge on the east, bordering Crumlyn Bog, was similar and the watered ponds of the large Glan-y-Wern Colliery of Sir Bussy Mansel still remain on the fringes of the marsh.

The probable intensity of usage of mining water-wheels in this area is suggested by the existence at the well-documented Llanelli coalfield of some 11 water-powered pits in the eighteenth century. The close proximity of the Llanelli and Swansea collieries also ensured that many individuals were active in the two adjacent coalfields. For example Thomas Price, the landowning coalmaster of Penlle'r-gaer, was also mining in the Bygyn area of Llanelli in 1729-31. One of the first recorded (1749) colliery 'water-engines' in the Llanelli coalfield was that pumping at Bynea from the 'water-level' of the Swansea Four Feet Vein. This was built and operated by Henry Squire, David Evans, John Beynon, all from Swansea. Another Swansea landowning coalmaster active in the Llanelli area is likely to have been the Thomas Popkin (probably of the Forest family) who married the daughter of Sir John Stepney, the Llanelli magnate, in 1747 and who mined coal at Cwmfelin before his death in 1770. Chauncey Townsend and two generations of his successors on the Llansamlet coalfield: John, Charles and Henry Smith, were also active in both areas. However, having acquired the mineral rights of nearly all the Llanelli coalfield (1752) and so having reduced the potential competition in the local coal market, concentrated nearly all their resources on the larger Swansea coalfield throughout the latter eighteenth century (except for a few years in 1766-69).

Even so the pit at Caemain Colliery, powered by a water-wheel, was provided by Charles Gwyn, Chauncey Townsend's agent or tenant, in 1762 and in 1766 Chauncey Townsend may also have had at least one water-powered colliery at the head of the Yspitty Canal.[153] There is no doubt that Swansea coalmasters were well-aware of the widespread application of water-powered technology to colliery production in the eighteenth century.

The landscape this produced was observed by the Swedish industrial reporter Eric Svedenstierna on his visit to Swansea in 1803:[154]

'To return to Swansea, one finds ... such a profusion of copper works, coalmines, ... ponds, canals, aqueducts ... that ... a stranger would be able to go around here for several days without being able to make order out of this chaos ...'

The picture of the widespread application of water-power usage in the early eighteenth century in the lower Swansea Valley is not totally consistent. In this rolling upland area where the thick Swansea Five Foot vein outcropped the southern area was worked by the Prices of Penlle'r-gaer by horse-winders and steam-pumping engines from 1731. Further north on Graig Trewyddfa the workings were operated by the landowner/coalmaster Thomas Popkins until supplanted by the Lockwood, Morris coppermasters. Robert Morris had taken over the bankrupt concern of the Llangyfelach Copperworks in 1726 but only managed to operate the works successfully from 1727 with a large injection of capital from the London mercantile fortunes of Lockwood and Gibbons. A picture of 1747 shows very primitive hand-operated windlasses in use on the partnership's Treboeth Common pits (possibly including their Pentre Pit at Landore) and in 1767 they were still building elaborate horse-engine operated pits like that at Tirdeunaw. By 1754, however, they did have a powerful water-wheel winder and pump at the Landore Coalpit and then a second at the Clyn-du Water machine Pit on the hillside overlooking the site of what became Morriston.

The landscape of power became transformed by the arrival of steam-engines on a large scale. At this stage in Swansea a limited volume of water available from tributary streams was fed onto fairly large diameter water-wheels by long circuitous valleyside leats. At Llanelli the tidal flats only allowed small falls of water which required broad wheels to generate enough power to power their collieries: 12ft wide at the Bres Pit. Replacing them were tall beam-engine houses with attendant boiling- and condensing-water ponds fed by small leats and watercourses.[155] Upland regions of Britain tended to remain dependent on water as a main power-source for some time after the widespread adoption of steam as the prime-mover in lowland districts. However, the rate differed in proportion to the extent of the capital expended in the development of the industry or coalfield. The low investment of the majority-operating Townsend/Smith dynasty at Llanelli may have had a critical effect on the slow development of steam-power in that coalfield. The Swansea coalfield, by contrast, had its first recorded steam-pumping engines installed in 1717 and 1731 whilst the first at Llanelli was not installed to c.1750. By 1800 the Llanelli coalfield is known to have had six steam-engines; only just over half the number (11) known to have existed at Swansea by that date. At Swansea new water-power ponds in the Clyne Valley and Cwm Burlais were built and used well into the nineteenth century for powering collieries but the amount of available water for all the increasing number of industrial uses was finite and the development in the scale and intensity of industry demanded a large-scale use of steam-power.

In the eighteenth century the Swansea

copperworks required a variety of types of water-supply leats to supply their needs. The earlier copper and brass manufacturing centre of Bristol had similar requirements but their impact on the landscape was rather different and it is instructive to examine why this was so. Of some 18 rolling and battery mill sites around Bristol, some six were on the main rivers of the Avon and Frome, involving relatively short but wide watercourses providing a low head and powering undershot water-wheels. The several mills on the tributary River Chew were of similar type with broad watercourses cutting across meanders. At all these mills Swansea copper was processed into goods and also used as a raw material for brass, which was in turn turned into vessels for the market. Indeed, it was a Mr. Padmore, 'a capital mason from Bristol' who laid-out the site of the Lower Forest Coppermills on a similar site on the River Tawe at Swansea. The earlier mills at Upper Forest, Ynys-pen-llwch and Ynysygerwn (the latter on the neighbouring River Neath) were all similar to the layout of the Bristol mills.

Generally the area around Bristol was gentler and the east Bristol coalfield was not noted for its wide use of water-power. Sited on the coalfield was the Warmley Zinc and Copperworks which impounded the small Warmley Brook to form a thirteen acre lake with a three-storey statue of Neptune in its centre and house over a bridge at its head, both built of copper-slag blocks. In the lower Swansea Valley there were only two large water-power lakes: at the 1720

Cambrian Copperworks and the 1784 Upper (iron-) Forge at Clydach. Neither had an ornamental aspect but the earlier Gnoll Park Lakes in the neighbouring Vale of Neath did.

A third and commoner type of waterpower installation was represented by the long and circuitous leats arranged around the sides of the steeper south Welsh Valleys. The precedent for this was the well-capitalised venture established by the former London barrister Sir Humphrey Mackworth (1657-1724) at Neath after he had married the local Gnoll estates heiress, Mary Evans, in 1686. His new lead and copper-smelter was at Melin-Cryddan but the water-powered complex he established was to the north of this and east of the town of Neath. The head of the system was two very large embanked reservoirs on the summit of Cefn Morfudd, east of Aberdulais and to the north of the Gnoll. Watercourses led south to 'the upper Great Pond,'[156] lying along the east of the Gnoll hill. A watercourse, or leat, led westwards from the head of the upper pond along the northern and western sides of the Gnoll to provide 'The Gnoll' mansion with water, it then fed into 'The great Pond' which lay along the foot of the Gnoll hill on its southern side. Two other watercourses fed into the south-eastern corner of the two large pools at their junction, that to the south had at least three smaller pools or reservoirs along its course. Another leat ran northwards, from a stream to the east of the Melin-Cryddan smelter to 'The great Pond.' Water from 'The great Pond' ran out

Fig. 169. John Morris I had a watercourse engineered in 1800 that ran along the eastern side of the Clyne Valley to drive water-wheels operating his Ynys collieries. A substantial part of this leat survives in the Clyne Valley Country Park. (9500051/19)

fig.170

Fig. 170. The Clyne River runs through the trees on the left with a track on its right fringe that was the formation of a colliery branch of the Oystermouth Railway. Mill Wood on the right-hand side of the valley formed the south-western fringe of the Morrises' Sketty Park Estate. Jernegan's 'Belvedere' survives just below the first two arcs of later twentieth-century housing on the right. (955062/51)

through 'The double Battery Mill' and into a lower pool that fed 'A single Battery Mill'. The pool below this then received an additional feeder from the north and in turn drove 'A Mill for Rolling & Slitting of Iron and Brass'. The tailrace water from this then flowed down to a small pond driving 'A Mill for making brass Wire.' Below this point the water discharged through the town of Neath and into the River Neath alongside. An added landscape feature of this elaborate system was the pump and pipeline from 'The upper great Pond' by which 'The Water [was] forc'd from ye Pond to ye top of ye Hill' and so to supply the domestic needs of the Gnoll mansion itself. This system would have been known to Dr. John Lane when he left his first works in the lower Vale of Neath in 1717 in order to found the Llangyfelach Works in Swansea.

One point apparent from the examples discussed above is the frequency with which a watercourse, water-power site and/or buildings or a water-wheel

actually changed function once, or several times, from the purpose for which they were designed. Quantification of the elements of the Swansea water-power is difficult because of the uneven nature of the surviving documentary and topographical record but there must have been well over a hundred water-wheels in the lower Swansea Valley area providing power for a range of industries in the later eighteenth and early nineteenth centuries.

Certain areas and side-valleys in the industrial landscape had a particularly complex and complicated development such as the areas examined in the Nant Rhyd-y-filais valley at Landore and the Clyne Valley. Swansea was not unique in this: there are similarly developed dense nodes of industrial development intermittently throughout Wales. The Greenfield Valley in Flintshire was one such area that was also linked to Swansea by the nature of its trade. It had a large water-power resource in a similar proximity to navigable water at its mouth, as was also the case in the lower Swansea Valley and this led to a huge concentrating of water-powered copper manufacturing capacity using the copper produced at the great Anglesey coppermines and often smelted at St. Helens in Lancashire. Later, much Swansea copper was also processed here and by concerns that had works in both places. Interestingly the form of the water-power installations, now very well conserved, forms a significant contrast with those of Swansea, or even the other great centre of British copper-manufacturing around Bristol. At the Greenfield Valley, a small but fairly constant stream of water from St. Winifred's Spring and the navigable mine-drainage Holywell Level was utilised through a series of large pools impounded by substantial earthwork and masonry dams. At Swansea, similar large pools only really existed at the mouth of the Burlais Valley and on the Lower Clydach River at Clydach with the difference that the gradient of these minor watercourses at Sansea was not so severe so that these pools were of relatively small capacity with low dams necessitated by the topography.

There were other areas of Wales that had a similar density of water leats and water-powered industry. The wire-mills built in the Angiddy and Whitebrook side valleys of the Wye in Monmouthshire from the sixteenth and seventeenth centuries, together with succeeding paper-mills have left a legacy of interconnected leats and ponds that equals that once found in the Nant Rhyd-y-Filais Valley at Swansea. The form of the water-power installations is between that of the high-dammed pools in the Greenfield Valley and the pondless leats commonly found in the Swansea area and is more analogous to those low masonry dams found in the seventeenth-century copper-mills of the associated Bristol region. In the Swansea area the only surviving dam of this type is that of the late eighteenth-century upper forge at Clydach.

Within south-west Wales the late nineteenth-century concentration of water-powered woollen mills at Drefach-Felindre is another dense concentration of water-power but here simple linear leats parallel to the tributary streams of the River Tawe served a number of small mills in succession.

The vast metal mining districts of Wales were generally remote from sources of coal that might have been used for powering, winding and pumping and continued to depend almost exclusively on water until the end of the nineteenth-century. The positioning of water-powered coal-shafts at both Swansea and Llanelli was largely determined by the location of the outcropping of the major coal-seams, the rate of dip of the seams and the patterns of land ownership and of mineral leases. Such factors were even more restrictive in relation to the metal mining-areas which were developed in relation to the narrow vertical veins of mineralisation and not the mostly horizontal seams of coal which characterised the working coalfields. The localised upland location of the shafts that this determined entailed the construction of a web of very long watercourses, by the later nineteenth century, across areas of upland Wales, such as north Cardiganshire, that were as large, or even larger than Chauncey Townsend's great mid-eighteenth coal-mining and metal-smelting leat at Swansea. A precursor of all these had been the great water-scouring leats at Dolaucothi gold mine in Carmarthenshire, generally attributed to the Roman period.

In general, water-power was the predominant source of power used in the earlier Industrial Revolution in Wales, just as it was in Scotland.[157] In the Swansea area the extensive use of waterpower necessitated the construction of a system of water-channels over the lower valley which helped determine the siting of industry. The evolving industrial landscape in the Swansea area was a microcosm of the development of transport and water usage in the south Wales industrial area.

References

1. The river was navigable for 3.7 miles from the 'bar' at its entrance (i.e. from the later mouth of the enclosed harbour of Fabian's Bay).

2. Jones, *Swansea Port*, 317.

3. Badminton MSS. 898-899.

4. See the illustrations of Beaufort Bridge.

5. A. H. John,, 'Introduction: *Glamorgan, 1700-1750'*, *Glamorgan County History, Volume V, Industrial Glamorgan*, ed. A. H. John and G.Williams (Cardiff, 1980), 1-46, 16.

6. Jones, *Swansea Port*, 34.

7. Ibid.

8. William Edwards was known as William Edward in contemporary documents.

9. Badminton MSS. various.

10. R. O. Roberts, 'The White Rock Copperworks', *Glamorgan Historian*, II, (1982).

11. See the illustration of the Birmingham Copperworks Dock.

12. Reproduced on p.56 in 'Mr. Price's Report', Reports on the Formation of a Floating Harbour at Swansea with reference to plans submitted to the trustees, Printed by Order of the Trustees, 1831, Swansea. Included in a composite volume entitled *Reports on the Harbour of Swansea*, 43-61.

13. Shown on the engraving of White Rock Works in 1744 reproduced opposite p.116 of Francis, *Swansea Copper*

Smelting and Griffith Price's and C. R. Jones's coal-spouts are shown on 'A Plan of the River Swansey in Glamorganshire, taken in 1771 by B. Jones (reproduced on p.54 of Jones, *Swansea Port*).

14. The Swansea Vale Railway was made into a more sophisticated public railway in 1852 and may then have had the more advanced types of boat loading stage.

15. Jones, *Swansea Port*, 184.

16. Ibid.

17. Ibid.

18. William Kirkhouse's Letterbook (Swansea Reference Library, Local History Collection).

19. According to the *Swansea Canal Case* ships of only up to 200 tons could reach Landore wharves.

20. Trewyddfa Canal Traffic Book.

21. E. Nance Morton, *The Pottery and Porcelain of Swansea and Nantgarw*, Plan V, 71.

22. Badminton MSS. 2080.

23. The 1841 census makes no clear distinction between the boatmen or bargemen employed either on the canal or river. The total numbers of boatmen for the respective waterways are presumed on the basis of a knowledge of their dwellings.

24. Jones, *Swansea Port*, 68; the Trustees first met in 1791.

25. Ibid., 191.

26. Badminton Collection, Group II, MSS. 2080. For more details of this and other railways at Landore see Chapter Four.

27. Lewis, *Wooden Railways*, 110 and 232.

28. Ibid., 314.

29. Ibid., 323-24.

30. John, *Llansamlet Mining 1700-1740*.

31. (Badminton MSS. 1285, d. 1768 and Swan. Mus., small maps and prints collection, 356 n.d. but between 1757 and 1770). The line was built by Calvert Richard Jones, a local landowner (labelled 'Mr. Jones's Waggonway'): Mr. Calvert Jones had the southern part of the Copper House Wharf ('C') in 1768 (Badminton MSS. 1285, d. 1768 and *B. Ferry estate plans*, 76, (before 1780 until after 1798).

32. A. Rees, *Rees's Manufacturing Industry*, Volume 2 (Newton Abbot, 1972), 32. Selective reprint of Rees's *Cyclopaedia* published in 1802-19.

33. Now the site of Manselton and not to be confused with the later Pentre Pit at Landore: Cowley, *Pentre Colliery*, 4.

34. Cowley, *Pentre Colliery*, 4.

35. Ibid.

36. The first 'railway tunnel' usually noted on Tyneside was a 1770 drainage tunnel also used for transport (Flinn, British Coal, 151).

37. Badminton MSS. 2136 & *Report from the Committee on the Petition of the Owners of Collieries in South Wales, House of Commons, 1810* (344) IV, 51.

38. 'Clyndu Colliery', 1838, and 'Map of Park Level Colliery', Sept. 1839, H. Richard (R 10493) and

39. 'Cwmgelly Colliery 5' Seam', Richard & Mathews, 1911 (Plan No. 5782) (T. Bryn Richard MSS., UWS).

39. The same was true elsewhere in south Wales. Lewis, *Wooden Railways*, 247-48 and evidence of the 1810 Committee on Collieries in South Wales, p.60.

40. 'In 1788 there was about 240 tons of Cast-iron Tram Plates in Landore Coll'y' (*J. Morris C. Book*).

41. Lewis, *Wooden Railways*, 262-67.

42. Trinder, *Industrial Landscape*, 148.

43. Ibid.

44. Ibid.

45. Hadfield, *S. Wales*, 67-8, the engineer was described as 'Evan Hopkin of Llangyfelach' in 1805 (Jones, *Swansea Port*, 137).

46. Lewis, *Wooden Railways*, 292.

47. Ibid., 293.

48. Ibid., 317; for the 1738 iron-rails in Whitehaven see J.I.C. Boyd, *Narrow Gauge Railways in North Caernarvonshire; Vol.2, The Penrhyn Quarry Railways* (Oxford, 1985), 16.

49. It would be interesting to know what the circulation was of Outram's two similar accounts. One was his *Observations* prepared for the shareholders of the Brecknock and Abergavenny Canal, printed for them in 1799 (one extant copy is N.L.W. Mayberry MSS. 383). The other was his *Minutes to be observed in the construction of railways*, reproduced in M.F. Outram, *Margaret Outram 1778-1802* (London, 1932), 350-353.

50. Lewis, *Wooden Railways*, 293.

51. N.L.W., Mayberry MSS. 383.

52. Lewis, *Wooden Railways*, 42.

53. Ste.-Marie-aux-mines (Markirch) in the Liepvre Valley (Leberthal) of Alsace (Lewis, *Wooden Railways*, 51).

54. Baxter, *Tramroads*, 18-19.

55. *J. Morris' Commonplace Bk.*

56. Ibid., each plate laid would have approximated to the size originally proposed, as three and a half miles of track would have consisted of some 9,240 plates of 4ft. in length weighing 58lb. each. If the plates conformed to measurements taken in 1826-27 (*English Railways 1826*) then 9,363 plates of 3ft 11.37 in. in length would have weighed 57.4lb each.

57. Lewis, *Wooden Railways*, 321 and 265.

58. C. Hadfield and A.W. Skempton, *William Jessop, Engineer* (Newton Abbot, 1979), 173.

59. Baxter, *Tramroads*, 18-19. Lewis, *Wooden Railways*, 311.

60. See also: P.R. Reynolds 'An Unusual Type of Tramplate', *S.W.W.I.A.S.N.* 20 (November 1978), 6-7. However, the plate provenanced to Cribarth was in fact found by Gordon Rattenbury and Stephen Hughes on the line of the Garth Pit Waggonway which localises the type to the lower Swansea Valley area. The detailed chronological sequence of the Landore Waggonways as worked out for this study also places the plate from Pentre Pit at an earlier date than the 'terminus ante quem' established by Reynolds. A standard

contemporary description of all types of rail in use in Britain is found in *English Railways 1826*. Other modern sources include: D. Pollard, 'Bath stone quarry railways 1795-1830', *BIAS Journal*, 15 (1982), 13-19; B.E. Osborne, 'Early Plateways and Firestone Mining in Surrey, An interim report', *Proceedings of the Croydon Natural History and Scientific Society Ltd.*, 17, part 3 (February 1982), 74-88; Riden, P.J., 'Plate Rails at Codnor Park', *Industrial Archaeology, The Journal of the History of Industry and Technology*, 10, 1 (February 1973), 77-82; R.G. Gilson and G.W. Quartley, 'Some Technical Aspects of the Somerset Coal Canal Tramways', *Industrial Archaeology*, 5, 2(May 1968), 140-155; K.W. Bean, 'Plate Rails at Godstone', *Industrial Archaeology*, 7,2, 184-189; D.E. Bick, 'Tramplates of the Gloucester and Cheltenham Railway', *Industrial Archaeology*, 3, 3 (August 1966), 193-200; J. van Laun, 'Excavation on the Hay Railway', *J.R. & C.H.S.*, XXIII, 3 (November 1977), 83-86; D.E. Bick, 'Cast-iron Tramplates', *J.R. & C.H.S.*, XXIII, 2 (July 1977), 69 and J van Laun, D.R. Steggles and P.G. Rattenbury, 'The Hay Railway', *J.R. & C.H.S.*, XXIV, 2 (July 1978), 74-77 and Baxter, *Tramroads*, 37-58. Plate-rails as used by Thomas Telford are illustrated in his *Atlas to The Life of Thomas Telford* (London, 1838).

61. One was the Oystermouth Plateway that was largely built by the former partners of the influential Lockwood, Morris Partnership and their landlord. Such influence can be seen by the fact that in about June 1807 the Swansea Harbour Trustees requested 25 tons of iron tramplates 'of the same dimensions as those used by the Oystermouth Tramroad Company' for the construction line along the Eastern Harbour Pier (Jones, *Swansea Port*, 136). The William Bevans who may have built the plateway to the northern of the two Graigola Levels (Collier, *S. Wales Coal*, 31), were in business near the Landore and Morriston Collieries of Morris and Lockwood (Thomas, *Swansea Districts I*, 67) and indeed did work on the waggonways of at least Lockwood and Co. ('Report Book of John Blenkinsop', 66 (Leeds City Archives, 1546)).

62. The weights of the Landore plates are given in two forms in the 1826-27 description (*English Railways 1826*) which work out variously at about 45 and 84 lb. The rail lengths of 3ft. 10in. are almost certainly among those other dimensions measured by Dechen and Oeynhausen in Prussian inches. If converted to British imperial units this gives a length of some 3ft. 11.37 inches.

63. Collier, *S. Wales Coal*, 28-29.

64. The rails used by Smeaton in 1756-59 were nine inches wide and formed a 'timber road, commonly called at the Collieries, where they are used, a Rail Road'.

65. See attached table of plate-rail data.

66. The Cwm Clydach and Cwm-twrch ('Palleg') Waggonways (*Canal Map of 1797-99*). The correct use of the terminology 'rail road' to mean an edge-railway is confirmed by the fact that the line was still laid with edge-rails of the early 'Losh' design in 1826-27 (*English Railways 1826*, 55-56).

67. Edward Martin did much canal and waggonway engineering and was also a businessman in his own right. He had permission to build the Cwm-twrch or Palleg Waggonway in 1797-98 (Canal Minutes, 4th July 1797 and 8th July 1798) in order to work coal from levels opened by him and his father-in-law (Thomas, I.A. *Upper Tawe*, 68 and Roberts, *Gower Coalmining*).

68. *Engineer's Report.*

69. The line is clearly marked as a 'tramroad' on the *Canal Map of 1796-99*.

70. J.D.H. Thomas, unpublished M.A. thesis on 'The Industrial Development of Llangiwg Parish', University of Wales, 1974, p.43.

71. Thomas Sheasby was Harper's executor.

72. *Outram's Observations.*

73. Lewis, *Wooden Railways*, 319.

74. Baxter, *Tramroads*, 46.

75. Lewis, *Wooden Railways*, 319.

76. At nearby Chesterfield was one of the first two places where plates of Curr's type were used on the surface (Lewis, *Wooden Railways*, 318). These plates at Butler's Wingerworth Ironworks weighed 24 lb. per 3 ft. (C. Hadfield and A.W. Skempton, *William Jessop, Engineer* (Newton Abbot, 1979), 173).

77. The Dadfords were busily constructing iron edge-railways to the Glamorganshire, Brecknock and Abergavenny, and Monmouthshire Canals in the early 1790's (Lewis, *Wooden Railways*, 16).

78. C. Hadfield, *The Canals of the East Midlands* (Newton Abbot, 1966) 51.

79. P.J. Riden, 'The Butterly Company and Railway Construction, 1790-1830', *Transport History*, 6, 1 (March 1973), 30-52.

80. C.D. Marshall, *History of British Railways down to the Year 1830* (London, 1938) 147-48.

81. Lewis, *Early Wooden Railways*, 261.

82. Ibid., 292-93.

83. J. Childs, 'Inside Penlle'r-gaer House Two Hundred Years Ago', *Gower*, 41, (1990), 22-38.

84. Ibid. quoting NLW Penlle'r-gaer MSS B 7:6.

85. Cowley, 'Pentre Colliery: A Short History', 3-7.

86. Ibid.

87. Childs, 'Penlle'r-gaer', 37.

88. Cowley, 'Pentre Colliery', 3.

89. Jones, *Swansea Port*, 105.

90. Ibid. 75-6.

91. Cowley, 'Pentre Colliery', 3.

92. Ibid.

93. Neath Abbey Ironworks drawings M/141 (WGRO).

94. D/Lous/W/Collieries 138 (Cumbria County Record Office, Carlisle).

95. Clyn-du adit site at SS 6698 9734; ventilation shaft sites at SS 6690 9740 and SS 6683 9753; site of Clyn-du steam-engine shaft at SS 6675 9770; site of Clyn-du water machine shaft at SS 6647 9715 and position of the end of the canal underground at SS 6571 9829.

96. See S. Hughes, 'The Development of British Navigational Levels', *Journal of the Railway and Canal Historical Society*, XXVII, 2 (July 1981) 2-9, for a full discussion of what 'mining' or navigation levels are.

97. The Gwauncaegurwen Underground Canal was one. The Rhandir-mwyn Upper Boat-level near Llandovery may have been built by the merchant/entrepreneur Chauncey Townsend (G.W. Hall, *Metal Mines of Southern Wales* (Westbury-on-Severn, 1971) and W.J. Lewis, *Lead Mining in Wales* (Cardiff, 1967) 164). Townsend had built a lead smelter ('spelter') fairly close to the Forest Copper and Lead Smelter and its Clyn-du Underground Canal in 1755 and second & third smelters in *c.* 1757 (Roberts, *Copperworks Chronology*) with the John Smith who was probably also the manager of Rhandir-mwyn by the 1760's. A letter from John Griffiths to the Earl of Powis on the 1st April 1767 states that someone had gone 'to Mr. Smith's Mines at Cerrig y mwyn near Llandovery'). Townsend was acquiring lead mines in the interior of Wales from the 1750s. He brought forward false claims on the Grogwynion and Logculas Mines in Cardiganshire and took a lease of them in 1758. He also took a lease of the great Cwmystwyth Mine in 1759 and perhaps significantly drove a deep haulage and drainage level, the 'Lefel Fawr' (= Great Level). Smith was also probably managing operations on that site (S.J.S. Hughes, 'The Cwmystwyth Mines', *British Mining No. 17*, A monograph of the Northern Mines Research Society (Sheffield, 1981).

Rhandir-mwyn had like Cwmystwyth been previously exploited (In a list of 'Lead Ores' in Wales in Lewis Morris' notebook is an entry of 1747 for the 'Cerrig y Mwyn Mine near Ystradffin.'). Townsend had proposed his first surface canal in 1752 (he later actually built two) and of course had built the very early railway tunnel at Landore between 1762 and 1768. John Smith later built the Llansamlet Canal with its substantial Tunnel through the White Rock Copperworks site.

98. Hughes, *Navigational Levels*'.

99. Ibid.

100. See the following section on the Lefel Fawr Mining Canal.

101. See the entry for the 'Rhydydefaid Colliery Canal'.

102. E. Martin, 'Description of the Mineral Bason in the Counties of Monmouth, Glamorgan, Brecon, Carmarthen and Pembroke', *The Philosophical Transactions of the Royal Society of London,* (1806), 342-347.

103. Francis, *Swansea Copper Smelting*, 101.

104. R.O. Roberts, 'The Smelting of Non-ferrous Metals since 1750', *Glamorgan County History, Volume V, Industrial Glamorgan* ed. A.H. John and G. Williams (Cardiff, 1980), 47-96.

105. Ibid.

106. Jones, *Swansea Port*, 308 and in 1835 'mills on the said ffarm called the fforest [had been]... lately erected' and ' a parcel lately inclosed on Trewidva Common'. (Badminton MSS. 898-899). Clarifying the purpose of the letter lease is a reference to 'a parcel inclosed from Trewyddva fforest, & all the buildings erected on the last mentioned parcel (Forest Copper Works) for smelting copper and lead...' in a 1746 lease renewed in 1754 (Badminton MSS. 1261).

107. Joseph Harris, 'Journal of two visits to Wales' in 1746 and 1748 (Ynysfor MSS., N.L.W.). My thanks to Peter Wakelin for bringing this source to my attention. It is unfortunate that Harris was only interested in the metallurgical processes at the works as befitted being the assay-master of the royal mint (see P. Wakelin and J. Day 'Joseph Harris in Bristol, 1748', *Bristol Industrial Archaeological Society Journal*, 15 (1982), 8-12) and hence ignored any possible innovations by his compatriots in coal-mining techniques. Both Llangyfelach and Forest Copperworks were in production in 1748 but only Forest was producing copper (N.L.W. MSS. 15, 103-B, pp. 25-29.

108. The White Rock Works opened in 1837 (Roberts, *Copperworks Chronology*) had such a level as did the Middle and Upper Banks Works opened in 1755 and c. 1757? (Roberts, 'Non-ferrous Metals', *Glamorgan County History*). ' A level was driven from the Middle bank Copper Works to the Warkey Vein & along its 700 yards to the east' and 'A Level was driven from the Upper Bank Copper Works to the Two feet or Cwm Vein'. The very high value based on these 'on-site' coal resources available for the prodigious consumption of the smelters is highlighted by the fact that both these levels were driven to seams only 2 feet thick (Logan, *S. Wales Coalfield*, 1837).

109. 'A Map of the Fee of Trewyddfa in the Parish of Clace within the Parish of Langevelach in the County of Glamorgan Belonging to His Grace the DUKE OF BEAUFORT. This map was drawn by Lewis Thomas in 1761, from a survey made by Whitterley some years before, with some additions thereto made by Lewis Thomas in 1761.' The largely rotted original is Badminton MSS.1278. A very good copy was made by Wm. Davies in 1857 and accurately copied by J.M. Davies in 1940. This appears as Map 6 in the latter's unpublished U.C.W. M.A. thesis of 1942, 'The Growth of Settlement in the Swansea Valley.'

110. J.E. Ross, *Radical Adventurer: the Diaries of Robert Morris, 1712-1744* (Bath, 1971), 177.

111. Information on waterlogged conditions of collieries from J.M. Davies.

112. '26 May 1845, Rees Morgan Evidce as to Lockwood & Co's Old Canal', 'taken at Clyndee level mouth, by me

J.P.H. present Mr. Redhead & Morris the Canal Agent' (Badminton MSS. 2136).

113. *Svedenstierna's Tour.*

114. My thanks to Dr. Fred Cowley for drawing to my attention a later copy of a map showing the canal (now in the T. Bryn Richard Collection, UW Swansea.). An 1838 original of that map of 'Clyndu Colliery' is in the Abandoned Mines Record Office as are a 'Map of Park Level Colliery' Sept. 1839, H. Richard (R 10493) and 'Cwmgelly Colliery 5' Seam', Richard & Mathews, 1911 (Plan No. 5782).

115. Shown on ibid. and on the map of Landore-Morriston (as '25' and '26') in c.1794 reproduced in this section. It seems quite likely that the Clyn-du steam-engine shaft partly utilised an earlier ventilation shaft to the canal.

116. Rees Morgan, 1845 (Badminton MSS. 2136).

117. Thomas, *Swansea Districts* I, 86.

118. The site of the mouth of Graig Level was at SS 6676 9672.

119. *Beaufort Plan & Case.*

120. v.i. Charles Hadfield.

121. A.M.R.O., 'West Glamorgan Collieries Ltd. Swansea', 'Plan of Workings, Rhydydefaid Colliery', Richard & T.M. Mathews. 1.9.'20. Plan No. 7190, Reg. No. 11/5386: Many late mining plans incorporate information from a long succession of available predecessors.

122. N.M.R. Mss. 'Burry Port, History of its Industries, Communications, and Coal Mines' by R.G. Thomas, M.E.

123. F. Mullineux, 'The Duke of Bridgewater's Underground Canals at Worsley', *Transactions of the Lancashire and Cheshire Antiquarian Society*, 71(1961), Fig. XXVII.

124. Trefor Ford and J.W. Rieuwerts (eds), *Leadmining in the Peak District*, (Derby, c.1979) 60).

125. Roberts, *Gower Coalmining*, 67 and Boucher, C.T.G., *James Brindley, Engineer, 1716-1772* (Norwich, 1968) 103. William Shaxby was Harper's brother-in-law. (Thomas, I.A. *Upper Tawe*, 69).

126. A close relationship between the two is probably indicated by Sheasby becoming an assignee of Harper's property after the latter's death.

127. 'Lower down (than Pentre'r Lamb) there is a small coalmine by the side of the river. Nearby is the canal head' (letter from Benjamin Malkin in June 1801, quoted from D. Watkin Morgan, 'An outline History of the Upper Reaches of the Tawe River', *Brycheiniog*, XII, 1966-67, 121-130. A date for the level of 1801 is also given in that article. A date of 1805 is given for driving the level of 1805 in Thomas, I.A. *Upper Tawe*, 69. Both accounts are based on rather garbled oral sources but the story of the underground canal cannot have made sense unless the level was at least planned in the 1790s as seems likely from what reliable evidence exists. The level mouth is at SN 8154 1247.

128. D. Watkin Morgan, 'An Outline History of the Upper Reaches of the Tawe River', *Brycheiniog*, XII, 1966-67, 121-130. If William Watkins had been in charge of the driving of 'Lefel Fawr' in 1801 he would only have been nineteen and at a time when he was supposedly busy in Edward Martin's new mines in the Twrch Valley.

129. Thomas, I.A. *Upper Tawe*, 69.

130. A.H. John, 'Introduction: Glamorgan, 1700-1750', *Glamorgan County History, Vol. V, Industrial Glamorgan*, 1-46, 12.

131. For detailed information on Swansea Canal powered mills see S.R. Hughes, 'The Swansea Canal: Navigation and Power Supplier', *Industrial Archaeology Review*, IV, 1, Winter 1979-80, 51-69.

132. D. Crossley (ed.), *Water-power on the Sheffield Rivers* (Sheffield, 1989) v-xv.

133. R.O. Roberts, 'The White Rock Copper and Brass Works, near Swansea, 1736-1806', *Stewart Williams's Glamorgan Historian*, 12, 136-51.

134. Ordnance Survey Drawings, c.1813.

135. West Glamorgan Archives Service, D/D B.F. E/1 'Survey of Briton Ferry estate lands, plans and reference tables', n.d. but a compilation of various dates, 71 and 83.

136. References to the 1840s topography in this section are drawn from the Tithe Map for Llansamlet.

137. *State of Llansamlet Colliery, 1771-72*, 250.

138. *Gwernllwynchwith Colliery Plan, 1808.*

139. The detailed references to installations in the Landore area have already been published by Stephen Hughes in the *Bulletin of the South-West Wales Industrial Archaeology Society* and the water-power section in S. R. Hughes, "Landore: a study of water-power during the Industrial Revolution", *Melin* 3 (1987), pp. 43-59.

140. For details of the Swansea Canal sites see S.R. Hughes, 'The Swansea Canal: Navigation and Power Supplier', *Industrial Archaeology Review* 4, 1, (Winter 1979-80), 51-69.

141. *Badminton G.II. MSS.*, 2134.

142. NLW, Brecon Q/RP 35-36.

143. For details of this and other remains in the Swansea area see: S.R. Hughes & P.R. Reynolds, *A Guide to the Industrial Archaeology of the Swansea Region* (Aberystwyth, 2ed., 1989), 17.

144. M.V. Symons, *Coal Mining in the Llanelli Area; Volume One: 16th Century to 1829* (Llanelli, 1979), 97 and Badminton G.II. MSS., 2134.

145. *Bulletin of the South-West Wales Industrial Archaeology Society* and *Melin*.

146. S.R. Hughes, 'The Swansea Canal: Navigation and Power Supplier', 56-8.

147. Wanlockhead leadmines engine, south-western uplands of Scotland.

148. B.G.mss II, 2080.

149. Vivian mss C3 (1852): UCL (Swansea); O.S.25", Glamorgan XV 13 (1876).

150. O.S.25", Glamorgan XV 13 (1876 & 1897).

151. For details of the early Llanelli coalfield see M.V.

Symons, *Coal Mining in the Llanelli Area; Volume One: 16th Century to 1829* (Llanelli, 1979).

152. Ibid. and John Williams, 'The Coal Industry, 1750-1914' in A.H. John & G. Williams (ed.), *Glamorgan County History, Volume V, Industrial Glamorgan from 1700 to 1970,* (Cardiff, 1980), 155-210, 157-58.

153. Symons, 40-42, 62-65, 74-82 & 315.

154. E.T. Svedenstierna, *Svedenstierna's Tour of Great Britain*, (Newton Abbot, 1973), 42.

155. Symons, ibid.

156. Most of the Melincryddan and Neath water-power system is shown on 'A PLAN of the TOWN & PORT of NEATH in the COUNTY of GLAMORGAN. Being part of ye Estate of Sr. Humphry Mackworth. Situated near ye Bristol Channel.' By M. O'Conner, 1720.

157. J. Shaw, *A History of Waterpower in Scotland* (Edinburgh, 1980).

CHAPTER 3

'COPPER-MEN, COLLIERS & MECHANICS'[1]

fig.171

Swansea Engineers and their Innovations

The dense industrial infrastructure which was established at an early date in the area of Swansea and other parts of Glamorgan provided a natural training school in which local artisans were able to develop into innovative engineers. Others were brought ready trained from early coal-mining districts in the north of England. The great bulk of engineering biographical literature has concentrated on the 'Heroes of the Industrial Revolution' whilst largely ignoring the lesser figures who contributed to what was a sizeable profession of pioneering and largely self-taught engineers. The more detailed archaeological and historical analysis of local mining and engineering infrastructures is producing much evidence of local artisan engineering expertise concentrated in the earliest centres of industrial development.[2] Within Wales, Swansea, and later Merthyr Tydfil, acted as nodes of economic activity that provided the opportunities for these self-taught innovators to develop their natural aptitudes.

In Glamorgan, the locally born artisan-engineers who were pre-eminent in the scale of their achievement were William Edwards (trained as a stone-mason); 'Mr Powell' (a blacksmith) and Watkin George (a carpenter). They all produced engineering innovations of importance to the establishment of the world's first industrial revolution. William Edwards and 'Mr Powell' both

Fig. 171. William Edwards was a well-known artisan-engineer whose 'hands-on' training ground was the dense industrial structure that existed in South Wales by the mid eighteenth century. He was also the first of an important group of Independent Minister and Deacon architects active at Swansea. (The National Library of Wales)

131

worked in the Swansea area. Edwards and subsequently William Jernegan both worked as architects and civil engineers to John Morris I, designing buildings and furnace structures, while 'Mr Powell' developed the machinery necessary to work John Morris's deep pits.

In addition to Edwards, Powell and George, the lives and achievements of a number of other architects and engineers, whose origins lay outside Glamorgan, are examined in this chapter. The Hopkin family represent a dynasty of early railway and civil engineers who gained experience in constructing the transport infrastructure of the Swansea area and who then spread their influence over a much wider field. William Jernegan is known only for his architecture and engineering works in the Swansea area whilst Thomas Sheasby (senior), an architect from Tamworth in the English east Midlands, was responsible for much of the largest eighteenth-century engineering project in the Swansea area - the Swansea Canal. Edward Martin from Cumbria and his locally trained assistant, David Davies, had a particular expertise in mining geology and engineering; while George Kirkhouse was brought in from the Great Northern Coalfield near Newcastle upon Tyne, to develop the infrastructure of the Llansamlet coalfield and made a spectacular misapplication of Newcastle practice to the south Wales coalfield.

One of the earliest engineers from outside the Swansea area was John Padmore who was called from the existing copper-industry centre of Bristol to design the water-power engineering of the Llangyfelach Copperworks in 1717. His next recorded involvement at Swansea is in designing the Upper Forest Mills of Lockwood, Morris & Co. in c.1735 and was also an engineering partner of Morris in the venture to develop collieries to the west of Swansea accessible from the Loughor Estuary by the Sluice Pill Canal. Padmore had learnt his craft on the eighteen or so water-powered coppermills of the Bristol-Bath area and was highly regarded by the internationally known engineer John Smeaton. He was a polymath whose range of internationally significant innovation included one of the first structural uses of cast-iron (in 1735 on the Bristol Great Crane); built the third British enclosed dock at Sea Mills Bristol and was engaged building steam-engines for the Bristol Waterworks Company and pumping-machinery for the Bristol Hotwells. In the portrait illustrated he is pictured with one of his smaller treadmill cranes that loaded large Bath-stone blocks from his pioneering Ralph Allen's Railway onto river barges on the Avon.

William Edwards (1719-1789)

'Architect ... Bridge-builder' and 'Minister of the Gospel.'[3]

The best known of the three identified native artisan-engineers of Glamorgan was the mason William Edwards (in contemporary documents he is referred to as William Edward) who built the largest arched bridge of the eighteenth century, at Pontypridd. Edwards was under the particular patronage of the Morris family in the Swansea area, who employed him as their architect/civil engineer for much of his working life, a relationship which culminated in the foundation of Morriston in 1778-9. Morris was more than a mere employer, being sufficiently interested personally in Edwards the man to commission a miniature portrait of him and to obtain English translations of Edwards' sermons.[4] There is little doubt that Morris's philanthropy, and his non-sectarian concerns, were deeply influenced by Edwards' wider compassion.

Edwards was born in the farmhouse of Ty Canol at Groeswen near Caerphilly in 1719. When William was seven years old his father, Edward David, was drowned while fording the raging waters of the River Taff on horseback (presumably near the Upper Boat Ferry) while returning from Llantrisant market. This may explain why William was later so determined to provide a secure bridging of the Taff near this point. The four children and their mother subsequently moved to Bryn Tail farm, situated near Eglwysilan parish church, and only a mile and a quarter south-east of the site of his later bridging of the Taff.[5] By the time he was 15 Edwards was repairing the dry-stone walls on his family's farm and busy earning extra income for his family by repairing the walls of his neighbours.[6] He later claimed to have often visited and studied the huge ruins of nearby Caerphilly Castle in order to examine the masonry techniques and the construction of the arches. Next to Ty Canol was the Caerphilly iron-furnace and nearby were the iron-forges of Bassaleg and Machen.

The building of the pillars for a horse-shoeing shed at a smithy nearby allowed Edwards to observe stone-masons at work and to acquire the skills necessary to dress stone; at this stage he was commissioned to build a workshop for a neighbour.[7] In 1738 two other events took place in Caerphilly that probably influenced the course of William Edwards' life. The Reverend David Williams, minister of Trinity Independent Church in Cardiff, brought Howel Harris, the great preacher and revivalist, to preach in the Caerphilly area on Whit Monday and Tuesday. Williams also established the first circulating school in the Caerphilly area at about the same time, teaching local people to read and write in Welsh. Subsequently William Edwards' name appeared on the successful petition presented to Glamorgan Quarter Sessions in 1739, seeking permission to build the Nonconformist Watford Independent meeting house at Caerphilly.[8]

In 1740, aged 21, Edwards was commissioned to construct the water-powered iron-forge situated on the western side of the castle at Cardiff, which was fed by a leat from the River Taff. While undertaking the construction he lodged with a blind baker, Walter Rosser, who taught him English. Benjamin Malkin, who had a long interview with William Edwards' son, David, noted:[9]

> 'After he had performed his engagement at Cardiff, he built many good houses, with several forges and smelting houses, and was for many years employed at works of this nature by John Morris, of Clasemont, Esq.'

Edwards' work for the concern of Lockwood, Morris & Co. started in the time of Robert Morris, John Morris' father. It almost certainly began at about the time of the construction of Edwards' first Pontypridd bridge in c.1746. The Beaufort Bridge of c.1747-52 is firmly attributed to him, and Robert Morris' Forest Copper and Lead Works, to which the bridge gave access, and which started to produce in about 1748, was the only smelting works constructed for Lockwood, Morris & Co. before the 1790s. In view of the deep respect that Morris developed for Edwards it may well be that Robert Morris stood as one of the guarantors for the young mason's famous bridge at Pontypridd (as rebuilt this was the largest span constructed in the eighteenth century). The longevity of this relationship and possibly its beginnings in c.1748 is suggested by another aside of Malkin's made after his discussions with Edwards' eldest son David:[10]

> 'The building of this new town [Morriston] was begun about the year 1768 [in reality the site of the town was laid-out in 1779], on a plan furnished by William Edwards, who superintended the construction of Mr. Morris's copper works for twenty years.'

In this period Edwards also built a water-mill in Eglwysilan parish which may have been the new mill on the Nant yr Aber, built at Graig-y-Fedw, Abertridwr, north of Caerphilly.[11]

William Edwards deeply felt spiritual life also developed concurrently. He said later 'I ... discard all secular concerns from my mind on the Sabbath, that nothing should disturb the duties of that day.'[12] In 1742 another nonconformist meeting house was built near Caerphilly, this time to the west at Groeswen, for the use of the Methodist followers of Howel Harris. In January 1743 Howel Harris and George Whitefield, one of the leaders of English Methodism, met at Groeswen and appointed officers for members of the Methodist connection living in a large part of south-east Wales. William Edwards was appointed supervisor of six societies, or groups, within the Groeswen church. In 1745 he was ordained as joint minister of Groeswen chapel.[13]

In 1746 the Hundreds of Senghennydd and Miskin decided to build a three-arched bridge over the River Taff, to connect their two areas. The contract, paying £500 and requiring securities insuring the bridge for seven years, was awarded to William Edwards, then 27 years old. The stone set into the side of the present bridge records a date of 1750, presumably when the first bridge was completed.[14] The scale of Edwards' engineering and architectural expertise may have induced Robert Morris to commission him to build the very similar Beaufort Bridge and the adjoining Forest Copper and Lead Works which were built in the years 1747-52.[15]

The Forest Works were built to a most novel design. The earlier Lower Forest rolling-mills on the east bank of the River Tawe were built before 1730 by Mr Padmore, 'a capital mason of Bristol'.[16] These copper rolling-mills were housed in a simple rectangular shed, as also were all the copper-smelting furnaces erected in the Swansea area before and after this, even those at the Landore and Birmingham Copperworks associated with Morris' later innovative architect/engineer, William Jernegan. The most elaborate smelting building hitherto had been that of the Llangyfelach Copper, Lead and Silver Works which remained in use as the Lockwood, Morris & Co. Works until 1748.[17]

The architecture of the Forest Copperworks achieved a careful balance between architectural embellishment and internal function. It comprised four symmetrically placed smelting-houses. Their unique circular form may have resulted from a concern to maximise the external wall area near the hot furnaces, thus providing ventilation for fumes and heat on all sides, by means of large circular openings on two levels, some of them above open semi-circular headed arches, and also with space between the various furnace 'pavilions'. This arrangement could have caused a congested and restricted interior for those stirring and pouring the metal in the four calcining furnaces (the south-eastern pavilion), or the eight ore and slag reverbatory furnaces (the other three pavilions), and or those stoking the fires necessary to maintain the great heat required to enable smelting to occur.

The historian George Grant Francis was present at

Fig. 172. John Padmore was called from the existing copper industry of Bristol to design the water-power engineering of the Llangyfelach Copperworks in 1717. He designed the Lower Forest Mills of Lockwood, Morris and Company in c.1735 and was also partner to Morris in developing collieries to the west of Swansea accessible from the Loughor Estuary by the Sluice Pill Canal. (Victoria Art Gallery, Bath and North-East Somerset Council)

the demolition of most of the original works in the late 1830s and noted that:

> 'the old buildings ... were formed of four large circular structures: the whole arranged so that the fire-places being on the outside of the circular walls, they [i.e. the reverbatory furnaces] were so placed within them , as to conceal all that was going on in the manufactory.'[18]

This innovatory arrangement is confirmed by the contemporary illustrations that show the single chimneys from each furnace arranged in an inner ring at the ridge point of the truncated conical roofs of the four smelting-houses.[19] This would then have allowed the heat generated from the coal, which was fed in from the outside, to be reflected down from the vaulted roof of the smelting-chamber before being exhausted through the tall chimneys circling the centre of the building. The searing, debilitating heat of the furnace stoke-holes added to that emanating from the furnace basins would thus be avoided. This would have eased the working environment of the skilled furnace-keepers, but whether the stokers preferred greater heat to exposure to the elements (common in the working environment of the time) is not recorded. The earlier Llangyfelach Works of Lockwood, Morris & Co. had five large furnace chimneys on the two side walls of the major smelting-house indicating that all the furnace coal had to be wheeled down the centre of the building into the combined heat and fumes of stoke-hole and furnace.

The new arrangement at Forest must have been a satisfactory means of heat control as William Jernegan was to re-use it in the Landore Copperworks of Lockwood, Morris & Co.

This concern for the working conditions and health of the workers reflects Edwards' social conscience which was to have a large influence on the development of John Morris I, managing-director of Lockwood, Morris & Co.

The four cupolas found on the earlier Llangyvelach Works were imitated in the Forest Works by two: one was located on top of the central refinery which may have been for ventilation; the other over the pedimented works entrance which may have housed a bell. The gable over the works entrance housed a clock with a lozenge-shaped face whose upper face parallelled the sides of the gable. Edwards' Beaufort Bridge connecting the two halves of the Forest Works was used as a public thoroughfare: the only other bridged crossing of the lower River Tawe was the roguish Thomas Popkin's ramshackle timber-built toll-bridge at Upper Forest to the north. The fact that a public thoroughfare passed through the centre of the works probably encouraged the partners in the concern to indulge in the architectural sophistication and innovative design which can be seen. The pedimented entrance-arch, spanning the public road, pierced a long range of offices and workshops which joined the two circular smelting-houses on the river side of the works. Flanking the north-western side of the Beaufort Bridge was the hipped-roofed 'Assay House' of 1747.[20] Its two storeys were both centred on striking three-light 'Venetian Windows' that would have appeared in all the builders' pattern-books of the period. Contemporary illustrations show that the assay house had a wide central chimney on its rear facade to vent its testing hearths, and that this, as with all chimneys in the works, had a broad-banded capping cornice; that illustrations show Edwards' bridge had two similar projecting bands on the parapet. The finely cut masonry of the triangular cut-waters remained visible on the bridge until its demolition in 1968.

At the centre of the works was the octagonal refinery which was smaller than the circular smelting-house pavilions but was capped by a domed venting cupola was surmounted by a rather tall ball finial. Reverberatory furnaces in gabled housings extended from four of the eight walls of the octagon. The public road through the works skirted the octagon on its south side and one wonders whether John Morris I's later architect/engineer, William Jernegan, remembered its elegant form when designing his octagonal lighthouses for the western pier of Swansea's new harbour and at Mumbles, and also the octagonal

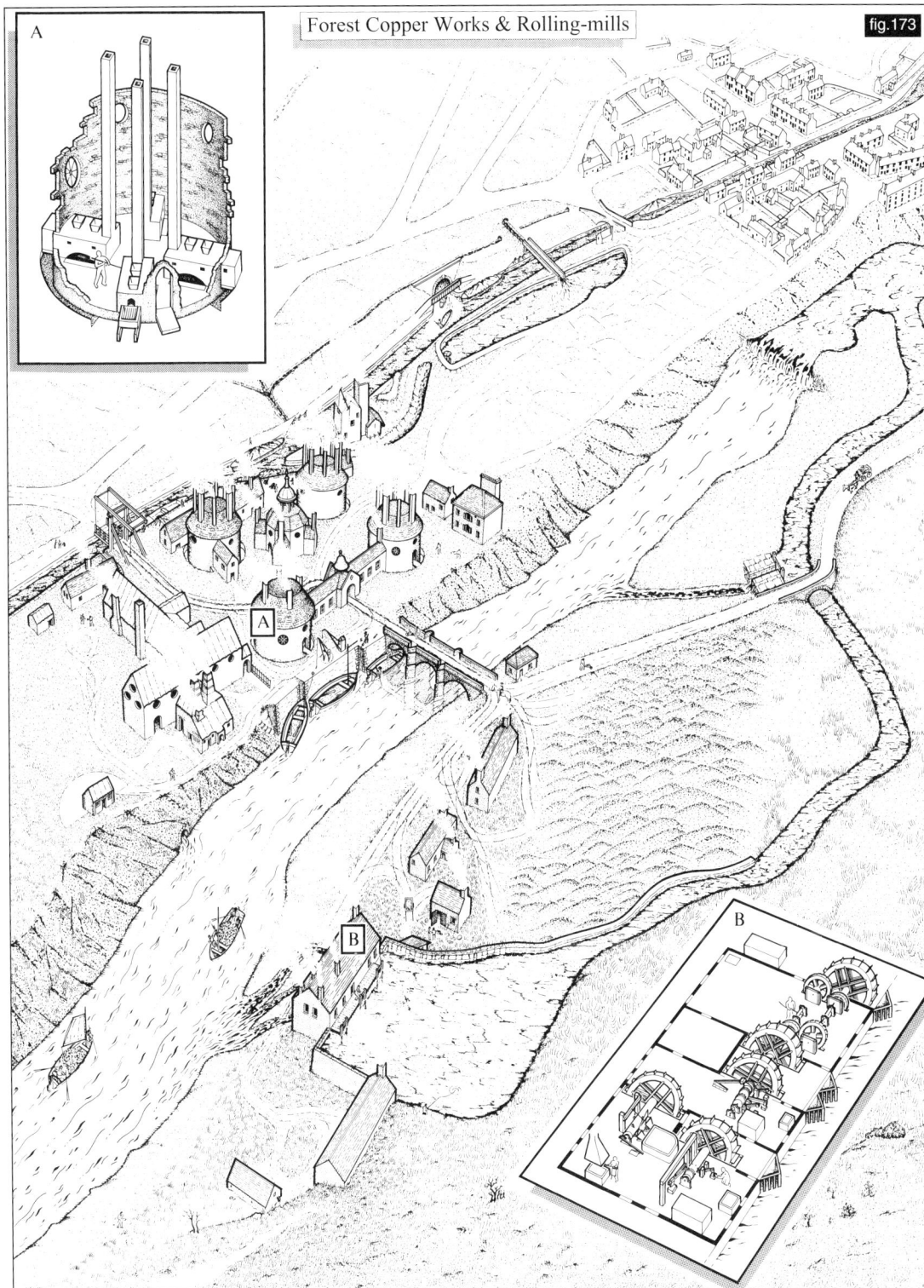

Fig. 173. John Padmore's Lower Forest Coppermills can be seen in the foreground (B) with a cut-away view of the interior. The second cut-away is one of the circular pavilions at William Edwards' Forest Copperworks showing the arrangement of four large reverbatory calcining furnaces. Behind the works can be seen the engine house of Copper Pit and the portal of the Clyn-du Underground Canal.

pleasure-villa of 'Marino' for the future husband of John Morris's sister.

A further architectural refinement at the Forest Copperworks was the provision of ornamental glazing bars on the circular windows. If the windows were glazed, then presumably they also opened. These windows consisted of an inner circular eye with eight radiating segments. The bars may have been of copper.

More conventional, rectangular smelting-houses with circular windows and banded chimneys which were

135

fig.174

Fig. 174. Forest Copperworks in the 1780s drawn from the north. Foreground: works pond; centre: central octagonal refinery; left: the cupola and weather-vane of the pedimented works entrance; front of Kilvey Hill: the Lower Forest Coppermills; to the right of the buildings: stack of the Copper Pit. (By permission of the British library)

added to the original buildings at Forest at a later date may also have been Edwards' work on the site, as Malkin noted later that William Edwards' 20 continuous years of service building smelting-furnaces for Lockwood, Morris & Co..

All the original buildings at the Forest remained standing until 1838.[21] Demolition took place soon after so that only the south-western of the circular smelting-houses remained after 1844.[22] The octagonal refinery was still standing in 1897.[23] All that remains today of the Forest Works is a later rectangular block on the west side of Beaufort Road (erected to the north of the original works by 1838). It is likely that substantial substructures of the Forest Works and furnaces remain underground.[24]

As the Forest Works and the Beaufort Bridge were being completed, events called William Edwards back to the Pontypridd bridge further east. The Taff was a temperamental river; Edwards' father had been drowned fording it when it was in spate and Leland

in the sixteenth century had noted, 'the Water of Taphe cummith so doun from woddy Hilles, and often bringgith down such Logges and Trees, that the Cuntery wer not able to make up the Bridges if they were stone they should be so often broken.'[25] The flotsam and scour induced by a seasonal spate two years and two months after the completion of construction (in *c*.1752) caused the first three-arched Pontypridd Bridge to collapse.[26]

Edwards persuaded his backers to allow him to build a single arch which would avoid the circumstances for collapse of the original bridge. The first single arch was nearly complete when a flood carried away the centring and the partially completed works. Work began again with strengthened centring and the bridge was successfully completed.

The opening in 1755 of the huge arch attracted much attention, but the large weight of material laid to ease the ascent over the sides of the bridge, compared with the shallowness of material on the

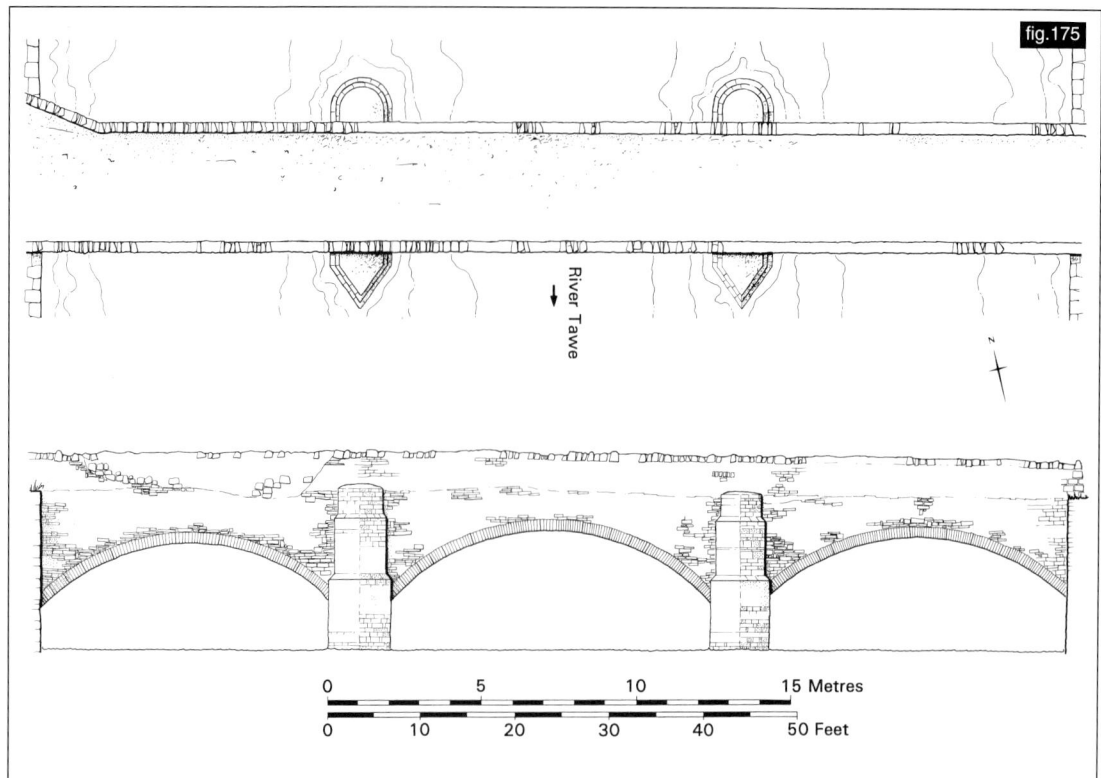

fig.175

River Tawe

Fig. 175. The drawing of Edwards's Beaufort Bridge (1747-52, demolished 1960s) shows what his early multi-arch bridge designs were like; it was a bridge similar to this that he first built at Pontypridd in 1746-50 and which was subsequently washed away (redrawn from Swansea City Highways Department drawings).

0   5   10   15 Metres

0   10   20   30   40   50 Feet

crown of the arch, caused the distortion and collapse of the arch after six weeks.[27]

Edwards' fellow local preacher, Thomas Morgan, was prompted to submit a full account of the building of the Pontypridd bridges for publication in November 1764 by an assertion in *The Gentleman's Magazine* that the Rialto Bridge in Venice (95ft in span, with a height of 24ft) was 'the greatest curiosity of its kind in Europe, having the widest and flattest arch.'[28] Morgan's subsequent article on the bridge in the *Annual Register* for 1765 gives a fuller account of the building of Edwards's fourth and successful bridge at Pontypridd:

> 'By this time the mason was greatly in debt, and greatly discouraged; but the lords Talbot and Windsor (who have estates in that neighbourhood) pitied his case, and being willing to encourage such an enterprising genius, most generously promoted a subscription among the gentry in those parts, by which a sum of money was raised that enabled him to complete the bridge in one arch for the last time. In order to lessen the quantity of matter in the abutments pressing upon the arch, and thereby to bring it to an equipoise with that on the crown, he has contrived three circular arches in the abutments; these pass through from side to side, like round windows, and gradually decrease in the ascent.'[29]

This was not a new idea, although its realisation in the large arch built at Pontypridd had a significant effect upon the design of subsequent bridges throughout Britain.[30] The lightening of the masonry load over abutments had been achieved in Roman times and again by innovative French civil engineers and architects, who had been trained in the modern period following the foundation of the *Corps des Ingenieurs des Ponts et Chaussées* in 1716 and the *École des Ponts et Chaussées* in 1747. The means by which this was achieved often involved the placing of conventional arches over piers and abutments. In 1735-6 two designs, by John Price and Batty Langley, for the proposed Westminster Bridge in London were published which utilised circular voids of various diameters to lessen their loadings over intermediate piers.[31] One intermediate pier, of the approved design by Charles Labelye, was sunk into the Thames river-bed in 1748 and the adjacent arches, which were rebuilt in 1750, incorporated voids spanned with large segmental arches of the type which were subsequently utilised in Edwards' bridge abutments of 1756.[32] The brilliant pioneering civil engineer John Smeaton was watching the work.[33]

From the quotation given above it seems likely that Edwards watched the crown of his arch being forced up by the pressure of the weight of the spandrels and the abutments pressing in on the lower sides of the arch, for the six weeks that it took for the arch to collapse. However, when he was pressed by his guarantors to justify the building of his new single-arched span at Pontypridd he turned to Smeaton for advice.[34]

The use of fully round openings on the Pontypridd bridge is of course also reminiscent of those so prolifically applied to the round buildings of the Forest Works. The new bridge was successful and still stands. However, the obvious lightening of abutment load by the visible device of the three round openings seems to have been augmented by three other structural features. Such features may also have been used in the large span of Wychtree Bridge at Swansea. In 1778 a very full account of the Pontypridd Bridge construction was published which noted that:[35]

> 'In order to lessen the quantity of matter in the abutments pressure upon the arch, and thereby bring it to an equipoise with that on the crown, he has contrived three cylindrical arches and one semicircular in each abutment, the semicircular are hidden by the side walls. It is also strengthened by very large pieces of timber laid in the work between the great arch and cylinders.'

A civil engineer making a study of the bridge in 1838 also had access to a model of the bridge and of its centring used in the erection of it. He claimed, in the publication of his structural analysis and survey, that the voids between the round openings were filled with charcoal rather than with gravel or rubble, in order to lessen further the previous loading on the sides of the lower arch which imposed by the weight of the abutments.[36] Recent restoration observed by Cadw: Welsh Historic Monumnets failed to find any charcoal in the abutments, Neither were such expedients found during the demolition of the Wychtree Bridge in Swansea, which had a lesser loading and where such weight saving devices would have been unnecessary.

These masonry arches and hollow spaces hidden in abutments and piers, later became more common as structural devices after the success of their use at Pontypridd, the span of which was not equalled for 40 years.[37] Edwards' two other large spans over 80ft, those at Wychtree Bridge in Swansea and Dolauhirion (near Llandovery), used weight reducing openings. Voids were subsequently commonly used by Thomas Telford in such aqueducts as Chirk and Pontcysyllte[38] and John Rennie in the huge masonry aqueducts of the Lancaster Canal.[39]

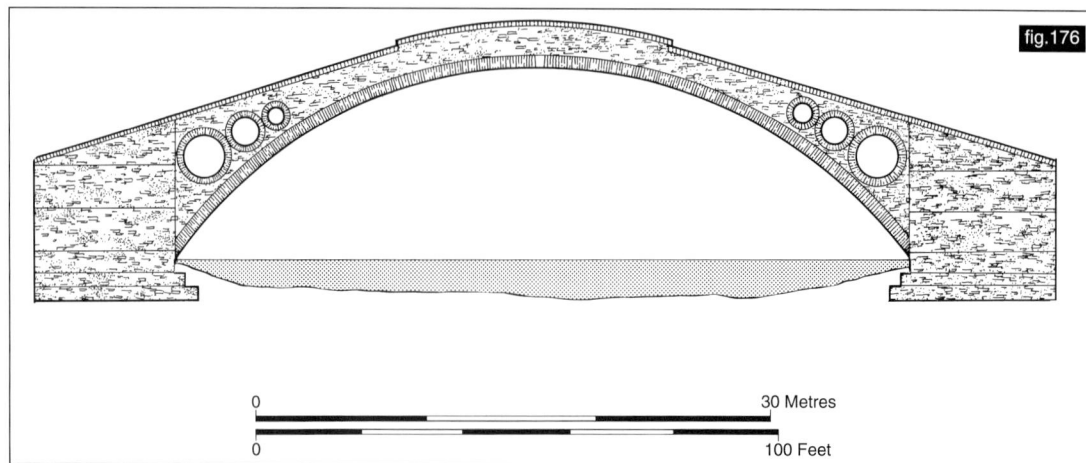

Fig. 176. Edwards's fourth attempt at building a bridge over the Taff at Pontypridd resulted in his receiving international attention as the constructor of the largest arched span to be achieved by man (140 feet) (redrawn from a contemporary illustration in the 'Gentleman's Magazine').

The arch ring of the Pontypridd bridge was very narrow for the large span involved, it looks almost as if Edwards was reusing voussoirs from the collapsed three-arched first bridge in order to save money. This point was noted by the most eminent engineers of their generation - William Jessop (ex-pupil of the great John Smeaton and mentor of the younger Thomas Telford) and John Rennie (who visited to the Taff Vale in 1806 when reporting on the Glamorganshire Canal's water-supplies and was the originator of the Melingriffith Pump).[40] They wrote:

> 'There is a bridge across the river Taff in the County of Glamorgan, of upwards of 135ft span with a rise of 32ft, and what is more remarkable, the depth of the arch stones is only 30 inches; so that in fact that bridge far exceeds in boldness of design that of Neuilly.' [i.e. Perronet's internationally known innovative bridge over the Seine in France].[41]

In fact, the most recent mathematical analysis has concluded that such a close correspondence to theoretical requirements as found in the Pontypridd Bridge suggests a mathematically determined design. However a sound mathematical theory of the mechanics of bridge construction was only evolved in two books, published in 1743 and 1758 by William Emerson and so would not easily have been available to Edwards during his increasingly refined constructional achievements.[42]

The span of the arch, 140ft, together with the rise that this entailed: 35ft (on a substantial segment of a 175ft semi-circle), produced a bridge that was very difficult for wheeled traffic to use.[43] Even after a raising of the abutments in 1826 the approaches to the crown of the bridge were still 1 in 4. A comment was made in 1838 that

> 'it is so steep ... that it is necessary to use a chain and drag so that when a carriage reaches the centre of the Bridge, one end of the chain is

attached to the hinder end of it , the other end being secured to the drag upon which a boy generally places himself; so that as the carriage descends on one side the drag is pulled up on the other, and this relieves the horse in descending. Without this contrivance it would almost be impossible for a loaded vehicle to pass over in safety.'[44]

The stability of the Pontypridd bridge had a lasting effect on William Edwards' bridge designs. The intermediate piers of upland river bridges were often subject to flood damage, even those whose design was advised on by the famous stonemason and engineer Thomas Telford.[45]

Nine bridges were attributed to William Edwards by Malkin who used information given by William's son David who placed them in roughly chronological order.[46] Of the eight, besides Pontypridd, three were multi-arch spans like the first Pontypridd bridge of 1745-6. For the Beaufort Bridge (c.1747-9) Edwards used the same scale of construction as he had used on the first Pontypridd Bridge. He could have spanned the Tawe in a single arch which he did 30 years later when he built the Wychtree Bridge very close by.

The other multi-span bridges were rather different cases, being erected over major rivers that could not have been spanned by single arches. One, over the Usk at the town of Usk, is attributed to the same early period as the Beaufort bridge and the first Pontypridd bridge. This five-arched bridge has segmental spans of some 50ft. Other sources, however, convincingly suggest that the four arch 'New Bridge' over the Usk at Tredunnock was that in fact built by Edwards.[47]

A seven-arched bridge over the River Wye at Glasbury was also built in 1776-80 with William superintending his sons in one of his last major works. His advocacy of the use, whenever possible, of large single spans was justified when the Glasbury

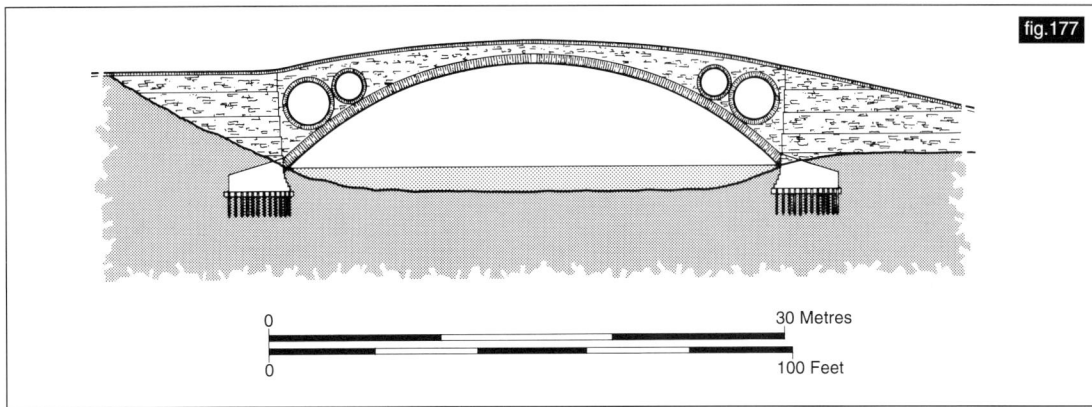

fig.177

Fig. 177. Edwards's Wychtree Bridge across the River Tawe was erected in 1778 alongside the future site of Morris Town. With a 95-feet span it was the second largest of Edwards's bridges, and with a flatter deck was much more practical for vehicular traffic (redrawn from a Swansea City Engineers' drawing).

bridge was one of several large bridges destroyed by the floods of 1795. There seems to have been nothing particularly deficient concerning the foundation of Edwards' bridges, those of his final bridge, the Wychtree Bridge *(c.*1778-80), were examined during demolition.[48] They reached down through sand strata to a raft 14ft below ground level. A concrete mass of boulders set in lime mortar constituted the 1ft thick building raft. The mortar would have been hydraulic mortar (setting under and being impervious to water) as used on Edwards' Pontypridd Bridge; it was obtained from sources at Aberafan and Aberthaw. Set in the building raft were the heads of 5ft long oak piles; 6in diameter. The bottom foot of these was bedded in a layer of stiff clay and gravel, which was considered to be firm enough on which to lay the new raft of the concrete bridge built to succeed Edwards' bridge in 1959. A firmer foundation could have been secured for Edwards' bridge by laying the building raft directly onto the firmer clay, but the foundations as built were sufficient for the fairly narrow bridge to cope with traffic flows until 1959. It was demolished in 1959 after refusal to sell the land adjacent which would have facilitated its retention.

The fourth bridge attributed to William Edwards lies to the north of Swansea on the River Tawe (at Pontardawe). Its high, elegant arch is believed to contain a single circular opening in each abutment, although these have now been filled in. The nature of the design suggests that it is probably not earlier than his pierced abutment, single-arch experiments at Pontypridd in 1755-6. A bridge on this site is certainly existed as far back as the beginning of the eighteenth century when it was recorded by Edward Lhwyd.[49] A bridge over the Afan at Aberafan was supposed to be the eighth by Edwards (*c.*1768), and was of similar span but had no visible piercing of the abutments although it may have had the hidden arches used in his fourth Pontypridd bridge.[50]

A bridge over the River Aman at Ammanford with a 45ft span, said to be Edwards's fifth could well,

from its size and form, have been earlier. The sixth bridge was again a substantial single arch with abutments pierced by single circular openings. This was built in 1773 over the River Tywi at Dolauhirion and has an elegant 84ft span that still remains.

The Wychtree Bridge, begun in 1778, over the River Tawe facilitated the foundation of Morriston to the west of the new bridge. The bridge was Edwards' greatest span after the later Pontypridd bridges, having an arch 95ft wide and 20ft wide. Other bridges attributed to William Edwards include Pontycymer[51] and Pontygwaith over the upper Taff.

While William Edwards was undertaking these bridges he was continuing in his Swansea-based role as architect and civil engineer for Lockwood, Morris & Co.. The new waggonway from Plasmarl Level to the River Tawe at Landore was constructed in 1769-71 and in 1772-4 a 'New Quay' was built alongside the river by a team of masons working under William Edwards. John Morris I considered it a considerable structural advance over earlier river quays.[52] Possibly this was because it was constructed of masonry throughout, bonded with hydraulic lime from Aberafan and set on a raft of mortar-bonded rubble sitting on oak piles, as was his later nearby Wychtree Bridge. The quay survives in good condition on the west bank of the River Tawe just to the north of the Landore Viaduct.

A substantial quay was also added to the Forest Copperworks, immediately south of the Beaufort Bridge, this was built in three stages between 1747 and 1791, for trans-shipping copper-ore from river boats into the works. It is likely that this too was built by William Edwards as part of the infrastructure of the works. Further modifications were carried out to the copperworks in order to increase its smelting capacity. A transverse annex towards the river housed two large furnaces: either brass-annealing or calcining the initial roasting of the ore, brought in from the enlarged quay alongside. The smaller reverbatory furnaces for further refining were ranged along the western side of the building. The main block of the new smelting-house had semi-circular arched openings for access. The upper sides of the main block had roundel openings and three were grouped in the gables: an arrangement later very common in the cast-houses and forges of the ironworks of south Wales.

What was strikingly innovative in the construction of the new smelting-house was the roof of corrugated-copper sheets.[53] These seem to have been long narrow sheets of rolled-copper which were formed into a convex cross-section and laid in parallel order down the slope of the roof. Those over the main entrance to the calcining furnaces were extended downward to form a projecting curved porch. Contemporary travellers commented on the use of copper-sheeting for roofing purposes in the Swansea copperworks and a view of the Melincryddan copperworks at Neath in 1792, may show similar forms of roofing on buildings that include a smelting-house with a three similar circular openings in the gable.[54] Morris and Mackworth (operator of the Melincryddan Works) used the same architect, John Johnston, for their houses, both of which had copper glazing-bars. Perhaps they both used the same architect (this time William Edwards) for their smelting-works which may have had rolled-copper roofs.

In 1778, the year that Edwards began his last recorded bridge at Wychtree, his standing was such that he was called to inspect the large bridge over the Wye, being built by James and James Parry (father and son).[55] In 1779 John Morris I acquired the farm to the west of the new Wychtree Bridge and in his 60th year Edwards laid-out the new town of Morris Town,[56] with its double grid of wide streets with two intended squares. According to Walter Davies, John Morris I then instructed 'His Architect, William Edwards' to design a church, a chapel and a standard form of house for the new inhabitants of the town that was to be developed. Morris, in fact, built at least three of the new houses himself and left them to Mary and Lewis Morgan, his daughter and son-in-law, and their three children at his death in 1789.[57]

The philanthropy shown in the expense of laying-out the new town may have had much to do with the influence of the charismatic Edwards on Morris and his late father.

Edwards had continued as a joint minister at Groeswen (near Caerphilly) from 1745. In 1752 his fellow minister, Thomas William, was excluded from the Methodist congregation and the chapel became firmly Independent. Thomas William died in 1765 and in 1766 the chapel had to be extended. William Edwards continued his ministry until his death on August 7th 1789. His chapel was noted as the richest in Wales for the qualities of its congregation. He was one of the co-leaders of a

religious revival in the Rhymney Valley and in 1772 a letter was written to Howel Harris, leader of the great religious revival in Wales noting that:

> '...there are about twenty remarkably gifted brethren from Groeswen Church doing uncommon work from Ystrad [Mynach] Bridge to Machen Forge. Blessed be God. All Caerphilly Town come to hear: the beautiful country about Caerphilly is become a Beulah!'[58]

Edwards was a very liberal puritan Calvinist and:

> '...during the last years of his ministry, he always avoided in his discourses those points of doctrine that were more peculiarly in dispute between the Calvinists and other parties. He frequently repeated and enforced a maxim ... that the love of God and of our neighbour is the ultimate end of all religions...
>
> of ... the ... minister[s] ... salary ... he distributed the whole among the poor members of the Church, and even added very considerably to this largess from his own personal property.'

He was well respected by the most intelligent and liberal of all sects and parties, and died, very much lamented by all who knew him.'[59]

The planning of Morriston reflects this non-sectarian benevolence. Morris and Edwards planned to lease one-rood plots of land for a period of three generations or 50 years at a cost of 7s. 6d. a year ground rent for workmen to build their own houses. The best workmen were also offered sufficient land at the Common to graze a cow.

The affection of the new population of the town for William Edwards was reflected in the colloquial name of 'Hewl (Heol) Neti': i.e. 'Ned's (Edward's) Road' for the main street in Morriston.

John Morris also established a workmen's compensation scheme to cover periods of loss of work by employees through accident and sickness; the scheme provided also for the payment of old age pensions and he built a Poor House in Morriston for necessitous persons.[60] Presumably this was the large, rather elegant block of accommodation on Wychtree Street, called the 'Wychtree Hostel',[61] which was demolished during the 1960s, the lower level of this accommodation was built substantially from cast copper-slag blocks. It can be presumed that this third communal building, like the church and chapel, was also designed by William Edwards, it had a central pediment with a half-round lunette window and split-level accommodation on four floors accessible by communal entrances.

Edwards' role as a nonconformist minister was also felt in the Morriston area. In 1782 he built the new Libanus Chapel on land provided by John Morris I which was later to form part of the settlement of Morriston. The chapel was intended for members of the Independent chapel at Mynydd Bach, whose local members had previously worshipped in the old thatched cottage of Ty Coch which was situated on the slope of Graig Trewyddfa above Morriston. The chapel was enlarged in 1796, became an independent congregation in 1829 and was further enlarged in 1831 and 1857, so that it is now difficult to see anything of Edwards' original chapel. Edwards became close friends with the Revd. Lewis Rees, the influential minister of Mynydd Bach and, frequently preached (in Welsh) in the chapels under the local minister's charge. It was agreed that the weekly Sunday morning services for the area should be maintained at Mynydd Bach with no clash in the times of services between there and Libanus.[62]

William's sons learnt their skills as masons and civil engineers from working alongside him on his various engineering projects; David Edwards' name actually appears on the parapet of his father's Dolauhirion Bridge. Edwards' sons are also recorded as having worked on the Glasbury Bridge and may also have been employed on the contemporary Wychtree Bridge at Swansea. Such was the experience accumulated that Thomas Edwards was already acting as consulting engineer by 1781, making many inspections of bridges constructed for the Justices of the Peace of Breconshire between then and 1791.[63] Edward and William junior appear not to have built any bridges in their own right and the latter died in 1791 after being wounded while on war service while the former does not feature in his father's will.[64]

David and Thomas Edwards were bridge builders in south Wales like their father but the exact number of bridges they built is unknown. It is not recorded if they built any smelters, forges or any other types of buildings in Swansea, Morriston or elsewhere. Like their father they both had dual occupations: David was a farmer at Beaupré near Cowbridge in Glamorgan and Thomas kept the Three Cocks Inn at Aberllynfi near Glasbury in Radnorshire;[65] presumably William would not have approved of his son's new occupation.

Thomas built bridges at Talybont (1781-82, River Usk, demolished); Pont Ithel (Hay, 1783, River Llynfi, arches are 20, 32 & 20ft) and Fforddfawr (1791-92). In 1793-96 his last recorded bridge work involved the widening of the old Wye Bridge at Brecon on its upstream side. This was damaged by floods in 1795 and his widow and David Edwards paid the £150 compensation due in 1801, this was in order to discharge Thomas's bond for seven years' maintenance.[66]

David Edwards was more prolific than Thomas but also seems to have restricted himself to multi-arch bridges with lesser spans than those achieved by William Edwards except for the fine medium span at Edwinsford which has unpierced spandrels.[67] The bridge over the Usk at Newport, however, was a very substantial multi-span bridge, it was built by David and Thomas Edwards and would bear comparison with some of the other major bridges erected elsewhere in Britain at the time.

David's five known bridges, 'among many others, ... built'[68] are as follows: Llandeilo-yr-Ynys near Carmarthen (1786, River Tywi, the three arches are 45ft); Pont Loerig near Whitland (River Taf, the three arches are 27ft); Newport (1793-1801, River Usk, the five arches were between 62 and 72ft); Edwinsford near Talley (1793, River Cothi, had a single arch) and Bedwas near Caerphilly (River Rhymney, two arches).[69]

David seems to have been the best in his family for establishing firm foundations for his piers. The Usk at Newport has a huge tidal range and the 14ft wide piers extended 57ft from low water to the top of the parapets, the bridge was not replaced until 1927.[70] In November 1931 all the bridges on the Tywi were either destroyed or damaged in heavy flooding. Heavy scouring of the bed collapsed part of the upstream section of one of the piers of David Edwards's Pont Llandeilo-yr-Ynys, but the quality of the masonry remaining allowed it to be rebuilt.[71]

Two of David's sons became doctors but he commented that his son William was:[72]

'a very skilful mason, and particularly so in all kinds of bridge and water-works. He now superintends many of the locks and bridges of the Kennet and Avon navigation from London to Bristol.'

At the time of David Edwards talking Malkin, William was also bidding for the contract to rebuild the Usk bridge at Caerleon.

Perhaps the most successful work of this engineering dynasty was the Wychtree Bridge, it combined a large arch with lightened spandrels with a roadway easily used by road traffic. However, there is no question that the Pontypridd Bridge was the most internationally influential product of these engineers, their contribution to the landscape of the world-centre of non-ferrous smelting at Swansea will merit further study.

Fig. 178. *The elegant and graceful structure of Edwards's Wychtree Bridge was a feature in the mixed industrial and rural landscape visible from Morris's new house on the valley side at Clasemont (engraving by Rothwell). (The National Library of Wales)*

Mr Powell (d. 1783)

'... he had only to perceive to comprehend, and from Comprehending to execute.'[73]

Perhaps the most vivid description of one of the engineering innovators in the Swansea area, in Mr Powell's case a former blacksmith, was written in 1783[74]. George Watson, a representative of Boulton & Watt, visited John Morris I, managing director of Lockwood, Morris & Co. at Landore, and:

'found him in deep affliction and real distress, having that day by a most melancholy and unfortunate accident lost Mr Powell his Engineer who being down repairing the Engine, by accident the plug of the Steampipe dropt out, when the Steam issued with such force as to scald him so dreadfully that he died in two days, - most Universally lamented, - It seems this man was a wonderful prodigy of nature, a person who had achieved such extraordinary (*sic*) feats that Mr Morris wish'd above all things your being acquainted with him, and very feelingly regrets that you can now only see his works, - He came there about twenty years ago a young man and common blacksmith, who from working for and attending on, the millwrights and engineers, was so expert and clever, that he had only to perceive to comprehend, and from comprehending to execute, insomuch that he undertook to construct an Engine, which he effected far superior to any before, made everything himself - and it went to work the very first day with ease, expedition and regularity, and so continued without alteration, from which time he has erected Several Engines with great improvements - and invented a very simple but consequential contrivance to draw the Coal from a Pitt about 100yds from an Engine, the Water from thence being carried to a Water Wheel, which being made double close together only revers'd receives the Water first on One which draws up, then is diverted to the other that turns it the other way and lets down, the rope being round the Shaft and runs on each side alternately as the Wheel reverses - One Man only attends the pitt, shifts the Stream, receives the Coal, empties and returns the baskets - and with such expedition that it draws 150 Tons of Coal in 24 hours, close to this is a small but curious Engine to throw the Water when it is scarce and they are short of it - They tell me that it is of singular simple Construction, his own Invention and without some Customary Movements, but I do not know enough of the Matter to describe it.'

The type of water-wheel winder which is described had been known for centuries and so was not unusual, it was probably the water-wheel winder known to have existed at Landore Colliery. The 'small but curious Engine' may have been one of the earliest hydraulic rams (otherwise attributed to Joseph Montgolfier or a Mr Whitehurst of Derby in

fig.178

Fig. 179. *'Mr Powell' had joined Lockwood, Morris & Company in c. 1763, and by 1778 John Morris I was proudly mentioning his abilities to the steam-engine manufacturer Matthew Boulton. (Birmingham City Archives, Boulton and Watt Collection, Box 2/18/2)*

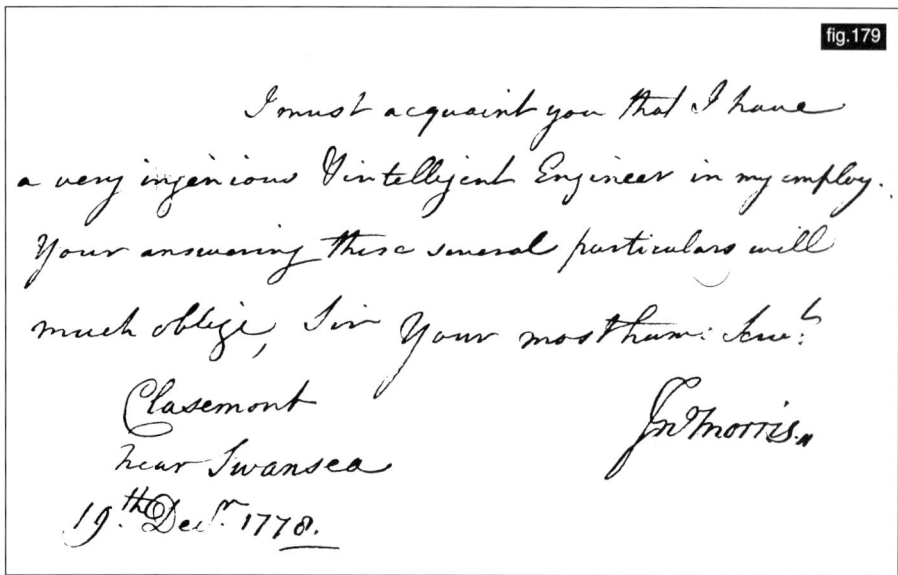

fig.179

I must acquaint you that I have a very ingenious & intelligent Engineer in my employ. Your answering these several particulars will much oblige, Sir Your most hum: serv.t

Clasemont near Swansea 19th Dec.r 1778.

Jn Morris.

1772) but the small proportion of water used that would have been thrown back over the water wheel by such a device suggests that it was something different. The remark that 'it is of singular simple Construction...and without some Customary Movements' suggests that it was of a type of engine already known, probably a simple type of atmospheric engine without a beam or pistons, such as 'Joshua Rigley's rotary steam engine' illustrated in Nicolson's *Journal*, vol.1, for 1797. This was little more than a version of Thomas Savery's steam-pump (invented in the seventeenth century) which was applied to keep a water-wheel turning in times of drought.

Powell also constructed steam-powered (or more correctly, atmospheric-powered) rotary engines which were among the first of this type built. George Watson's letter of 1783 discusses the unfortunate Mr

fig.180

Powell's steam-winders. One was a small dedicated winder and the others combined pumping and winding engines. Such combined engines were common in the early nineteenth century when they were low-pressure steam-engines rather than the atmospheric- engines that Mr Powell was presumably developing. The description suggests that this deals with the other deep pit at Landore Colliery:

'And the Otherside the Engines, upon one other pitt he has erected a Small Steam Engine that an apparatus very singular of his Invention is fix'd that draws up the Coal and only one man necessary to receive and deliver the Coal and attend the apparatus and one other to the Engine for the Coal being drawn up the basket strikes a small Contrivance that shifts the barrel itself. lowers down the basket on the bank - the man

having only to empty the basket and return it to the Mouth of the Pitt. when by his pulling down a handle [i.e. a clutch mechanism] the barrel is thrown into contact with a reverse wheel and it descends again, but its movements are made by the beam - the most ingenious part being that whether the beam ascends or decends [*sic*] the Barrell is made to go round in one direction [i.e. by a crankshaft or equivalent] - This small Engine draws 120 Tons of Coal in 24 Hours and another apparatus of the same kind is fixt to the beam of One of the large Engines over another Pitt and draws about 130 Tons of Coal in 24 Hours - This large Engine has another Close adjoining [i.e. Plas Marl Pits], One has a 40 Inch Cylinder, 10 In pumps and 6 foot Stroke upon a Mine 30 fathom - the other has a 46 In Cylinder 12 In pumps 6 foot Stroke upon a Mine 30 Fathom and both move from 10 to 12 Strokes a Minute - very easy regular and Constant, being both unceasingly at Work night and day, except any accident or repair, because these two Engines can hardly and bearly relieve the Mines, as Mr. Smith does not work his Engine near to them [i.e. 'Chauncey Townsend's Fire-engine (later Calland's) Pit] by which they have some of the Water to draw off therefore these Engines are kept so much to the heighth [sic] of their powers. That they cannot Stop a day for fear of the water increasing upon them - It was from this reason that the Engineer was repairing (for he did everything himself) a fault while the Engine was in a dangerous State, because he would not slacken her Fire'g these two Engines consume immensely of Coal full Two Hundred Weigh per Month - and worth to Mr. Morris I find £1200 per anno and I suppose three times as much if in Cornwall, for each Engine has four boilers, and the fires are fed from pits above them into which the Coal is thrown as it is drawn up - and so it keeps constantly and gradually sinking upon the fires as they burn away, a Man only now and then raking or clearing the way below. Mr. Morris wishes for One of your Engines [eventually this was bought and was the 'stupendous' 66 inch Landore Boulton and Watt Pump installed in 1802] and means to have One as soon as he has an opportunity - he hopes and expects Mr. Smith will soon Work again, and if so, he must work his Engine - and when that works - Mr. Morris can relieve himself with only one of his present Engines - and then he will put One of Yours in the room of the other, by which he will be secure in future whether Mr. Smith works or not, This being his Situation he can do nothing decisive at present - yet whenever you

Fig. 180. At Gwernllwynchwith the distinctive tall, long and narrow shape of the building suggests it was the casing of one of the early winding-engines developed at Swansea. View looking west towards the end wall which probably has a shaft beyond. (920049/4)

or Mr. Watt were at leisure, he would be very happy to see you that you might survey his Works - see what was necessary to be done and in fact fix and conclude everything [Powell's atmospheric winder was replaced by a Boulton and Watt steam winder (an ex-Cornish pumping-engine) in 1787] - so that there might be no delay on that account - when an opportunity enabled him to begin the Operations - This late Engineer has left a Son abt. 24 years of Age - who must now look after the Engines, he was with his Father - but his abilities have never been tried and therefore is not known whether he thoroughly understands his Fathers principles - I told Mr. Morris that it was your intention of waiting upon him in your way to Cornwall, when circumstances prevented, as they did also on your Return, or you would have accompanied me down - but I told him you would shortly, he said he should be very glad to see you, that he only went from home Once a Year to London - and that was sometimes in March or April, I told him he shou'd be apprised [sic] before it was undertaken, and I could only get off from spending a week more by promising to do myself the pleasure of accompanying you.'

This information may have helped spur James Watt on application of rotary motion to his higher pressure steam-engines. Watt first applied his rotary engines to colliery winding in 1784. With their separate condensers these Watt engines were much more economical in their use of fuel. However, in coal-producing areas, large consumption of fuel by relatively simple and easily maintainable early atmospheric-pressure machines was a bearable cost. In the event, the young Mr Powell may not have been such a successful innovator as his father for the Morrises were soon buying in more advanced machines from elsewhere, the Plasmarl winder in 1787, the Landore pumping-engine in 1802 and Trevithick high-pressure winders from 1803.

Even so, a surviving survey of the steam-power in use in the Swansea area in c.1840 shows the survival of these simple but effective prime-movers in the Smiths' coalfield to the east of the River Tawe.[75] Two of the Llansamlet collieries still had 'common atmospheric-engines', one used for both pumping and winding. These may have been machines originally built by Chauncey Townsend in the mid to late eighteenth century. In c.1840 an atmospheric-engine had even been newly erected to power the pumps at the nearby Llanwern Colliery; it was presumably the re-use of a much earlier engine.

Powell (senior) and his fellow engineers working for Morris, seem to have innovated to both improve the efficiency and meet the challenges of running the large copper-smelting and coal-mining complex of Lockwood, Morris & Co. As has been noted, the archaeological and historical evidence suggests that John Morris introduced cast-iron plate-rails in 1776, some 11 years before the generally accepted date for the introduction of such rails by John Curr at Sheffield.

Edwards, George, and Powell, mason, carpenter, and blacksmith, had all acquired their basic training and experience by being artisans. Three other routes into the early engineering profession were available in the Swansea area, by those who had originally been architects, contractors or engineers.

The Hopkin(s) Dynasty of Engineers

The intense economic activity of the Lower Swansea Valley led to the emergence of other dynasties of locally born engineers: some of whom went on to use their engineering experience in other areas both of Britain and in other parts of the world. Perhaps the most notable of these was the Hopkins family of Llangyfelach (a village on the north-western fringe of Swansea).

Evan Hopkin is the first member of the family who is known to have been involved in engineering. With his sons David and Roger, he is reputed to have been the contractor for the Penydarren Tramroad nine and a half miles long, between Merthyr Tydfil and Abercynon, under the engineer George Overton in 1800-02. He was thus already an experienced railway builder in 1803-5 when he constructed, what is generally believed to be, one of the earliest powered railway inclined-planes at the head of the Vale of Neath, on the tramroad from the Neath Canal to the Abernant Ironworks near Aberdare.

Hopkin's next contract was as resident engineer and surveyor of the Oystermouth Railway seven and a half miles long in 1804-7, Edward Martin (from Martindale in Cumbria) was the consulting engingeer. In the General Assembly Minutes of the Railway Company, it was noted 'That Mr. Martin be requested to ascertain from Evan Hopkin upon what terms he will undertake the laying the whole tramroad and build the culverts.' However, it seems to have been David Hopkin, Evan's son, who actually carried out the building work, for the Minutes continued: 'That Mr. Martin and Mr. Williams [John Williams, land surveyor from Margam who also had produced the large-scale surveys made during the construction of the Swansea Canal in 1794-99][76] be desired to mark out the intended line and that David Hopkin attend them.'[77]

The sum of £95 0s. 9d. was then paid to David Hopkin on 3rd November 1804, the first of many such payments.[78] The committee subsequently decided to dispense with the services of Evan and David Hopkin and chose to make Benjamin French their manager, but by the end of 1805 the company was also employing Evan's other son, Roger Hopkin,[79] he was referred to as 'the engineer' while still in employment in June 1807.[80]

In August 1805 a committee of the Swansea Harbour Trust, which was then building the western pier at the entrance to the new harbour in Fabian's Bay, decided that the existing supervising engineer

Fig. 181. The Pennant sandstone quarries along the southern face of Kilvey Hill from which stone was taken to the harbour pier by a railway incline. Its formation can be seen running towards the right-hand lower corner of this view. (935054/50)

was not competent to complete the work and recommended the employment in his place of 'Evan Hopkin of Llangyfelach, who has been used to the laying down tram roads, and engaged in the management of other works where large numbers of workmen have been employed.'[81] This work was approaching completion in July 1809.[82] Concurrently with this (in 1806) Evan Hopkin was also 'the surveyor in charge' of constructing Dr. Bevan's Tramroad from the Dinas limestone quarries to the head of the Neath Canal.[83]

In 1807 Evans' two sons, David and Roger Hopkin, ventrued into colliery operation with the lease of Cathelid Colliery in Cwm Clydach to the north of Swansea. They moved from their previous home at Penllanau Farm (probably to the west of Landore) to live closer to the colliery at Cathelid Ganol.[84] Roger subsequently moved away from the district altogether, in 1810 he was engineer to the Severn and Wye and Monmouth Railways in the Forest of Dean[85] It was probably after he moved to England that the final 's' was added to his surname. David, subsequently described as a 'land surveyor of considerable ability', seems to have remained in south Wales with his father. Between 1810 and 1812 they acted as contractors for the construction of the Aberdare Canal, the course was surveyed by Edward Martin of Morriston and the resident engineers were successively Thomas Sheasby junior and George Overton, it was completed in August 1812.[86]

In 1819 Evan Hopkin was declared bankrupt and his sons were forced to give up the lease on their colliery in Cwm Clydach.[87] Roger Hopkins was subsequently recorded as working in the west of England where he joined William Stuart as assistant engineer of the Plymouth and Dartmoor Railway (25 miles long, completed 1827) in April 1821. He was required to report on a part of the railway that Stuart had already constructed, and his findings were so detrimental that Stuart was dismissed in October 1821 and Roger Hopkins was appointed in his place.[88] His older son, Rice Hopkins (1807-57), also learnt his engineering on the construction of this line.[89] Roger was said to have been 'a bridge builder of repute, who had obtained many commissions on the continent' but this is likely to have been Swansea Valley folklore relating to his activity as a bridge builder in Devon in the 1820s, where the Teignmouth and Shaldon Bridge of 1827 was one of his larger commissions.[90] In 1830 Roger Hopkins was noted in the *Plymouth Directory* as 'Civil Engineer.'[91]

Roger, with his son Rice and his younger son Thomas (1810-48), next engineered the first steam railway in the west of England. This, the 14 mile long Bodmin and Wadebridge Railway, was built between 1832 and 1834 and used two locomotives built at the Neath Abbey Ironworks, Camel (1834) and Elephant (1836).[92] In 1836 Roger appeared again in the *Plymouth Directory,* this time as on 'architect'.[93]

In 1831 Roger was tendering for work in Swansea again, with the submission of a scheme to build the first non-tidal dock in the port (the later North Dock).[94] He was living at Gwauncaegurwen, at the head of the Amman Valley in the 1830s, when he sank a coal-pit which was aborted when waterlogged strata was encountered. A sizeable partly-completed tramroad down to the Swansea Canal at Pontardawe was constructed in 1838.[95] A successful pit was sunk in 1837, shares were sold in 1839 and by 1842 the pit was producing 30-40 tons a day to be sent down the new Llanelly Railway via a short connecting line.[96] Roger Hopkins was recorded in the 1841 Census as a 'Civil Engineer' living near Caegurwen (Gwauncaegurwen).

Roger Hopkins, again with his two sons Rice and Thomas, started the Monmouthshire Iron and Coal Company's Victoria Ironworks at Ebbw Vale in 1836.[97] All three are said to have been Directors with Roger Hopkins as the Managing Director in the period *c.*1837-39. A writer to the *London Mining Journal* in 1845 was obviously questioning Roger Hopkins' all-round competance when he wrote:

'A member of the first board was appointed at the commencement resident directing engineer of all operations above and beneath the surface: this gentleman was undoubtedly clever in his profession of a civil engineer - but you well know, Mr. Editor, a fish is not amphibious, and according to my notion of things, constructing bridges, harbours, and railroads, in the West of England, is not the best method of studying the geology of the South Wales coal-field, or the manufacture of iron.'[98]

There was obviously some substance to the reporting of difficulties in the company and the operating company went bankrupt in 1849.[99]

Rice Hopkins, in partnership with Thomas, carried-out engineering works in south Wales and the west of England. At the time of his death in 1857 he was engineer to the Llanidloes & Newtown Railway; the West Somerset Mineral Railway and Watchet Harbour Commissioners.[100]

Roger's third son, Evan (born near Ystradgynlais in 1806) became a solicitor and then sole proprietor of the *Plymouth Times*. Roger's grandson, Rice (born in Plymouth in 1842), started a successful woollen warehouse business in Australia, going out as agent for Scottish woollen mills after 20 years service as an evangelist in Orkney and in England.[101]

William Jernegan (1750/1-1836)

'...has a turn for mechanical inventions.'[102]

The self-taught engineers of the late eighteenth century acquired their initial experience from a variety of sources. There were the artisan engineers who had been trained and worked as blacksmiths, carpenters or masons. Contractors, such as members of the Bevan or Hopkins families went on to undertake engineering works. Another group were the architect-engineers such as William Jernegan, or Thomas Sheasby senior (the engineer of the Swansea Canal). By the end of the eighteenth century the growing industrial wealth of Swansea had generated enough work for there to be at least five architects practising in the town.[103]

Of these, William Jernegan was pre-eminent, the architect who designed the principal structures built in the leading town of south Wales and hence the outstanding architect of the region. Of the five Swansea architects of the period he was the only one who was not also a builder and therefore the only true professional.[104]

He was an innovative architect who was responsible for many of the new mansions built for the coppermasters and others just beyond the western boundary of Swansea. These include Sketty Hall (remodelled *c.*1790 and 1802); Morris's Sketty Park House (1806); Brynymor (*c.*1820) and Clyne Castle (remodelled 1802 and 1820) where there was not a single flat primary surface in the front elevations; his Marino (1783) was a unique octagonal structure.[105]

Jernegan's known works are confined to Swansea and the surrounding area and he is commonly said to have originally been a Channel Islander. However, his family cannot be traced there. It seems more likely that he accompanied the Leicester architect John Johnston to the area in order to supervise the building of the coppermasters' houses Clasemont (Morriston), and The Gnoll (Neath) and the associated castellated dwellings of the Ivy Tower (Aberdulais) and Morris Castle (Landore). The Jernegan, or Jerningham, family were resident in Norfolk and included an architect in this period. Jernegan's first known commission was that of 1783 for Marino, the residence of the future brother-in-law of the coppermaster John Morris owner of Clasemont.[106]

He is known to have been associated with a series of pioneering industrial buildings. In 1793 he designed the coal-fired Mumbles lighthouse as a double octagon with two lights.[107] He probably also designed the first quay, hotel and customs house at Milford Haven.[108] In 1798 he converted the lighting arrangements at the Mumbles light to a single cast-iron and glass lantern, which he designed and had made by the Neath Abbey Ironworks.[109] He also designed the innovative cast-iron lighthouse for the West Pier of the new harbour that he was building for the Swansea Harbour Trustees.[110]

A plan by Jernegan for 'A Copper Works on a Declivity' was recently found among the papers of Matthew Boulton,[111] Boulton is known to have been associated with the Rose Copperworks, but this drawing is perhaps more likely to have been made by Jernegan for John Morris I, who built his new copperworks in Landore in 1793. The plan fits what is known of the Landore works better than the Rose

Fig. 182. Plan and elevation of the Mumbles Lighthouse. William Jernegan was the leading Swansea architect employed by the Morrises. His characteristically octagonal structure had a cast-iron and glass lantern.

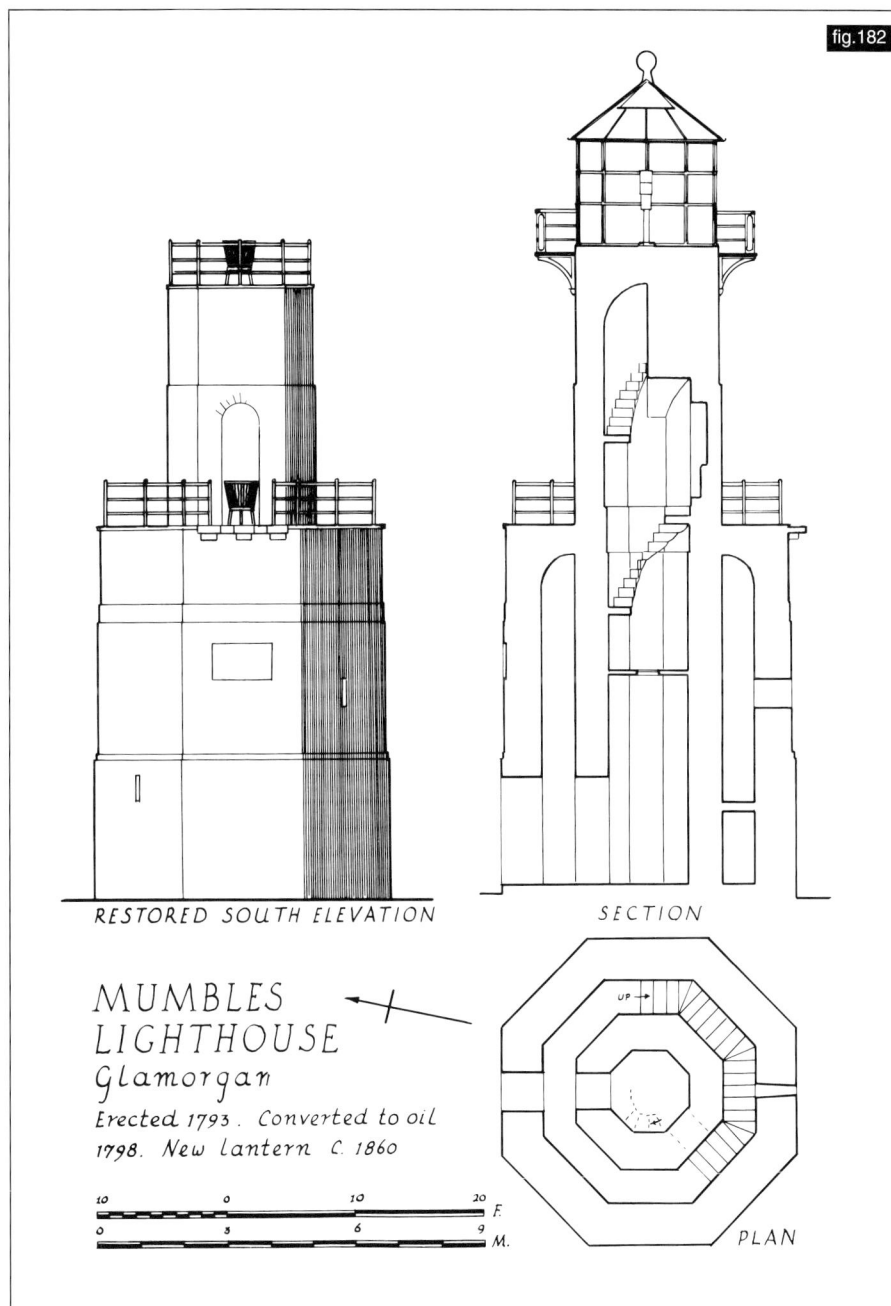

RESTORED SOUTH ELEVATION   SECTION

MUMBLES LIGHTHOUSE
Glamorgan
Erected 1793. Converted to oil 1798. New lantern C. 1860

PLAN

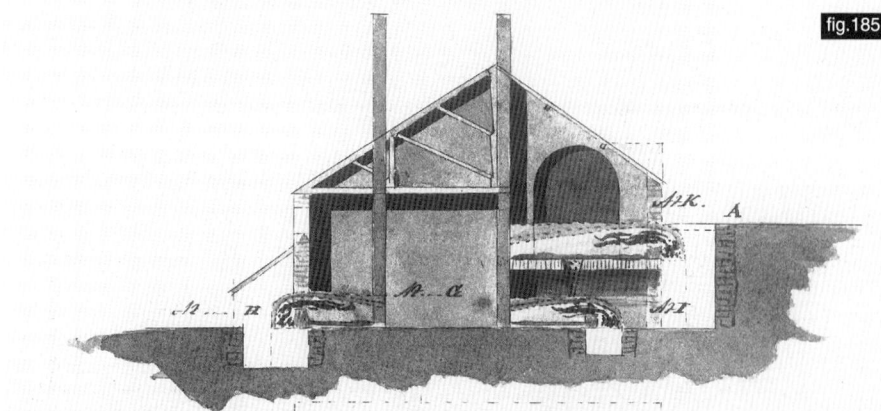

Fig. 183. Drawings for a copperworks that William Jernegan sent to Matthew Boulton are almost certainly of the Landore Copperworks of 1793. The first drawing shows the characteristic round ventilation openings for the furnace-hall at an upper level, with a lower lean-to roof covering the hearths and coal stoking-points for the furnaces (drawings for 'A Copper Works on a Declivity'). (The National Library of Wales)

Fig. 184. Uphill elevation showing the very large ventilation openings on the sheltered side of the building, the lower part of which was built facing a retaining wall on the hillside with the Trewyddfa Canal on top. (The National Library of Wales)

Fig. 185. Jernegan's use of split levels in a furnace-hall was innovatory, although the upper furnace floor would have needed iron reinforcement to have worked. The sloping nature of the valley sides meant that coal arrived by canal at the upper level and could be used to fuel the large calcining furnaces at that level. (The National Library of Wales)

Works, which was also located on a sloping site. It appears that, following the death of Morris' previous architect, William Edwards, in 1789, Jernegan had taken over this role. That the plans were found among Boulton's papers can be explained by the fact that Morris introduced Jernegan to Boulton in 1800, when the latter was enlarging the Rose Works. On the plan 'various sheds' are noted on the south side of the smelting-house, which suggests that the drawing was of an existing building, not one prepared specifically for Boulton. At Landore Copperworks, sheds did stand in the position indicated on the plan.[112]

The plan for the new copperworks at Landore was strictly functional and was essentially an asymmetrical large shed in which to house 14 reverberatory furnaces of various sizes. The building was designed to facilitate the process of successive roastings and meltings of the copper-ore and not *vice-versa*. This fits with Morris' later comment (see below) that Jernegan had 'a turn for mechanical inventions.'

The building was innovatory in having furnaces on two levels: ore arriving at the upper level was roasted in the two large calcining furnaces and then fed directly to the four small ore-furnaces below via

gravity. At least four units of coal were needed at this period to smelt one unit of copper-ore. Consequently, to ease the flow of the main raw material, again by gravity, the new smelter was built into the retaining bank of Morris' Canal, which itself led directly underground into the mines. On the bank of the canal was also a railway from Morris' other collieries at the head of the Nant Rhyd-y-Filais.[113] Coal from all of these sources could be fed directly into the stoke-holes of most of the furnaces which were on the outside of the smelting-house as with Morris' earlier Forest Copperworks. A lean-to on the lower eastern side of the building which faced towards the River Tawe gave shelter to the workers stoking the fires on that exposed side of the building.

Two other expedients were used to ventilate the searing heat of the interior of the smelting-house. The two-storey chamber containing six reverberatory furnaces on the lower eastern side of the building, was given a continuous row of circular windows on the higher level as had previously been used in Morris' Forest Copperworks. The other expedient was to place each of the reverberatory furnaces in large open arches. These would have been a particularly striking device on the uphill western side of the building, but would not have been seen to full aesthetic effect, as they were mostly below the level of a masonry retaining wall which ran parallel to the western elevation of the building. These openings were grouped to facilitate maximum comfort for those stirring and attending the beds of the reverberatory furnaces and were not, in any case, designed for artistic effect.

The northern bay of the west elevation also contained a large calcining furnace on the ground floor of a two-storey chamber. This sat in a large semi-circular arch which extended almost to the eaves level of the outside wall, appearing above the large retaining wall that bordered the long narrow well between the building and wall on its western side. This large chamber was further ventilated by tall narrow semi-circular arched openings that symmetrically flanked the main furnace arch (semi-circular arched openings one storey in height were used in Morris's earlier smelting-house at the Forest). A similar arrangement vented through the main furnace chamber on the east, to three close-set circular windows that terminated the run of six widely spaced circular windows, which ran along the upper eastern wall of the smelting-house.

The central five bays of the west (uphill) elevation of the smelting-house were on two storeys with the fundamental layout determined by the need to house two large calcining furnaces on the upper floor of the chambers. Both of these large furnaces were set in the upper part of very tall and wide arched openings that extended almost to the top of the west wall of the smelting-house. The eaves cut back and gables placed over these openings so as not to block any of the ventilation available in the three-quarters of the arch which extended above the level of the adjacent retaining wall. The large opening in the ground-floor wall in the lower part of the arch gave onto the two smaller reverbatory ore-furnaces below each calcining furnace. Water wells in each side chamber flanking the furnaces received the molten metal tapped from the adjacent furnaces in order to granulate the material. It could then be wheeled over to the metal storage area in the south-western corner of the building, before the final refining was done in the three pairs of furnaces flanking the eastern wall of the smelter.

The section of the building shown in the drawing makes it clear that the intermediate floor was of firebrick and actually constituted the floor of the first-floor calcining furnaces. Smaller chambers flanked the upper furnaces and the furnace-rooms were vented into them by large transverse arches. These side chambers were themselves vented by substantial openings, headed by semi-circular arches that extended through the two storeys of the building. These openings also accommodated steps down which the furnace-charge of copper-ore could be brought. The upper floors were supported across these large openings in the walls by shallow segmental arches. A rather similar arrangement of arched openings extending through two floors, with intermediate arches interspersed with circular windows is found in the large stable block built by George Haynes as part of his 'gothick garden-village' at Clydach in *c.*1803. This also seems likely to have been designed by Jernegan.

Morris, in 1800, introduced Jernegan to Matthew Boulton who was expanding his Rose Copperworks which was in a similar situation, as a result of which Jernegan gave Boulton his drawings of 'A Copper Works on a Declivity.'[114] Significantly, it was to Boulton that Morris made his comment about Jernegan having 'a turn for mechanical inventions.'[115] Thomas Jernegan spent enough time discussing his plans with Matthew Boulton for the former to become familiar with the latter's niece, Mary Mynd; however, she was 50 years his junior and turned him down. Whether this meant a change of architect for any extension to the Rose Copperworks is not clear.[116]

Morris was heavily involved in the scheme for the new Assembly Rooms in Swansea (1805-24) and

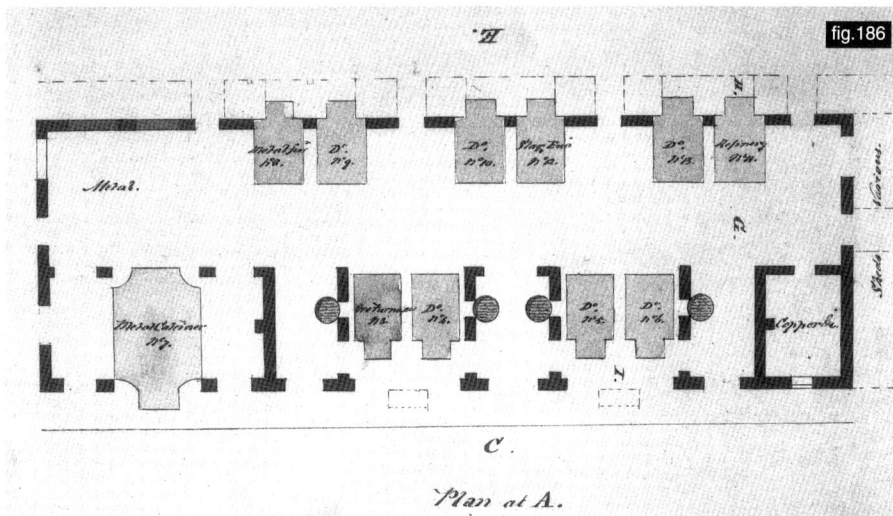

C.

Plan at A.

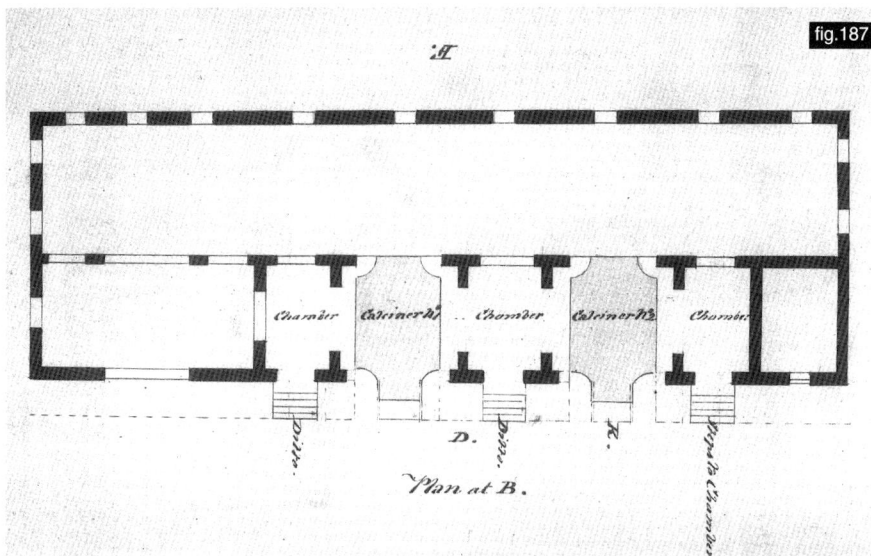

fig.187

Plan at B.

Jernegan was the architect for this (still extant) building that was used for all the ticketings for the Swansea copper-trade.[117] During the troubled course of their scheme, in 1811, Jernegan was declared bankrupt. He later worked his way back to reasonable affluence and his gravestone can still be seen in St. Mary's churchyard.[118]

fig.188

Fig. 186. Plan of the lower, main furnace-floor showing the range and variety of reverbatory furnaces used for repeated roastings of copper-ore in the 'Welsh process'. Alongside some of the furnaces are sunken shallow wells of water which would have been filled from the adjacent canal. (The National Library of Wales)

Fig. 187. Plan of the upper level with two larger calcining furnaces at the upper (canal) level. (The National Library of Wales)

Fig. 188. Jernegan designed the Assembly Rooms in Swansea (completed in 1824) to house a main social venue for the coppermasters and other local notables. The large rooms inside were also used for the important copper ticketings. (960211/1)

Charles Roberts and the Thomas Sheasbys; father and son.

The Swansea Canal (1794-8) was the largest engineering task undertaken in Swansea and its hinterland in the eighteenth century. The canal was built by direct labour and Charles Roberts was appointed engineer for the first nine months.[119] Roberts' origin is not clear, the course of his career is vague and his engineering ability was questionable. He seems previously to have been engaged on the Manchester, Bolton and Bury Canal authorised in (1791) where 'Charles Roberts of London' was appointed engineer at the first meeting of the Canal Company. However, in September 1793 Roberts was found 'not to have acted in the Execution of his Duty with proper discretion and Economy' and was dismissed.[120] In his report on the Swansea Canal, Roberts seems not to have understood what the water requirements might be and his water-storage basins on the hillside overlooking the canal were subsequently abandoned and allowed to silt up. On April 1st 1796 Thomas Sheasby senior replaced him as engineer of the Swansea Canal.

Charles Roberts subsequently surveyed the line of the Newcastle-under-Lyme Canal in 1797.[121] He may then have become the Mines Agent of Lord Dudley and his representative on the Dudley Canal Committee.

Thomas Sheasby, 'architect of Tamworth in the County of Warwick' was appointed contractor for the construction of the Atherstone-Polesworth section of the Coventry Canal in August 1786. His contract was extended to Fazeley in 1787-8 and here he met James Brindley's former assistant Thomas Dadford (senior) who was advising on the design of the Tame Aqueduct.[122] On the 31st March 1789 he was given the contract to raise the section of the Coventry Canal from Coventry to Atherstone to the level of the Oxford Canal by Michaelmas. On 19th July 1790 his working partnership with Thomas Dadford senior started, when he was awarded the contract for the construction of the Cromford Canal together with Thomas Dadford.

Sheasby's entry into south Wales came when he and the Thomas Dadfords (father and son) were appointed on 19th July 1790, as joint contractors to build the Glamorganshire Canal for a fixed sum.[123] One at least of the Merthyr ironmasters may have been aquainted with Dadford by reason of their common west Midlands origins and interests. In the succeeding few years the Dadfords and the Sheasbys were to be responsible for the engineering of all the major canals in south and mid Wales.

On 16th August 1791 the Committee of the Neath Canal noted that 'Mr. Sheasby, a surveyor who has been employed on the Birmingham Canal, informed the Committee by letter that he has nearly finished his work and should be happy to find employment on the Neath Canal.'[124] In the first instance, however, it was Thomas Dadford junior who was engaged to build the Neath Canal, but he left in the middle of 1792 in order to build the Monmouthshire Canal, just as the Neath Canal had reached the southern side of what was to be a major aqueduct over the River Neath at Ynysbwllog. Subsequently on 5th July 1792 the Neath Canal Company agreed that Thomas Sheasby should build the aqueduct and finish the canal according to his estimate by 1st November 1793. There was a sub-committee formed to oversee him and it was resolved that 'if it is found that Mr. Sheasby goes on improperly the General Committee are to stop payment till the same be properly done'. There was obviously disagreement between the two parties and Sheasby failed to complete the canal by the agreed time. To compound matters there was a breach on the Glamorganshire Canal in 1794 which the Dadfords and Sheasbys had completed and where they were still acting as managers. They refused to repair the breach and whilst the arbitration was proceeding on Sheasby's Neath canal accounts, he and the elder Dadford were arrested at the instance of the Glamorganshire Canal Company. He was not found to be liable for the large sum claimed by the Glamorganshire Canal Company and was subsequently also able to claim some additional money from the Neath Canal Company.[125]

While this was going-on Thomas Sheasby senior carried-out the survey and the majority of the engineering on the Swansea Canal. On the 5th April 1793 he was appointed to survey a canal route as far as Defynnog by 12th June.[126] On 2nd December 1793 he reported a survey (assisted by James Cockshutt, later engineer of the Trewyddfa Canal) of another possible canal route between Loughor and Llandovery.[127] Sheasby was appointed surveyor for the Trewyddfa Canal in 1794. The Swansea Canal Act was also passed in 1794 and as has been noted, Charles Roberts, was appointed engineer for the first nine months with Thomas Sheasby senior as assistant engineer from the 1st July 1795 until Roberts was dismissed on 1st April 1796 when he became the engineer of the undertaking: a post he held until after completion on 2nd July 1799.[128] His business interests alongside the canal included the opening of the Brynmorgan Coal Level in Cwmtwrch, in partnership with George Haynes. He was then taken-on as surveyor and engineer of the Wolverhampton and Birmingham Canal company on 17th August 1801 and retired on health grounds on 11th June 1804.[129]

From 1796 Thomas Sheasby junior was helping his father engineer the Swansea Canal. By the 3rd July 1798 he was working for the canal company

Painting of White Rock Copperworks with Upper Bank beyond (Gastineau, 1830). (City & County of Swansea: Swansea Museum Collection).

Side elevation of Tabernacl Chapel, Morriston, showing how the substantial contribution of local nonconformist entrepreneurs allowed the building of a complete classical temple design with the developed side elevations missing in the chapels of poorer worker congregations. (The National Library of Wales).

Painting of Smith's Canal Tipping Staithes on the east bank of the River Tawe at Foxhole. (Glynn Vivian Art Gallery).

Roger Hopkins surrounded by some of the symbols of his profession. On the table to the right is a partly unrolled section of the strata in a mine and behind it are surveying instruments. (Dr D.G. Hopkins)

preparing estimates, and when his father left on 2nd July 1799 he was appointed in his place. On the 7th July 1801 Thomas junior was allowed time to carry-out other engineering commissions in exchange for a reduction in salary. He then left the employ of the Swansea Canal Company in July 1802.[130] On the 17th January 1810 he was taken-on to build the Aberdare Canal but left on 20th August 1811 after being offered a job as clerk and superintendent to the Severn and Wye Tramroad Company as it constructed the short Lydney Canal and Dock. He stayed on, enlarged the harbour there in 1820 and retired on a pension in 1847 as the new locomotive railways helped make the old horse-drawn tramroad obsolete.[131]

Other Swansea Engineers

There were other locally born engineers in early nineteenth century Swansea. One such was Rhys Jones, Sir John Morris II's engineer, who gave evidence at the King's Bench over the legality of Morris' Cwm Burlais Tramroad in 1839.[132] Others included William Bevan of Morriston, who was a partner in the Penfilia Vein Coal Company in the early nineteenth century and his son, also named William, who worked as an engineer on the Saundersfoot Railway in 1832.[133]

Engineers were also brought in from other coalfields. The career of Edward Martin (d. 1818) from Cumbria has been discussed elsewhere.[134] He may have applied Powell's rotary atmospheric engines to John Smith's neighbouring Llansamlet coalfield judging by the surviving engine house remains at Gwernllwynchwith. Unfortunately, his pupil, son-in-law and partner, David Davies of Morriston, died when aged 33 and so was unable further to apply Martin's developed geological and other knowledge of the Swansea area.

The Kirkhouses were another dynasty of engineers who worked in the Swansea area. George Kirkhouse had been brought down from the Great Northern Coalfield on Tyneside to develop the Llansamlet Coalfield of Chauncey Townsend, a London merchant who had leased the eastern side of the Tawe Valley at Swansea in July 1750. Kirkhouse made a major blunder by sinking a multitude of shafts and building many miles of wooden railway to them in order to develop the Llansamlet Coalfield in accordance with the best Tyneside practice. A potential profit was turned into a loss and Townsend's successor John Smith, used Edward Martin's ideas on hillside drainage and transport tunnels to completely change the infrastructure of his coalfield operations.

George's grandson, William Kirkhouse, was rather better versed in local geology and practice and was one of the best known of the early nineteenth-century engineers working in the Swansea area. His career has been described elsewhere.[135]

The Significance and Influence of the Swansea Engineers

The intense economic activity of the Swansea landscape in the eighteenth century acted as a catalyst in producing new skills needed to produce the sophisticated infrastructure in which smelting and mining activities could prosper and grow. Much of this activity was away from the large market and population centres of the world's first industrial revolution and it is difficult to know to what extent this, locally developed, expertise spread on to the national and international stage. It was only at the end of the eighteenth century that English tourists 'discovered' and reported on this seaside arena of intense activity, while it was not until the beginning of the nineteenth century that our continental European cousins came to examine and write on this scene.

References

1. B. Morris, 'Swansea Houses - Working Class Houses, 1800-1850', *Gower,* 26, 53-61 citing the quotation 'but the town is not merely inhabited by Copper-men, Colliers and Mechanicks; very few of whom live in the Town' in the petition supporting the Swansea Paving Bill submitted to Parliament in 1787.
2. v.i. Professor Michael Flinn, quoting the findings of the research undertaken for the British Coal Research Project.
3. 'Obituary of considerable persons with biographical anecdotes, Aug. 7', *The Gentleman's Magazine and Historical Chronicle,* (1789), 767-68; will of William Edwards, April 8, 1789 (proved Aug. 10).
4. H.P. Richards, *William Edwards: Architect, Builder, Minister* (Cowbridge, 1983), 67 & 82.
5. B.H. Malkin, 'New Bridge', *The Scenery, Antiquities and Biography of South Wales (*London, 1804), 83-95 and Richards, *op.cit.,* 11.
6. Ibid.
7. Ibid., 85 and S. Smiles, 'William Edwards, bridge builder', *Lives of the Engineers* (London, 1862), 266-75.
8. Richards, *op.cit.*, 11.
9. Malkin, *op.cit.*, 91.
10. Ibid., 594.
11. Richards, *op.cit.,* 17-8.
12. Richards, *op.cit.,* 33, quoting *A Portrait* (of William Edwards, MS. 9927D, d.1782-83, NLW).
13. Ibid., 12-3.
14. Ibid., 21. Some accounts give the number of spans of the first Pontypridd Bridge as four arches.

15. G. Grant-Francis, *The Smelting of Copper in the Swansea District* (London, 1881), 111; R.O. Roberts, 'The Smelting of Non-ferrous Metals since 1750', *Glamorgan County History*, vol. 5, 47-96, p.87 gives 1748 as the last date of recorded production in Lockwood, Morris & Co.'s Llangyfelach Copperworks and the first recorded production of the Forest Copperworks in 1752. The Beaufort Bridge connecting the new smelter to the earlier rolling-mills on the opposite bank of the River Tawe would have been needed from the beginning of production and is attributed to Edwards by Malkin, based on conversations with the latter's son David. The land for the Forest Copperworks was leased from the Duke of Beaufort in 1746, and renewed in 1754 when 'all the buildings ... for smelting copper and lead' had been erected and allowance for constructing further buildings given (NLW, Badminton Collection Group II, 1261).

16. Lockwood, Morris & Co. came to a resolution to erect mills at Forest in March 1728 and Mr Padmore 'a capital mason of Bristol, was to come to Swansea to view a situation R.M. [i.e. Richard Morris] had in his eye to make a draught or model and estimate - or to send his brother-in-law, Wm. Sawyer.' In July 1728, Robert Lockwood 'agreed to let ye erection of the intended Mills sleep a little while' while on a visit to Swansea (Morris. MSS. (UCS) quoted in Jones, *Swansea Port*, 309). Subsequently ...'Richard Lockwood and Company have by Lease or Grant from the said Duke [of Beaufort] built and erected a wear across the said river upon or near the foundation of an ancient Wear that was upon the said river near fforrest yssa (isaf) aforesaid for the diverting the water of the said river for the service and use of certain Battery works by them erected on the ffarm and lands called fforrest yssa' (NLW Badminton Collection Group II, 1496, d.1732).

17. R.O. Roberts, 'The Smelting of Non-ferrous Metals since 1750', In: *Glamorgan County History: Industrial Glamorgan,* vol. 5 (Cardiff, 1980), 47-96, 87.

18. G. Grant-Francis, *The Smelting of Copper in the Swansea District* (London, 1881), 111, describes the demolition of the buildings, but probably dates demolition a few years too early as they still appear on the Llangyfelach Tithe Map of 1838.

19. The Forest Works were illustrated on an engraving by Rothwell in 1790 and a watercolour by J.W. Smith in 1792 (the latter reproduced in D. Moore, *The Earliest Views of Glamorgan* (Cardiff, 1978)).

20. Grant-Francis, *op.cit.*, 111. Visible on an engraving of Forest Copperworks published in 1791, reproduced in M. Gibbs and B. Morris, *Thomas Rothwell; Views of Swansea in the 1790s* (Cardiff, 1991), 33-5.

21. Tithe Map of Llangyfelach, 1838 (NLW).

22. Badminton Collection, Group II, 1286, d.1844 (NLW).

23. O.S. 1/10,560 Gl. Sh. XV S.W., 1897.

24. Tithe Map of Llangyfelach, 1838 (NLW).

25. E. Jervoise, *The Ancient Bridges of Wales and Western England*, (London, 1936), 95.

26. The first two hastily written accounts of the building of the Pontypridd Bridge were prepared by William Edwards' fellow preacher Thomas Morgan as a rejoinder to the London periodical asserting (1764-5) that the Rialto bridge was the largest span in Europe. This account was subsequently rewritten and 'whatever was incorrect in this printed account, has been ... carefully set aright from the Architect's own mouth ...' (NLW 2049 (Panton 84) d.1778; discussed in Richards, p.25).

27. Richards, *op.cit.*, 26-7.

28. *The Gentleman's Magazine and Historical Chronicle*, 34 (October & November, 1764).

29. T. Morgan, 'An Account of a Very Remarkable Bridge in Wales', *The Annual Register; or a view of the History, Politics and Literature for the year 1764*, (1765) 147.

30. T. Ruddock, *Arch Bridges and their Builders 1735-1835* (Cambridge, 1979), 50.

31. Ibid., 5-7.

32. Ibid., 16-17.

33. Ibid., 18.

34. Ibid., 48.

35. NLW 2049B (Panton 84), provenance discussed in Richards, *William Edwards*, 25-9.

36. T.M. Smith, 'The conditions of the stability of the Arch of the Pont-y-ty-Pridd Bridge', In G. Snell, *The Stability of Arches* (London, 1846).

37. T. Ruddock, *op.cit.*, 48.

38. R. Quenby, *Thomas Telford's Aqueducts on the Shropshire Union Canal* (London, 1992), 37..

39. Design drawings for the Lancaster Canal (PRO).

40. C. Hadfield, *Canals of South Wales and the Border* (Cardiff and London, 1960), 98.

41. Richards, *op. cit.*, 44.

42. Ruddock, *op.cit.*, 47, 49-51; W. Emerson, *Doctrine of fluxions* (London, 1743). Arches of equilibrium were established by W. Emerson, *The Principles of Mechanics* (2nd. ed. London, 1758)

43. Ibid., 45.

44. Ibid., 47.

45. Thomas Day, 'Telford's Aberdeenshire Bridges', *Industrial Archaeology Review,* XVII,2, (Spring 1995), 193-207.

46. Malkin, *op.cit.*, 88-89.

47. Ruddock, *op.cit.*, 52; quoting Malkin and J.G. Wood, *Rivers of Wales Illustrated* (London, 1813). See also Richards, *op.cit.*, 58.

48. Richards, *op.cit.*, 53-4.

49. Jervoise, *op.cit.*, 84.

50. Richards, *op.cit.*, 54.

51. Ibid., 58-9 quoting M. Williams, *The Making of the Welsh Landscape.*

52. Badminton Collection, Group II, MS. 2082, d. early nineteenth century (NLW).

53. Gibbs and Morris, *op.cit.*, 33-5.

54. D. Moore, *op.cit.,* 43.

55. Ruddock, *op.cit.,* 51 quoting Brecon Q/S Records (NLW).

56. Will of William Edwards, quoted in Richards, *op.cit.,* 72-73.

57. Ibid.

58. Richards, *op.cit.,* 13.

59. Ibid., 31.

60. Ibid., 62-63.

61. A. Scoville, *Morriston's Pictorial Past* (Cowbridge, 1988), 21.

62. Richards, *op.cit.,* 64-5. Lewis Rees was the father of Abraham Rees whose Encyclopaedia is still a major source for industrial archaeologists.

63. Ruddock, *op.cit.,* 51 quoting Brecon Q/S records (NLW).

64. Richards, *op.cit.,* 73.

65. Ruddock, *op.cit.,* 51.

66. Ibid., 52.

67. T. Lloyd, *The Lost Houses of Wales* (London, 2nd. ed. 1989), 61.

68. Malkin, *op.cit.,* 93.

69. Ruddock, *op.cit.,* 53.

70. Ibid. and Richards, *op.cit.,* 59.

71. Jervoise, *op.cit.,* 76.

72. Malkin, *op.cit.,* 94.

73. G. Watson to M. Boulton, 4 Jan. 1783 (M. Boulton Papers, Birmingham Central Library, Box M).

74. Ibid.

75. 'Statistical Table of Steam Power, Swansea', n.d. but *c.* 1840 (Morgan MSS., Swansea Museum (RISW)). For a fuller discussion of the steam-power in use in the Swansea area at this time see S. Hughes, B. Malaws, M. Parry and P. Wakelin, *Collieries of Wales: Engineering & Architecture* (Aberystwyth, 1994), 69 and 87-8.

76. J. Marshall, *A Biographical Dictionary of Railway Engineers* (Newton Abbot, 1978), 118. p.c. Gerald Gabb, quoting Jones, *Swansea Port,* 137. Several of the following references have been kindly supplied by Gerald Gabb, *Swansea Canal Survey 1794-9* (British Waterways Estates Office, Gloucester, G.W.R. Swansea Canal, Plan No. 20) 77; S.R. Hughes, 'Aspects of the use of water power during the Industrial Revolution', *Melin: Journal of the Welsh Mills Group,* 4 (1988), 23-37.

77. UWS Library, Oystermouth Railway General Assembly Minutes, 6 August 1804.

78. UWS Library, Oystermouth Railway Accounts Committee, 3 November 1804.

79. UWS Library, Oystermouth Railway, Management and Accounts Committees; 18/2/1805 & 8/11/1805 respectively.

80. UWS Library, Oystermouth Railway Management Committee, 1 June 1807.

81. Jones, *Swansea Port,* 137.

82. Ibid.

83. *The Cambrian,* 16 May 1807.

84. T.V. Evans, *Clydach a'r Cylch* (Tonypandy, 1901), quoted in R.P. Roberts, *History of Coalmining in Gower from 1700 to 1832,* unpublished MA thesis, University of Wales Cardiff (1953), 74.

85. Marshall, *op.cit.,* 195-6.

86. C. Hadfield, *The Canals of South Wales and the Border* (Cardiff, 1960), 121-2.

87. Oystermouth Railway Committee Minutes, 26 September 1819 and Evans, *Clydach a'r Cylch.*

88. R. Handsford Worth, *Early Western Railroads,* (Weston-super-Mare, n.d. reprint from *Trans. of Devon & Cornwall Nat. Hist. Soc.* 1887-88); p.c. Paul Reynolds quoting K.S. Perkins, 'Opening up south Devon - the Hopkins connection', *The Devon Historian,* 45, (October 1992), 9-17.

89. Marshall, *op.cit.,* 117.

90. Perkins; J.H. Davies, *History of Pontardawe and District* (Llandybie, 1967) 91-2.

91. p.c. Donald Hopkins, great-great-grandson of David Hopkins.

92. Handsford Worth, *op.cit.,* 13.

93. p.c. Donald Hopkins.

94. R. Hopkin, 'Report on the Formation of a Floating Harbour at Swansea', *Reports on the Harbour of Swansea* (Swansea, 1831), 15-42.

95. *Cambrian* 17.3.1838; P.R. Reynolds, 'Gwauncaegurwen Tramroad', *R. & C.H.S. Tramroad Group, Occasional Paper 39* (November 1986); and. Davies, *Pontardawe,* 44.

96. Davies, *Pontardawe,* 44, 104.

97. Marshall, *op.cit.,* 118; L. Ince, *The South Wales Iron Industry, 1750-1885* (Birmingham, 1993), 110.

98. Ibid., & p.c. Donald Hopkins; K.S. Perkins, 'Opening up south Devon - the Hopkins Connection', *The Devon Historian,* 45, (October 1992), 9-17 citing the *London Mining Journal* (1845), 22.

99. Ibid.

100. Marshall, *op.cit.,* 118.

101. p.c. Donald Hopkins.

102. T. Lloyd,' The Architects of Regency Swansea', *Gower,* 41, (1990), 56-69, 59

103. Ibid., 56.

104. Ibid., 58.

105. Ibid., 59.

106. Ibid., 59

107. D.B. Hague, *Lighthouses of Wales* (Aberystwyth, 1993), 86-89.

108. Lloyd, 'Architects of Regency Swansea', 60.

109. Ibid., 61.

110. Jones, *Swansea Port,* 68.

111. W. Jernegan, 'Plan of a Copper Works on a Declivity' (NLW).

112. 'Plan of Morfa Slebech, Morfa Arglwydd, etc. showing the Landore Copper Works & Trewyddva Canal etc (NLW, Badminton Group II, 1280-81, *c.*1844), The Rose (1780), and Ynys or Birmingham Copperworks (1793), are shown to have different plans on the deposited plan of the Swansea Vale Railway (NLW, Brecon Q/RP 35-36).

113. Badminton MSS Group II, 1295, n.d. but *c.* 1793.

114. Lloyd, 'Architects of Regency Swansea', 62.

115. Ibid., 59, quoting Boulton Papers MB/M2/187 (Birmingham Reference Library).

116. Ibid., 62.

117. Ibid., 60.

118. Ibid., 62.

119. Hadfield, *op.cit.,* 49.

120. C. Hadfield and G. Biddle, *The Canals of North West England* (Newton Abbot, 1970, Vol. 2), 246.

121. Ibid., 206.

122. Charles Hadfield's index of canal engineers and C.T.G. Boucher, *James Brindley, Engineer, 1716-1772* (Norwich, 1968), 103.

123. Most of the information in this section is drawn from Charles Hadfield's index of canal engineers.

124. Neath Canal, Committee Minute Book, 16 August 1791 (GRO).

125. Ibid. for the appropriate dates and Hadfield, *The Canals of South Wales*, 65 and 95.

126. *Hereford Journal*, 17 April 1793.

127. Ibid., 27 November 1793.

128. p.c. Charles Hadfield.

129. Ibid.

130. Ibid. quoting the Swansea Canal Minutes.

131. Ibid. and Hadfield, *Canals of South Wales*, 121-22 & 212-15.

132. Quarter Sessions, 16 April 1839, Richard Mansel Phillips v. Sir John Morris II.

133. B. Baxter, *Stone Blocks and Iron Rails* (Newton Abbot, 1966), 225.

134. S. Hughes, *The Archaeology of an Early Railway System: The Brecon Forest Tramroads* (Aberystwyth, 1990), 112-13.

135. Ibid., 127-31

CHAPTER 4

MASTERS & MEN
DYNASTIES & DWELLINGS

Fig. 189. Workers at what was still one of the world's largest copperworks watching King George V, Queen Mary, Captain Hugh Vivian and other members of the Vivian family on a royal visit to Swansea on July 19, 1920. (City & County of Swansea: Swansea Museum Collection)

Introduction

In and around Swansea there still remain substantial workers' houses, settlements and former mansions from the period when the city was the copper-smelting capital of the world. The number and form of each of these buildings and the way they relate to each other in the landscape can be used as a source of historical information on several different levels.

The most numerous physical survivals from the copperworks era are the hundreds of dwellings that once housed the copperworkers in semi-rural communities. Most evident, and most important to any understanding of the coppermasters' motives in social provision, are the regular layouts of workers' houses built by the coppermasters themselves. It is important to examine how these houses changed in the eighteenth and nineteenth centuries.

To understand what the remains of this social landscape mean, it is necessary to compare it to what had gone before, and to what was provided in those other substantial industries of Swansea that interacted with the dominant copper industry. It might, for instance, be possible for the coppermasters to have left housing to the market. Only an examination of the remains of the housing provided by the copper-masters will show whether it was better, worse, or average when compared to that provided by others.

For any conclusions to be arrived at in assessing motives of enlightened self-interest and philanthropy, it is necessary to try to find out what proportion of the industrial housing stock was provided by the coppermasters, and hence how important was the existence of their housing in meeting the needs of the copperworkers.

In order for judgements to be made on any of these

topics, it has to be ascertained whether enough remains survive in Swansea to support conclusions. It needs to be seen to what extent houses were just simple individual structures for living in, to whether they had a character that formed part of distinct settlements with recognisable types of plan, and with communal buildings for education, religious worship and communal use.

Workers' settlements, like the copper-smelting works, formed nucleated clusters of structures in a semi-rural landscape.

Perhaps they also related to the linear systems already considered. These networks of water-power and transport served the dispersed nodes of copper-smelting and coal-mining and may also have directly influenced where workers lived. There is also the question of whether the separate mansions of the industrialists related to the workers' settlements and the linear systems of power and transport. The evolution of the landscape of this first industrial period depended on how the growing profits from established copper-smelting works were expressed in terms of evolving mansions, parklands and philanthropy.

Perhaps the most basic question to ask on the transformation of the living conditions of the population in the Swansea area during the Industrial Revolution is, to ask how the lot of the Swansea copperworker compared to that of his contemporaries who had remained in rural employment.

At a more complex level perhaps, information may be gleaned concerning human motivation within trends of economic and social history. The buildings, themselves physical manifestations of the historical process, may tell us about the aspirations of internationally powerful entrepreneurs and managers who effectively developed booming businesses, generated employment for hundreds, and later thousands, and who, in the process, founded wealthy dynasties. It may be possible to begin to understand issues that concerned these influential men and their wealthy business partners by examining the bricks and mortar of domestic and institutional buildings. An examination of the surviving industrial housing, and of complete industrial settlements, can be expected to clarify the intentions of the coppermasters. They may have been motivated by a combination of factors: to provide accommodation for key and other workers in order to ensure a labour supply in what were then fairly remote rural locations; or to satisfy an urge to ensure social control and to encourage the docility of a large workforce. It could be that there were more altruistic motives of philanthropy and paternalism at work. All of these factors may have manifested themselves in material displays of enlightened self interest, and the

consequences may have contributed to the good financial investment of a smooth-running coppersmelting enterprise in eighteenth-and nineteenth-century Swansea.

To understand the broader context of this industrial society, it is also necessary to understand its relationship to what is known of the character of the copper and other industries during the world's first Industrial Revolution. It may be instructive to place this society in a European context, with its new international level of affluence.

It might also be argued that the Swansea industrial settlements have a distinctively Welsh character, with copious provision of large and elaborate nonconformist places of worship; this is examined in detail in the next chapter.

In the following sections the surviving domestic fabric, and some of the available written sources, are studied to see which of these important questions can be answered. The more significant facts to emerge will be summarised at the end of each section and in the general conclusion.

Early Workers' Housing in Wales

Providing housing for industrial workforces in Wales has a very long history stretching to a period well before the Industrial Revolution. There has always been the need to provide accommodation for specialist workers concentrated at centres of economic activity which were sited in areas remote from existing agricultural, mercantile or urban settlements. In prehistory the stone-axe factories of Aberdaron, Penmaen-mawr and the copper-mines of the Great Orme (now Llandudno) would have required settlements nearby to accommodate the workers who operated them. Later, in the Roman period, the two great legionary fortresses at Chester and Caerleon were serviced by substantial industrial settlements, at Holt and Bulmore respectively. The Roman gold-mines at Dolaucothi would have been similarly served. With the Edwardian conquest of North Wales in the 1280s, specialist German miners were employed to tap more effectively the mineral wealth of Flintshire and other lead-and copper-mines further to the west. These specialists would have been provided with a reasonable level of accommodation to attract them to employment in a remote area and then to keep them there - factors that were also to exercise the minds of later industrialising magnates in Swansea.

In the medieval period, the Cistercian monasteries of Wales, at Strata Florida and elsewhere, derived much of their income from the exploitation of local mines, such as the great copper and lead mine at Cwmystwyth. In the sixteenth century the monopolistic Society of Mines Royal leased its rights to lead-and silver-mining activity in mid-Wales to Thomas 'Customer' Smythe and his manager, Charles Evans. They employed 40 miners and concentrated their activities nearer Aberystwyth (and available sea transport) at Cwmsymlog, where two German miners from the great technologically advanced mines in the Harz Mountains of Saxony (now a World Heritage Site) were employed to improve techniques. Here is one of the first references to actual workers' houses in Wales, for at Cwmsymlog in 1586 a new blacksmith's shop was erected and houses were repaired for the miners.[1] Such recycling of the available building stock to house workers was to emerge later in Swansea. Both German and Belgian experts in copper, silver and zinc smelting were also later to be provided with special accommodation at Swansea.

By 1565-66 it was royal policy to establish centres of manufacturing that were independent of the more industrially developed industries on the continent of Europe. Of all the British rivers examined for their suitability as a power source, the Angiddy at Tintern was chosen as a site for a new wire-producing works, using ductile Osmond iron produced from the Monkswood Furnace at Pontypool. In 1568 the Company or Society of the Mineral and Battery Works was set up to license the manufacture of such goods and it took over the Tintern Works. Again, a number of key experts were brought from Europe to provide technical expertise and they would have needed to be housed near the site. By 1763 there were no fewer than eight main water-powered metalworks on the river and two had terraced rows of workers' dwellings attached: at the Lower Wire Works and Pont-y-saeson Forge. The earliest surviving dated house, however, appears to be that of the the seventeenth-century works' owners', the Foleys.[2] To the south of the former works is a two-storey house with a datestone bearing the inscription 'E.F. 1699', evidently built by Edward Foley. Undated two-storey workers' terraces survive alongside several of the former works' sites including Pont-y-saeson Forge. The Tintern Wireworks was by far the largest industrial installation in sixteenth and seventeenth-century Wales, employing some 600 people.[3]

A subsidiary manufacturing site of the Tintern Wireworks was established in the Whitebrook Valley, to the north, in 1606. This is one of the earliest recorded instances of housing being newly built specifically for workers for by 1630 14 houses had been constructed for workers on the 'waste' (i.e. common land). The wire industry ceased in c.1720, but an early paper-making industry was started in the valley 30 years later and much workers' and managers' housing seems to survive. In Wales much of the workers' housing, and indeed the industrial installations themselves, were built on unenclosed waste, for until the 1790s about a quarter of Wales consisted of such land. The proportion of commonland and marsh within the lower Swansea Valley was much the same and much of this was utilised for coal-mines, copper-smelting works and workers' housing.

The early workers' housing in the Angiddy and Whitebrook Valleys forms one of the early types of workers' housing found in Wales, namely that built at remote rural iron-forges. One of the other early references to houses newly built to house workers, may be to a charcoal store and workmen's dwellings built at Machen Upper Forge by, or for, John Greenhough of the Van (Caerphilly) before 1662.[4] Without more wide-ranging fieldwork, it is impossible to be absolutely sure of the antiquity of present housing remains on these sites.

Elsewhere, the forge at Monmouth was in existence by 1628 and most of its workers' housing is still intact.[5] A large two-storeyed terrace that certainly dates from the eighteenth century or earlier, still exists. Another survival of considerable age is the terrace of workers' houses attached to the Llanelly Furnace and Forge in the Clydach gorge near Abergavenny, which probably dates from c.1730.[6]

At Swansea the most significant source of employment at the world centre of the copper trade was that driven by the great copper-smelting works, but two earlier types of industrial site existed which were also important to the copper and, later, the tinplate industries.

Housing at Water-powered Forge Sites in the lower Swansea Valley: 1647-1784

The development of workers' housing in the Swansea area fits well into this general picture of workers' housing in Wales. There were four substantial forge sites in the lower Swansea Valley built during the seventeenth and eighteenth centuries; they were rather similar to those found elsewhere in Wales. The first was the Ynys-pen-llwch Forge, built in 1647 by Bristol merchants and which certainly had substantial workers' housing by the time of the 1841 census, but of which nothing remains. Sir Humphrey Mackworth, the philanthropic builder of the terrace of workers' housing at Esgair Hir Copper and lead mine (in Ceredigion), was in control of the Ynys-pen-llwch 'rolling, slitting and wire mills' by 1726 and may have built or extended the housing there.

The second lower Swansea Valley forge was

founded at (Upper) Forest in 1696 by the Crowley ironmasters alongside the mansion of the landowner and later forge master, Thomas Popkin, and whose ruins remain. By 1841 the works' manager, Charles Conway (described as a 'Copper Agent' i.e. works' manager), almost certainly lived in the old Popkin mansion.[7] Four other dwellings had been built by then and are also recorded in this first detailed census.

The third forge site in the lower Swansea Valley was built for copper rather than iron and was the Lower Forest Forge, built for Lockwood Morris & Company c.1730. These coppermills had 10 houses on site by 1841, and probably were created by 1793 when two blocks of buildings existed to the north of the mills. These were enclosed on the island by the works' headrace and the River Tawe (five blocks of ancillary buildings surrounded the mill at this date). Forty-seven people lived here, including one collier, four coppermen, two smiths, two hammermen, two rollermen and two bargemen.[8]

The fourth large forge site was the Clydach Upper Forge constructed c.1784; the first detailed map of the site in 1841 shows a terraced group of houses for the forge workers. At Swansea the four large early forge sites all had workers' housing attached. This is not surprising, these sites were in relatively isolated rural locations determined by the ready availability of substantial water-power sources, and a nucleus of skilled workmen had to be attracted to them to operate them successfully for the forge owners. By contrast, the first copper-smelting works of 1717 and 1720 at Landore and Swansea were either in, or within relatively easy reach of, Swansea town.

There are also structural remains of workers' dwellings at one of the forge sites in the lower Swansea Valley. The houses at the Upper Forge at Clydach were probably built in 1784 by the ironmaster Richard Parsons. By 1813 two blocks of housing had been built for workers at the forge on top of the cliff to the south-east of the forge site.[9] The houses of the western row, on the edge of the cliff, survived in their original form until they were modernised (and largely re-roofed and gutted) in about 1975. Chronologically, the Clydach Upper Forge houses may be the next surviving workers' housing in the lower Swansea Valley industrial area after the Morriston houses. In form, they are like a terraced adaptation of the Morriston cottages with double frontages and a one-room depth giving onto a blank rear wall.

The façade-width of one of the houses, 15ft, is comparable to the 16ft of town terraces in Swansea such as in Pottery Street. However, in Swansea and in the copperworkers' settlements the houses are all single-fronted, whereas Forge Row at Clydach has four double-fronted houses, 23ft wide, and only one single-fronted house, which is 15ft wide. The dimensions are similar to those of the Morriston houses (26.5ft by 17ft and 25.5ft x 15.5ft), which were single-aspect, double-fronted houses also. This is also the normal type of workers' house in the iron belt at the Heads of the Valleys. The height of the Clydach houses is comparable to that of the Morriston houses and of those in the iron-working district: their front and back walls are little more than a storey and a half in height, with the first-floor windows set close to floor level and directly under the eaves. The floor area of 480 sq ft was less than that of the Morriston houses and approximated to that of the head colliers' houses at Tirdeunaw (see below).

By contrast, houses in Swansea invariably had two full-height storeys with windows on their front façades, and copperworkers' housing in Morriston and Trevivian (from the 1830s) was also higher. In Swansea town this was the case as early as 1813-23, judging by the houses in Pottery Street which were surveyed before their demolition. In fact, all the windows of Forge Row, both upper and lower, were small, almost square, openings and this characterised the more remote, late eighteenth and very early nineteenth-century houses of the south Wales ironworkers in the Heads of the Valleys region, such as at Gellideg, Merthyr Tydfil (built between 1769 and 1813), Stack Square and Engine Row, Blaenavon (built in 1789-92), and Forge Row, Cwmavon, near Pontypool (1804-6).[10]

It would be profitable to compare these houses with any that the ironmaster Richard Parsons built for his other workers at the Ynysgedwyn Ironworks. However, the houses at Ystradgynlais known as Gough's Buildings were probably built by the landowner, the Revd Fleming Gough, for his colliers in the 1820s and they had a front block with a slope behind. A further block was added by George Crane for his ironworkers in 1829. The long sweep of College Row at Ystradgynlais was built for the Ynysgedwyn ironworkers from 1837 onwards and consists of double-fronted houses. The houses built by George Crane, who was Richard Parsons's successor at Ynysgedwyn from 1823, have two full storeys, fairly large windows on both storeys and balancing, blind window recesses over the front doorways of each house.[11]

The nearest comparison to the Clydach Upper Forge houses is in Stack Square and Engine Row at Blaenavon, where the houses had two rooms on each floor and a larder. They were quite large for their time and were probably intended for foremen and craftsmen employed on full-time contracts and essential to the working of the furnaces. The Clydach cottages were probably built for the skilled craftsmen who turned the cast-iron pigs from Ynysgedwyn into malleable wrought iron at their hearths.

Clydach Upper Forge can be dated to 1784, and there is no reason to doubt that the houses of Forge Row at Clydach are among the earliest extant industrial workers' houses in Wales. Like Blaenavon, and the eighteenth-century terrace and manager's house at Monmouth Forge and Tintern Upper Forge, the row is immediately adjacent to the place of work rather than some distance away as at the later lower Swansea Valley settlements of Morriston, Trevivian, and Grenfelltown. The walls and even the chimney-stacks (which were brick on all other sites considered in this section) were of local Pennant sandstone rubble; casement windows were used, rather than sash windows; timber lintels were installed over openings rather then the brick arches found in other settlements; and a red, pantiled roof (probably pre-dating the availability of north Welsh slate that only came with the partial opening of the Swansea Canal in 1796) survived until the 1970s. However, refurbishment of this terrace has removed many of its original features.

Housing at Seventeenth- and Eighteenth-century Colliery Sites at Swansea

Some of the first mining at Swansea, possibly in the Roman period and certainly by the thirteenth century, took place on the outcrops of coal where the substantial Swansea Five-Foot seam was exposed high on the western and eastern slopes of the lower Swansea Valley. Rudimentary tracks developed to give pack-horse access to many neighbouring adits on the exposed seams running along the lips of the valley, and the colliers' settlements began to appear. A few simple cottages seem to have been erected alongside these tracks by the miners for themselves and their families.

To the west, the outcrop lay along the top of the open commonland slope of the fee of Graig Trewyddfa: the road northwards from Swansea town to Clydach at this time ran along the crest of the common and just above the row of mining tunnels that ran into the outcrop on the southern part of the unenclosed land. On the northern part of Graig Trewyddfa a long straggling settlement of colliers' dwellings, set within small gardens enclosed from the common, developed along both sides of what has become Graig Road on the hillside above, and to the west of, what is now Morriston, that is just below the more northerly outcropping of the Swansea Five-Foot seam and its adits. One of these houses, Ty-coch, is recorded as being in existence by 1682,

fig.190

when it was being used for Welsh Independent (i.e. Congregationalist) religious services.[12] Eleven separate plots of land are shown enclosed from the common alongside the seam in 1761, and by 1793 these seem to have included 17 cottages with another four added to the north and east.[13] The landowner was Thomas Popkins, and later (from 1730) Lockwood, Morris & Co. were leasing the thick seam here from the Duke of Beaufort, but the direct provision by employers of housing here seems to have been limited to the building of the 24 apartments for colliers and their families at Morris Castle, sited over the southern end of the outcrop. The proximity of this seam explains the founding of

Fig. 190. The western valleyside of Graig Trewyddfa looking south-west from Morriston. By the seventeenth century, groups of squatter cottages housed colliers on small plots enclosed from the common; the successors to these cottages can be seen at bottom right flanking the early coal road now known as Graig Road. (935056/41)

Fig. 191. Early postcard of Morriston viewed from the west showing some of the thatched squatter cottages on Graig Trewyddfa in the foreground. The Clyn-du Coal-pit engine house and chimney are visible to the left with the central church of St John visible behind.

the alum works on adjacent land above the site of Morriston, the existence of its water tanks now marked by the present Cwm Bath Road, Bath Road and Bath Avenue.[14] Presumably workers in this manufactory may also have lived in the houses on the common.

At Bon-y-maen, on the eastern side of the valley, by the beginning of the eighteenth century, the Mansels seem to have been the local land-owning family most directly involved in large-scale coal-mining. One quarter of their whole estate income was derived from coalmining.[15] However, these collieries were on land remote from Swansea town and across the river to the north-east. By 1722 the Mansels had built houses for no less than 75 of their colliery workers.[16] These were probably traditional, detached cottages set in small roadside gardens (shown on later maps) enclosed from adjoining farmland on the Mansel's Briton Ferry estate at Bon-y-maen: Cefn Road actually follows the line of the Swansea Five-Foot seam outcrop down towards Crymlyn Bog.

The development of coal-hungry smelting-works at Swansea from 1717 created the incentive to invest in more substantial collieries to feed this trade. The simple short mining tunnels on the outcrop of the Swansea Five-Foot seam on the western side of the valley were largely worked-out by this period. A succession of shafts were sunk to work the deeper outcrop to the west in the first half of the eighteenth century. The Prices, squire-coalmasters of Penlle'r-gaer, appear to have had a worker's house alongside

one of their coal-shafts at Pwll-y-domen (now Manselton) by 1738. Other shafts and levels are likely to have had at least a 'head collier' living adjacent.

By 1768 Lockwood Morris & Co. had three houses, for head colliers alongside their Tirdeunaw Colliery shaft while there was a cluster of cottages alongside the mouth of the same firm's Plasmarl Level. The eighteenth-century Pentre-gethin Colliery tunnel mouth (on what is now St. John's Road at Manselton) also had a colliery manager's house alongside its entrance cutting (which stood until living memory).

The 24, or so, flats built for colliers at Lockwood Morris & Company's Treboeth Level in the early 1770s, and known as Morris Castle, were unusual in their multi-storey flatted form (they are discussed below).

Thus, in the eighteenth century houses for head colliers were being built next to individual coal shafts and colliery levels to the immediate north of Swansea town, both by the colliery operators who were drawn from the local gentry class and by those who were the proprietors of the adjacent copper-smelting works. Other colliers working in these nearby pits would have lived in the northern part of Swansea town or in cottages carved from roadside farms or encroachments on the local commons.

Different factors were at play in the more remote areas where the Swansea Five-Foot coal-seam outcropped high on the sides of the lower Swansea Valley, across the river on the ground around Bon-y-maen and on the summit of Cnap-llwyd at Landore. Both were places where large-scale housing did not exist. Here housing provision might be made on a more lavish scale; in this period both by a colliery proprietor drawn from the local gentry (e.g. Price) and by one who represented a partnership of whose capital was drawn from the south-east of England (e.g. Morris').

Swansea Copperworkers' Housing

Swansea Copperworkers' Housing- 1717-1800

In the eighteenth century there were nine copper-smelting works constructed in the area of modern Swansea, eight in the previously largely rural lower valley to the north of the town, five to the west of the river and three to the east. The first was Dr Lane's Llangyfelach smelter of 1717. This lay over a mile north of the old town in what was then a relatively uninhabited area next to the Neath Road on the west bank of the River Tawe. The workforce of this early copperworks was relatively small, reaching 40 workers in 1727. However, there is little evidence of even key workers having special accommodation built for them.[17] The manager, John Phillips, and his assistant, who were in charge of building the works over a period of four months, had to lodge in Swansea throughout this period as there was nowhere nearer where they could stay.[18] Dr Lane's venture was under-capitalised (he went bankrupt in 1724), but in 1717 'a small house with one room above and below' (i.e. an advance on a cottage but still retaining the rural mores of communal living and sleeping spaces) was recommended for the foreman of the proposed copperworks.[19]

In fact, local rural wages were much lower compared to those available in the new copperworks;[20] thus the mass of potential unskilled recruits needed no extra incentive of housing provision to escape the fairly basic subsistence living that was all that was available in agricultural employment on local landed estates.

The workforce itself may also have been responsible for the building of occasional cottages around the edges of the commonland of Treboeth, Cnap-llwyd and Graig Trewyddfa, but there is no evidence of large-scale squatter housing on the substantial commonland existing, around the works in this period. In fact, by the 1760s there appear to have been no more than five houses on the fringes of the commonland in the Landore area and proximity to coal-mines suggests that these were for colliers rather than for copperworkers.[21]

Two of these cottages occupied small enclosures from the Cnap-llwyd common on its south-western fringe flanking the Trewyddfa road that led from Swansea town to Clydach. This was adjacent to the site of Cwm Pit and was also near Cwm Level. Four more cottages occupied narrow strips of land on the fringe of the junction between Cnap-llwyd and Graig Trewyddfa Commons, alongside what are now Cnap-llwyd and Trewyddfa Roads and close to the outcropping of the Swansea Five-Foot seam and its mining tunnels. From the map evidence it is obvious that these cottage plots had been divided off from the larger fields of Trewyddfa Farm (now the area of Llewelyn Park). In fact, this process of the growth of linear settlements on roadsides, which continued in intensity in all the industrial settlements of early nineteenth-century Swansea (in the proximity of coal-mines, copper and tinplate works), was found elsewhere in industrialising south Wales.[22] It allowed landowners to retain control of settlement expansion rather than tolerate a larger measure of haphazard encroachment on the commons. It would also have been a valuable source of extra income from rents for the landowners. Some of these cottages survive in a much rebuilt condition around the high ground surrounding Morris Castle at Landore.

The second copperworks, the Cambrian, was built by local Quaker merchants and industrialists in 1720, but it was the only copper-smelting works built within the old borough and housing convenient for the workforce was at hand.

Fig. 192. An eighteenth-century croggloft cottage, 'Allt-y-screch', Coed-cae, at 535 Llangyfelach Road sited immediately west of Pentre (Penfilia) Pit (still standing in the 1960s). (GL0270)

Three of the four copperworks to be built in the mid-eighteenth century were situated on the opposite (eastern) bank of the River Tawe, just to the north east of the town but across the river and hence remote from existing available accommodation. Another four works were between one and two miles north of the then edge of Swansea town and on the west bank of the river.

What evidence survives suggests that copper-smelting works operators felt obliged to provide houses for a sizeable proportion of their workers in these fairly remote locations.

Four main types of dwellings might be provided for workers: works, gate keepers might live in a lodge attached to the works office; semi-detached houses might be provided for managers; terraced blocks might house a fair proportion of relatively few workers; and on the works site or nearby in a converted farmhouse or mansion would live the copperworks' manager or 'agent'.

Lockwood Morris and Company may not have had the working capital to build extra workers' housing after taking over the Llangyfelach Copperworks at Landore after Lane's bankruptcy in 1727, but they had doubled their capital to £20,000 by 1845.[23] In the 1730s they began their relocation to the riverbanks at Lower Forest near Morriston. The 13 workers resident in the 10 or so houses at Lockwood, Morris & Co.'s Lower Forest Coppermills have already been mentioned. Forest Copper-smelting works (1747-52)[24] had an 'Office House' with a gate-keeper on site.[25]

In 1757, one of the copper-smelting concerns, probably Lockwood, Morris & Co., had no less than 40 workers' dwellings on land leased from the Duke of Beaufort at two shillings a year, with permission to build more.[26] This total would include colliers and coppermills' staff in addition to those at the actual copper-smelting works.

Much of the evidence for the other works only appears in the first detailed census of 1841, although a few blocks of early workers' accommodation can be identified from a combination of map, architectural and written evidence. More personnel than the refiner, watchman and managers were resident on the works' sites; detailed evidence for two of the works to the east of the river is available from the 1841 census. There were 11 workers' dwellings noted at the Upper Bank Works with 50 inhabitants, including 13 coppermen and one bargeman. At the neighbouring Middle Bank Works were 10 houses, also with 50 inhabitants, including 11 copper-workers and six labourers. It seems possible that up to a quarter of the workforce for each of the smelting-works might have been resident on site.

Where accommodation blocks were to one side of the main works or nearby, it is possible to identify them from the map evidence. One block existed to the south of the White Rock Works (1737) alongside Smith's Canal, and two were built on the opposite (west) bank of Morris's Canal from the Birmingham Works (by 1793).[27] The latter, known as 'Barracks Row', consisted of 10 workers houses in two terraces of five, a number that might well have been normal for on-site accommodation provided at each of the eighteenth-century works.[28]

One block of workers' accommodation on the edge of Morriston, ('Wychtree House') survived to be photographed and partially recorded in the modern period. The coppermasters building it, Lockwood, Morris & Co., had already built one large block of accommodation for their colliers in the form of 24 flats in three-storey blocks linked by the four-storey towers known as Morris Castle.

The concept of providing a whole model township (Morriston) in two successive grids of roads laid out to the north-east of, but asymmetrical to, the earlier 'Wychtree House' was new. Here workers funded the construction of their own houses to a standard plan on fairly generous plots provided by their employers. This was an innovation of c.1779 introduced by John Morris I, resident managing partner of Lockwood, Morris & Co. which bought farmland to carry this out.

Morris had also built three pairs of semi-detached houses (with substantial gardens) in 1775 on the commonland of Graig Trewyddfa near Lockwood, Morris & Co.'s Forest Copper-smelter; these may have been for skilled specialists and managers at the works.

Other workers wanting to live near the works would have had to lease a plot of land and build a cottage. The largest concentration of such activity in the eighteenth century was probably at Foxhole, to the east of the River Tawe and south of the White Rock, Middle and Upper Bank Copperworks. In 1793 there were four blocks of housing at the foot of the steep Kilvey hillside on the east of the road leading north from the Swansea Ferry.[29] By the early nineteenth century these roadside blocks had merged into two almost contiguous terraces.

Each copper-smelting and rolling works would certainly have had a fairly substantial residence for the works' manager or 'agent' on or near the site; some were re-used local mansions and farms. For example, the old Cnap-goch Mansion of the Popkins family became available for use as a managerial residence. The main industrialists and land-owning family to the east of the lower River Tawe, the Mansels, had bought the Popkins estate on the east of the river so that they could provide a site for a smelting-works (1737), whose operators would buy their coal. The old mansion was then available for the White Rock smelting-works, manager, Edward Jones, to use as a residence. It was later subdivided for workers' use after the substantial White Rock House was built for the works' manager on the riverside to the south of the works in *c*.1771.[30] Another manager's, or agent's house was situated by the (Smith's) canal access bridge to the nearby Upper Bank Copperworks; at the time of the 1841 census it was occupied by 50-year-old Thomas Brown, 'Agent [i.e. manager] in Copper Works', and his 20-year-old son, Thomas, 'Agent in Patent mills'.

The total number of people employed in the lower Swansea Valley copperworks at the end of the eighteenth century was not large. Coppermills such as those at Lower and Upper Forest may have employed five to ten workers (while most copperworks also had smaller-scale water-powered plant). There were 40 employed at the Llangyfelach lead and copper works and mills at Landore in 1717. There were 25 copper-furnacemen employed in a copper-smelting works such as that at Middle Bank in 1796, but 25% of the Hafod Copperworks' labour force in 1810 comprised additional ancillary staff, suggesting that the total labour force may have been about 35 at Middle Bank.[31] The total employed in the larger White Rock Copperworks and mills in 1755 was 55.[32] There were eight copper-smelting works in the lower Swansea Valley at the end of the eighteenth century. If they employed an average of 40 then the total workforce employed directly at the smelting works might have been some 320, with another 15 or so employed at the two separate coppermills on the river: that is, a total of about 335 in all. An average of 4.8 people later lived at the workers' houses at the Middle and Upper Bank Copperworks, and as successive generations tended to be employed at the works, an average of 1.5 people in each household were employed at the works. Therefore, at the end of the eighteenth century about 1,100 people were directly dependent on the copperworks, living in some 225 houses or cottages, the great majority in the industrial satellite communities to the north and north-east of Swansea. By 1801 these workers would have formed just over a quarter of the population of this semi-rural area of the Swansea Valley, and some 40% of these might have lived in houses built by the coppermasters. For the rest, a partnership such as Lockwood, Morris & Co. might provide land on which workers could construct their own houses. A number would have lived in roadside cottages on small plots of land rented from the local landowners; others lived on similar sized cottages and land enclosed from the local commons. Others again would have trekked daily from the north Swansea suburbs a mile or so to the south, or caught the White Rock or other ferry across the river from the town.

Each of the nine eighteenth-century copper-smelting works required prodigious amounts of coal to roast and refine the imported ore. Colliers were a major element of the lower Swansea Valley population as they must have been for centuries. To the west of the River Tawe, Lockwood, Morris & Co. became an integrated mining and smelting concern from the 1730s, supplying the other smelting concerns between Landore and Morriston and building houses for their colliers near the mines. Other of their mines were near the new township of Morriston (1779) and colliers came to form an element in the population of that town. Chauncey Townsend started by owning another integrated and mining concern to the east of the River Tawe from the mid-eighteenth century. His miners already had large-scale housing that had been provided by the Mansels earlier in the century. From the late eighteenth century his successors' the Smiths, were again solely running a large-scale colliery concern as had the Mansels early in the century.

The main determinant in the location of the lower valley population was that each of the nine copper smelting works in the eighteenth century had to be situated on the riverbank in order to receive its ore. The workers needed, in turn, to be near the works but the piecemeal deployment after 1737 and the foundation of Morriston in 1779 were on such a scale that they heralded the construction of several townships in the lower Swansea Valley where available land and accommodation attracted workers locally but also from several of the adjacent works, along with colliers, craftsmen and professionals and their many dependants. The building of Morriston and other early housing at Swansea is of international importance, because of its early date. The metal-smelting complex at Swansea predated that at Merthyr Tydfil by some 40 years. The ironworkers' townships of Georgetown and Dowlais around Merthyr (both now largely demolished and rebuilt) were the result of more recent industrial activity.

However, the location of the semi-rural copper-smelting works with blocks of workers' accommodation and managers' houses, and the first larger-scale workers' settlement at Morriston, was outside Swansea. Swansea town itself at the end of the eighteenth century was still a fairly genteel resort with no copper- smelting works left within the old borough boundaries. The influences on the two distinct parts of the urban and semi-urban conurbation were complex. Coppermasters had developed the town's port, although it was the town's

Fig. 193. Two of the straight main streets of the second grid of eighteenth-century Morriston can be seen running diagonally across the lower left. The Anglican church of St John's, with its tower, remains in the central crossroads where Sir Morris I had built its predecessor. The spire of the later Tabernacl Chapel, 'the Cathedral of Welsh Nonconformity' is visible just below it. (983515-09)

merchants and the upper Swansea Valley landowners who developed the upper valley anthracite coal trade which exported via the improved tidal port and town wharf's from the end of the eighteenth century onwards. The coppermasters and industrialists resident nearby also contributed to the general trade and social life of the town. Despite all this trade-driven activity, the *Gloucester Journal* on 14 August 1786 was able to justify a comment that: 'Swansea, in point of spirit, fashion, and politeness, has now become the Brighton of Wales'.[33] The town was still largely occupied by the sort of tradesmen found in any market centre, and the mariners who were engaged in its extensive sea trade (substantially in

copper ore and return cargoes of coal); in 1787 it could still be said that 'the town is not merely inhabited by Copper-men, Colliers and Mechanics; very few of whom live in the Town'.[34]

Swansea Copperworkers' Housing-1801-51

By the beginning of the nineteenth century, Swansea's position at the centre of the international copper industry had made it the nucleus of the largest population centre in Wales. The copper trade was now very firmly the dominant economic driving force of both the town and its surrounding communities. The town's population in 1801 was just over 6,000 in the area of the ancient medieval borough, but its nascent industrial suburbs (in the northern and north-eastern parts of the present city) made this the largest centre of population in Wales, with over 10,000 people. What became the iron-making capital of the world at Merthyr Tydfil was then smaller, its 7,700 inhabitants making it the second most populous place in Wales. Complementing Swansea at the other end of Wales, and partly dependent on it for smelting its copper ore, was the third largest centre of population: this was Amlwch on Anglesey, with a population of 5,000, serving what was then the world's largest copper-mine at Mynydd Parys, and with a port busy exporting ore.[35]

Large-scale growth of both internationally important metal-smelting towns of Glamorgan was accelerating dramatically. In the lower Swansea Valley basic patterns of urban growth already established in 1801 still existed in 1851, in that nearly all copperworkers and those involved in mining coal continued to live in a series of well-defined satellite communities close to their places of work, with a sizeable minority living in the northern part of the town. What was different was the scale of that settlement.

By the mid-nineteenth century, the great expansion in the existing copperworks, and the foundation of larger new ones, had boosted the total population of the Swansea conurbation by a factor of almost four. The rates of growth in the old town, however, and new industrial satellite communities were greatly different and for the first time the population of the old borough was eclipsed by that of the surrounding area. In 1849 the population estimated to be living in Swansea town itself was about 17,458, with a further 19,000 living in the industrial villages around the town. Swansea borough had almost trebled in population in 50 years but the new satellite communities had grown by a larger factor of almost

five. There were estimated to be 3,166 houses in the borough and no less than some 3,724 in the industrial villages and townships.

At the end of the eighteenth century, as has already been noted, there had been some 335 workers employed by the copper-smelting works. In 1851, the census returns allow us to be more precise: there were 1,146 workers identified as working in the copper industry, an increase of 342% in the first half of the nineteenth century and, as might be expected, this was the period when workers' settlements were enlarged, some newly built, and most of the new houses for copperworkers constructed.

The 1841 census allows the distribution of the copper-smelting population to be determined with some accuracy, although only some 72% of a copperworks' labour-force might be identified as 'coppermen' (generally workers tending the rows of reverbatory furnaces), whilst others at the works were ancillary and transport staff such as carpenters and masons.[36]

The long-developed centre of Swansea town, well to the south of the copperworks themselves, was not where large numbers of copperworkers might be expected to live and only four copperworkers lived there in 1841. There were three communities where more than 100 copperworkers were concentrated: northern Swansea (113 copperworkers); Foxhole, to the east of the river (119), and Morriston (278). Other lesser centres of the copper-smelting workforce were the early-nineteenth-century Grenfelltown at Pentre-chwyth (about 100); Graig Trewyddfa, including part of the recently enlarged Landore (97); the newly built Trevivian or Vivian's Town at Hafod (67), and mostly rural north Llansamlet to the north-east of the town and river (20). In all, that is a total of some 800 coppermen in 1841, although the total employed by the copper-smelting works, including ancillary and transport workers (but excluding seamen), would be nearer

Figure 194

Summary of Swansea workers' settlements and employment (drawn from contemporary census returns)

Morriston	1841	2,187 people	456 houses	70 colliers	278 copper workers
Graig Trewyddfa	1841	1,023 people	210 houses	140 colliers	97 copper workers
Foxhole/Whiterock	1841	902 people	174 houses	19 colliers	167 copper workers
Trevivian (Hafod)	1841	256 people	46 houses	0 colliers	67 copper workers
Trevivian	1851	–	46 houses	–	–
Trevivian	1861	508 people	94 houses	3 colliers	104 copper workers
			12 houses building		
Grenfelltown*	1871	386 people	71 houses	1 collier	82 copper workers
N. Swansea	1841	–	–	80 colliers	113 copper workers
Cen. Swansea	1841	–	–	4 colliers	4 copper workers
N. Llansamlet	1841	–	–	161 colliers	20 copper workers

*Grenfelltown only occurs as a distinct entity in the 1871 census.
(The term 'collier' denotes all coal workers).

Fig. 194. Table showing size of workers' settlements in Swansea.

1,100. By 1851 there were 1,146 copperworkers, and from then the number of copper-workers in the Swansea area was fairly static for the 60 years until 1911, remaining at an average of 1,200 workers.[37] The average number of people in a household situated in a copper workers' settlement like Foxhole in 1841 was five and an average of 1.1 within the household were 'coppermen'. If the 28% of copperworks, employees who were ancillary and transport workers is included then by 1851 more than a third of households (6,505 people in 936 houses) in the semi-rural surroundings of Swansea included a copperworker as breadwinner or other member of the household. The situation in Swansea town was quite different and here only some five per cent of households (737 people in 147 houses) included, or were dependent on, a copperworker.

Seamen, carrying copper ores in and return cargoes of coal out, were the largest group of workers in the town itself and in Swansea as a whole largely outnumbered copperworkers from 1861 onwards.[38]

The question arises of what proportion of the total of some 884 copperworkers' dwellings built in the lower Swansea Valley by 1851 were constructed by the coppermasters. In this period, the eight existing copperworks were joined by two much larger ones: the Hafod (1809 smelting and 1821 rolling) and the Morfa Works (1828 rolling and 1835 smelting). There is evidence both of existing copperworks' concerns constructing more houses and of the new concerns building on a substantial scale, but precise figures are hard to ascertain. By 1849 the Vivians (Hafod) had 91 workers' houses at Trevivian and the Grenfells (Middle and Upper Bank Works) probably had some 83 at Grenfelltown and Foxhole. The earlier Lockwood, Morris & Co. could have constructed some 42 dwellings specifically for copperworkers, the White Rock Company is known to have constructed about the same number, and the Morfa Company probably about the same too. The proprietors of the smaller Birmingham and Ynys Copperworks may have constructed 10 houses each.

It seems possible that by the mid-nineteenth century, some 320 houses occupied by copperworkers had been built by copper-smelting concerns, out of a total for the copper industry workforce of about 884. Overall, that would mean a 360% increase in employer-built housing in the first half of the nineteenth century. Even so, the proportion of the workforce provided with housing by the coppermasters had remained substantially static: some 40% of the copperworks' workforce were housed by employers in the eighteenth century and about 36% by the mid-nineteenth century. Much of the rest of the copperworkers' housing was built in

the two major copperworkers' settlements at Morriston or Foxhole by the workers themselves on land leased or sub-leased from the copperworks' owners and other copperworkers. Others were built on roadside plots rented or leased from local landowners who were also often industrialists in their own right.

In fact, an important national report in the 1840s highlighted cottage-building as a particular characteristic of the Swansea area:

'Accommodation for the Poorer Classes - The number of small or cottage houses inhabited by the labouring classes is striking; and Mr Bevan, surgeon and registrar of Swansea, attributes much of the comparatively healthy state of Swansea to this circumstance. 'Dr. Bevan himself remarked: 'The practice of cottage-building prevails at Swansea to an extent seldom witnessed in the manufacturing districts of England, and as a natural consequence, we seldom find more than one family located in each house.'[39]

By the early nineteenth century, Swansea was internationally important for its copper-smelting role. The industrial communities were of a substantial size but as yet did not include a large element of the modest but growing tinplate-industry workforce. At this point it is worth noting the main copper-smelting workforce settlements as indicated by contemporary sources in the mid-nineteenth century.[40]

By mid-century. the biggest of these settlements was still Morris Town or Morriston, which had a population of some 2,000 and was connected to the other settlements that had grown up around the copperworks and pits on the western side of the valley 'by a range of straggling cottages'. A lesser, but still substantial, village with a total population of some 1,500 people was Mile End and Landore, situated alongside the Neath Road between the Morfa and Landore smelting-works on the western bank of the River Tawe; and, from 1838, just to its south, Trevivian or Vivian's Town (Hafod). On the eastern bank of the Tawe some 1,200 people were now concentrated in Foxhole and Pentre-chwyth (including Grenfelltown).

It is worth examining the economic and employment balance of at least the largest of these industrial settlements in some detail. This will show to what extent a fairly mature community of this type was dependent on the copper industry on the eve of the foundation of the first tinplate works in the lowermost three miles of the Swansea Valley (1845). The population structure of Morriston will be examined as it was in 1841, when the first detailed census is available. The town then had a population of 2,187 (living in 443 houses: an average of five people per house), of whom 25% were working (including one 'poorwoman', presumably on relief). Of this 557-strong working population, exactly half are easily identifiable as copperworkers including 193 'coppermen' (and one 'copperwoman') presumably attending the furnaces; 24 workers in coppermills (including 20 'rollermen', three 'hammermen' and one 'shearer'); seven 'agents' (i.e. managers) and six 'engineers'.

The only other large group placeable in a particular industry are the 50 colliers and 20 other workers in the coal industry, which itself fuelled the copper-furnaces: that is, one eighth of the working population (a much larger proportion than in the other industrial settlements actually built by coppermasters).

There were also 97 general artisans living in the town (17% of the working population), of whom 34 were masons; 27 [black]smiths; 15 carpenters; 11 sawyers and eight joiners. It seems likely that this group would be working in either the local coal or copper industries. Therefore, up to 73 might work at the copperworks alongside the other 278 copperworkers identified above (i.e. equivalent to an additional 21% of a probable total of 351 workers at the copperworks living at Morriston). In total, 63% of Morriston's population might have worked at the copperworks.

Another 16 workers (about 3% of the working population of Morriston) were involved in transport, probably mostly of copper but also of coal: six [canal] boatmen; three [coal] cartmen; three ostlers; two [river] bargemen; one [canal] gauger and one haulier.

In 1841 there were only 13 other industrial workers living at Morriston and operating possibly unrelated industries. However, four working in iron-founding could well have been employed in a copperworks, while the rest may substantially have been service industry workers: six [handloom?] weavers, one maltster and one tanner. The exception to these, from an industry that would potentially rival the copper industry, was the one 'tin man', the first tinplate works at Morriston may have been Lower rather than the Upper Forest Works, both converted from water-powered coppermills.

Most of the rest (13%) of the working population served the mainly dependent populace copperworks in a service capacity. Of these 66 (12% of the working population), some 37 were craftspeople or retailers: tailors (12); shoemakers (6); bonnet-makers (6); dressmakers (4); milliners and hatters (3); sadlers (2); curriers (2); a wheelwright and a cabinet maker. There were also six publicans and 17 other retailers: grocers (5); cordwainers (3); shopkeepers (2); tallow chandlers (2); a victualler; a baker; a butcher; a cheesemonger and a fruiterer. There were, too, another six workers in the service sector: washerwomen (3); a charwoman; a watchman and a gardener.

Interestingly about 6% (32) of the working population of Morriston were of the 'professional' classes. Of this number, the seven copper agents (managers), three colliery and six copperworks' engineers have already been mentioned. There were four ministers of religion (two 'clergymen', one curate and one Methodist minister); three teachers (two 'schoolmasters' and one 'schoolmistress', presumably at the Birmingham Copperworks Schools); three accountants; two surgeons; one Officer of Excise; one mineral surveyor; one stocktaker and lastly, for entertainment, one 'harper'.

Of the three major centres of settlement for industrial workers in the Swansea area in the mid-nineteenth century, two were a continuation and expansion of those founded by eighteenth-century proprietors and their workers. The location of these works, and therefore the near proximity of their workers' housing, had been determined partly by localised and direct horse-worked railway and canal access into the nearest available sources of coal under Graig Trewyddfa and Llansamlet.

The situation with regard to the settlement at Mile End on the Neath Road and the adjoining area of Landore was different. The expansion of surface canals down the western side of the lower Swansea Valley (c.1780 and 1794-96) allowed for the possibility of the foundation of new works. These were then to the immediate north of Swansea town and the river-mouth, but still retained adequate bulk transport for their coal from further up the valley via the Morris and Swansea Canals. The three new works built along their line were at Landore (1790), and Hafod (1809) and Morfa (1828 and 1835). All were within a mile of the northern edge of Swansea borough where some of their large labour-force could live. There is no evidence that Lockwood, Morris & Co. attempted to lay out a planned township for their workers at Landore, as they had previously done at Morriston.

In fact, there is little evidence for any substantial settlement at Landore before 1826.[41] Away from the main Neath Road at that date there were a few cottages on the old road that ran, sandwiched between the River Tawe and the Trewyddfa Canal, across the south-eastern foot of Graig Trewyddfa Common, between Plasmarl Coal-level and Pits on the one side, and the Landore Copperworks, Pits and wharves, on the other.

The only beginnings of denser settlements

between Landore and Morriston at that date were three clusters of cottages, largely occupied by colliers and strung out longitudinally on the fringes of roads along the outcropping of the Swansea Five-Foot seam between Landore and Morriston, generally on the higher land away from the river. By 1826, the most southerly part of this long ribbon development was almost contiguous (only some 400yd to the north) with the most northerly houses of Swansea town, extending along Carmarthen Road and Llangyfelach Street from the end of High Street. This southerly settlement was Pentre'r-ystwyth (the winding village or Pentre-estyll), along the southern part of the Llangyfelach Road and the northern part of Cwm Burlais, and centred both topographically and economically on the Millbrook Coalpit (later the site of the Cwmfelin Tinplate Works). Four hundred yards to the north of these houses was the southern end of a second north-south ribbon settlement: Brynhyfryd/Treboeth. Cottages had been built along the Landore (east) side of the Llangyfelach Road at Brynhyfryd (sub-divided from the westernmost fields of the Morris's Mysydd and Pwll-yr-oer farms). However, the lands on the opposite (west) side of the Llangyfelach Road, under separate ownership at this point, were not divided into building-plots until 'Eaton Town' was constructed later in the nineteenth century. A fairly continuous row of houses then continued around the intersection known as Brynhyfryd Square (adjacent to Pentre and Cwm Pits and Brynhyfryd and Cwm Coal Levels), northwards along the west side of Llangyfelach Road and eastwards on the web of roads that accompanied the piecemeal enclosure of Treboeth Common. At the north of Treboeth Common was the Tirdeunaw Pit and to the east was the Treboeth Coal Level. The most regular part of this settlement was an oval enclosure of land ('Waun Gwn')[42] on the northern part of Treboeth Common, owned by John Dillwyn Llewellyn, which had been bisected by a north-south boundary and 11 plots, mostly neat rectangles, laid-out on either side. By 1838, there seem to have been 12 cottages built on these and more irregular adjoining plots.

In general, the multiple roadside plots, enclosed from adjoining fields, show little sign of the systematic sub-division for building-plots as used at Waun Gwn. This particular case of orderly planning must have resulted from a very deliberate initiative to realise income from rent from colliers on the part of the industrialist-landowner John Dillwyn Llewellyn, who had no direct interest in the running of the nearby mines. The uniform boundary-line between many roadside plots and the adjoining larger agricultural fields, from which they had been divided by 1826 and 1838, does suggest that the initiative for enclosure came from the landlord rather than from individual cottagers, who nevertheless may have been responsible for the fewer number of irregular individual plots on the roadside and also for enclosures from the commons. By 1838 the settlement extended for a mile alongside the Llangyfelach Road with some 53 houses in 41 plots: the average size of building/garden plot was a substantial 4,500 square yards.

By 1825 another straggle of houses extended northwards for a second mile from Swansea town, to Morriston along the Trewyddfa and Graig Roads; they were built on irregular and tiny plots along the crest of Cnap-llwyd Common, then on roadside sites enclosed from Cnap Llwyd and Trewyddfa Farms, and they continued northwards in an intensification of the earlier settlement (enclosed from Graig Trewyddfa Common) flanking Graig Road above Morriston.

These settlements presumably had a sprinkling of copperworkers from the end of the eighteenth century and certainly had a significant copperworking population by 1841, when the 210 houses on the hillside of Graig Trewyddfa between Landore and Morriston were homes to 97 copperworkers, as well as the 140 colliers who largely made up a working population that, with dependants, formed a community of 1,023 people.[43] These diverse self-built cottages had a usual occupancy of 4.9, which approximates to that in the other industrial settlements around Swansea at that date. As with the other industrial suburbs it is possible to see (from successive maps), a process whereby additional cottages gradually infilled each of the fairly generous family plots. By 1825 it is also possible to see from the map evidence that the first of three terraces that stood in the area by 1838 had been formed by joining cottages together on the Llangyfelach Road.[44]

Between 1826 and 1838[45] the form of the largely copperworkers' settlement at Landore had begun to appear, linked to the earlier (largely colliers') settlement at Brynhyfryd by an extension of the farm-drive of Pwll-yr-Oer (the farm site was later used to build Seilo Newydd Chapel). It was tripled in length eastwards to the Neath Road by a road (now Siloh Road) that was to form the spine of the new semi-urban settlement of Landore. The Landore, Morfa and Hafod Copperworks were all accessible along the Neath Road and copperworkers' houses began rapidly to in-fill the roads. The first of the north-south cross-roads (Byng Street, named after the family of Sir John Morris II) had appeared half-way along the extended drive, with a terrace of some 13 workers' houses to the south of the Siloh Road and seven to the north, built between 1826 and 1838.

All of this development was on the farms of Mysydd and Pwll-yr-Oer that were owned by the Reverend Thomas Morris and tenanted by his brother, Sir John Morris II. These first two terraces and another three short ones (with another 16 houses along Siloh Road) to the east on Millbrook Ironworks land were probably built for the Ironworks staff by Sir John Morris II and his two partners after they had bought the works in 1830. Sir John Morris II also gave the local Welsh Independents (at what became Seilo) the land for a schoolhouse and then their chapel (1829).[46] The strength of the copperworkers in the community is perhaps indicated by the fact that Robert Monger, the manager of the Hafod Copperworks, was one of the founder members of that congregation.

The first terrace had appeared on the west side of Neath Road by 1826, to the east of Seilo Chapel: the chapel congregation itself acted as a focus for the formation of building clubs or societies as must have been the case elsewhere in the lower Swansea Valley.[47] By 1838 the proprietors of the Morfa Copperworks (rolling-mill 1828, smelting-works 1835) had bought land to the west of the Neath Road, just to the north of their copperworks, and the longest terrace (16 houses) in the settlement, at that date, was built for their workers. A further short terrace of three houses and four detached residences to the south, opposite the by-product chemical works, may have been for specialists and managers. A further house in grounds to the south of Heol y Glo, now Station Road and four other buildings on Morfa land may have been the main manager's house and other dwellings.

Landore was not a regularly designed company town: it may have had elements laid out by the Morrises and the Millbrook Ironworks Company and by the Williamses of the Morfa Copperworks, but it did not have the grid-plan of a 'Morris Town' or Trevivian. That said, there were distinct differences between the regular company terraces built for the iron and copperworkers near the Swansea Canal and Neath Road, and the irregular upland cottages mostly built by colliers a few hundred yards west at Brynhyfryd.

Interestingly, terraces had even appeared in parts of those more westerly colliers' settlements by 1838. Most of these, like the three terraces already mentioned, were part of a long piecemeal process of infilling gaps between cottages along the street frontages of garden/building plots. By 1838 another two terraces stood high on Cnap-llwyd Common, alongside the south-eastern side of Trewyddfa Road, to the north east of its junction with what is now Salem Road. That nearest Morris Castle to the north had comprised two joined cottages by 1826, a four-

dwelling terrace by 1838 and was to be a row of six houses by 1899.[48] The site of the longer terrace to the south west was a triangular, empty-plot at the road junction on the common in 1826, but an eight-house terrace of two sizes of house had been constructed by 1838. Before 1899, extensions to that terrace had absorbed another separate dwelling already extant in 1838 and the row then, as now consisted of 14 houses with a second terrace on nine houses later added to the lower part of the plot.

Another fairly irregular settlement was at Foxhole, across the river and north east from Swansea, where accommodation came to be provided for large numbers of works. Two very long, almost joined, terraces of workers' houses were built along the main valley road east of the river, to the south east of the White Rock Copperworks. In 1841 there were 50 copperworkers and 11 colliers living in the 77 houses along the main valley road. The settlement, as a whole, seems to have consisted of a mixture of housing built by the copperworks' proprietors, and houses built on land leased to their copperworkers.

Most additions to the industrial settlements of the lower Swansea Valley followed this piecemeal development with a great deal of self-building. Business leaders facilitated the framework in which workers could securely accumulate savings and in which the workers' own building clubs and societies could operate; they set up savings banks to encourage the thrift of small savers, with boards of trustees who were forbidden to profit from their trusteeships under an Act of 1817 (other contemporary banks did not usually receive deposits of less than £10). The first savings bank in Glamorgan was set up in 1816 in the copper-smelting centre of Morriston by the Quaker philanthropist William Bevan, he acted as its first treasurer, with Sir John Morris I as its first president.[49] By 1829 this institution had accumulated balances from depositors worth no less than £9,209. The Morriston bank was early in the field in Wales (the first such institution in Britain was founded in 1804); in Glamorgan others followed at Swansea in 1816 (abortively as most of the potential worker savers were in the industrial townships such as Morriston), Bridgend in 1817, Cardiff in 1819 and Dowlais in 1852.

Sir John Morris II chaired a meeting in 1827 in which the largest of the savings banks in Glamorgan was founded, the Swansea Savings Bank, which absorbed that at Morriston in 1829. This bank had 1,104 depositors with savings of £37,663 in 1837 and 10 years later 2,322 with deposits of £92,434. The partners of general banks in Swansea also included some of the wealthiest industrialists: the 'Glamorganshire and Swansea Bank' of 1771-1804

(Sir John Morris I); 'Sir Herbert Mackworth Bart. & Co' (Swansea and Neath) of 1783-91; William Bevan at Morriston in c.1800 and Haynes & Co. at Swansea and Neath in 1813-25.[50] Building clubs became commoner later in the century.

There is little evidence for speculative building in the semi-rural industrial settlements away from the town. One particular point of interest was the different approaches shown by the owners of the two large new works at Hafod and Morfa. The Cornish proprietors of the latter works seem to have built 26-28 houses in close proximity to the works about the time of its construction. Their earlier compatriots at Hafod (1810) do not seem to have built on a large-scale until the 1830s, then they did so, possibly encouraged by the example of the proprietors of the adjoining new and competing works at Morfa. However, once the Vivians had started making provision for their workers, they developed the most comprehensive planned settlement made by the coppermasters at Swansea.

In the later nineteenth century, the neighbouring Hafod and Morfa Copperworks, founded in the first half of the nineteenth century, became the centre of enormous industrial complexes requiring yet more housing for the workers needed in a range of related industries.

Swansea Copperworkers' Housing, 1852-1901

It was in this period that the term 'Copperopolis' might be regarded as losing its validity as a name for Swansea and its surrounding settlements, for by 1891 the number of tinplate-workers outnumbered copperworkers for the first time and the lower Swansea Valley became the centre of the world's largest tinplate-making area. As has already been mentioned, the number of workers in the copper industry remained fairly constant from 1851 until 1911. It might be assumed that the numbers of copperworkers housed by the copper-smelting concerns would also remain static. However, in this period there was consolidation in the industry, with production becoming concentrated in the new larger works that were sited nearer the coast on the lower west side of the Swansea Valley, between Swansea town and Landore and, to a lesser extent in the eastern valley works. In 1851, the owners of the Morfa and Hafod were each importing some 27-28.5% of the total of copper ore brought into the port of Swansea, while the owners of the Middle and Upper Banks Works and the White Rock Works to the east of the river were each smelting about half that (14.3-16%).

In the later nineteenth century, the copper-smelting concerns and their workers dealt with the need to enlarge the workers' housing nearest the surviving bigger works in diverse ways. However, the picture is somewhat complicated by the fact that each of the larger works diversified and became part of integrated complexes often smelting other non-ferrous metals such as silver, lead and nickel, producing sulphuric acid and other materials as by-products, and servicing works such as iron-foundries, gasworks and sawmills. Therefore, the enlarged industrial settlements housed workers involved in all these tasks and also in the newer tinplate industry centred at Landore, Morriston and to the west of Hafod. In the 1891 census the tinplate industry workforce outnumbered copperworkers in the Swansea area for the first time, but even then by a large margin 3,442 to 1,146, and this change in the dominant industry continued from that period onwards.[51]

It is instructive to look at the structure of the most developed settlement built by a copper-smelting concern near the zenith of the industry in 1861 and to compares with that recorded in 1841 at Morriston.

By 1861, the number of houses recorded at Trevivian (or 'Vivian's Town', now known as Hafod) in the census was virtually double what it had been in 1841 and 1851, an increase of 48 from 46 houses to 94 and 12 additional houses were also being built. The density of occupation was slightly higher than in the other industrial settlements; perhaps a better standard of housing allowed larger families to survive. In 1841 the population of 256 had an average of 5.6 people living in each house, whilst in 1851 a population that had almost doubled to 508 had a slightly less denser occupancy of 5.4 people per household.

The working population of Trevivian in 1861 was 179;[52] interestingly this represents 35% of the total population which compared with 25% of the population working at Morriston 20 years earlier in 1841. Possibly in view of the Vivians' sanitary housing, more of the second generation of a household survived in improved health to find jobs. It does not seem to represent a significantly larger number of women employed at the Hafod Copperworks, for whereas there was one 'copperwoman' identified at Morriston, there were still only two women employed 'in the Copper Works' who were living at Trevivian 20 years later. Of the working population of 106, 59% (compared to 50% at Morriston) seem to have been fairly clearly attending the copper-furnaces, mills and engines and to have been dependent on the copperworks for their income. They included 71 'coppermen' and four 'smelters of copper'; seven probably in the coppermills (including three 'rollermen', one 'shearman', a stoker, a fireman and an engine driver); two 'agents' (i.e. managers), one chargeman, three engineers, three fitters and two brasiers, two women 'in the copperworks', and a copperman's widow and nine labourers.

The colliers who represented a significant group at Morriston barely existed at Trevivian, where the owners of the Hafod Copperworks and the houses did not own or operate the coal-mines on its western fringe. Two colliers lived in the Trevivian houses with their families and there was also a coal weigher who more probably worked at the copperworks. The other major industry of the lower Swansea Valley was represented by two tinplate-workers who would have been employed in one of the tinplate works to the west of the community.

The group of 21 general artisans of the working population; (12% living at Trevivian; 17% at Morriston) probably all worked at the copperworks: nine masons, five blacksmiths, three carpenters, two sawyers and two joiners. In all 128 (72%) of Trevivian's working population were probably employed at the Vivians' Hafod Copperworks, compared to 63% of the Morriston population who worked at the three nearby copperworks, and coppermills. In addition, another worked at the Hafod Silverworks (1846), which occupied the northern part of the main copperworks site and extracted silver from copper ore. The two transport workers a [canal] boatman and a [river] barge-man, were probably also directly employed by the copperworks company.

Twenty seven people (15% of the working population; 13% at Morriston) were employed in the service sector at Trevivian. Nine were craftspeople or retailers: seven (female) dressmakers, two (female) milliners, a shoemaker (male) and a grocer's assistant. Also employed in the service sector were seven house servants, two watchmen, four charwomen, two housekeepers and a laundress.

Interestingly, the percentage of professionals in a company-owned town like Trevivian was the same as in Morriston, where individuals owned the houses. There were 10 members of what could be termed professional status. Of this number, the two 'agents' (managers) and three engineers have already been mentioned. There were also two school mistresses, a schoolmaster and one student (teacher?) employed at the Hafod Copperworks Schools, and an 'Independent Preacher'.

The total workforce of Vivian & Sons probably peaked at 3,000 in 1886, with some 1,000 employed at the Hafod Works complex. By 1877 there were 263 houses at Trevivian, housing perhaps some 322 workers, suggesting that by c.1880 32% of the Hafod workforce lived in company housing at that date.

The large-scale accommodation for the Vivians' workers at Hafod may have been begun a considerable time after the foundation of the works, but this was not unusual. What was unusual was the eventual large size of Trevivian and the institutional and educational provision made for it. By comparison, the Morfa Copperworks proprietors seem to have made some immediate provision for housing their workers, but although the works grew considerably in the nineteenth century, this housing did not expand on a significant scale. Their workers used building clubs to continue the development of Landore at their own initiative and expense.

The workers at Whiterock Copperworks continued their development of the fairly irregular settlement at Foxhole to the north east of the town. It was a development in some ways not dissimilar to that at Landore, but with some institutional and religious buildings provided by the coppermasters. By the mid-nineteenth century, individual workers obtained 60-year leases from the proprietors of the White Rock Copperworks for plots on the upland waste of Kilvey Hill, high on the east side of Kilvey Road. Two leases were granted retrospectively in 1846 for encroachments made by workers in 1840 and 1844. From then on the pace of building quickened with seven more leases of building granted land in the four years between 1850 and 1854; the annual rent charged was about £1 for five of the plots but three others cost £3-4 and possibly housed more than one related household.[53]

The steep site and haphazard arrangement of Foxhole did not compare well with the more orderly planned settlements laid out by the coppermasters at Morriston, Grenfelltown and Trevivian, or the gentler landscape upon which Landore had been constructed. Most of these planned settlements were also west of the prevailing winds which blew the copper-fumes eastwards; pollution settled on the inhabitants of Foxhole in much the same way as the sulphurous pollution affected the contemporary, industrial centre of Stoke-on-Trent. In his 1849 Report on the Public Health of Swansea, G.T. Clark had noted that:[54]

> 'Across the Tawe, and upon its steep bank opposite to Swansea, are built the long and irregular villages of Fox-hole and Pentre-guinea. The houses are for the most part niched into the hillside, and are old, damp, and very dirty, and have no back premises. The water-springs used by the lower cottagers are defiled by those living above them on the hill, and the roads and gutters, when I visited them, were in a filthy state. The copper smoke here affects' the vegetation and the glass in the windows ...'

Further away from the copperworks and adjoining Foxhole to the south was the grid-iron plan settlement of St Thomas. This was a later general suburban construction of the 1850s and 1860s, resulting from access to the town for the lower eastern valley *via* the new Tawe bridge near the river mouth. However, this new settlement was given some communal facilities by the local resident coppermasters, the Grenfells. In fact, the Grenfells, proprietors of considerably smaller works than the Vivians, had expanded their housing provision around their three earlier rows at Grenfelltown to the east of the River Tawe, so that by 1871 there were 71 terraced dwellings housing 386 people.

Building societies and clubs, often arranged around the membership of the many respective chapels, became common in this period. By 1856 the Swansea Savings Bank had 165 friendly and charitable societies as members and five years later no less than 223.[55]

Both of the largest copperworks of the nineteenth century (Hafod and Morfa) were within about half a mile of the northern part of Swansea town. By the time of the first detailed census in 1841, there were 113 copperworkers living there; with 80 colliers also resident, this made it into one of the largest of the industrial settlements. The dense population of this urban area meant that speculators provided some of the housing.

By 1876, the Borough of Swansea Improvement Scheme had targeted five streets and nine infill courts to the south west of St John's Church as slums to be demolished in order to enable a broad new approach (Alexandra Road) to be made to the main High Street Station. Most copperworkers, in fact, lived north of this in the less densely developed northernmost part of the town; nearer the river in this clearance area lived workers involved in the maritime trade. The clearance involved 319 units of property around Back and New Streets. Only 9% were owner-occupied and 91% were owned by property entrepreneurs and let for profit. Four prominent members of the local community owned 68% of this condemned property: these were Lewis Llewellyn Dillwyn, MP; J.D. Llewellyn; the Reverend W.M.D. Berrington and the Reverend Thomas Thomas. Of the four, the local MP held 21.6% of the housing and the Reverend Thomas Thomas the smallest proposition at 10.7%. The Reverend Thomas Thomas owned one of the five courts covered by the clearance scheme. He had 34 houses in Regent Court, arranged in two long terraces which were on either side of this alley and with gardens mostly about the same size as these houses, the 26 of which were one-room deep, with an earth-closet at the foot of each of the tiny yards. The average rent which Thomas Thomas charged for each property was 1s. 6d. a week. However, the rent charged to the tenants would have been marginally greater than this as the landlords usually let to agents who sublet the properties, thereby acting in effect as rent collectors. In Thomas Thomas's case, his agent was Mary Parks, who was also one of the agents acting for Lewis Llewellyn Dillwyn. Thomas Thomas's houses were regarded as of very low standard or value compared to the others, for while he received compensation in 1879 worth 16 times the annual income from his houses, three of the four main landlords received compensation worth 29-43 times their annual income.[56] Thomas Thomas, who was the eminent Welsh Independent minister of Siloh, Landore, and one of the most prolific chapel architects in Wales, resigned unexpectedly in 1876, quite possibly as a result of the publicity attached to this clearance of Regent Court.

As in many other early-nineteenth-century towns, there was a tendency for the growing population to be accommodated by the insertion of these 'courts' behind the existing street frontages. The tiny constituent cottages had little space around them except for the alleys leading to passageways in the neighbouring street frontage. These workers' dwellings often had only one room on each floor and no windows in their rear elevations. By 1852 there were over 900 such houses out of a total of 3,500 in Swansea.[57] Shared sanitation in the courts drained to adjacent cesspits. Consequently, the cholera epidemic that spread across the urbanised areas of Britain also broke out in Swansea in 1831 and 1849; many deaths occurred in these crowded houses and even in better dwellings unfortunate enough to be sited near insanitary watercourses. None of the northern part of the town was served by a piped water-supply in 1849 when cholera also killed many around the houses of copperworkers in places like Greenhill Street and Bethesda Court. Swansea's first building by-laws were introduced in 1860.[58]

Generally, however, conditions in the old town and dispersed rural industrial suburbs were not as bad as those in some other new urban settlements. The various public health investigations of the 1840s in Swansea had only managed to find one example of what might be called a cellar-dwelling. Infant mortality at nine per 1,000 was double that of the rural district of Gower, but the general rate of mortality was less than that of more rural market towns such as Haverfordwest and Carmarthen and comparable to that of Aberystwyth. The rate of 15.5 per 1,000 was considerably less than the 27 per 1,000 found in Merthyr Tydfil, the unhealthiest town in south Wales.[59]

The general trend in the late nineteenth century was for copperworkers' settlements to move nearer the town along with the two larger nineteenth-century copperworks. That move had been facilitated by the Swansea Canal being able to supply coal to the works from the north; the canal itself was a source of drinking and other water for Trevivian and the other copperworkers' settlements. Later on in the nineteenth century, both works also had locomotive railway connections, but of more relevance to the industrial settlements was the arrival of passenger tramways from Swansea, which helped to provide a new cohesiveness to the urban area. The new works were not dependent on water-power on their main sites although the Millbrook Ironworks, which produced some machinery for the copperworks and their port and was a contributor to the growth of employment at the nineteenth-century community at Landore, was driven by large water-wheels. Both Morfa and Hafod had steam-driven rolling-mills from the 1820s, so the relationship between the concentrations of works and settlements and the old linear features of water-transport and power was largely broken. Interestingly further urban expansion between Pentre-estyll, Landore and Morriston on the west side of the lower Swansea Valley, often took place along old disused horse-worked railway formations that had been converted to roadways. The old copperworks' township in Morriston became largely a settlement for tinplate-workers in the last few decades of the nineteenth century.

Agents' and Managerial Housing

The works' 'agent' or manager was the key staff member at each of the copper-smelting works. His importance was reinforced by the fact that nine of the 11 copperworks built between 1717 and 1828 lacked a locally resident managing partner of the operating company for most of the works' operating life.

As already noted, in 1809 when the Hafod Works was opened, John Vivian advised that a 'house for the agent and few cottages for the most valuable hands' were required, or the company would 'stand a bad chance of getting and keeping the best men at Swansea'. The location of the agents' houses at Upper Bank, White Rock and Hafod Copperworks and the Upper Forest Copper-rolling mills are known. In two of four cases, a suitable house near the new works was adapted for the use of the agent or manager. In the eighteenth century, at White Rock Copperworks and the Upper Forest Coppermills, the former mansions of the local Popkins family (themselves former colliery and forge operators) were adapted as managerial dwellings. In time, the old Cnap-goch mansion and its outbuildings near the White Rock Copperworks were converted to form part of 24 workers' dwellings, and the agent or manager was given a newly built residence near the

Fig. 195. 'Aberdyberthi House', early nineteenth-century dwelling of the general manager of the Hafod Copperworks. With four rooms on each floor, it had double the rooms of the earliest workers' housing and in addition had a large attic, where servants lived, and a cellar. At the rear was a coach-house and extensive gardens, walks and a glass-house.

works, and on the riverside to its south probably by 1771 and certainly by 1805.[60] By the 1820s the general manager of the Hafod Copperworks was given a fairly roomy, if undistinguished, new house at Aberdyberthi on the later site of Trevivian. A new outbuilding, or coach-house, of cast copper-slag blocks, was also provided (and still survives).

The house of the Upper Bank agent is the only one known to have been newly built on the works' site, and its occupant may also have managed the adjacent Middle Bank Copperworks. The managers' houses from the Hafod Copperworks and Upper Forest Coppermills survive; the former main Hafod manager's house, submerged amongst later terraced housing at Trevivian has been recorded.

The very large later-nineteenth-century works, such as Hafod and Morfa, came to have a much larger managerial structure, but only two other large managerial residences are known to have been constructed, 'Hafod' and 'Brynnant' for the Hafod Copperworks at Trevivian, and only one of these remains. Each of the three managers' houses at Trevivian had extensive gardens with conservatories and greenhouses. Each had nearly four times the floor area of the average worker's house at Trevivian.

To the south of the Hafod Copperworks was a diverse industrial complex: first, the Hafod

Fig. 196. The remaining one of the two later nineteenth-century Hafod Copperworks departmental managers' houses situated at the south-west of the Trevivian settlement on Morgan Street. (960209/3)

Phosphate Works and to the south of this, successively, the Hafod Iron Foundry, the Hafod-Isaf Nickel and Cobalt Works, and the Hafod Saw Mills. The newly built residence of the superintendent of the Phosphate Works (37ft by 42ft, over three times the floor area of the average Trevivian worker's house) stood in grand isolation on a mound on the east bank of the Swansea Canal, immediately adjacent to the works, and looking towards the workers' housing at Trevivian.

There is little information available on the architectural features of the main managers' houses. The house intended to be built at the Llangyfelach Copperworks in 1717 was planned as a very humble affair with only one room on each floor. The early nineteenth-century Hafod Copperworks general manager's residence, Aberdyberthi House, is a much larger double-pile Pennant sandstone rubble-walled box, with a simple pilastered doorcase. However, the latter feature is the only sign of ostentation, besides its size, to distinguish it from the later workers' houses at Trevivian. By contrast, the surviving later-nineteenth-century departmental manager's house at Trevivian is distinguished externally by its greater three-storeyed gabled height, its rendered walls and the use of round-headed windows.

The other local industries in the Swansea area also, of course, had managers' houses. There was, for example, a fairly large manager's house at the Clydach Upper Forge which was built at right-angles to the smaller workers' dwellings in Forge Row.

The Architecture of Swansea Workers' Housing

The Form of Swansea Workers' Housing

Pictorial depictions on eighteenth-century maps show that the cottages in use as squatters' and colliers' dwellings on the small roadside plots and commons to the north of Swansea town were similar to the traditional cottages of rural west Wales. One map of 1736 shows one of these cottages, serving as head collier's house, on a small enclosure of land at a coalpit on or near Pwll-y-domen and the Llangyfelach Road, the pit being represented on the map by a hand-windlass.[61] This and all the surrounding cottages were single-storeyed with two small windows flanking a fairly central front doorway, with a chimney at a gable which was slightly further from the central doorway in order to accommodate the large cooking-hearth area.

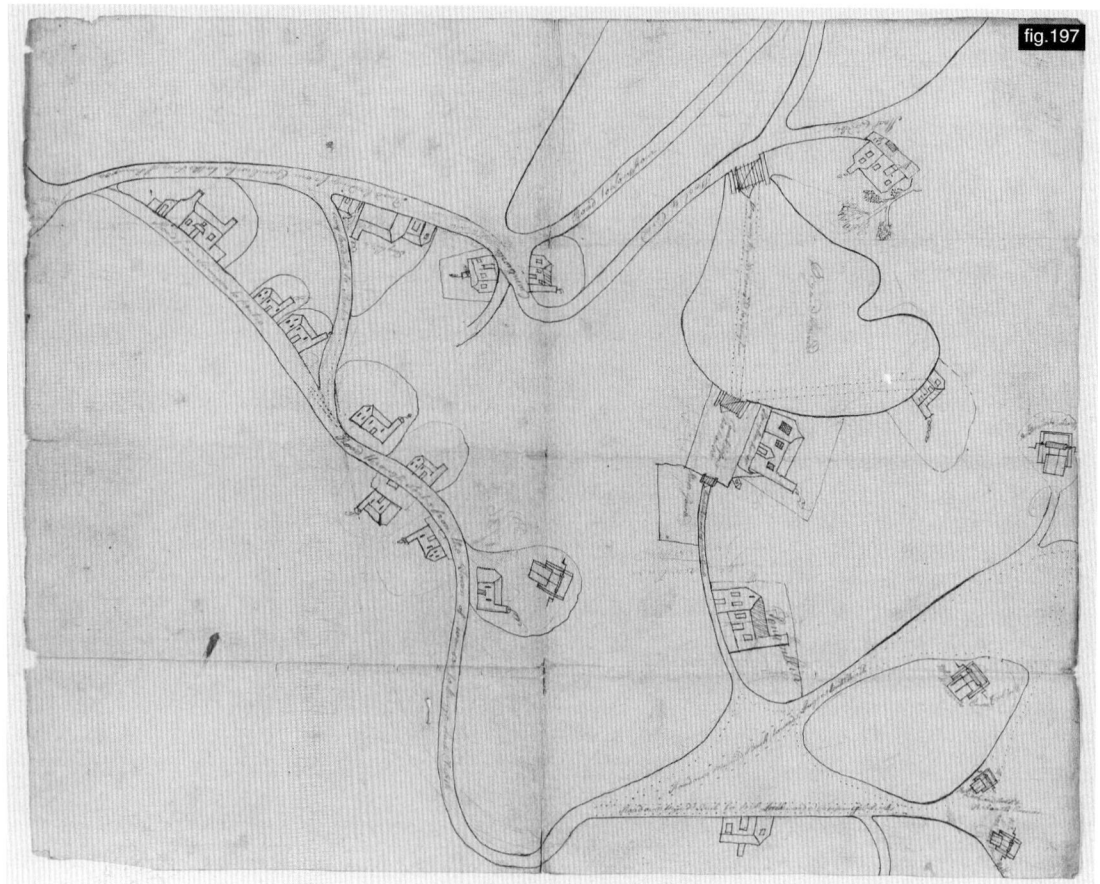

Fig. 197. Diagrammatic map of 1736 showing cottages and coal-pits in the area immediately north of Swansea town (the High Street/Dyfatty Street junction is at left) to Landore and Tirdeunaw. A head collier's cottage is shown at Pwllydomen alongside a representation of the Llangyfelach Road running along the bottom of the map. (The National Library of Wales)

Internally, the door opened onto the cooking and living area with a sleeping room/parlour at the unheated end, sometimes with a small half-loft (*croglofft*) over this, accessible by ladder, for at least some of the usually numerous children to sleep in.[62]

The Pwll-y-domen colliers' cottages had no gable windows, only those flanking the front doorway. Some of its neighbours had a single centrally-placed gable window lighting the bedroom/parlour, while others had two ground-floor windows at this end of the house with a single upper window to the *croglofft* over.

A local landowner/coalmaster such as Thomas Price (*c*.1687-1763) of Penlle'r-gaer House would use this local building tradition when building the collier's house at Pwll-y-domen before 1736, and Mansel in the 1720s had probably also used this type when building large numbers of cottages for his workers in the Bon-y-maen area.

Despite examples being built on the orders of the gentry/industrialists, it was all very much in the building tradition generated by the meagre economic resources available from the poor subsistence farming of west Wales. This only generated enough income for cottagers to build rudimentary single-storey dwellings of local rubble-stone or clay walling, and this type of dwelling continued to be used throughout the nineteenth century in west Wales by both agrarian and industrial workers. In the Swansea area, the early use of this type of dwelling was displaced by two-storey houses as the industrial entrepreneurs increased in wealth and were able to draw on the carpenters and masons already active in one of Wales's largest towns. Elsewhere, rudimentary cottage designs continued in use, as with the workers' housing built in the 1820s for workers on the Brecon Forest Tramroad (at the top of the Swansea Valley) by local entrepreneurs.

Early eighteenth-century maps of the lower western side of the Swansea Valley have graphic depictions of a scatter of single-storey cottages among the coal-shafts of the unenclosed commonland of Graig Trewyddfa and Treboeth. Other miners' cottages of this type seem to have stood near the multitude of mining tunnels into the outcrops of the thick Swansea Five-Foot seam at the crest of Graig Trewyddfa and on the east side of the valley at Bon-y-maen. These lacked any substantial division into rooms and carried on traditions of communal rural living.

There are accurate plans and elevations of some of the houses built by coal (and copper) masters near Pwll-y-domen, on land enclosed from Treboeth Common in 1767 alongside the Llangyfelach

Fig. 198. A rare elevation sketch of a working eighteenth-century colliery with the proposed new collier's housing sited to the north of the walled pit-prop storage area (right) ('d'). The earlier head collier's cottage was to the south of the shaft. In May 1767, stone was brought to the site to erect the new semi-detached head colliers' houses. The formal layout shown of another three dwellings to the north was never completed. (The National Library of Wales)

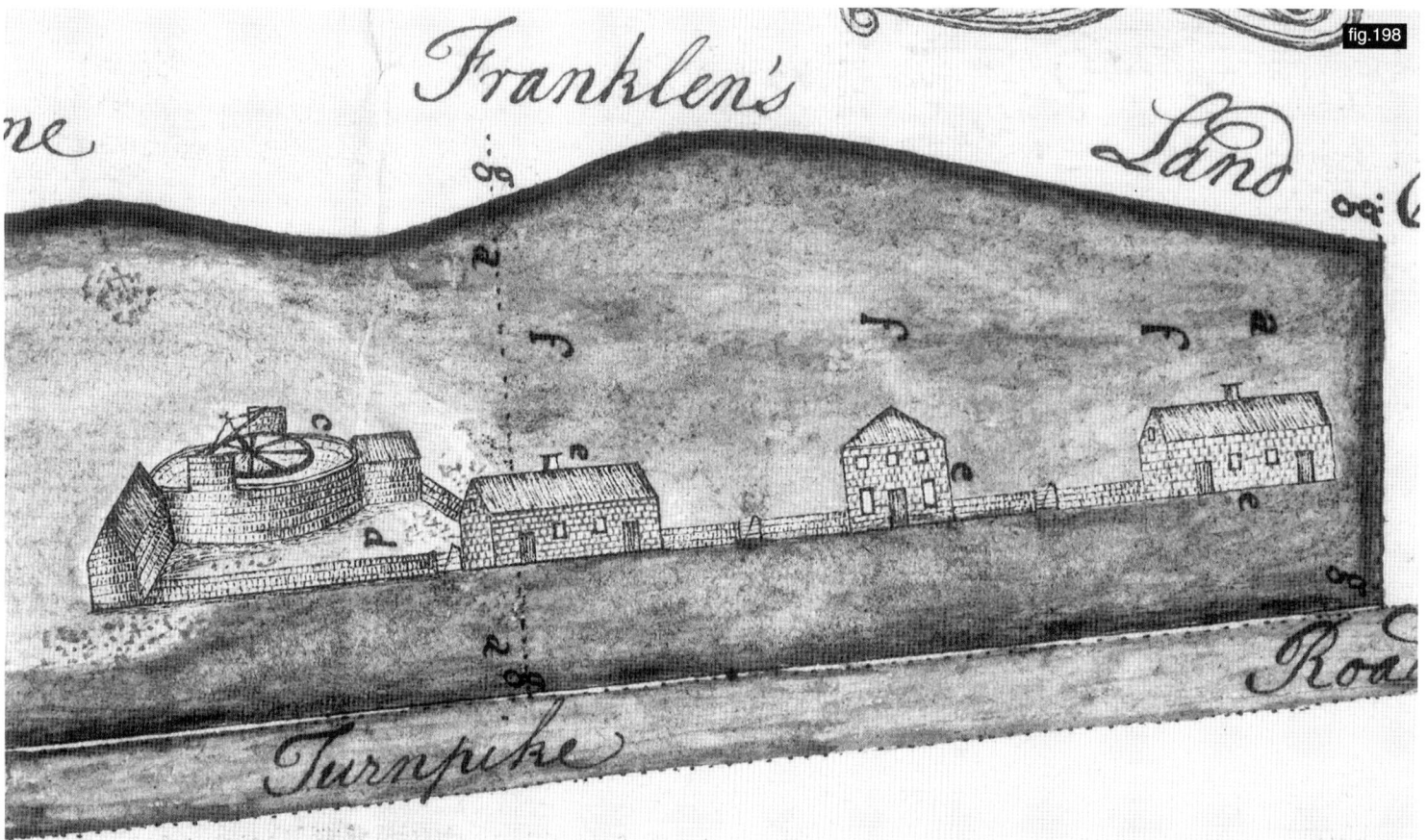

175

Road.[63] The Tirdeunaw shaft was sunk in 1758 by Lockwood, Morris & Co. and a 'cottage and garden belonging to the colliery' added on enclosed land immediately to the south. The buildings surrounding the pithead were added nine years later, together with the semi-detached pair of 'head colliers' houses and gardens' to the immediate north.

These were superficially of the earlier vernacular type in that they were single-storeyed with a window in each of the upper gables indicating the existence of a partial sleeping loft. The chimney and fireplaces were on the shared partition wall between the two cottages. (It is worth noting that one of the earlier cottages illustrated around Treboeth Common in 1736 also had a central chimneystack.) The doors were fairly central to the internal living space in each cottage (i.e. set slightly towards the outer gables because of the large internal cooking-hearth), with a single window between the door and central partition in order to illuminate the hearth area. The front block (which probably had a partitioned sleeping area) measured some 18ft x 15ft internally, with a low half-loft some 8ft x 15ft over this. Interestingly, the two houses had a shared and unheated projecting back wing, each with a room measuring some 12ft x 8ft internally. These unheated rear ground-floor rooms were presumably for food preparation and storage as was the case elsewhere or possibly they provided an extra bedroom, and form an interesting precursor to the 'tunnelback' (shared back-wings of a dwelling) common in late nineteenth-century terraced housing, a distinctive industrial housing type. The total internal living space for each of these cottages was some 486sq ft (including 120sq ft of partial height loft space). The cottage walls were almost certainly constructed of Pennant sandstone rubble, which was the universal building material for the area. The cottages stood until the early twentieth century.

These semi-detached houses mark the earliest recorded involvement of an urban housing type, produced in the Swansea area under the pressures of the need to house more than one family on a limited area of available land, and built with finite business capital. This is significant as the first known move away from simple single-storey one-cell cottages and the first recorded indication of a separation of function in such workers' dwellings around the town.

Those two cottages were originally designed to be part of a larger layout, for by 1768 at Tirdeunaw the copper-smelting company of Lockwood, Morris & Co. had developed an idea for the first planned layout of workers' housing, and for the first time this included a full two-storeyed house as its central feature, flanked by two semi-detached single-storey cottages.[64]

Little detail is known about the terraces of five to 11 dwellings that each of the eighteenth-century copper-smelting works' owners seem, to have built to house about a quarter of his workforce on or near the works site. They may well have resembled the three-storey, single-room deep, Pennant sandstone, Kelston and Warmley terraces of the 1760s known from the associated Bristol copper industry (which in the former case still stand).

If terraces represent an urban form of housing making greater use of limited available land then this is even more true of multi-storey flats such as those erected by the coppermaster John Morris I in the 1770s. These may well represent some of the first multi-storey workers' flats built since Roman times, although their prominent positioning and castellated appearance suggest that their primary function may have been to ornament the landscape near Morris's new mansion (David Dale had already built huge ranges of apartments for his textile workers at New Lanark).

The application of ornament to workers' housing was a new departure, mainly because few of the cottages built for workers in the lower Swansea Valley had been sited on or near coppermasters' parklands and so generally were built as simply and cheaply as possible. Morris Castle was a block of 24 multi-storey apartments built to accommodate John Morris I's colliers at Treboeth Level; it was erected on the crest of Cnap-llwyd common. It is known to have been substantially complete by 8 August 1773[65] and rents were paid on it from 29 August 1773.

Twenty-four families were housed in Morris Castle (sources mentioning 40 families probably

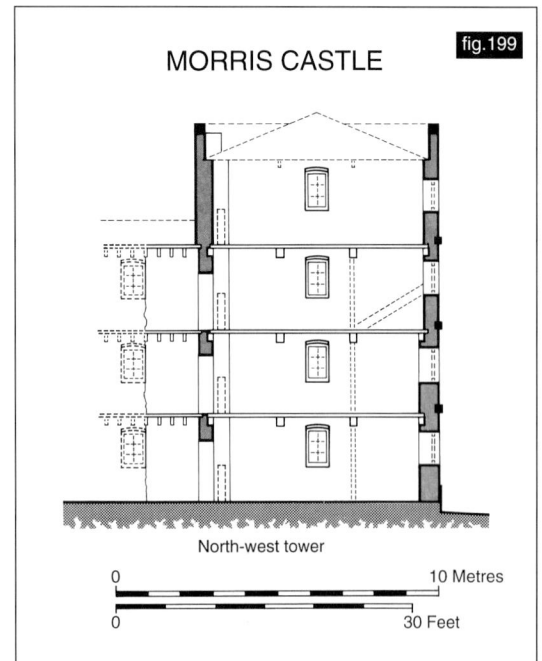

Fig. 199. Morris Castle, one of the first blocks of multi-storey workers' housing built. North-south cross-section of north-western tower of Morris Castle showing three-storey courtyard sides and trace of internal stair into fourth floor of new tower (after a survey by Bernard Morris).

fig.200

Fig. 200. Remains of the four-storey corner towers of Morris Castle, seen from the south-west before severe storms toppled half of the far tower in 1987. The larger upper tower flats were probably the residences of head colliers. (GL1368)

include accommodation provided in nearby cottages), a block 'of collegiate form' with an internal quadrangle. The position of the access stairs is problematical but they probably gave onto timber galleries around the internal courtyard, giving access to the 24 flats.[66] The boundary banks of the attached garden plots to the north flank the remains of the entrance road. These were mentioned in an advertisement in *The Cambrian* in 1811, when John Morris I seems to have been trying to sell the castle.[67]

One other block of workers' accommodation ('Wychtree House') survived to be photographed and partially recorded in the modern period (intact until 1964 with the lower two storeys still in place in 1975). This split-level housing with two-storeys of accommodation provided at higher and lower levels was built into the hillside near the western River Tawe bank at what became Morriston. Forest Copperworks (1749-51) lay some 1,000ft to the south west and the block was built into the south-eastern side of Wychtree Street, which was the old road from Swansea town to the Forest

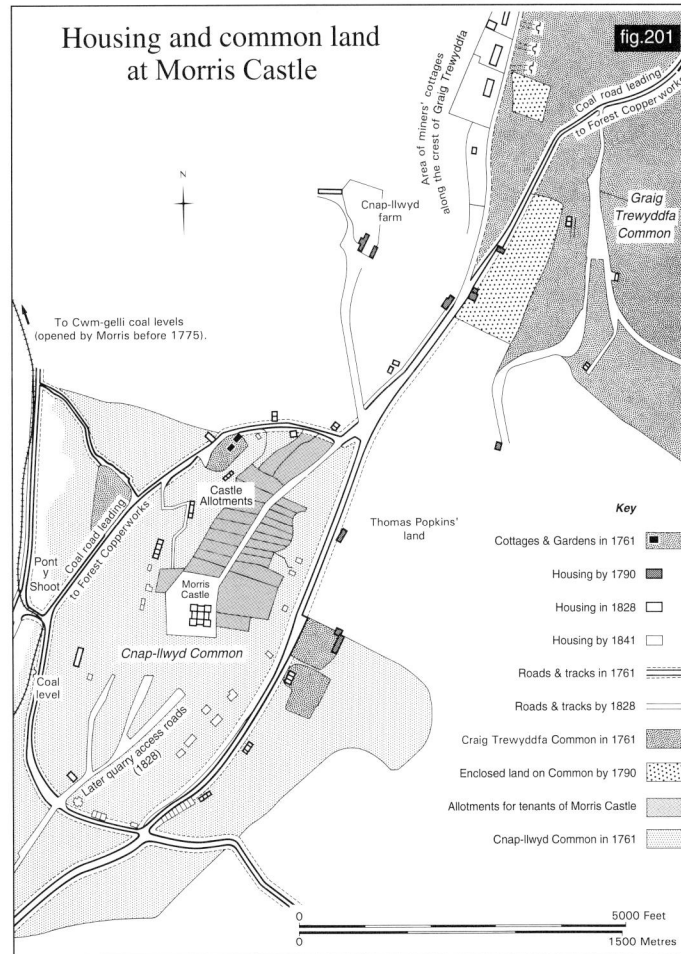

Fig. 201. With the agreement of the Duke of Beaufort, Lockwood, Morris & Company placed their colliers' flats of Morris Castle, with their extensive gardens, in the centre of Cnap-llwyd Common. In this location, it was also a landscape feature for John Morris's new house at Clasemont.

177

fig.202

Fig. 202. Plan of the vaulted basement of 'Wychtree House' or 'The Hostel' on Wychtree Street, Morriston. This may have been built soon after Morris Castle by the architect John Johnston as almshouses for workers of Lockwood Morris & Company. (Bernard Morris)

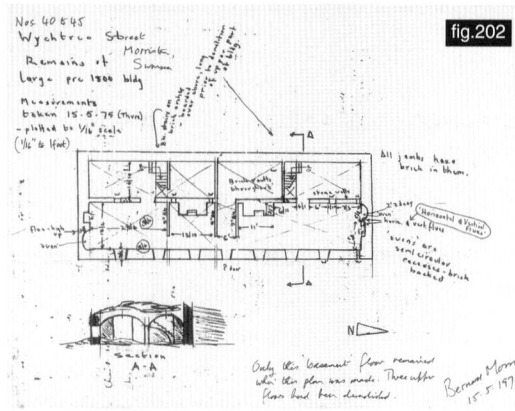

fig.203

fig.203a

Figs. 203, 203a. 'Wychtree House' or 'The Hostel', Morriston, photographed before demolition in the 1960s.(GL8376)

which faced onto the riverbank and the Lower Forest Coppermills weir. Over the lowermost dwellings was a brick vault separating them from the more highly finished upper storeys which mainly faced onto the public road.

In 1775 Morris built another group of three semi-detached houses for his copperworkers: his memorandum book for that year notes that he had had six cottages built. A group of three semi-detached houses is shown on the 1794 map of Graig Trewyddfa; they were prepared for John Morris I on the commonland on the hillside above the Forest Copperworks. These may well have been identical to the Tirdeunaw design of eight years previously but presumably housed key personnel from the nearby Forest Copperworks. They had garden plots approximating in size to those later adopted in Morriston. These houses which stood on what became Pentremalwod Road have been demolished and their detailed form is unknown.

In 1779, Morris's engineer/architect, William Edwards, laid out what was called Morris Town (later Morriston) with plots on which copperworkers and colliers could build using standard plans produced by him. The plots were rapidly divided between progeny and only two late eighteenth-century houses survive in Morriston from this phase. They seem to conform to a standard type and probably follow the Edwards design for dwellings in the new township. In these first architect's designs for industrial workers' housing, symmetrical layout replaced the irregular vernacular tradition. This is the 'Renaissance' symmetrical plan with a central doorway found in cottages and farmhouses from 1750 onwards. These two houses had oblong plans, 25.5-26.5ft long to the street and extending back about 12-13.5ft, and probably divided into two rooms. In 35 Morfydd Street, Morriston, there was almost a third more living space (634.5sq ft) than in the colliers' houses at Tirdeunaw designed 12 years previously, and unlike those it was almost to full height except for the small areas next to the back walls where the roof intruded down. The house at 91a Woodfield Street was slightly smaller (552sq ft) but subsequently had a 'slope' (or rear lean-to) added which provided a living-area of 710 sq ft, almost a 50% increase on that of Tirdeunaw.

As in so much industrial housing, the upper rear walls were blank, but the back door on the ground-floor did provide an element of through ventilation. The comfort of the parlour on the ground floor would have been considerably improved by the addition of a second hearth at this end of the house. A vestige of the vernacular tradition of south-east Wales (where Edwards came from) may be evident in the narrow, confined hearth-stairs at 35 Morfydd Street, where

Bridge, existing before Morriston was laid out on land to its north-west after 1779.

By comparison with other buildings, Morris Castle can be shown to have been designed by the Leicester architect John Johnson, who had been brought into the area to design coppermasters' residences. The same can be inferred in the case of the pedimented 'Wychtree House', which resembles another of the multi-occupied blocks called 'White Houses' which Johnson designed in Leicester for the occupancy of his relatives.[68] 'Wychtree House' incorporated substantial areas of cast copper-slag block walling and quoining in its lower two storeys

such stairs succeeded stone stairs built into the walls. The workers' terrace probably built in *c*.1784 for the Upper Forge at Clydach is also only one-room deep but two rooms wide and the houses were of similar size but standing in terraces.

Up to this point the form of workers' accommodation for two industries has been described - coal and copper. The former tended to be individual dispersed accommodation constructed on waste or commonland. The latter tended to be focused on larger individual units of production and, as with earlier Bristol practice, consisted of grouped rows or terraces. The exception was sponsored by the locally-based managing directors of Lockwood, Morris & Co., who built semi-detached and detached colliers' houses and later grouped colliers in an ornamental apartment block. Subsequently, however, they built copperworkers' accommodation on their earlier model of semi-detached and detached housing.

Elsewhere, detached houses were largely the prerogative of managers and owners. By the later nineteenth century, housing in Morriston had been joined into continuous rows, and the only other instance of detached workers' houses in the Swansea area was alongside the Swansea Canal, where married workers with responsibility were given housing at dispersed locations along the waterway. Even then, only one of the four ordinary lock-keepers or lengthsmen (there were no fewer than 36 locks on the Swansea Canal) was given a new house when the canal was built in 1794-98 and that was in a dual-purpose Swansea building which had a committee room enlarged by 'Mr. Powell the architect' in 1807.[69] Other purpose-built dwellings erected at the time of the canal's construction were 'Fountain Hall', reserved for the canal manager/engineer (eight rooms and designed by the canal engineer/architect William Sheasby), and another smaller dwelling built for the head of the canal maintenance yard.

In addition, at Clydach, a building that was a rented 'counting-house' for assessing traffic on the canal was subsequently also used as the dwelling of one of the lengthsmen. The local tradition was that this had been a thatched cottage or a school-house, and its thick, irregular walls of rubble-stone may indicate that this was so. The canal company bought the cottage in 1834 and doubled it in size to be a four-roomed dwelling.

As has already been noted, each of the eighteenth-century copperworks probably had at least terraced housing for some 10 workers on or near the works site. The workers' housing in (north) Swansea town built from *c*.1800 was also constructed on the terraced model. As with the early copperworkers' housing, this was usually only one room deep but was two full

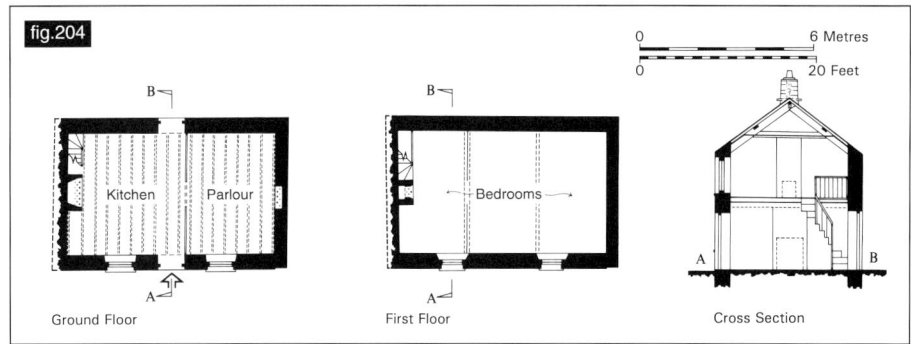

Fig. 204. William Edwards produced a standard design for Morriston and 35 Morfydd Street is one of the earliest surviving, with a façade that may approximate to the original appearance of the houses. (9500357/2)

Fig. 205. The two floors of 35 Morfydd Street provided fairly generous accommodation for the period.

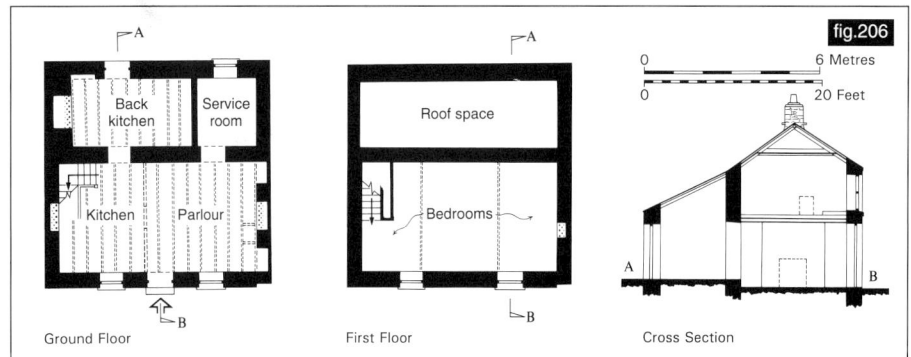

Fig. 206. The second early house at Morriston is 91a Woodfield Street; the ground floor fenestration has been altered subsequently.

Fig. 207. The plan of 91a Woodfield Street is similar to 35 Morfydd Street but has a simple outshut extension at the rear.

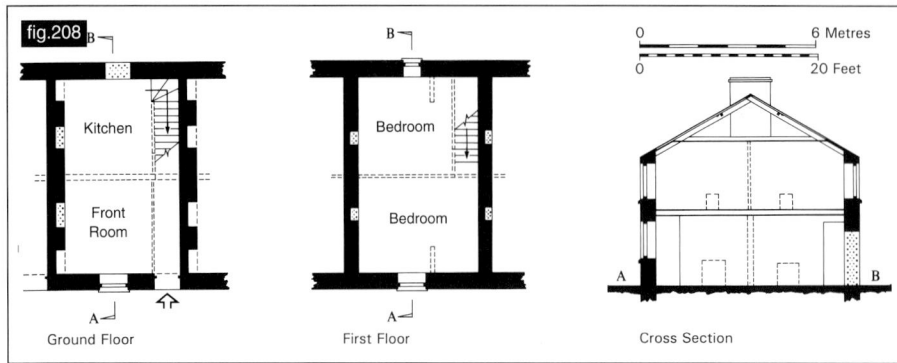

Fig. 208. By 1799, a dwelling, or dwellings, stood on roadside land in the position occupied by the present 29-32 Pentrechwyth Road, opposite the present entrance to the original Grenfelltown terraces. The present roof to the four houses, with purlins abutted and tenoned into the principal rafters, is of a type found in the houses of Swansea Canal workers from 1810, to shortly after the mid nineteenth century.

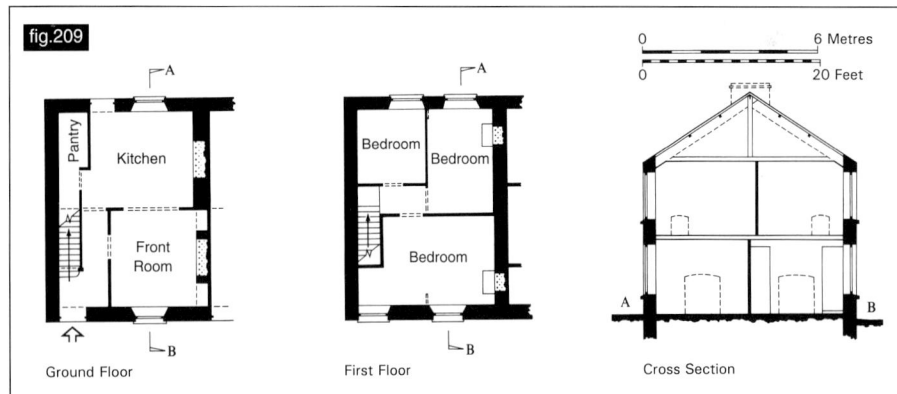

Fig. 209. 1 Grenfelltown, part of three terraces of workers' houses built in 1803-13 by Grenfell and Williams. These houses have been re-roofed in the late nineteenth century, although at least the lower walls are probably of the early period.

Fig. 210. Rifleman's Row, the southerly of the three original Grenfelltown rows looking towards the middle of the terraces set at a lower level. Sources of water available to the inhabitants were a well sunk at the eastern end of the terraces and the leat of the White Rock Copperworks. (9500350/1)

storeys in height. In that respect the typical Swansea town house of the early nineteenth century was of a rather better standard than was much of the ironworkers' housing in the Heads of the Valleys area. Houses from this period on Princess Street and Pottery Street, which were surveyed before demolition, had full-height, sixteen-pane sash-windows, rather than the reduced-height, eight- or twelve-pane windows often seen in ironworkers' cottages.[70]

The smaller town houses, with only one room on each of their two floors, had what was considered adequate lighting from the large windows on the front facades and so had windowless rear elevations that were only pierced by a single door. On the 1852 large-scale public health plan of Swansea, there are shown 970 houses of this two-roomed type, approximately 12ft to 15ft square. The 1845 *Report on the State of Bristol, Bath, Frome, Swansea, Merthyr Tydfil and Brecon*, presented to the Health of Towns Commission by Sir Henry De La Beche, stated that there were 3,369 houses in Swansea at that time, so two-roomed houses would have formed about 29% of the total housing stock in the 1820s.[71]

Swansea craftsmen almost certainly worked on houses both in the main large urban centre and in the nearby satellite villages. This was a different situation from that facing entrepreneurs in the isolated Heads of the Valleys area, where there were no large pre-existing centres of population, including artisans, so industrialists there are thought to have brought in craftsmen from their home areas to design accommodation in a variety of forms.

In the early and mid-nineteenth century there was another larger design of workers' housing common in the Swansea area that was usually narrower than the detached two room wide house and was built in terraced rows. Like the earlier *croglofft* cottage this had partially-heightened sleeping space and is often known as a 'catslide' roofed house, with a 'slope' extending back from the roof of a full two-storeyed two-roomed house to cover a partial-height rear bedroom over a back kitchen extending to a low third wall of the house.

A house of a design that may be transitional between the two types was already in use for workers' houses by 1812, when the Swansea-based management committee of the Swansea Canal specified that a new lengthman's house at Ystalyfera (in the upper Swansea Valley) should have '4 rooms and a slope'.[72] The double terraces built for the colliers and ironworkers on a landowner's estate in the upper Swansea Valley at nearby Ystradgynlais in 1826-41, and known as 'Gough Buildings', seem to have also consisted of a two-room front block with a rear slope.[73] The type was certainly in use in Swansea town by 1830-37[74] and was adopted by the Vivians for their first houses in Trevivian in about 1837-40.

It may have been a type widespread in industrial south Wales. Certainly the second great concentration of industry in the eighteenth and early nineteenth century at Merthyr Tydfil had many examples of a similar type of house, although with interesting differences of detailed design. The Crawshay family, owner of the world's biggest ironworks at Cyfarthfa, used a catslide design for their workers' housing between 1795 and 1830;[75] their Rhydycar houses of this type are on display at St. Fagans. A major design difference was that the Merthyr houses were lower overall and therefore did not have a rear low-ceiling bedroom, as was found in Swansea houses. Instead, the rear slope was occupied by an unheated small bedroom and pantry, side by side, on the ground floor. The Swansea house slopes were heated and could have contained a back kitchen and pantry with bedroom above.

Sir Henry De La Beche recorded the evolution of the catslide house from an earlier, simpler and smaller type of workers' house in use towards the end of the first half of the nineteenth century; he wrote that in 1845 '...houses having only two rooms are numerous in this place [i.e. Swansea]. Houses of this description have frequently been built here within the last ten years'. But significantly he then went on to say that '...there are some such in course of erection now, but commonly of a better sort; latterly, in erecting these houses of two rooms, a slope behind is, generally speaking, added; some of these slopes contain a small bedplace'.

The 'Crawshay' type of house was centred on the various ironworks of the family in west Merthyr, Hirwaun and Nantyglo, although other examples of the type have been located in Aberdare, Rhymney, Blaenavon, Pontypool and Rock (Monmouthshire).[76] Two of the other ironmasters of Merthyr, the Homfrays at Penydarren and the Hills at the Plymouth Ironworks, provided a few four-room 'catslide' houses by subdividing the main upper floor; again a quite different design device from the upper rear bedroom provided in the taller Swansea workers' houses.[77]

The area of each room in Swansea town workers' houses *c*.1800-30 would have been about 13ft by 14ft and into this intruded a straight wooden stair leading up from the front door and ending above the rear door. In this respect, Swansea area workers' houses differed from those of south-east Wales and the iron-making belt, where semi-circular stone stairs were placed in the thickness of the wall. The internal floor area of these typical two-roomed

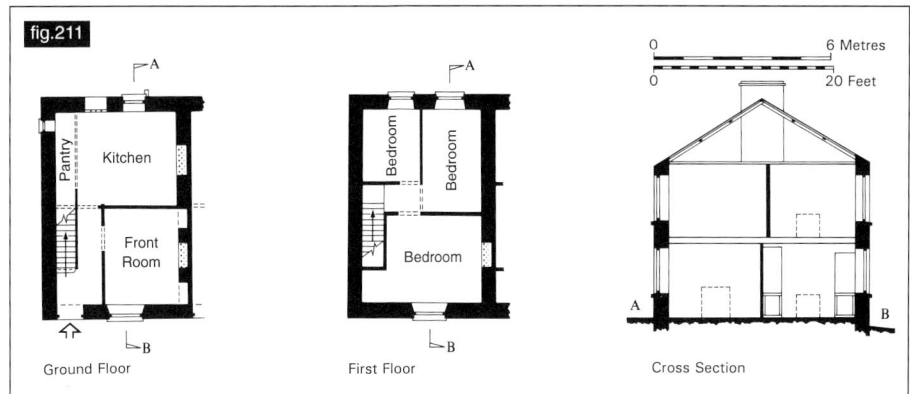

Fig. 211. In the later nineteenth century, Grenfelltown was expanded with new terraces at right-angles to the old, flanking a new vehicular access to the settlement from the Pentrechwyth Road. The houses were provided with three bedrooms, and the ceiling was at full bedroom height.

Fig. 212. Later houses built between 1851 and 1877 on the western side of the new steeply-inclined access to the older houses at Grenfelltown. (852239)

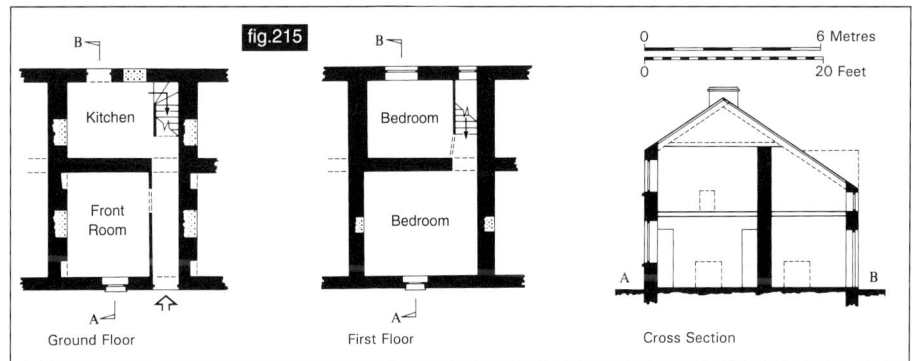

Fig. 213. Front elevation of one of the Trevivian houses of c. 1837 showing the windows illuminating the main rooms on each storey. The quoining is of rectangular, cast copper-slag blocks; the garden walls of rough copper-slag and the garden wall-cappings of half-round copper-slag blocks. The house walls are of squared Pennant sandstone. (9500055/24)

Fig. 214. The facade of 15 Vivian Street, one of the few to remain unrendered from the 48 houses built at Trevivian by 1844. The front of squared Pennant sandstone is toothed to the west indicating that it was originally intended to continue these original catslide-roofed houses, with long gardens and pig-sties, towards Aberdyberthi Road. (9500055/27)

Fig. 215. A house from one of the original four terraces of Trevivian (43 Vivian Street). Externally similar to the ironworkers' houses of south-east Wales, the Trevivian and Swansea examples had an extra children's bedroom sited under the 'catslide' roof at the rear.

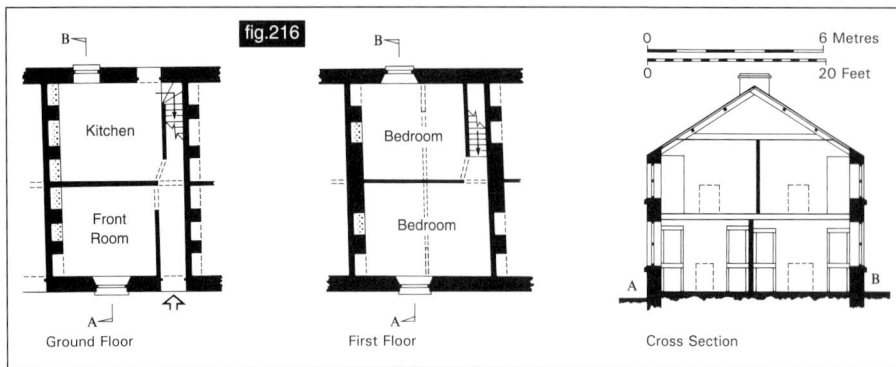

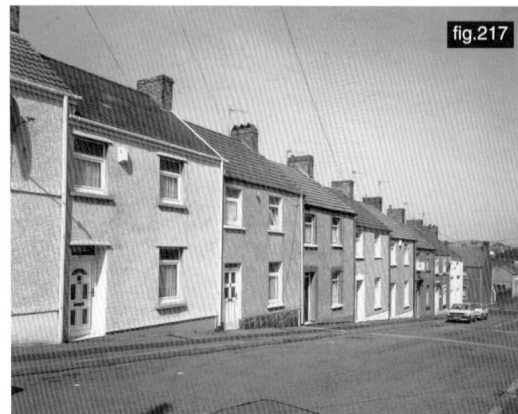

Fig. 216. The houses on the north side of Bowen Street infilled one of the original blocks of Trevivian in the 1860s. The cross-section shows how contemporary expectations for workers' housing had changed. Four heated rooms rather than two or three were provided, and both bedrooms had a nearly complete full ceiling height (22 Bowen Street).

Fig. 217. 28 Vivian Street and other houses of late nineteenth-century date looking east to the original houses of Trevivian east of Aberdyberthi Street. (9500354-1)

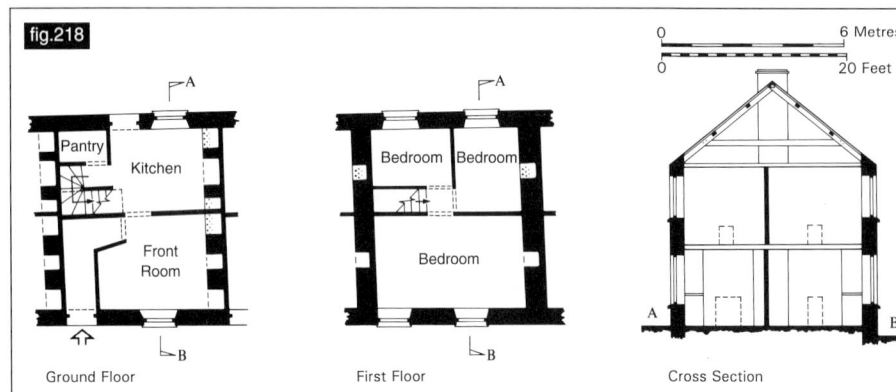

Fig. 218. One of a terrace of 18 houses built on the north side of Vivian Street. Houses in the later nineteenth century showed advances in their social provision. Three bedrooms allowed segregation of parents, boys and girls, and bedrooms were also completely full-height for the first time. However, the two smaller rear bedrooms were unheated (28 Vivian Street).

Fig. 219. Later nineteenth-century houses on Aberdyberthi Street, the more distant forming part of the sixth phase of development at Trevivian. These quite large houses had the tunnel-backed, partial-width rear extensions found in almost all the late nineteenth-century urban housing in Britain. (920057/2)

Swansea houses was 364ft^2, which compares with about 800ft^2 for an average modern house.[78]

The next stage in the evolution of workers houses at Swansea was to have terraced houses built with upper storey rooms of a symmetrical cross-section, being completely, full-height (or almost so) with a depth of two rooms. This involved a greater financial cost for the copper concern building them: rear walls were higher than with cottages or catslide houses, and much larger spans and widths of timber were needed for the trusses and rafters of the new roofs; imported pine from the Baltic could be used for the more substantial spans.

It is difficult to be precise about this when evolution into what is recognisably a modern house-form took place. By 1813 Pascoe Grenfell and Owen Williams of the Upper and Middle Bank Copperworks had built three large terraces of housing on the hillside above their works, at what became known as Grenfelltown. Whether the original form of these houses was as a catslide design, or of full-height and two-rooms deep, is impossible to say, as the houses seem to have been substantially rebuilt in the late nineteenth century; the rubble-stone walls of the three nearly contiguous Grenfelltown terraces may be original, but the roofs seem to be of a later imported red-wood with continuous purlins (characteristic of late nineteenth-century roofs) resting on the principal rafters. The walls are of two full storeys and the generous double windows to two of the three bedrooms are likely to be a later rebuilding.

Of the four original blocks of housing built by the Vivians at Trevivian (i.e. 'Hafod') in 1838-44, that nearest the Hafod Copperworks survives as full-height houses rather than in the catslide form of the other three rows; it was almost certainly also rebuilt in this form in the later nineteenth century.

The transitional stage to almost full-height houses may have taken place in the mid-nineteenth century, probably in the 1840s or, 50s. The management of nearby Cwmavon Copperworks built housing at Ty-Maen, near Port Talbot, in about 1848 with full-height upper floors.[79]

At Swansea, the second phase of Trevivian in Bowen Street has houses of this date and type with large pegged and tenoned purlins butted against very substantial roof trusses. The upper ceilings only slope slightly at the edges against the outer walls of the house. At Grenfelltown an earlier block of housing lay on the north side (i.e. opposite the three *c*.1813 terraces on the hillside above) of the main Pentre-chwyth Road up to Bon-y-maen, on land which the copperworks' proprietors held on a long lease from the Mansels. This survives as a terrace two rooms in depth, with the outer sides of the two upper rooms again having ceilings less than full height, due to the principal rafters intruding below the general ceiling level.

It is worth noting that the number of rooms in the average Welsh industrial worker's house built between 1775 and 1875 varied between two and four. Houses with only one room, or with five or more, were uncommon.[80] The three-bedroom house, which was the ideal of housing improvers who sought to segregate parents, boys and girls, may have come into use for Swansea workers in Grenfelltown when it was extended in the 1870s.

The terrace of workers' housing was often used at Swansea; however, it was a particularly common device to maximise the number of dwellings, and hence deployment of scarce capital resources, on a given piece of land. It was not restricted to Swansea and seems to have been used from an early period, large-scale urban and functional use must have evolved naturally. In Britain in the medieval period it can be seen in such complexes as the close of the Vicars' Choral at Wells Cathedral. In the early modern period terraces of storeyed dwellings were built by the Venetian state near their Arsenal, which was the largest industrial works in Europe by the seventeenth century.[81]

Terraced housing was in use in the non-ferrous metals industries of Wales by *c.*1700. The houses at Esgair Hir Coppermine, built *c.*1700 by the Neath copper-smelting magnate, Sir Humphrey Mackworth, appear to have been a row of eight single-storey dwellings.[82] By 1850 the 'barracks' there had been extended to a longer terrace of 10 dwellings, and four had been made two-storey (of which substantial ruins remain). The nearby Cwmystwyth copper and lead-mine was under the same management and by 1704 had a single-storey terrace of miners' houses which had been extended from one dwelling to four, with hearths at one end of each dwelling. This was flanked by three similar detached dwellings, one with a lateral fireplace and three having upper gable openings, probably indicating the existence of a sleeping loft.

House Walls

The typical copperworkers' houses were not intended to enhance their surroundings or the lower valley landscape, none being close to the large landed estates of their owners, with the notable exception of Morris Castle and to a lesser extent Morriston. For this reason the houses at Grenfelltown, Trevivian and Morriston were not designed to present any elaborate architectural display but rather to provide adequate living space for workers at an acceptable or minimal cost. Mass-walling consisted of undressed coursed rubble from the local quarries (only rendered in the later twentieth century).

Works' proprietors could generally quarry the locally available Pennant sandstone from the coal measures situated in the hillside behind their works and their landlords generally agreed to this under the terms of their leases. For example, under the terms of the 1737 lease, the White Rock Copperworks partners could freely take stones and limestone, together with 'earth and clay' for making bricks and tiles, from Bussy Mansel's lands in Llansamlet. The remains of their quarry are still on the western side of Kilvey hill above Foxhole.

Window-dressings and chimneys consisted of coarse red-brick and the main quoins or cornerstones of the houses consisted of cast slag blocks. The architectural ornamentation in a copperworkers' settlement was reserved for its institutional elements of churches, chapels, schools, markets and sometimes a large block of housing (at Morriston, 'Wychtree House', a pedimented block possibly designed by John Johnston). Cast copper-slag blocks were used to cap garden walls which were often made of rough copper-slag pieces set in slag mortar.

Ornamental Housing

Most workers' housing was fairly severe in appearance, with no unnecessary ornamentation that would have diverted essential capital from the main productive business. The only exceptions to this occurred when the housing was visible from the works owner's house. The castellated Morris Castle, home to 24 colliers, stood visible from Sir John Morris's Palladian mansion at Clasemont, and both were almost certainly designed by John Johnson. The castellated design and hilltop siting provided a dramatic landscape feature at the southern end of Graig Trewyddfa. The Pennant rubble-stone of which the castle was built was covered in white lime plaster and dressings were provided by glistening blocks cast from Morris's own copper-slag. The function of this castellated 'eye-catcher' is directly analogous to the castellated Ivy Tower in the neighbouring Vale of Neath. This banqueting house (complete with copper glazing-bars), on the northern part of the Gnoll Castle estate, was designed for another prominent copper magnate, Sir Herbert Mackworth, in 1776-8.[83] A castellated fort was added to the skyline of the Vale of Neath at about the same time.[84] The architect of Gnoll Castle and the Ivy Tower (and presumably the 'fort' and extant 'ruined castle' and castellated Gnoll lodges) was John Johnson (1732-1814), who had been associated with Mackworth since at least 1767 and may have been introduced by him to John Morris I.

Similarly, the gothic mansion of the Swansea entrepreneur George Haynes, on the canalside at Clydach, was flanked by a matching block of workers' dwellings, all of these possibly designed by the architect William Jernegan.

Such examples of ornamentation that was applied to workers' housing, visible to landowners and industrialists from their houses and parks are paralleled elsewhere; for example, the bow-fronted canal worker's house visible from Pontypool Park, near Pontypool; and ornate estate-workers' houses of the later nineteenth century, such as those of Powis Castle and Leighton near Welshpool.

The generality of austere workers' housing in south Wales contrasts with the more ornamental workers' villages, such as Saltaire (Bradford) and Akroyden (Halifax), provided in the textile belt of northern England in the third quarter of the nineteenth century.

However, the Swansea houses were more substantial than the cottages which had preceded them, and the grim houses built for copperworkers at Cwmavon with all their dressings cast from copper-slag rather than using softer stone. In the use of mass sandstone rubble, they equate in ostentation with the eighteenth-century copperworkers' rows at the Greenfield Valley in Flintshire or those at the Warmley and Kelston works and mills on the eastern fringe of Bristol.

Even so, the houses in rural workers' settlements such as those outside Swansea town, or in the Greenfield Valley, were undoubtedly plainer than houses in the more genteel urban townships nearby. As with the houses of the entrepreneurs, housing built for the lower classes in the town of Swansea during the period 1800-30 generally had ground-floor openings set in blind-headed, semi-circular arches, a form of ornamentation also seen in other reasonably affluent urban centres such as Holywell in Flintshire and Liverpool.

House Roofs

Some of the earliest dated roofs of workers' houses in the Swansea area are of houses built alongside the Swansea Canal from *c*.1810 onwards. These houses had the wooden-pegged roofs, with purlins tenoned into the principal rafters, that are characteristic of late eighteenth- and early nineteenth-century workers' housing in the Swansea area. This type of roof was used at Trevivian from the 1830s until the 1860s and is found in the block of housing situated below the level of the main Pentre-chwyth Road at Grenfelltown.

The secondary cottage surviving on the main street at Morriston has a light roof with tie-beams bolted onto the purlins. The main three rows at Grenfelltown have had roofs rebuilt with red wood and passing purlins, of a later type, not butted onto the principal rafters as with most earlier workers' housing. The housing at Clydach Upper Forge, which probably dates from the 1780s, is still partially roofed with curved red-clay pantiles.

There is no evidence of any of the workers' housing being roofed with the copper-sheeting that covered many of the copper-smelting works houses, or of being provided with copper glazing bars that graced some of the coppermasters' mansions.

The terraces at Grenfelltown suggest that by the first decade of the nineteenth century houses in the Swansea area could be much wider than those built in the more remote iron-working region at the Heads of the Valleys. Before this period, there was no real divergence of type: the houses were largely one room in depth with roofs mostly 12ft in span. This was the standard that continued in the Heads of the Valleys region until the 1820s and it has been suggested that the houses of ironworkers there may have had a narrow main roof span because of an absence of carpenters.[85] However, there was no such shortage in the long established and fairly large population of Swansea, and substantial lengths of imported softwoods would have been easily available through the docks, so making it easier to build house roofs of two-room depth.

Gardens and Allotments

Some of the large gardens provided for workers by various of the coppermasters seem to have been material expressions of ideals of partial self-sufficiency. There is a comparative wealth of contemporary map evidence and material remains of gardens. These may reveal how the holdings provided by the industrialists changed over a period of time. As has already been seen, many workers' houses were not provided by coppermasters at all. Therefore the evidence for gardens and allotments provided by employers needs to be related to the size of plots upon which many colliers and copperworkers were constructing their own cottages.

Some of the earlier industrial housing at Swansea consisted of squatter occupation on common land and certainly part of the cottage settlement around what is now Graig Road, on the northern part of Graig Trewyddfa above Morriston, was there by the end of the seventeenth century. Most of the 20 cottages to the north west and above the road were certainly there by 1826, (ref. 1826 OSD) and 1838 (ref. 1838 tithe) in the irregular plots enclosed from the common had an average size of 487 sq yds. By

1838 there were also 11 cottages on the (lower) south-east part of the Graig Road and much of the northern part of Graig Trewyddfa Common, along the western side of Clyn-du Road, had been enclosed into a series of large gardens and small fields, with a substantial average size of those attached to cottages of 1,708 sq yds.

Most, but not all, of the multitude of self-built cottages in the lower Swansea Valley were given gardens. In 1838, for example, seven of the cottages on Cnap-llwyd Common to the north west of the Trewyddfa Road seem to have been built without gardens on unenclosed land, and may have been recently erected, some on the platforms provided by Pennant sandstone quarry tips, however, by the end of the nineteenth century gardens for these houses had been enclosed from the waste.

A more serious long-term deficiency in this regard was evidenced by some of the houses built at Foxhole to the east of the River Tawe by workers in the White Rock (1737) and Middle Bank (1765) Copperworks. In the northern part of the very long southern terrace built along the main Foxhole Road, and along the whole of the northern row of houses, there was an absence of garden provision that was still evident in 1849 when G.T. Clark was carrying-out his general enquiry into the sanitary condition of Swansea:[86]

> 'Across the Tawe, and upon its steep bank opposite to Swansea, are built the long and irregular villages of Fox-hole and Pentre-guinea. The houses are for the most part niched into the hill side, and are old, damp, and very dirty, and have no back premises.'

This tradition of gardenless building on the steep hillside of Kilvey Hill carried on into the third quarter of the nineteenth century.

The first known gardens provided by a Swansea coppermaster were those built for two of the head colliers at Tirdeunaw Colliery in 1767; these were only some 22 yds long and 10 yds wide (217 sq yds in area), that is less than half the 487 sq yds of the earlier gardens made by workers themselves at the squatter cottages on the north of Graig Trewyddfa Common.

At Morris Castle the fact that the 24 workers' dwellings were in flats precluded the dwellings standing within individual garden plots. The 24 or so rectangular and square allotments given to each tenant flanked the approach drive to the castle from the north. The earthwork banks defining the area of several of the gardens can still be seen. The hilltop site dictated that these were somewhat irregular but 20 of the plots fell within the range of 213 to 678 sq yds (an average of 433 sq yds). The four other allotments seem to have been two to three times greater (871-1,271 sq yds) than the average and presumably equate with the four larger tower apartments in the Castle that would have been let to head colliers.

By about 1789 there emerged at Morriston the view that the right sort of worker might mature and become a more dependable member of society if he were given a greater degree of independence with enough land on which to grow potatoes or even to keep an animal. Map evidence indicates the development of this idea in three phases, from John Morris I starting to build on the land used for Morriston in 1779 until the first available map showing the developing township in 1793.[87] The large split-level four-storey housing block of Wychtree House seems to have been built on the land nearest the Forest Copperworks alongside, the pre-existing road to Forest Bridge. There is no indication of any gardens being provided either for this property or for the block of housing that stood alongside the nearby Forest Coppermills (c.1737). The plot occupied by Wychtree House was some 2,033 sq yds in extent.

The first gardens at Morriston (c.1782) varied in size and shape because the edges of the blocks of the township were mostly developed within the irregular confines of the pre-existing fields, resulting in a jigsaw-like layout of interlocking garden blocks.

At Morriston the earlier and smaller grid of roads to the south east of the township (Libanus Chapel was built on the grid in 1782) nearest to the Forest Copperworks, was developed first. Here the two blocks (16 houses developed by 1844) nearest the works had relatively small plots of an average size of 444 sq yds (i.e. almost the same size as those of the earlier squatter houses on the adjacent common and almost exactly the same as the average for the plots built some 10 years previously at Morris Castle).[88] The larger blocks of land laid out, perhaps some seven years later, around the church of 1789, enclosed much more substantial gardens averaging about three times the size (1,277 sq yds) of the earlier ones; by 1844 some 2.6 of the four blocks (30 houses) around the church had been developed in this way. Interestingly, the expansion of the township in ribbon fashion along main roads beyond the confines of the Morris landholding between 1826 and 1844 was carried out using plots of a similarly large size, enclosed both from adjoining farmland and the Graig Trewyddfa Common (10 houses eastwards along Woodfield Street and Sway Road and five houses along Martin and eight along Clyn-du Streets to the south-west).

Indeed, the very large numbers of roadside houses built by colliers and copperworkers in the semi-rural area between Swansea town and Morriston in the nineteenth century were often sited on reasonably sized plots. The eight cottages, mostly built before 1826 (five before 1768) on the roadside fringe of Cnap-llwyd Farm (now Llewellyn Park), above Graig Trewyddfa, on land that was owned in 1838 by the industrialist/landowner John Dillwyn Llewellyn, had gardens with an average area of 911 sq yds. As already mentioned, the same land owner also had an enclosed area of Treboeth Common called 'Waun Gwn' that was fairly regularly divided into 10 cottage plots by 1826 with an average area of 2,344 sq yds. Llewellyn was only one of several landowners (including Sir John Morris I's son, the Revd. Thomas Morris) who enclosed roadside plots for renting in this way. Along a mile of the Llangyfelach Road from Pentre-estyll north through Brynhyfryd to Tre-boeth there were some 53 houses on 41 plots by 1838. These were built on fairly generous holdings of an average of about 4,500 sq yds. Several already had two or three dwellings on each plot by 1838 and two of these had evolved into roadside terraces.

The idea of partial self-sufficiency developed over a period of time. No contemporary business could afford the generous outlay of capital to build detached houses for each of the workers, unless the workers were building their own accommodation as at Morris Town; but instead would build compact terraced housing. The gardens which were provided for the workers in the two blocks of housing at Upper Forge Clydach (1784) were not quite as generous as those at Morriston.

At Trevivian the first two phases of houses in the 1830s and 1840s saw the provision of long gardens and combined privy and pig sty blocks at the end. The keeping of a pig to feed off unwanted scraps was certainly encouraged, and was in line with more enlightened ideas elsewhere.

Fig. 220. The long gardens ascending the hillside at the rear of the Grenfelltown terraces are of the large size characteristic of late eighteenth- and early nineteenth-century workers' housing in Swansea. (9500052/34)

At Grenfelltown (c.1803-13) the houses fronted directly onto tracks, but at their rear earthen banks of long gardens survive, 5.6 yd wide and extending up to 57.9yd on the slopes of Kilvey Hill (324 sq yds).[89] The length is variable, with a minimum of 19.8yds (i.e. 111 sq yds) at the western end where an old field road curves round the backs of the houses up to Llwyn-heiernin Farm. The 1877 Ordnance Survey map shows considerably shortened gardens terminated by other tracks cutting across the present garden earthworks and leading up to a colliery drift-mine on Kilvey Hill.

Landore was developed as a settlement on a large scale in the 1820s and 1830s. Between 1826 and 1838 two terraces were built on land owned by the Revd. Thomas Morris and rented by his brother Sir John Morris II, almost certainly for the Millbrook Iron Company, in which the latter was a shareholder from 1830. Of these houses on and to the north of, Byng Street, 11 were single and six double-fronted. The six to the north of Siloh Street had gardens split between the back and front and those to the south had gardens all to the rear, the former at least had latrines at the far end of their plots. The average gardens were some 55 yds long and some 4.1 yds wide, for single, and 9.6 yds wide, for double houses; the total garden areas were some 224 and 523 sq yds respectively.

In 1835-38 the proprietors of the Morfa Copperworks built a terrace of 16 houses along the Neath Road at Landore and these seem to have had rear gardens about 36 yds long and 5.5 yds wide, with garden areas some 194 sq yds in extent.

All the 46-48 original houses in Trevivian (c.1837-41) had gardens 4.9 yds wide and between 29.5 and 32 yds long (145-157 sq yds); those on Neath Road were set back from the street frontage with front plots over 39ft long and back plots 56ft long, while the other houses on the quieter back roads were set directly on the street frontage, with back plots some 89ft long. The second phase of 36 houses in the later 1840s, largely completing the two blocks nearest the Hafod Works had gardens slightly wider and shorter than those in the first phase (150 sq yds).

Most of these early nineteenth-century terraced houses have very long gardens with the provision of pig sties in the first two phases of Trevivian, which, with the generally large gardens at Grenfelltown, means that some notion of self-sufficiency persisted into the early and mid-nineteenth centuries. However, their comparatively narrow width means that the plots available generally do not compare to those provided in the second phase of Morriston (1787 onwards), nor to the plots provided by the owners of roadside fields, for colliers and copperworkers to build on. It must also be remembered that the worker lessees of most of these Morriston plots and a proportion of later roadside plots had themselves sub-divided the property blocks to form continuous terraced housing with narrow garden plots.

By the mid-nineteenth century there seems to have been an end to more general notions of self-sufficiency. Pig sties were not provided beyond the second phase of Trevivian, although generally long gardens did continue to be provided during the third quarter of the nineteenth century. The nine higher terraces, and two semi-detached blocks, some 62 houses in all, built along Kilvey Road in Foxhole during this period, tended to be terraced into the hillside without gardens at all.

At the end of the nineteenth century houses built at Trevivian also had small gardens, but then space was made on the hillside slopes to the west of the settlement and above Cwm Burlais for the laying-out of allotments as well as for the drilling of the Hafod militia force.

In some ways, the type of provision of gardens provided by the industrialists for their workers from the end of the eighteenth century to the end of the nineteenth had gone full circle, with a revival in the provision of allotments rather than large gardens.

It has been suggested that providing the workers with gardens was,[90] 'a subtle, yet potent way of exercising control over the labour force outside working hours as well as a means of enhancing the Vivian's reputation as philanthropists.'

Judging by the fact that most of the workers had an evident determination to provide themselves with substantial garden plots this seems to be an over-elaborate hypothesis. The first known garden plots provided by a coppermaster were those laid-out by Sir John Morris I in the 1770s and comparable in size with those enclosed by the workers themselves on encroachments from Graig Trewyddfa Common (some 433-487 sq yds). The larger plots provided by Sir John Morris I from c.1789 were almost three times as large as these early gardens and this change does seem to have been done intentionally. In the last two years of his life, in 1819, John Morris I recorded what he thought he had intended some 40 years previously in giving[91] 'artificers and labourers ... a sufficient plot laid on, for raising potatoes....I need not add that such persons never become burdensome to their Parish but continue to hold out the best example of prudence to their neighbours.' The agricultural commentator Walter Davies also wrote in 1815 that Sir John Morris I[92] ...'encouraged some of the more steady labourers, and others, with a quantity of land each, sufficient to keep a cow.'

Whether any of the workers kept a cow seems uncertain; the Vivians were later to provide pig sties for their workers, but there is no evidence that the Morrises expected their workers to build cow-sheds, and in the second generation the plots were generally sub-divided and would have been too small to keep animals. That this urban animal husbandry actually happened at Swansea is only evidenced in the 84 or so dwellings built in the first two phases of Trevivian in the 1830s and 1840s. It did take place elsewhere, for instance, on the houses provided for workers on the Powis Castle Estate (1851-61) and the Montgomeryshire Canal in mid-Wales.[93] In 1872 the enlightened model 'Plans for Labourers Cottages' sent-out by the 'Inclosure Office' included combined pig sty and toilet blocks similar in function to those already built at Trevivian.

There was obviously some preoccupation to prevent the parish poor from increasing the parish rate for industrialists, as was noted by Morris. At Swansea this did not cause industrialists to build their workers' housing in the neighbouring parish as happened at Blaenavon. Internationally the most famous and largest workers settlement at the end of the eighteenth century was New Lanark in Scotland. Like Morris Castle, it was developed largely on plots without gardens. However, the long rows of mostly two-room tenements in four-storey blocks for some 2,000 inhabitants, largely built by David Dale at New Lanark in the 1790s and continued by his son-in-law Robert Owen, do seem to have had some allotment provision made to one side of the settlement.[94] As on the continent, a tradition of a tenement living without a garden became a cultural normality in urban Scotland, but it was not a feature developed in industrial south Wales.

The tendency to replace large house gardens by allotments was a general one in Great Britain. The houses in the elaborate west Yorkshire model township of Saltaire had minimal gardens from its foundation in the mid-nineteenth century, but the settlement was surrounded by allotments on all four sides.

It is interesting to note that, similarly, employer-provided workers' housing built during the European mainland's first Industrial Revolution in the southern half of Belgium at Le Grand Hornu (1820s) and Bois du Luc (1830s), had small gardens, only about twice the size of the floor area of the houses. These were built originally in rural locations, but a tradition of gardenless urban dwellings quickly took root in French-speaking Europe.

Water-supply and Sewerage

As in other urban areas, adequate standards of public water-supply and sanitation were not attained until after the passing of the Public Health Act of 1849.

The report by G.T. Clark, Superintending Inspector of the General Board of Health, under the terms of the Act of 1849, noted that in Green Row (at the northern end of Swansea town) 'there were three houses of two rooms each without toilets or sewers' and that he had been informed that as many as 30 have lodged the same night in one of these cottages. 'They are a nuisance to the neighbourhood.'

The provision of an adequate water-supply in the northern part of the town was no better. In Dyvatty Place, Mr Davies (who compiled a report on the worst parts of the town) noted that he had seen '...as many as 36 pitchers waiting to be filled at these two spouts about two weeks ago; this day, May 18, after much rain, I saw 16 persons waiting'. In Mariner Street '...the residents buy most of their drinking water at this time of the year, as they can rarely get any at Dyfatty spout without waiting for hours'.

Water supply was a general problem in Swansea at this time. In 1849 'The chief supply of Swansea [was] derived from public and private springs, from wells and pumps, and [for those few who were connected] from the Swansea Water Works. The public springs [were] numerous, though ill managed, and rendered foul by privies and for want of being collected in proper cisterns.' There were 'five public pumps, and very many of the better class of houses' were 'also furnished with pumps and cisterns for rain-water'.

In Morriston and Trevivian some water was used from the Swansea Canal as well as from local watercourses. The following is Clark's assessment of Morriston in 1849:

'[The houses] are almost all ill provided with privies, and the people draw water, often contaminated with sewage, from various springs and brooks within two to ten minutes' walk, or from the canal.

The position is good; so good that, although the houses do not lie very closely together, they are capable of being cheaply drained and supplied with water. The people are chiefly employed in the adjacent collieries and copper-works, and, although they have suffered under the temporary depression of trade, I found the interior of their houses particularly clean and neat, and they lamented that their landlords would not provide them with a tolerable drainage and with privies. I was accompanied by Mr. Evans, surgeon to those [copper-] works, who pointed out ...[that the]...2,000 inhabitants, who are, for the most part, colliers and copper-men.... are, on the whole, a well-fed and well-clad population (a description which will apply to all the other places mentioned in report)...[However]...The place is scarcely ever without fever of some form or another, and which frequently breaks out epidemically, and often with great virulence and fatality.

This unhealthy condition of the place is, to the best of our knowledge and belief, mainly owing to its possessing, comparatively speaking, no drainage whatsoever; no privy accommodation. The weir situated immediately below Morriston [i.e., the Lower Forest Mills weir] is unquestionably another very fertile source of disease to the population; it receives nearly the whole of the animal and vegetable filth of the town, as well as the foul water from the collieries [the tunnel under Water Street, Morriston, still disgorges iron-oxide-stained mine-water into the River Tawe above the weir]; in the summer months it is stagnant (excepting now and then a freshet clears it), and at these times especially the foetid exhalations arising from the weir are highly offensive, and as the supply of water for drinking purposes in Morriston is insufficient, and at a distance, the bulk of the inhabitants drink their tea and beer brewed with this filthy water.

It is, therefore, plain that fever food and cholera food are unremittingly supplied to the lungs and bowels of the inhabitants of Morriston.'

A detailed investigation in the same report of the sewers and drains of the Swansea area, which were found to be non-existent, states:

'In Morriston the sewage is left to find its own way along the open street-gutters, which at the time of my visit were in divers places clogged up with it. In hot weather the nuisance thus created is considerable, and the Medical Report, already cited, shows the ill effects not to be confined to the mere production of an inconvenience.'[95]

The three Grenfelltown Rows of 1799-1813 were immediately east of the former gentry house of Cnap-goch, on land just above the water-supply leat leading from the east side of Kilvey Hill to the White Rock Copperworks. This probably provided much of the settlement's water-supply in the same way that the Swansea Canal was used by the inhabitants of Morriston and Trevivian. In addition to this relatively clean water source (the property of the neighbouring works), the philanthropic Grenfells provided a well at the north-east end of the rows.

In the 1850s Henry Hussey Vivian incurred considerable expense in an attempt to sink wells for his workmen at Trevivian but the operation proved a failure; presumably the water-table drained by the

Cwm Burlais ravine and neighbouring mines was too deep to be readily accessible. This lack of water was in contrast to the adequacy of the other features in the housing. 'Here, from the general personal cleanliness induced by the good arrangements of the houses, the inhabitants felt severely the want of water.' They had 'none soft enough for washing nearer than the canal, a good ten minutes' walk with a two-gallon water jar'.

> 'These people prepare for washing-day by bringing up a part of the water the day before. All agreed that 1d a week would be a very light burthen for water; and one woman whom I found busily engaged in washing clothes said she would pay 3d. a-week with pleasure to save the labour of bringing up water from the canal.'[96]

Henry Hussey Vivian built drains when asked to do so by his workers and was criticised by Clark for making them structurally over-elaborate.

Roads in Workers' Settlements

Some of the workers' housing at Swansea, such as the original three hillside rows at Grenfelltown (1799-1813), were completed without a vehicular road network. Indeed, Grenfelltown seems to retain its original narrow Pennant sandstone pavement along the valley side of its three rows. Industrial workers carried what they needed for themselves and it was only the better-off members of the community who received their household necessities by horse and cart. Retail street traders used hand-carts or ponies to carry their goods.

Even so, most workers' houses were built alongside tracks accessible by wheeled transport. Even the original seventeenth-century squatter community on Graig Trewyddfa Common, above and to the west of the site of Morriston, was bisected by a winding roadway which has survived and evolved into the surfaced but narrow line of Graig Road.

The larger settlements laid out by the coppermasters at Morriston and Trevivian were arranged alongside grids of streets but generally the roads were not structured or surfaced to a high standard. Morriston was originally designed with roads some 15.4 yds wide, but the north-eastern part of the second grid was completed in the early nineteenth century with roads narrowed to only 6.6 yds in width. G.T. Clark noted in his 1849 report that 'The streets and roads [of Morriston] are ill made, and no care is taken to keep them clean'. In the later nineteenth century, Grenfelltown was given unmade roads and along with Trevivian still has substantial lengths of unmade road.

The workers' houses of north Swansea town fared no better. Dyfatty Street had 35 houses, 17 without toilets. Davies observed, in his mid-nineteenth century health report, that:

> '....it was so bad in the winter that it could not be crossed by the inhabitants but with great difficulty, the dirt being knee deep. The surface of the road was maintained with 'loose stuff' from the nearby quarry but scavengers only came to clear dirt from the road every six months. In these conditions, these townships were probably no better or worse than other urban industrial areas of the time.'

Other development at the end of the nineteenth century was sited alongside the ballasted formation of former horse-drawn railways such as those that became Eaton Road, Dinas Street and Pentremalwod Street, along the western side of the lower Swansea Valley.

It was quite usual for some, or all, of the roads of a workers' settlement built by an industrial concern to be named after members of the founder entrepreneur's family. This certainly happened at Landore (Byng Street) and Trevivian; it occurred even at what seem to be settlements provided by the most philanthropic industrialists.

Significant Issues in Industrial Housing

Workers' Rents

Rents for accommodation in the copperworkers' settlements compared well with those in the town of Swansea. A two-roomed house in the town, often in crowded courts with small yards and no sewers, was rented at £4 to £4 10s per year in 1845.[97] Most of the early Trevivian houses, including the pre-1844 rows, had two rooms upstairs (including the low-ceilinged bedroom situated under the long rear roof, 'slope' or 'catslide') and two down, and the rent of these was 1s 6d per week (£3 18s per annum), just below the rent paid for only two rooms in Swansea town. Trevivian houses also, of course, had their long gardens with each pig sty privy and coal hole, the Vivians and other coppermasters also provided some free coal.[98]

In 1849 yearly rents in Trevivian for the five-roomed cottages recently completed near the schools were put at £5 4s (i.e. 2s per week). G.T. Clark, in his Public Health Report on Swansea, noted:

> 'Each house contained two rooms below and three above, and the windows were made to open. Nothing could be cleaner or more comfortable. For each house the tenant paid 2s weekly, a sum not so large as I found elsewhere

to be often charged for tenements of the worst description, and yet these improved cottages are stated to yield a fair return for their cost. The road in front of each row was neatly formed, and the footway pitched and flagged.'

Marginally higher rents were charged for the houses built by the Grenfells for their copperworkers on the east side of the River Tawe at Foxhole and Grenfelltown. Here a weekly rent of 6d a room, was charged in 1847 and 2s a week was charged for the usual 4-roomed Grenfell cottage, 25% more than may have been charged by the Vivians for similar houses.

Comparative housing stocks in the two main metal-smelting industries of south Wales.

Very few total stocks of workmen's houses built by individual companies in Wales have been surveyed; the houses of the Blaenavon Ironworks Company and the textile area of Penygloddfa (Newtown) are two exceptions.[99] The results of the former survey have been analysed in detail and are worth comparing with the fullest housing stock surveys undertaken for the present study in Swansea. Significant differences are the larger scale of the Blaenavon housing in what was a more remote location, reflecting the comparative size of the works in the iron and copper-smelting industries.

At Swansea the largest copper industry housing stock, that of the Vivians, survives remarkably intact. There have been very few losses and the wholesale clearances proposed before World War II never materialised, improvement schemes being adopted in the post-war period instead. Three main types of Hafod copperworkers houses have been recorded, types that constituted at least 74% of the total housing stock built by Vivian & Sons before 1877 (195 houses out of 263). The recording rate of the three main types of the houses provided for the Middle and Upper Bank copperworkers is even higher. Of the 83 cottages owned by Grenfell & Sons in 1892, 67 survive and representatives of each type have been recorded, 81% of the total housing stock at that date. Data on a significant proportion of the earlier, eighteenth-century housing stock provided by Lockwood, Morris & Co. have also been recoverable. Many of the plans of 25 of 42 dwellings built by the company near Landore has been reconstructed: that is, 60% of the total housing-stock built by the company and 100% of the dwellings known to have been built by the company to four apparently different designs. In addition, it has been possible to determine the form of the design prepared for the company in 1779 for

their Forest copperworkers, to build themselves on land leased from the company. It has also been possible to reconstruct 100% of the ironworkers' accommodation at the Parsons's Upper Clydach Forge, built in c.1784.

The Blaenavon partnership of 1788, Hopkins, Hill & Pratt, differed from the Swansea copper entrepreneurs in being compelled, by the much greater remoteness of their works, to use significant parts of their initially scarce capital to build large numbers of workers' houses from the start of their industrial activities. No less than 64 (64%) of their initial housing stock built in the period 1788-1800 had a floor area of only 240sq ft. Even in the eighteenth century workers' houses in the Swansea area were much larger, though much fewer in numbers. The two colliers' houses built by Lockwood, Morris & Co. at the Tirdeunaw Coalshaft in 1758-67 were 486 sq ft; i.e. more than twice as large, although some 140sq ft of this was almost certainly sleeping-loft. The 24 colliers' dwellings built at Morris Castle in 1763 were also generous in size. The eight apartments in the lower two storeys of the corner towers would have had an area of at least 441sq ft, whilst the 12 dwellings in the three storeys of apartments in the 'curtain wall' between the towers would have been some 400 sq ft. Presumably, the head colliers lived in the four flats that seem to have existed in the top two floors of the corner towers, which had an area of no less than 924.5 sq ft. It was only the key workers at Blaenavon who had houses of similar dimensions to those built for ordinary colliers in the Swansea area; in the period 1788-92 there were four houses of 450-500sq ft built and five were the largest at sizes above 500sq ft.[100] Lockwood, Morris & Co. also specified generous dimensions for houses to be built by the workers themselves on ground laid out for the township of Morriston: the surviving early houses there have dimensions of 552 sq ft - 634.5 sq ft, and 43 of these model dwellings had been constructed by 1790. Similarly, the houses provided in the Swansea area for the ironworkers of Clydach Upper Forge c.1784 were mostly a substantial 480 sq ft. The vast majority of the Blaenavon eighteenth-century houses consisted of two rooms, while the much more spacious Swansea workers' houses seem to have generally been of four rooms, allowing the sleeping accommodation of parents and children to be separate.

The new houses built both in Blaenavon and Swansea grew in size during the nineteenth century and it is instructive to compare what was happening in the workers' townships of the two industries by the late 1830s. There were 97 new houses built at Blaenavon in the period 1828-39, 37 of 400-450sq ft and 60 of 450-500 sq ft: about half had three or four rooms. The size of houses had grown substantially and the more adequately financed company could now build houses that approximated more closely to those provided for workers at Swansea.

The 46 houses built by Vivian & Sons for their copperworkers at Trevivian in Swansea c.1837-39 were marginally larger than those, their Blaenavon counterparts of 518.8 sq ft, with four rooms that allowed a separation of the bedrooms of parents and children in all the houses. The only qualification of this more generous provision of space at Swansea was that the children's accommodation was provided in the contemporary equivalent of the sleeping loft, 431.3 sq ft of the house being at full height and most of the rear bedroom being under the sloping roof of the rear 'catslide'.

The fairly generous workers' housing provision at Swansea continued to increase in individual house size through the middle and later nineteenth century. There are problems in dating the first copperworkers' houses provided by Grenfell & Sons at Swansea. It is known from elsewhere that workers' houses could be rebuilt within the original shells on up to two occasions: over half of the smallest early housing stock at Blaenavon was rebuilt about 1810.[101] It appears that there was rebuilding of the early Grenfelltown houses, but a standard Grenfell & Sons worker's house size is apparent both from the four houses built on the north side of Pentre-chwyth Road (nos. 29-32, apparently rebuilt from a building already on the site c.1799-1812) and from the 55 houses in the three contiguous rows of Grenfelltown, also first constructed in the period 1799-1812. These 59 houses of standard dimensions were of a fairly generous floor plan of 555 sq ft which may well date from before 1812, even if the roofs and fenestration of the dwellings have been altered. The degree of separation in the houses, with the earlier surviving roofs situated on Pentre-chwyth Road, is the same as in the earlier Trevivian houses (i.e. with four rooms allowing the separation of parents' and children's bedrooms) but with the whole house built to a full height.

The second stage of housing at Trevivian is more closely documented and datable than that at Grenfelltown. Here the 119 houses built, in and around Bowen Street, for copperworkers between c.1851 and 1867 have a more generous floor area of 589.4 sq ft. Four full-height bedrooms were provided.

The third type of housing exhibited by 30 houses built at Trevivian in the period 1867-77 is more generous again in floor area, height and the degree of separation between the generations. A floor area of 606.9 sq ft allowed the creation of five rooms and the separation of the female and male children. This was the first of the Trevivian houses to have a completely full-height upper floor. The front walls of the houses at Trevivian had risen to 16.5 ft over 30 or 40 years

from an original height of 14.7 ft (the second type of Trevivian houses had walls 15 ft in height).

Very similar trends were visible on the, eastern side of the valley at Grenfelltown. The height of the front wall of the four houses that were originally built on the west side of Pentrechwyth Road between 1799 and 1813 is 13.6 ft, compared to the 14.7 ft of the first Trevivian houses built c.1837-41. The 55 houses of the three contiguous rows at Grenfelltown, although originally built between 1799 and 1813, have obviously had their roofs rebuilt and their walls heightened. The closeness of their wall heights of 16.6 ft to the 16.5 ft high walls of the second of the Trevivian house types, built between 1867 and 1877, suggests that these Grenfelltown houses were rebuilt to a similar size in the same period and certainly well before the sale of the company in 1892. Like the third type of houses built at Trevivian, these have five rooms, allowing the complete separation of the sleeping quarters of parents and both sexes of children.

A third type of Grenfelltown house was built, lining the road access to the Pentrechwyth Road, between c.1877 and 1892 to exactly the same plan-form as the three rebuilt long rows there. However, there are important modifications to the structure of these houses which suggest the increased status of ordinary workers. For the first time in a copperworkers' settlement, houses were built with first-floor windows to the same height as ground-floor openings. Interestingly, this architectural device was already in use 40 years earlier, when the workers' houses of 5-14 Princess Street were built in Swansea town and when the Hafod Copperworks manager's house was built in the same period. The third type of workers' houses built at Trevivian in 1867-77 did not have first-floor windows as large as those on its ground-floor. However, both groups of houses were the first in their respective settlements to have completely full-height ceilings on their upper floors.

At Blaenavon there is evidence for the construction of houses in neighbouring parishes in the late eighteenth and early nineteenth centuries in order to avoid additional poor-rate costs falling on the manufacturer.[102] At Swansea there is no evidence of this, but rather of coppermasters, such as the Vivians, being concerned that the authorities in Swansea town were moving poor rate dependants into the parish of Llangyfelach to be dependent on the rates raised from the copperworks.

The large-scale anthracite-fuelled ironworks that developed from the 1830s in the upper Swansea Valley used the two approaches to provide workers' housing that had long been used by the coppermasters in the lower valley. George Crane of

fig.221

Fig. 221. A surviving three-storey block of workers' housing at Kelston copper and brass mills. The earlier copper-industry centre at Bristol was building blocks of housing for its workers on relatively remote sites before most of the smelting companies re-located to Swansea. (960176/34)

the Ynysgedwyn Ironworks provided terraced rows for his workers' housing, whilst James Palmer Budd (whose father had been one of the managers at the Hafod Copperworks) of Ystalyfera Ironworks sub-let building plots.

The rows of ironworkers' housing that were laid out across hillsides, without vehicular access, by the ironmasters in the fairly remote Heads of the Valleys region, never became a strong feature of the larger industrial settlements at Swansea. The three staggered rows at Grenfelltown are the only large, formalised group built by the coppermasters, although the two adjacent 'barracks' blocks at the Birmingham Copperworks and the multiple rows built by the workers themselves across the hillside at Foxhole, at least superficially resemble the characteristic terraced rows of the Heads of the Valleys region. Generally, rows such as the large formalised layout at Grenfelltown had to be developed in sequence and this also indicates that the entrepreneur intended to be the house builder from the outset.[103]

Copperworkers Housing in other industrial centres

As the Swansea copper industry was closely linked with other copper-smelting and manufacturing centres it is worthwhile comparing the workers' housing built in those areas to that at Swansea. As already noted, Robert Morris II had seen copperworkers' housing built by a copper-working concern (Roe & Co) at Macclesfield a year before his brother is known to have built houses near the Forest Copperworks in 1775 and five years before the laying out of Morriston.

The early eighteenth-century Swansea copper industry was derived from that at Bristol and remained closely linked to it for many years. It is instructive to examine the housing policy of one of the Bristol area's largest copper-smelting companies, that of William Champion. Its manufacturing sites, like those of Swansea, were established in rural locations on a coalfield and/or in proximity to available water-power resources. Champion had built three-storey blocks of housing 'for the Mechanics Families &c' between 1748 and 1761 at each of his main copper-smelting and rolling sites, although these housed only the more skilled of his 2,000 workers.[104] The now demolished 'Row' at the Warmley smelter and mills formed part of the '25 houses and tenements' provided on the main smelting site. The similar terrace at his isolated Kelston Coppermills on the River Avon is still inhabited. The largest Bristol copper producer, the Bristol Brass Company, took over the Warmley Works in 1762 and in 1790 switched all its copper-smelting operations to the Forest Copperworks at Swansea.

The provision made for the workers of the Bristol-based John Freeman and Company, which ran the White Rock Copperworks, was not as extensive as that some of the other companies at Swansea. The company obviously had some philanthropic concern, for in the 1860s and 1870s it was paying monthly pensions of between eight and 18 shillings to at least seven retired copper-rollers from the old water-powered mills at Publow on the River Avon between Bristol and Bath.[105]

By the end of the eighteenth century, two other copper-working concerns (besides that of William Champion) were already housing their workers in terraces. Thomas Williams of Anglesey, 'the Copper King', was a leading partner in the first of the Banks Copperworks from 1778 and in the second from

1785. Williams came to run both of the two Anglesey copper-mining companies, each of which had a smelting-works near St Helen's in Lancashire, operated by many experienced copperworkers from the earlier south Wales works. The Ravenshead Copperworks, set up at St Helen's by Williams in 1780, was one of the largest in the country, with 48 reverberatory furnaces. Key workers, from among the 200-300 employed, were housed in the 16 cottages of 'Welsh Row'. William Morgan from Margam Copperworks was appointed manager of the largely English workforce, and in addition to his £60 annual salary he was given a rent-free house, free coal and candles and the keeping of a cow.[106]

Thomas Williams also built a large house alongside his Thames-side copper-rolling Temple Mills at Marlow, (which was formerly owned by Townsend, builder of the Middle and Upper Bank Works, and Pengree, and was purchased in 1788) which were subsequently taken over by Pascoe Grenfell, the major Swansea coppermaster who had originally been Williams's employee. The terraces of brick-built copperworkers' houses lining the Thames at Temple Mills still survive and have been converted into fashionable housing.

Pascoe Grenfell was a polyglot Cornish copper merchant who advanced the trade in Williams's copper goods throughout Europe. He was also a partner in the Greenfield Company (a subsidiary of Williams's other concerns) which in 1785 took over Thomas Patten's Battery Works (latterly Newton, Lyon & Co's works) in the Greenfield Valley and greatly increased the capacity of its water-storage in order to drive the enlarged works that was then provided with 35 houses for the workers required to operate it. These houses were largely sited in the very long terrace of 'Battery Row' (almost 500 ft long) which was built along the side of the valley overlooking the works, reservoir and was demolished in the 1960s, some foundations are still visible. These were spacious double-depth (25 ft deep), double-fronted houses with two windows on each floor flanking a central doorway. Both floors had three-pane, high-casement windows capped by segmental arches with the roof coming down to the top of the upper arches.[107] They were on a very constricted site between the valley road, the works' access and the works' reservoir, with little or no room for any gardens to be provided.

The proprietors of the 'Birmingham Copperworks' (in Swansea) were copper and brass manufacturers of that city. They had also built two rows of 'Barracks' for their workers in the 1790s on the west side of the Swansea Canal. Matthew Boulton was a principal partner in the Birmingham and Rose Copperworks and had already built a row of cottages for key workers as part of his model Soho Manufactory (ornament and jewellery works and mint) founded in fields at Handsworth near Birmingham.

Nearer at hand in the copper-smelting area of south-west Wales, there were other examples of industrialists providing land for their workforce to build on. The Nevilles of Llanelli, for example, provided land at a moderate long-term rent for their coppermen and colliers to build their own houses.[108]

Comparisons with the tinplate industry

In the later nineteenth century the heart of the lower Swansea Valley was an area that became the international centre of a second metals industry. This was the tinplate trade, which initially required less capital than the copper industry and attracted entrepreneurs of relatively humble and local origin. The industry in the lower Swansea Valley became centred on Morriston, and to a lesser extent on Landore, both of which were substantial existing settlements based on the earlier local industries of copper-smelting and manufacturing, coal-mining and, to an extent, on iron-forging.

Circumstances had been different when the copper industry was established in rural areas lacking any accommodation for the workforce. However, there were exceptions when tinplate-works were established away from earlier copperworks in the north-eastern part of the lower Swansea Valley at Llansamlet. Here, the Aber Tinplate Works of the Foxhole Tinplate Company, (which had been built across the disused but still watered line of the Llansamlet, or Smith's, Canal), built cottages for their workers. The rents charged in 1902 were 4s a week for a cottage with three bedrooms and 2s 6d for a two bedroomed one.[109] These rents for this tinplate-workers' industrial housing seem quite high when compared with copper industry housing, even when inflation between the main periods of the two industries is taken into account.[110] At 1847 prices the tinplate industry rents worked-out at about 1s a room, whereas the Grenfells were charging 6d a room and the Vivians between 8-9d a room. Copper industry workers' housing rents therefore were between 25% and 50% cheaper than those of the later, probably undercapitalised, tinplate industry.

Industrialists' Motives in Building Workers' Housing

Industrial workers' housing was being built at Swansea by the early eighteenth century and what occurred was largely consistent with what happened elsewhere. Housing had always been built, and continued to be provided, to allow industry to function in rural areas with no large stock of existing accommodation. It is a mistake to see it as a purely philanthropic gesture.

The famous Welsh industrialist and visionary socialist, Robert Owen has often been seen as the founder of the workers community at New Lanark, when in fact it was his father-in-law, David Dale, who built nearly all the flats there (to house no fewer than 1,300 workers) in the late 1780s and 1790s, in order that Highlanders and others could come there to operate the largest-scale textile mills.[111] It is equally a fallacy to claim that the nine three-storeyed houses built at Cromford in 1771 by the earlier textile innovator Richard Arkwright (also one of the early partners at New Lanark) were the earliest planned industrial housing in the world,[112] when there is evidence from the lower Swansea Valley and elsewhere in Wales, for workers' housing built by employers in a much earlier period.

It has been concluded, from observations on the cotton industry, that there were two kinds of workers' housing provided by employers. 'Managerial' or necessary housing, provided by, or for, an industrial enterprise, consists of dwellings of a kind, or in positions, that would improve the productive efficiency of the business. Speculative, or 'entrepreneurial', housing was built with the aim of securing the maximum monetary return direct to the provider.[113] There is strong evidence that the workers' housing provided by the Swansea copper industry entrepreneurs was primarily built to provide accommodation to attract and retain workers necessary to operate the adjoining businesses and not for speculative purposes or for reasons of social control or philanthropy.

Conversely, Morris had commented in retrospect, on his intentions to help his workers to be responsible citizens. However, that was 40 years after the events of c.1779 and he had not been providing accommodation directly, but merely fulfilling a basic requirement of providing his workers with building land for an annual rent. By contrast, the Vivians provide much more evidence for their intentions in providing the fabric of houses on a large scale rather than simply leasing building land. This evidence survives as both contemporary written records and as buildings.

John Vivian, the Cornish founder of the Hafod Works and the first of the Vivians to invest in Swansea, had been a prime mover in the partnership that had built workers' houses for those employed in the company's copper-smelter at Hayle in Cornwall. That Cornish mining partnership had even built and provided an octagonal nonconformist chapel (from cast copper-slag blocks) for its workforce. In the light of that experience, when the Hafod Works was founded (in 1809), John Vivian advised that a 'house for the agent and a few cottages for the most valuable hands' were required or the company would 'stand a bad chance of getting and keeping the best men at Swansea'. However, the Hafod Works manager, William Morgan, did not consider this necessary and replied that 'the Hafod will command its full share of workmen' without them.[114] These were common and tenable views when the earlier relatively small-scale copperworks - and most remained relatively small - employed only a few specialists. Similar views had been expressed by Dr Lane when founding the first lower Swansea Valley copperworks in 1717.

In addition to this, John Vivian commented in 1810 that he looked for a return of 12% from any houses built 'in order to reimburse its costs in a reasonable time and keep it in repair'. In other words, what he intended with this fairly low annual rate of return was 'necessary' and not 'speculative housing'. This rate of return was within, but towards, the upper part of the range of charges made by colliery companies to their workers for housing in the Rhondda in 1908.[115] The Glamorgan Coal Company charged 16-17s a month for houses at Llwynypia valued at £80, that is an annual rate of return of 12-13%. Other houses of the same age valued at £180 only yielded £1 6s a month, an annual return of some 9%. Scarce capital resources were mostly taken-up in building these large, new 'model' copperworks such as the Hafod, or the Morfa, but some was set aside to house key workers at a rent that would pay for the costs of constructing the accommodation within about eight years.

If John Vivian's first intentions had been followed then, possibly only four dwellings would have been completed, for in 1812 Vivian was able to run the Hafod Works with a total workforce of 80 and with only five specialists (including the manager): two clerks, a refiner and his assistant. In fact, it looks as if William Morgan's contrary advice not to provide any housing was largely followed at first. However, a general manager's house was built before the 1830s and a building was also erected near the works' gates, called the barracks, that accommodated some workmen (and possibly their families as the Birmingham Copperworks 'Barracks' did).[116] It was either decided not to build houses for other key workers because Morgan's advice was taken, or alternatively that the workers in question were already satisfied with their existing dwellings, possibly in the nearby settlements of Swansea town, Landor and Pentre-estyll.

A copy of the 21-year contract signed in 1811, soon after the completion of his works, between John Henry Vivian and one of his five specialists (his refiner) survives and makes it clear that this particularly skilled and experienced worker, William Howell, lived in Swansea town and received the generous sum of 10 guineas a year in lieu of a house and garden.[117] That sum of 4s a week was equivalent to a rent for a house some 10 rooms in size for this one crucial worker (as already noted a five-room house in Trevivian in 1849 cost a 2s. a week rent); this sum was additional to his weekly wage of 30s., two pounds of candles a week, and one wey of coal a year.

John Henry Vivian seemed initially to lease building plots on the partnership's landholdings adjacent to the Hafod Copperworks to speculative builders, but then turned to developing a complete township at Trevivian or Vivian's Town in 1838-39. He was sensitive to local public concerns, being MP for Swansea from 1832 to 1855 and Sheriff of Glamorgan.[118] This local political involvement demanded the provision of decent housing after the Swansea cholera epidemic of 1831. Another factor was the bringing into production of the equally large adjoining Morfa Copperworks in 1835, which included a sizeable number of houses for workers. This must have increased competitive pressure for key and skilled workers considerably and there were hints that the prime motive for starting Trevivian was probably a mixture of self-interest in Trevivian's political career and the necessity of maintaining his skilled workforce in the face of greatly increased competition.

Interestingly, in September 1838 the Memorandum Book of William Jones, Manager at the Hafod Copperworks, noted:[119]

> Margam I believe there are upwards of 108 Cottages adjoining the Margam work & ab. 24 for Colliers.

This noting of the housing already provided for a copperworks' workforce arose from Vivian & Sons taking over the Tai-bach or Margam Copperworks of the 'Company of Copper Miners or English Copper Company.' It appears that all 73 of the then workforce (26 'underhire' [probably labourers]; 9 'coke'; 5 'fillers'; 4 'roasters'; 1 'refining'; 28 at the 8 ore furnaces and 6 calciners [14 of them boys]; 6 'masons, joiners & smiths' and 8 watchmen), except for the 14 boys employed, had accommodation provided by their employers in this more rural location. Jones then noted at the bottom of this list of employees that '73 Cottages would be required'. In a cryptic note written sideways is a further note that reads '70 cot 5 old would do. there are 14 cot', hinting at a decision to rationalise the workers' accommodation at Margam.

However, the fact that Vivian & Sons came into possession of a works, only eight miles east of Swansea, where more than the entire workforce was provided with housing by the management, must have increased pressure on them to make better provision for their Swansea workforce, especially in the light of the increased competition.

By 1841 the first phase of Trevivian was complete, with 46 dwellings housing 67 workers from the Hafod Copperworks workforce. From that delayed start (and that delay in constructing housing was common in the earlier copperworks), Trevivian became the most developed of any for the copperworks' workforce at Swansea.

As has been noted, G.T. Clark, the philanthropic and scholarly manager of what was then the world's largest ironworks at Dowlais, commented in his report on the public health of Swansea that at Trevivian: 'For each house the tenant paid 2s weekly, a sum not so large as I found elsewhere to be often charged for tenements of the worst description, and yet these improved cottages are stated to yield a fair return for their cost.'

Such a well-informed external commentator noted how comparatively cheap were the rents charged by the coppermaking concern of Vivian & Sons, and that these were still judged sufficient to refund the capital reserves of what was after all, primarily a business and not a philanthropic concern. Similar comments on the fairness of the large (and also resident) coppermaking concern to the north of the river were made by the local Welsh Independent Minister W. Samuel Williams, in a Welsh-language parish history that the industrialists would not have read. He said that Pascoe Grenfell and his family might be churchgoers but that they 'inspired parents and children by being strong in religiosity and care' in commenting on a concern that charged 20% more than the Vivians for each room in a worker's house.[120]

There is further evidence that cottages were built by the Vivians at their various works and mines because the need for the maintenance of a sufficient workforce demanded it and that further profit from house rents was not a primary motor for house-building activity. In 1854 an acute demand for new workers arose when the furnaces and the new mill at Margam Copperworks were made idle and Vivian & Sons decided that cottages were needed to aid recruitment.[121] This well-established, mature and adequately-funded company was also building houses for its workers whenever remoteness and a lack of existing accommodation demanded it. At Amlwch in Anglesey, for instance, a group of houses known as 'Pentre-Vivian' was built for the copper-miners in the local Vivian Mines near Parys Mountain.

Again remoteness of a particular mine or works' location could determine that the Vivians had to house a complete workforce. Later in the nineteenth century, when Henry Hussey Vivian opened and worked (1872-81) a nickel mine on the remote Norwegian island of Senja he equipped the newly settled community of up to 600 with about 50 buildings, including a shop, school and church and houses for the miners and their families.[122]

The Vivians are also the only coppermasters known to have actually built a nonconformist chapel for their workforce in the lower Swansea Valley when they built houses for their colliery workforce at Pant-lasau, one and a half miles north-west of Morriston.[123]

The Vivians' main workers' settlement at Trevivian underwent large-scale expansion towards the end of the nineteenth century in response to the continuing increase in their workforce, which peaked at 3,000 in their various mines and smelting-works in 1886.[124] The key core-staff in 1909 included 64 general management and 48 technical management staff.[125].

However, at the end of the century the Vivians were also involved in speculative housing. A.P. Vivian had become sole proprietor of the Aberavon estate in 1866 and was subsequently advised by T. Gray, a manager at Margam, that building houses on the estate and selling them to his men would be 'a very good speculation; nine new houses were built in 1896. In 1910 a further expenditure of over £3,000 was made and rentals of £1,300 were received in 1914.[126] This was very much the era of urban speculative building on a very large scale for those with the capital to do it. Vivian & Sons was then a huge mature company with interests all over Britain and abroad. Perhaps the last few phases of the Trevivian development should be viewed as approximating to nearby speculative estate development in an expanding Swansea (such as the adjoining much larger grid of 'Manselton' to the west), rather than as a continuation of the earlier policy of 'necessary housing' to maintain the workforce in an era of increased competition in the industry.

As elsewhere, it is probable that the construction of industrial housing by employers at Swansea was carried out as a matter of necessity in what were then rural locations. If the site was at a considerable distance from the urban settlement of Swansea, as at Margam, then all the workforce might be provided with housing. If the site was very near the existing town, as at the Hafod Copperworks, then there might be little incentive to provide housing immediately on a large scale. Unlike the situation in the remote iron-smelting area of the Heads of the Valleys region, there was generally no need to provide housing in an unpopulated area. By contrast, the late eighteenth-

and early nineteenth-century workers' housing had to be built as cheaply as possible in the isolated inland iron-making districts in order to reserve capital. It was only later that houses could be improved to sizes equivalent to those provided in the copper industry housing of the Swansea area.

It has been suggested, with reference to Swansea and the Vivians, that the provision of workers' housing with gardens 'was a subtle, yet potent way of exercising control over the labour force outside working hours as well as a means of enhancing the Vivians' reputation as philanthropists.[127] Indeed, research into the provision of colliers' houses has concluded that those colliery owners providing accommodation for their workers also had a more contented workforce. There is supporting evidence elsewhere in south Wales for the idea that the provision of better workers' housing might shield the workforce from the influence of militant tendencies. At Merthyr Tydfil the provision by ironmasters of larger and more extensive housing can be related both to the larger output of the various works and to outside events like the Chartist Rising and the march of 1839.[128]

However, there is strong evidence from elsewhere in south Wales that company housing in fact freed workers from the constraints of conforming to the wishes of their employers. In 1894, during a strike affecting the South Dunraven Colliery at Blaenrhondda, about 30 colliery company houses were found to be empty after two weeks of the dispute and 46 after a month, suggesting that miners and their families were ready to leave their tenancies to seek work elsewhere. This suggests that company tenants were less tied to their work and place of abode than house owners.[129]

It could also be argued that such larger scale housing provision was a natural product of the more concerned and paternalistic employers' attitudes, or that it was a natural inclination of employers seeking to maximise the efficiency of their business, especially where the place of manufacture was remote from existing settlements. Trevivian was constructed at the same time as the larger houses built at Merthyr, though the origins and design of the houses probably predate the Chartist disturbances. There is no evidence at Trevivian of an intention to use housing provision to pacify the workforce, although in combination with other factors, such as the provision of children's education, it would have contributed to producing a more contented workforce.

Indeed, the Vivians regarded their relations with their workers in the first three quarters of the nineteenth century as being generally good. But, there is little support for such a cause and effect after

1875, when strikes became more common and relations between the employer and workers at the Hafod Works deteriorated, despite the expansion of Trevivian with the building of Odo Street.[130] Yet by that date much more alternative housing was available and the reason for the expansion of Trevivian on such a large scale may have been much more speculative. The predominant factor causing this latter unrest may have been higher wage rates paid by the booming tinplate industry. There is no evidence for the extreme social control in the Swansea company settlements that was exercised by the patriarchal Robert Owen at New Lanark with his patrols organised to control drunkenness.

It is worth examining the motives of the second large provider of workers' accommodation in the Swansea area, Grenfell & Sons. Again there is much evidence of a generally benevolent attitude, of which the provision of 83 workers' cottages is one manifestation. The production of a fairly contented workforce is indicated by the fact that there were no strikes or lock outs at the Bank works in the period of management of Pascoe St. Leger Grenfell (*c*.1844-79).[131]

The evidence is that most of the copperworking concerns were inclined to build houses for some 40% of their workforces in the late eighteenth and early nineteenth centuries. A strong local tradition in building has shown that many of the other workers were themselves inclined, or needed, to build simple single-storied rubble-walled cottages with thatched roofs. At Morris Town after *c*.1779 it seems that workers were themselves building two-storey houses and certainly the stonemason William Edwards built several for his own family there. By the mid nineteenth century there is evidence that workers at the other Glamorgan metal-smelting metropolis of Merthyr Tydfil were commonly building or buying their own houses without recourse to building-clubs or societies, so this tradition was not restricted to the lower Swansea Valley, even if contemporary commentators noted how common it was in the area.[132] It has been noted how the first single houses on large plots multiplied so that the cumulative effect was eventually to form continuous terraces along roadsides. It has also been noted that, more generally in south Wales, colliers and ironworkers, by the mid-nineteenth century, would often build two houses with a small mortgage so that one dwelling could be lived in by the builder and the other rented, so that the proceeds could help to pay off the total mortgage.[133] What seem to be leases taken from the White Rock Copperworks proprietors in the 1850s for building land for two or three properties almost certainly refers to this kind of development. Terraces built in this incremental way lacked the uniformity of those constructed by industrialists, building clubs or speculators.

The many chapels also built by the workers themselves provided the focus for the many building clubs that were formed in the industrial settlements of the lower Swansea Valley. By such means those who could not raise the money to build their own houses became householders by forming small building societies in which the members pooled their savings and sometimes their spare time too. The precise origin of such clubs among the industrial communities of Britain is not clear, there are known to have been terminating building societies in the Midlands from *c*.1775 and building clubs in Lancashire from the late 1780s.[134] It has been suggested that the clubs spread as workers moved from Yorkshire to south Wales.[135] The earliest such building club or society recorded in south Wales is the Pent-wyn Benefit Building Society, founded in 1838 at Abersychan near the iron-smelting works of the Monmouthshire Heads of the Valleys area.[136] Its members built 30 four-room houses with rather irregular internal partitions, perhaps showing evidence of the work of less-skilled society members. Terminating societies, like this, were dissolved when all the houses had been finished and paid for. Building clubs were particularly active during upswings in the economic cycle when money could be borrowed more easily.[137] Building clubs in the south Wales coalfield as a whole were very active for a short period in the late 1850s and at intervals until the 1880s. This was when trends for housing finance in the industrial communities of the south Wales coalfield and those of northern England led to the establishment of *permanent* building societies (and offices were also established in Swansea and south Wales); the 1874 Act that made balloting for advances illegal stopped their use in the north whereas they flourished in south Wales from the 1880s until the end of the great steam-coal boom just before the beginning of the First World War.[138]

Such clubs in the close-knit communities centred on the nonconformist chapels of the area were responsible for building more than a quarter of the workers' housing of the central valleys of Glamorgan after 1878. [139] In south Wales they attracted a cross-section of such communities, but some of the wealthier club members might have had between two and eight shares in the terraced houses and could rent them out to other workers in the industrial settlements.[140] Speculative builders were largely restricted to the larger urban areas such as Swansea and Cardiff towns.[141] Houses were often built in terraced units of 20 or more for sale, except in times of economic downturn when they might be rented. Some small-scale building of this type was carried out during the nineteenth century in the small semi-rural industrial communities of south Wales, in such places as the lower Swansea Valley and the Rhondda.[142] Such housing was one of the factors which initially allowed the Vivians to think that the housing needs of their workers were already being met. The two short terraces built by the 1840s, along the Neath Road on the Vivian's land at Hafod, may have built by the landlord of a public house attached to one block of houses.

Small-scale private landlords have earlier been noted alongside those who built additional houses on their large building plots. Very large-scale private landlords seem to have been uncommon in the industrial settlements and this was a situation common elsewhere in south Wales where it was usual for a landlord to have just three or four houses.[143]

In the lower Swansea Valley some accommodation already existed in Swansea town, but this was not the case in remote new industrial areas, such as Blaenavon and the Rhondda, where industrialists had to make some effort to house most, if not all, of their workers in order to operate industries in previously rural upland areas. As each of these industries matured, there was much greater conformity over the whole of industrial south Wales, in that other sources of accommodation in growing semi-urban areas could provide shelter for the majority of workers. In the coalmining area it has been pointed out that, in this respect south Wales differed from the Scottish and north-east of England coalfields.[144] This may be because south Wales, in the nineteenth and twentieth centuries, had great wealth generated by being the world centre of no less than three metals' industries and also by being the greatest international coal-exporting area.

Conclusions concerning workers' housing

The large-scale provision of workers' housing in the lower Swansea Valley had already been pioneered before Swansea became a centre of the copper industry, with the housing being provided for colliers in the lower eastern valley. Both types of housing provision were prompted by the presence of the underlying coastal coal deposits attracting works which needed operating by workers in what was a sparsely populated rural area.

Over a period of 150 years, when coppermasters provided housing for their workforce, some factors seem to have remained surprisingly constant. It is likely, in the later eighteenth and the nineteenth century, that all copper-smelting concerns accommodated about a third of their workforce in purpose-built housing. The housing that the copper-smelting concerns constructed generally compared well with the insanitary courts and dwellings built for workers in the northern part of Swansea town. Most rentals were lower and substantial gardens were provided in most copper industry housing. By comparison, little housing was provided for the dispersed, shorter-lived and smaller-scale centres of employment provided by later eighteenth and-early nineteenth-century collieries.

The coppermasters did take advantage of their capital resources, and even more the large amounts of capital generated by the industry in its heyday, to provide a substantial amount of housing which is still good enough to be occupied. As far as is known, nothing was produced that was as small, or built of temporary materials or as short lived as the timber houses built for the early workforce on the Severn Tunnel at Sudbrook. This was true also of the timber houses built by local under-capitalised entrepreneurs in the early days of the Rhondda coal industry (for Fernhill and Dunraven Collieries, Blaenrhondda and the 22 wooden dwellings at Baptist Square and Mountain Row, Blaenllechau, by the innovative entrepreneur David Davies and his partners) and the turf-walled houses for workers employed on the early railways in the late 1820s of the Brecon Forest.[145]

There were identifiable types of Swansea houses built by the coppermasters. Long-term specialist recording of industrial housing in the more isolated iron-smelting areas at the head of the east Glamorgan and west Monmouthshire valleys has shown that there, a primary reason for the type of housing was the policy of the company in undertaking the building and not usually the building type predominant in the locality before industrialisation.[146] The examination of housing built at Swansea shows some exceptions to this rule, in that the early and mid-eighteenth-century single-storeyed colliers' houses built by Price and the Lockwood, Morris & Co. partnership were obvious derivatives from the west Wales *croglofft* cottages already being used by the local collier builders on roadside verges and by squatters on the plentiful commonland.

Later, the original 48 catslide-roofed houses built by the Vivians in *c*.1838 were modelled on pre-existing housing types in the town. Even so, both companies almost immediately developed distinctive types of their own, as did all local copper-smelting concerns in the immediate Swansea area.

However, the development of workers' housing in the lower Swansea Valley was not entirely a consistent evolution in type. There were all types of accommodation existing at any one time, including isolated single-storey cottages (with sleeping-lofts)

- single-storey terraces - detached two-storey houses - two storey terraces and three-to-four-storey apartment blocks. There were also diverse approaches which other Swansea industrial entrepreneur might adopt in providing houses for his workforce. A works' owner could convert existing structures to provide habitations for his workforce, build houses for a varying core of his workforce directly, or else provide land for his workers themselves to build on.

Planned Settlements for the Workers

What has been considered so far has been the individual workers' dwellings and terraces and not the assemblages of some planned and partly planned settlements, of which they formed a part. The earlier coal industry, generally, was on a small dispersed scale except perhaps for Mansel's Collieries at Llansamlet. However, Mansel's colliers houses consisted, as far it is possible to tell, of cottages sited in fairly irregular plots along winding pre-existing roadsides at Bon-y-maen (Mansel and Cefn Roads), rather than in the regular grids found later at Morriston and Trevivian.

The founding of the Forest Copperworks some two miles to the north of Swansea prompted a development on an altogether more ambitious scale: the laying-out of a completely new settlement, arranged on two adjacent grids of streets and provided with a church and market. This may be one of the earliest metalworkers' settlements founded in the world's first Industrial Revolution. The workers from two other adjacent late eighteenth-century copperworks (Rose and Birmingham) for the most part also lived in the same town, and one of the works provided a school for the children of employees of the three nearby works.

Another full-scale settlement with a grid of streets was not laid out for another 60 years, with the construction of Trevivian. This was in open country nearer the town and its construction may have been prompted by concerns that sub-standard housing built by speculators during a time of cholera fears, would have produced circumstances severely damaging to coppermasters who were also MPs.

Morriston

The construction of Morris Castle by John Morris I in 1773 formed an 'eyecatcher' for his new grand mansion at Clasemont and for a mixed 'industrial picturesque' landscape that the Morrises were developing in the lower western Swansea Valley. Subsequently, 'Morris Town', or Morriston as it has become known, was laid out on the valley floor

below the parkland setting of Clasemont and visible from the banqueting or summer houses dispersed around the valleyside.

John Morris I's determination to make more adequate housing provision for his workers may have been inspired by his elder brother Robert's visit to the Warrington Copperworks of Roe & Co in May 1774. He noted that the works had originally stood on the outskirts of the town, but having been 'convicted of nuisance was compelled to be removed' a mile from the town. He then noted that 'Close to the Copper-works, there is [a] neat Row of small houses, to the number, as far as I can now guess, of about 12 or 14 for ye workmen. There did not seem to be above 20 workmen employ'd at a Time in the whole works.'[147]

Fig. 222. The first building on the site of Morriston was the seventeenth-century squatter housing on Graig Trewyddfa Common. The three semi-detached houses to the west of Forest Copperworks followed next, and then the smaller grid of houses on the south-east of Morriston proper. The grid to the north was developed in stages throughout the nineteenth century and subsequently sub-divided and infilled.

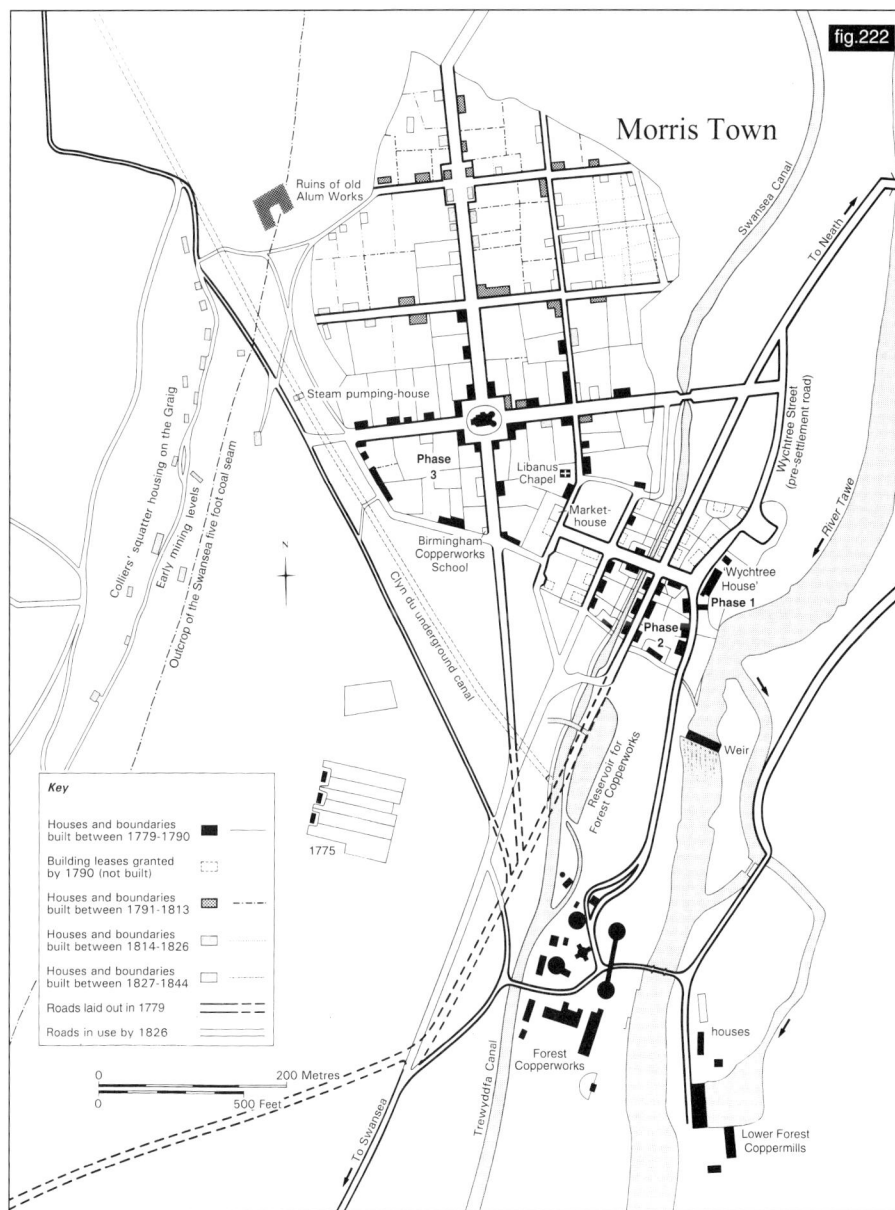

199

In the following year, John Morris I built three small groups of houses on the commonland on which his industrial installations were also sited, but even he may have been hesitant about building a full township on 'waste', even with the consent of the powerful lord of the Fee of Trewyddfa, the Duke of Beaufort. By the 1760s and 1770s, John Morris I, as a full and resident managing partner of Lockwood, Morris & Co., was deriving huge profits from the business and was acquiring vast tracts of land over the mineral deposits of the lower Swansea Valley and west Swansea.[148] He asked the engineer/architect William Edwards to begin laying out the township on four fields of his farm on the north side of Trewyddfa Common, about a year after the latter had built the Wychtree Bridge (with Morris supporting the new turnpike) across the River Tawe to its immediate south west in 1778.

A large building of workers' flats or 'barracks', not dissimilar in function to Morris Castle (1772-73), may have been the first building erected on the site, on the river side of the old road leading from the earlier Forest Bridge (Wychtee St) and on the part of the site nearest to the Forest Copperworks. It was later known as 'Wychtree House' and had lower storeys towards the river of cast copper-slag blocks, but a symmetrical facade with a central pediment towards the road suggesting it may have been designed by the architect John Johnson, who was very likely also the architect of both Morris Castle and Clasemont. Its asymmetrical siting in relation to the smaller and earlier grid of Morris Town suggests it was built before this was laid out.

This building has sometimes been called the 'Hostel' and may have been the 'Poor House' that William Edwards supposedly prompted John Morris to build at Morriston.[149] However, its lower two storeys towards the river were recorded and certainly did not form communal accommodation.[150] This large rectangular building lay on a steeply sloping site between Wychtree Street and the River Tawe. Seven windowed bays in its two upper storeys faced westwards onto Wychtree Street, with a third attic storey in the central three-bay pediment lit by a lunette window. The three openings on this side were accessible by bridges over a passageway which admitted light to the two storeys situated below street level. Four full storeys (with much cast copper-slag block construction) gave onto the eastern elevation, with access at the lower level. The wide double door in the centre of the west wall was for pedestrians, but the north door was much wider and could have provided access for coaches.

Edwards laid out the adjoining township, to the immediate north of Morris's Forest Copperworks, in the form of two adjoining plots infilled with interconnected streets on a gridiron plan. In 1819 John Morris I wrote down what he had intended in the laying out of this model settlement:

'About 40 years ago I appropriated a farm for the purpose of inducing the artificers and labourers of the County, to build thereon giving them a long term of years, at a nominal acknowledgement with a sufficient plot laid on, for raising Potatoes. The scheme at the time was thought a visionary one, by all around me; but I have lived to see, about 300 stone cottages, with tiled Roofs, built by this class of persons, & I need not add that such persons never become burthensome to their Parish but continue to hold out the best example of prudence to their neighbours.'[151]

Walter Davies wrote in 1815 that:[152]

'...he laid the foundation of Morriston, so called from the founder's surname; where dwellings have been erected for colliers and manufacturers, in well-formed and spacious streets: with a church, containing an organ for such of his workmen as prefer the established religion. This chapel of ease to the church of St. John's in Swansea he endowed. Being somewhat of a latitudinarian he erected also a chapel for non-conformists. His architect, William Edwards, the celebrated bridge builder, being himself a preacher, to distinguish between heterodoxy and orthodoxy, placed the protuberance of a tower at the west-end of the church, and at the east-end of the chapel, which is elsewhere considered as a badge of distinction.

The founder of Morriston's plan was to grant ground leases on plots of land, about a square rood each, for three lives, or 50 years, at 7s 6d a year each plot. The lessees were to build according to prescribed plans. He moreover encouraged some of the more steady labourers, and others, with a quantity of land each, sufficient to keep a cow. Only two or three thatched cottages once occupied where this neat little town, containing a population of about 1,100, now stands, a prominent feature of increasing wealth and comfort. Mr. Fox, in the year 1796, says, that in the year 1780, there was not a single house where Morriston now stands: that in 1796 there were 141 houses inhabited by 619 persons; so that in the last 16 years, the population has nearly doubled.'

Morriston was the earliest and largest of the planned settlements built by the copper magnates. The most remarkable survival is the street layout of two adjacent gridirons, with each laid out on slightly different alignments resulting from the use of different fields for their construction.

The small size of the garden plots on the smaller grid (filling just one field) nearer the river, its proximity to the Forest Copperworks, the placing of the first (nonconformist) religious building in its grid (1782) and the relationship of the larger, north-west grid to it, all suggest that the smaller riverside grid was laid out and developed first. It included part of the new straight main road to the Wychtree Bridge (1778) and Neath. The winding first main road, now known as Wychtree Street, originally followed the river bank to the Forest Bridge and subsequently became a quiet back road.

The smaller grid filled the first of four fields that Morris's town eventually covered. The new main road from Wychtree Bridge was laid down to run north to south, to the west of the earlier road and roughly parallel to it. Two roads of the new settlement, now Morris Street (its name probably indicating primacy) and Market Street, in turn ran parallel to this road on its north-western side. The centre of these parallel roads was bisected by the east-west axis of Castle and Globe Streets. The housing layout on the northern part of the grid (i.e. the further side from the Forest Copperworks) was never infilled to the regular layout planned (and shown projected on a map prepared for Morris in 1793).[153] In fact the first grid was partly destroyed by the Swansea Canal being constructed through the centre in 1794-96, largely removing three gardens and the projected sites for three more. This was despite the first proposed line and width of the canal being modified at the insistence of John Morris, to minimalise damage at this point. This

section of the Swansea Canal line, known as 'The Narrows', was consequently always a cause of congestion. As part of the Trewyddfa Canal, it was built and remained under the direct management of Lockwood, Morris & Co. and the Duke of Beaufort. By 1793 only the two blocks of this first grid nearest the Forest Copperworks had been built up on with houses.[154]

The second grid of Morriston was then laid out in three fields to the north, with roughly rectangular blocks modified in shape where it abutted against the smaller grid and also against former boundaries abutting common and field boundaries. Gardens three times as large as those on the first grid were laid out and the Anglican Church of 1787 added at its centre.

Most of the early housing has disappeared, although the market and rebuilt chapels and churches remain. The main street (now on the second grid and known as Woodville Street) was dominated by Tabernacl, which replaced Libanus, the Independent chapel built by William Edwards in 1782 (on the first grid). A map drawn for John Morris I in 1793 shows a second square laid out along Woodville Street (to the north of the existing square) on the larger grid.[155] The centre of the square may have been the original site proposed for the market. Whatever happened, the market was subsequently moved to the castellated market hall built by Sir John Morris II on the first grid, and by 1828 the site of the second square had been realigned to form extra housing plots.[156]

By the standards of the eighteenth century, a planned street layout with large garden plots and a religious building provided in a rural setting,

fig.223

Fig. 223. The architectural focus of Morriston was the coppermaster's church in the middle of the central square. The present church replacing William Edwards's smaller design lacks the western aisle and is dominated by the great steeple of John Humphrey's Tabernacl Chapel. (9500356/4)

fig.224

Fig. 224. Morriston may have originally had an open-air market, but in 1827, Sir John Morris II provided the covered area illustrated, which has four openings towards the street. The large room in the upper floor over the market was used for a dame school and the early Wesleyan congregations. (9500352/4)

represented an enlightened scheme of housing. However, by the middle of the nineteenth century the settlement had become much more densely developed and issues of public health were coming to the fore. At this time it was noted that many of the roads were unpaved, sanitation was lacking and people drew water where they could; but in fact conditions were no better in Swansea town itself.[157]

The Old Market on Market Street was built on the site of two small cottages to the east of the church square which had been converted into an Independent chapel in the years 1795-8. The Old Market was built as a communal structure of some pretension in the Morris tradition of ornamental castellation; four large arched openings on the ground floor of the Pennant sandstone building are capped by an upper floor with three sash windows and retains wooden gothic tracery; this housed a 'twopenny dole school', i.e., a charity school. The block is flanked by two small towers with gothic windows and was originally surmounted by battlements.

Several houses (besides the two surveyed in Morfydd and Woodfield Streets) at the southern end of Market Street may be of early construction, as is a small shop on the corner of Banwell Street.

Foxhole

The main area of settlement for Swansea copperworkers east of the River Tawe was to the south of the Middle Bank and White Rock Copperworks, where the narrow neck of the Swansea Valley widens out towards the sea at Foxhole. The long terrace of houses alongside the main road at the foot of the steep slope of Kilvey Hill was in existence by 1813. It was mainly built on land that was owned by the Grenfells, owners of the Banks Works by the time of the tithe survey; the Freemans of White Rock Works were only in possession of the 'upland waste' of Kilvey Hill itself, above the level of Kilvey Hill Road.

There was a larger workforce that needed new housing in the copper-smelting area on the east side of the lowermost Swansea Valley than on the west. These hundreds of people could not easily live in Swansea town, which was only accessible across a bridgeless three miles of tidal river. The only link was the White Rock Ferry plying from the south end of the works. Neither could they build squatters' cottages on nearby commonland. In these circumstances, it was always likely that they would develop roadside cottages of the sort that were such a prominent feature of the western valley side, between Landore and the site of what later became

Morriston. The growth of a large local settlement at Foxhole was assured in these circumstances.

Building land was at a premium at the foot of the steep slope of Kilvey Hill and two-storeyed houses, without gardens, seem to have been common here, rather than the cottages in large plots that were characteristic of the industrial settlements across the river to the west. The initial development of the Foxhole terraces was to the immediate south east of White Rock House and Cottages in an isolated location below the valley road and between Smith's Canal and the River Tawe. A small Baptist chapel (a denomination that never made an impact equal to the numerous Welsh Independent congregations), stood here. Later housing was on a higher level along the eastern side of the main valley road, on what was later the Grenfells' land at the foot of the steep slopes of Kilvey Hill, and consisted of two very long terraces of several different periods of construction. In 1841 there were some 50 copperworkers and 11 colliers living in these 77 houses along the main valley road.

Initially, long and steep rear gardens were only attached to the seaward end of the southerly terrace. These extended up the severe slope to Kilvey Road on the hillside above, along which the Kilvey schools and the Anglican church were later built by the Grenfells, rather than by the 'absentee' Freemans of the White Rock Works, although the latter gave the land for the first school. In the mid nineteenth century a series of terraces was built alongside the upper road and across the hill slope, above the northerly houses in the lower terrace. Those terraces below Kilvey Hill Road were built on land leased from the Grenfells in the 1840s and 1850s. Those built above were on the Freeman's land.

In total, the settlement at Foxhole seems to have consisted of a mixture of housing built by the copperworks' proprietors, with most houses built on by the copperworkers themselves on land leased by the companies to their workers.

About 1806, Pascoe Grenfell and Owen Williams, of the Middle and Upper Bank Copperworks, and John Freeman with his partners at White Rock Copperworks, jointly established schools on Kilvey Road (the first copperworks schools in Swansea) for the children of their respective workforces, these were sited on the hillside waste above this terrace and on Freeman's land.

In his evidence to the 1847 Inquiry into the State of Education in Wales, Pascoe St Leger Grenfell stated that 'The workmen's houses are generally pretty well built, but small, and deficient in the means of separation'. The houses which the Grenfells built had two bedrooms which were enough to separate the parents and children, and this was the standard of 'separation' also achieved in Trevivian.

The Grenfells' workers' houses at Pentre-chwyth and Foxhole were rented at the rate of 6d. a room and most had two rooms both upstairs and down. When the firm was offered for sale in 1892, they owned 83 cottages worth just under £100. If this total included the 55 houses at Pentre-chwyth, then there must have been only 28 Grenfell houses at Foxhole.

High on the hillside, to the south of the church and schools stood two rows of workers' houses. Freeman's Row was built on the east side of Kilvey Road just north of Jericho Road, between 1813 and 1826.[158] This row was clearly built by the Freeman Copper Company, which was the owner of the White Rock Works between 1764 and 1870-71. It consisted of a row of 10 houses ranged along the hillside with fairly short gardens, 30ft in length. Each of these dwellings was 21ft deep and 15ft 5in wide, slightly smaller than the houses of Trevivian and Grenfell Town which were on average 24ft deep. Belonging to this period also was a short row of three larger houses looking down the hillside from the junction of Kilvey Road and Jericho Road. These were 26ft deep, 18ft wide and had irregular gardens stretching between 35ft and 75ft to the road in front: these may have been for managers at the White Rock Copperworks. It is possible that they were built in the period 1804-26, when Owen Williams was a partner in the Middle and Upper Bank Copperworks, for the terrace was named Owen's Row. Another 13 houses were added between 1852 and 1877.[159] These were deep and narrow, 25ft by 13ft, and they had short back gardens 15ft to 40ft long; access was up a track which curved round the north of Freeman's Row. Facing the three original houses, on a short stretch of road linking Kilvey Road and Jericho Road, was another terrace of four smaller houses, 23ft deep and 16ft, these may have had no gardens at all.

All the houses in this interesting settlement were demolished as slums in the 1930s and a wooded slope now covers their site north of the suburb of St Thomas and below the surviving church and works' schools (of 1806 onward - now converted to housing north of the church, and the earliest school to the east now called the 'Gwyn Mission Hall'). John Humphrey's fine gothic revival school for girls was demolished in the 1970s.

Grenfelltown

This was a second, and later, settlement to the east of the river, situated three-quarters of a mile north-east of Foxhole. There was earlier housing already at Pentre-chwyth on this north-western flank of Kilvey Hill; it was scattered on the slopes above the Middle and Upper Bank Copperworks, where the road snaked up to the dispersed colliers' dwellings of

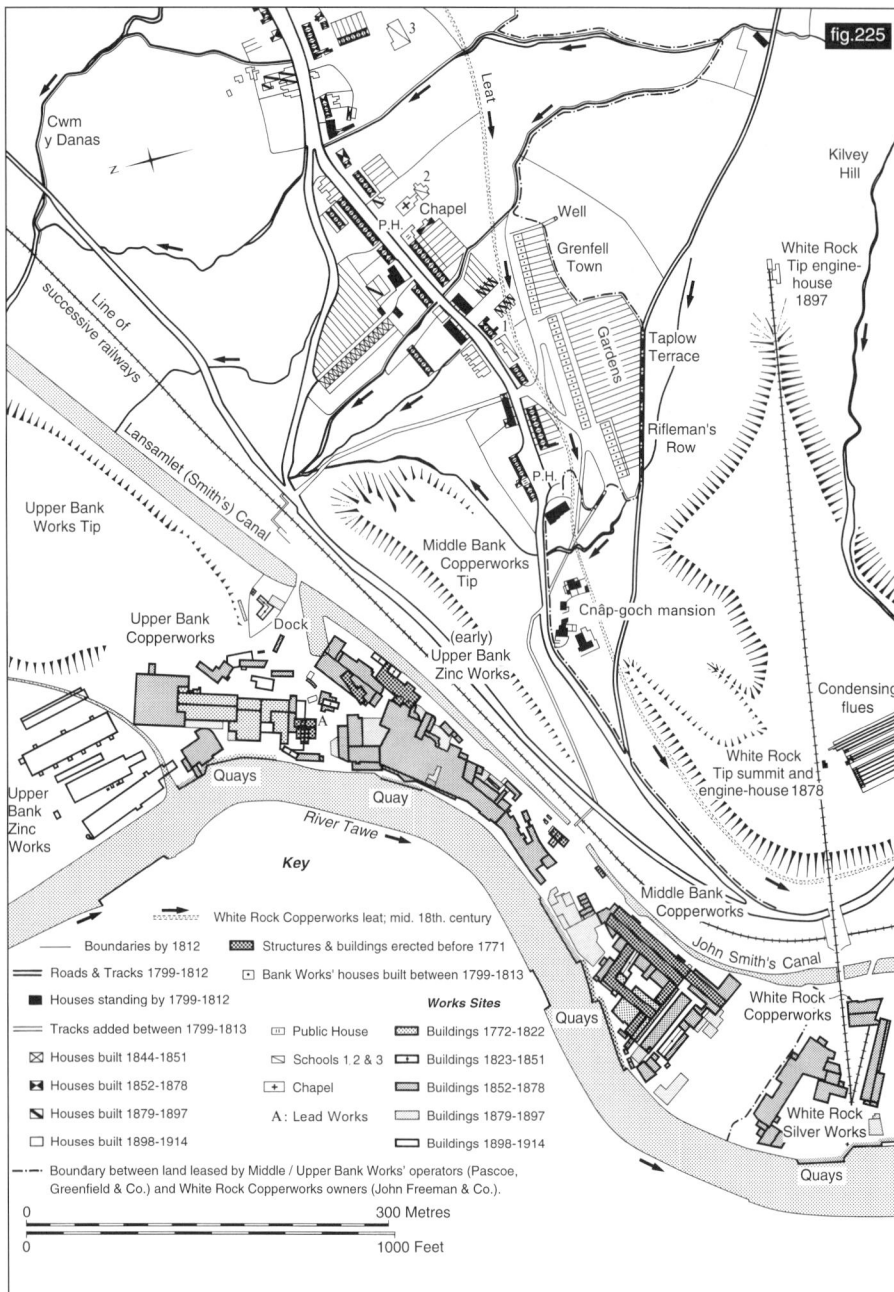

fig.225

Map labels:

Cwm y Danas

Line of successive railways

Chapel
P.H.
Well
Grenfell Town
Gardens
Taplow Terrace
Rifleman's Row
P.H.

Kilvey Hill

White Rock Tip engine-house 1897

Lansamlet (Smith's) Canal

Upper Bank Works Tip

Upper Bank Copperworks

Dock

Middle Bank Copperworks Tip

(early) Upper Bank Zinc Works

Cnâp-goch mansion

Upper Bank Zinc Works

Quays

Quay

River Tawe

White Rock Tip summit and engine-house 1878

Condensing flues

Middle Bank Copperworks

John Smith's Canal

White Rock Copperworks

Quays

White Rock Silver Works

Quays

Key

- →→→ White Rock Copperworks leat; mid. 18th. century
- —— Boundaries by 1812
- ═══ Roads & Tracks 1799-1812
- ■ Houses standing by 1799-1812
- ═══ Tracks added between 1799-1813
- ⊠ Houses built 1844-1851
- ▣ Houses built 1852-1878
- ◩ Houses built 1879-1897
- □ Houses built 1898-1914

- ▨ Structures & buildings erected before 1771
- ⊡ Bank Works' houses built between 1799-1813

Works Sites
- ⊞ Public House
- ◿ Schools 1.2 & 3
- ⊕ Chapel
- A: Lead Works

- ▨ Buildings 1772-1822
- ▤ Buildings 1823-1851
- ▦ Buildings 1852-1878
- ▥ Buildings 1879-1897
- ▭ Buildings 1898-1914

- —·—·— Boundary between land leased by Middle / Upper Bank Works' operators (Pascoe, Greenfield & Co.) and White Rock Copperworks owners (John Freeman & Co.).

0 ————— 300 Metres
0 ————— 1000 Feet

Fig. 225. The first workers' housing above the Middle and Upper Bank Copperworks was the occupation of the gentry house of Cnâp Coch by the manager of the White Rock Copperworks, which was subsequently subdivided for worker occupation. Between 1803 and 1813, the Grenfells built three terraces of 40 houses on the hillside ('Grenfelltown'). The settlement expanded and acquired a works school, a coppermaster's Anglican Church and workers' nonconformist chapels.

Bon-y-maen.[160] Just north of the converted mansion of Cnâp-goch (used by White Rock Copperworks' employees) were the Upper and Middle Bank Copperworks workers' houses. Three rows of houses, which still survive, were sited along the hillside above Pentre-chwyth Road as it curves up to Bon-y-maen. The settlement was labelled 'Pentre-wyth' in 1851 and 'Pentre-chwyth' in 1879, when the most easterly of the rows had the appellation 'Grenfell's Town', but the whole of the planned layout of houses owned by the Grenfells seems to have been called by this name while the family was in charge of the houses and the works.[161] The three

rows were built probably after 1799 and certainly before 1813,[162] presumably when the adjacent Middle Bank and Upper Bank Copperworks were under the joint ownership of Owen Williams and Pascoe Grenfell between 1803-4 and 1825-6, the latter being the more active partner in the business.

However, the form of the planned housing is distinctive for the Swansea area. The steep hillside situation of the settlement at Foxhole determined a development of long lines of dwellings across the hillslope, as was found elsewhere on the eastern side of the valley at Grenfelltown. This was a type common in the late eighteenth and early nineteenth centuries in the ironmaking centres of the Heads of the Valleys (and in the local ironworking centre at Clydach Upper Forge in 1784), and later became universal in the countless south Wales coal-mining centres of the mid-and late nineteenth centuries. In the lower Swansea Valley itself, terraced housing was already common in the cramped confines of Swansea town, where land was valuable, and at the foot of the steep hillside at Foxhole, but this form of housing was seen by the new entrepreneurs as a means of providing houses for a substantial number of their workers for the minimum outlay of capital.

The 40 houses built as 'Grenfelltown' in 1803-13 are arranged in three rows. Their roofs seem to be of late nineteenth-century form and the houses have probably been heightened with extra windows added. Their floor areas may be original, in which case they are not greatly different from the workers' houses that had been built earlier in the Swansea district. The north-eastern end house, no 1 Grenfelltown, is a fairly typical example of the rows. The house is 20ft 10in deep internally and 13ft 6in wide across the facade. The houses in Swansea resemble in size both the earlier Grenfell Williams houses in Flintshire and the smaller terrace that survives on the opposite, north side of the main road at Pentre-chwyth, where there were houses before the three Grenfelltown terraces were constructed.

The 40 original houses of Grenfelltown are arranged in three flanking but staggered rows across the hillside and have full-size sash-windows on both floors. The use of full-height sash windows on the first floor was common in the Swansea town area but not in the iron-producing districts, where the upper windows of workers' houses were of reduced height. However, the insertion of a second full-height window above the front doorway, as in the early Grenfelltown rows, was not usual in early nineteenth-century Swansea workers' houses and is likely to be an insertion, as are the large rear windows of the houses. At the rear, the peeling of

Fig. 226. Grenfelltown from the air: three original staggered rows of houses across the hillside stand at the top left with long gardens ascending the hillside. The works school, now a church, stands below the middle terrace with later housing to the lower left. The new valley road and roundabout cut through the former Middle Bank Copperworks tip. (983515-17)

the rendering (which covers all the terraces) around one of the windows has revealed window-dressings of bright orange brick set in a wall of Pennant sandstone rubble that seem to match the brick used in the later tunnel-back extensions. The houses may have quoins of cast copper-slag blocks throughout; the partly demolished outbuilding at the western end of Taplow Terrace has them, but the rendering on the terrace makes it difficult to be certain of their extent.

The names of the rows at Pentre-chwyth can be dated to the period prior to 1892-93, when Middle Bank and Upper Bank Copperworks were owned by the Grenfell family. The middle row is known as Taplow Terrace, after the village where the Grenfells had a home near their copper-rolling mills on the Thames. Taplow was also the name of one of their copper ore clippers that doubled as a yacht during the summer. The western row is Rifleman's Row; the obituary of Pascoe St Leger Grenfell in *The Cambrian* on 4 and 11 April 1879 recorded that 'In 1859 ... Mr. Grenfell gathered around him 200 volunteers, and formed the 6th Glamorgan Rifle Corps, which he commanded as Lieutenant Colonel, and the Corps was well known for its efficiency'. The eastern row now forms more of a nucleated settlement, with the half-row on the opposite side of a later road formed in front of the row with a steeply inclined access down to Pentre-chwyth Road. This later access is itself flanked by houses that had also been added to the settlement between 1851 and 1877, the whole being known as Grenfell Town. The total number of Grenfell houses, including Taplow Terrace, Rifleman's Row and Grenfell Town, was 55. In their present form they are more typical of the later nineteenth century in detailed layout, with tunnel-back extensions and short gardens on average about 40ft in length. A condition of the tenancy was that all urine from the houses be kept for collection for use in the copperworks![163]

The original access from Pentre-chwyth Road ran through what is now the church rooms attached to the early twentieth-century St Peter's Church. The former, a simple Pennant sandstone building built by the Grenfells in 1851, was used for an infants' school until 1877.

Landore/Brynhyfryd/Mile End

This is a settlement on the west side of the River Tawe, midway between the northern suburbs of nineteenth-century Swansea and Morriston to the north, largely lying on the south side of the Nant Rhyd-y-Filais side valley. The first copperworks in the lower Swansea Valley was built here in 1717 and remained in production until 1753. The works was initially poorly capitalised and there is no evidence of any other accommodation besides a small agent's (or manager's) house. With the bankruptcy of the proprietor Dr Lane, Robert Morris took over the works with the sponsorship of outside investors and had little cash available for building workers' housing. There is no map evidence of houses on any scale until 1826-38. Even after that date there is no grid of streets or large-scale formal layouts of terraces. Instead, there was a disparate terrace development by two industrial partnerships (John Morris II, and the Millbrook Iron Company and the Williams of the Morfa Copper Works) the landlord was the Anglican priest Thomas Morris, John Morris's brother. However, the settlement was initially mostly composed of detached cottage developments built by the workers themselves along a spreading web of irregular roads. Landholdings were divided up successively by families, possibly some for rent to other workers, and detached cottages grew into terraces which linked up alongside the roads; by 1838 there were two more such terraces. There were no Anglican churches or large works schools at the centre of the settlement as there were at each of the other four copperworkers' townships that were built, or added to, by the coppermasters. Instead, there was a flourishing growth of nonconformist congregations aided by the copperworks managers, both by providing works' accommodation for starting each cause and by raising funds and running the chapels as deacons.

Land was given by the landlord for Sunday and day schools attached to Siloh Welsh Independent chapel and a satellite Sunday school was built at Brynhyfryd, supported by Robert Monger, manager of the Hafod Copperworks.[164] The manager of the neighbouring Morfa Copperworks was one of the founders of the Wesleyan Methodist Chapel. The Siloh congregation was one of the most influential in Wales, appointing and supporting a famous architect-minister, Thomas Thomas, who in the 1860s popularised the use of the giant arch in the pediment as one of the most striking features of Welsh chapel architecture; he then preached the first sermons in the buildings he constructed.

In 1849, according to the surgeon of the copperworks - Mr Evans, the population was mostly of colliers and copperworkers. The district comprising Landore and Mile End was:

'A populous and scattered place, inhabited by the same class of people as Morriston, situate about a mile and a half north of Swansea. There is always a good deal of illness in this neighbourhood, and we are seldom without cases of fever. Drainage in the same neglected condition as in Morriston. Within an area of half a square mile are upwards of 200 houses without privies. The few that may be found are confined almost exclusively to the shops and public-houses.'[165]

The settlement continued to grow in the later nineteenth century under the impetus of what were probably the two largest copperworks in the world - Hafod & Morfa, along with the Landore, and Little Landore Copper and Arsenic Works, the Millbrook Iron & Steel Works and the Landore Tinplate Works.

Trevivian

In 1809 most of the initial workforce of the Hafod Copperworks was probably housed in the northern end of Swansea, half a mile to the south beyond the Cwm Burlais valley. Other workers lived to the north of the works at Landore and, as has been noted, conditions there were poor when Trevivian started to be built.

John Henry Vivian owned much of the land around the Hafod Copperworks and initially used the land on the east side of his works for tipping the vast amounts of waste material produced by the repeated roastings of the copper ore. As with the other works in the valley, these river side dumps were soon filled to capacity. The land on the west side of Neath Road was then bought and used as a new dumping ground, and a powered incline was constructed to raise the copper-slag up the hill and to the top of the new dump. This expansion of the dumping area explains why Trevivian was built to the south-west of the works and not on the land directly to the west. This mountainous dump, over 200ft high, towered over the northern end of Trevivian until the slag was cleared. During World War II some of it was used in the construction of aerodromes for the United States Air Force. The residue was removed in 1973 to fill the ravine of Nant Rhyd-y-Filais at Landore. Pentrehafod Comprehensive School now occupies the site.

The area on which the roads for Trevivian were laid out by 1844 was 1,640ft long and 1,150ft wide, an area approximately half the size of Morriston. The settlement occupied a triangular site, defined on its north by Pentremawr Road, on its east by Neath Road, and to the west by Cwm Road and

Fig. 227. 'Trevivian' or 'Vivian's Town' (Hafod), the most complete surviving copperworkers' township in Swansea. Its importance, with the complete remains of the works schools, is enhanced by the adjacent remains of what was once the world's largest copperworks and the survival of Vivian's mansion of Singleton Abbey.

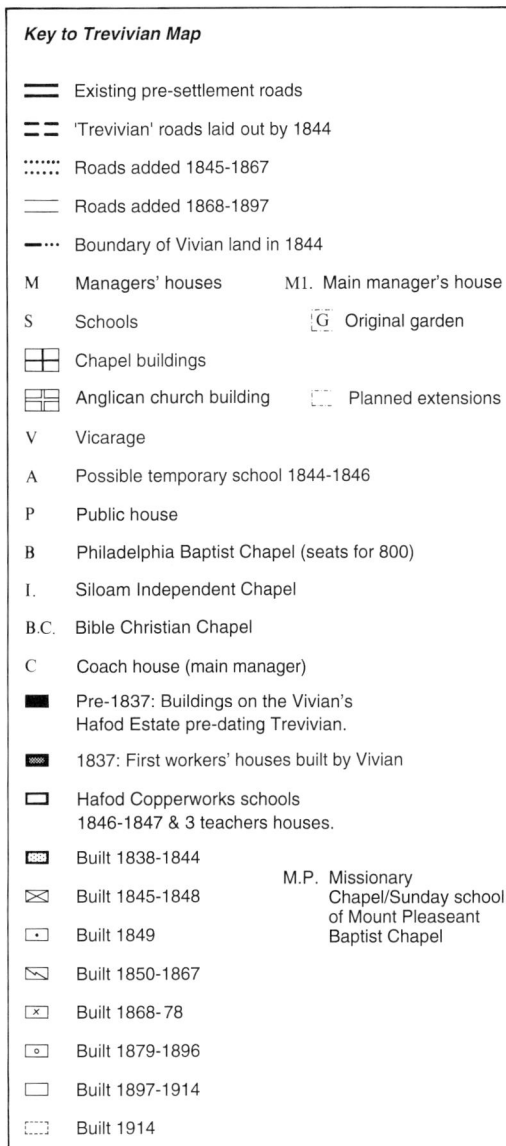

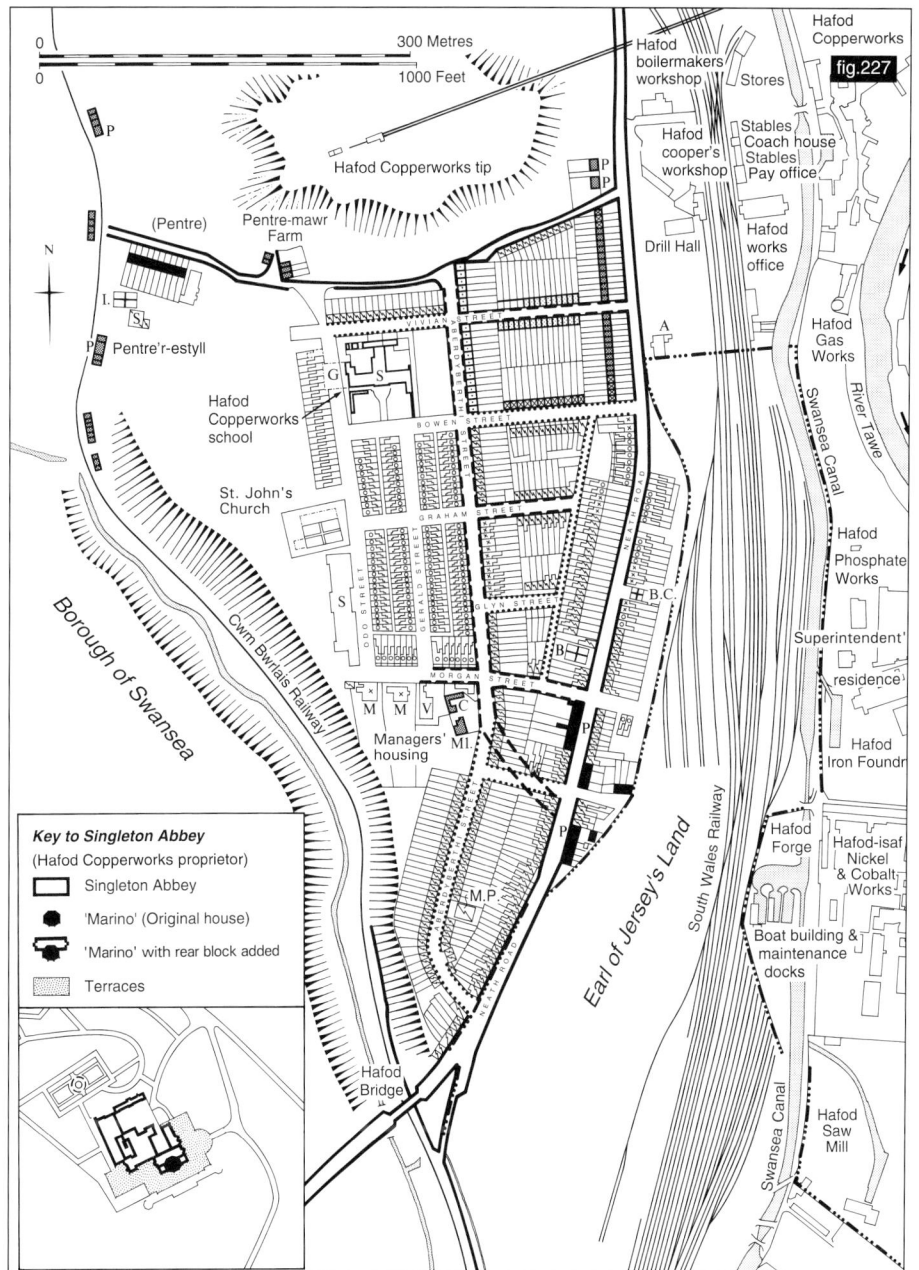

Llangyfelach Road in the Cwm Burlais ravine. At the southern end of the settlement, Neath Road crossed Llangyfelach Road by means of the Hafod, or Aberdyberthi, bridge. Aberdyberthi House lay in the southern part of this area, and gave its name to Aberdyberthi Street, which formed the spine of the southern part of the settlement, roughly parallel to Neath Road. It existed before the workers' settlement was built and was the house of the main manager of the Hafod Copperworks. Originally, Aberdyberthi Street joined Neath Road opposite Maliphant Street, but by 1876 it had been extended further south to its present junction. The final stretch of the road became known as Monger Street after Robert Monger, the manager of the copperworks.[166] Later, Odo Street, one of several named after members of the Vivian family, was added parallel to, and to the west of,

Aberdyberthi Street. The most northerly cross-street of the grid was developed first, predictably known as Vivian Street. Graham Street in the middle had also been laid out by 1844, together with Morgan Street to the south. Gerald Street and Bowen Street infilled the grid later on.

The next and final stage in the development of Trevivian began with the laying out of Odo Street in 1879-80 along the western edge of the existing settlement. It ran from Vivian Street in the north, beside the Hafod Copperworks School buildings, to the western end of Morgan Street in the south. The new thoroughfare was named after Odo Richard Vivian (1875-1934), later the third Lord Swansea.

207

Fig. 228. The second great tip of the Hafod Copperworks necessitated by the prodigious amount of waste produced by repeated roastings of the ore. The French artist's initial sketch must have suggested 'Provencal' roofs for the workers' houses at Trevivian in the foreground. (Le Tour da Monde, c.1860) (Reproduced by permission of Birmingham Library Services)

Fig. 229. One of the two original rows of Trevivian facing the main Hafod Copperworks across the Neath Road and the Swansea Canal. The nearest eleven dwellings were built with the long, shallow, rear roof pitches ('catslide roofs') common in small Swansea houses of the period. (950055/23)

The most important early central feature of Trevivian - and its outstanding survival - is the huge block of the gothic Hafod Copperworks School. On the western side of Odo Street, the much-delayed new church of St John's, Hafod, was constructed and this church, in a position that was obviously an afterthought, still remains but plans for partial adaptive reuse have already been made.

Workers' Settlements in Britain and Europe

It is worth considering how the nature of the provision of workers' housing in Swansea, differs from that generally provided in the best known of rural factory workers' settlements in Britain and, later, in the rest of Europe. The housing provided by the enlightened factory owner, Robert Owen and his predecessor David Dale, at New Lanark in south-west Scotland, was austere, as was the textile

housing provided by Richard Arkwright at the southern tip of the Pennines. Two of Europe's best-known 'model' workers' settlements were in the woollen textile belt of the Yorkshire Pennines: at Saltaire (near Bradford) founded by Sir Titus Salt, and Akroyden (near Halifax) built for James Akroyd.[167] Both Colonel Edward Akroyd and Titus Salt started by building very plain houses for Copley Mill (Halifax) workers between 1847 and 1853 and for the workers at Titus Salt's seventh textile factory at Saltaire between 1854 and 1857.

In 1856-59 James Akroyd employed the prominent Victorian architect Sir George Gilbert Scott to design All Souls Anglican Church near his Haley Hill Mills, and the high standard houses at Akroyden were built in 1859. The greatest group of textile (carpet) mills in Great Britain were also built at Halifax (Dean Clough) by the Congregationalist John Crossley in 1841-69 and the profits generated allowed the

Crossleys to house some of their 6,000 workers in the Italian-style development of Gibbert Street in West Hill Park.[168] In fact, it seems that the general embellishment of a workers' village was dependent, in these cases, on two factors: the generation of large profits from an active and successful enterprise and an appreciation of architectural embellishment in all the buildings of a commercial enterprise. Bradford had become the world's wool textile capital and Salt's innovation in using long fibre alpaca wool allowed him to generate huge profits.

The grand architectural design of Salt's seventh mill (at Saltaire), in the Italian Renaissance style, built by the prolific Bradford warehouse architects, Lockwood and Mawson, had a profound influence on the design of his workers' settlement. The distinctive ground-floor arch-headed openings, prominent banding and corbelled cornice tables are found both in the Saltaire Mill of 1851-53 and in Salt's later workers' housing, built particularly for overlookers and managers from 1860 to 1872. The huge demand for his products caused Salt to build his 'New Mill' in 1868 with a much grander chimney in the form of an ornate tower which was a copy of the campanile of the Venetian church of Santa Maria Gloriosa, complete with copied bell-stage. This prosperity, in turn, allowed the building of 45 almshouses around 'Alexandra Square' in 1868-71, with occupants carefully selected from the 850 households in the settlement. The provision of almshouses in an industrial settlement was most unusual. However, the dependence of industrial benevolence on commercial prosperity was vividly shown when a depression forced the Salts to cease trading in 1892.

There are of course differences of scale to the Swansea settlements: the infilled dwellings of Morriston by the 1840s only numbered about half those of Saltaire, and Trevivian had only about a third of the population of Salt's model settlement on the River Aire in the 1860s, whilst Grenfelltown was only a twelfth the size. The buildings of Swansea copper-smelting works were generally large, semi-open well-ventilated furnace sheds that did not lend themselves as readily to ornamentation as the multi-storey spinning-mills of the textile districts. Therefore, prominent architects were not generally brought in to design entire commercial schemes with linked domestic settlements.

The exception in the eighteenth century was the buildings associated with the resident copper-smelting Morris family which did construct buildings decorative in form and fenestration at its Forest Copperworks. The Morrises in turn provided nearby workers' housing with decorative banding and a pedimented barracks (possibly a provision for

Fig. 230. The second of the original two Trevivian terraces was twelve houses long lengthened to fifteen. It seems to have been enlarged in the later nineteenth century to the dimensions of the 1860s houses built for the settlement there. (950354/4)

their poorer workers) a hundred years before Titus Salt made similar but more famous provision.[169]

The largess seen at Saltaire was not standard in either Britain or in mid-nineteenth-century Europe. The first Industrial Revolution on the European mainland took place in the coalfields of French-speaking southern Belgium (*Wallonie*).[170]

The first workers' town here was built for textile workers in 1808 at Verviers, where 160 small and austere dwellings housing 800 people were built (and still remain) along four axes, each with two small low bedrooms over a single (12ft by 18ft) ground-floor room (and cellar) and a small garden 36ft long.

An exceptionally lavish scheme on the continent was the famous Le Grand Hornu, again an unusually comprehensive architectural scheme, and designed by Bruno Renard, built in phases between 1820 and 1852 with 450 dwellings housing 4,500 inhabitants, and centred on the grandeur of the classical colliery workshops and offices flanking a huge oval space with a statue of the founder, Henri De Gorge. Relatively plain dwellings had one heated ground-floor room with a lower heated bedroom over. The smaller rear bedroom and kitchen were unheated and small gardens only twice the depth of the houses lay behind.

More typical, because built by a company rather than a wealthy philanthropic individual, is the well-preserved colliery settlement at Bois-du-Luc, with a grid of austere dwellings, each of which was originally 'a slope' almost straight from a two-storey front to a rear single-storey and not the more lavish 'house with a slope' (i.e. a catslide roof) found in contemporary Swansea. A large ground-floor room (of 340sq ft) had partial low sleeping accommodation over, until rebuilt with two bedrooms in a full upper storey in 1880.[171] A total of 162 of these small dwellings were built to a common plan between 1838 and 1853, with an octagonal 'salle de fêtes' at the central crossroads added in 1854.

Most of the rest of Europe industrialised considerably later than Britain. Catalonia was one of the first areas in the Iberian Peninsula to do so, with large textile mills and settlements set near rural water-power sources, and steam-engines fuelled by coal imported from south Wales. The industrial Colonia Güell was built near Barcelona mainly between 1890 and 1918;[172] the terraced workers' dwellings of 1915 have two full storeys and an average floor-area of 950sq ft. These are fairly plain compared to the riotous Art Nouveau architecture employed for the managers' houses, the school and Antoni Gaudi's unfinished church.

The workers' settlements in Swansea and elsewhere in Europe have much in common. It is notable that the owner's residence did not form part of the layout but usually was set in parklands away from the grid of the streets, whereas the managers' houses were often part of the new settlements. It is also important to remember that the eighteenth-century workers' houses at Swansea are at the origins of what became an international social model.

Conclusions: Settlements

The considerable amount of available information concerning the communities surrounding the 12 main copper-smelting works in operation in Swansea between 1717 and 1860 can now be evaluated[173] The nature of the industry meant that from the very beginning attempts were made to keep the works away from the existing urban core of Swansea. Therefore, all the works needed to make provision to attract and house at least key workers in what were relatively isolated sites.

Five of these settlements provided the occupants with large gardens to keep pigs and grow potatoes and vegetables. Beyond this, schooling provision started in 1806 and at least four of the works' owners are known to have made provision for the education of their workforce, with possibly three more jointly contributing to what had originally been the Birmingham Copperworks School at Morriston. Four of the works' owners provided, in part or in full, Anglican churches. Sites were made available for building nonconformist places of worship. One coppermaster also provided a secular meeting hall for the use of the employees of three of the local works and one provided a barracks block that may have been intended for poorer workers.

The larger settlements laid out on a sizeable grid-plan and provided with churches, were only established by second generation resident managing-partners of copper concerns at Morriston and Trevivian. The involvement of the entrepreneurs with the local communities socially and politically, as well as economically, was a vital ingredient in making sure that their philanthropy was dispensed locally rather than elsewhere. Local political involvement was in their interest and once they were involved it is true that they had to be seen to be responsive to the needs of the local community: this was enlightened self-interest.

This process is very apparent in the housing and social provision made by the owners of the two large-scale and neighbouring Hafod and Morfa Copperworks founded in the early nineteenth century. The Cornishman Michael Williams, owner of the Morfa Copperworks, with his partners provided some housing for key workers but never a large settlement with educational and religious facilities or large-scale culverted drainage. He reserved most of his philanthropy and enlightened self-interest for the West Country, where he was MP for north Devon and supported charitable causes there and around his Cornish home - Caerhays Castle, the coastal mansion designed by John Nash.[174]

This was in stark contrast to his Swansea-based compatriots, the Vivians, at the neighbouring Hafod Copperworks. The large Morfa workforce at times was equal to or greater than that at Hafod. However, interest in providing amenities for the local workforce depended to a large extent on a managing partner having a home in Swansea and being involved in local politics and administration. A subsidiary residence could be enough to stimulate considerable local philanthropy, as with the Grenfell's Maesteg House on Kilvey Hill. If the primary residence of a coppermasters' dynasty was elsewhere, that is where much of the family profits would go. In the Grenfells' case it went to maintain and develop the main family mansions (near their copper-rolling mills) at Taplow in Buckinghamshire which still dominate the landscape. There, the resident family, involved in local public life, felt obliged to build workers' housing and fund the construction of local churches. Principal family members might have lived in Swansea for a substantial part of the year, but the family gathered at Taplow and the vault where they lay in death

Figs. 231, 231a. The main family home of the Pascoes remained near their Thames Valley coppermills at Taplow where they successively owned the mansions of Taplow House (top) and Court (left).

remains, alongside the River Thames and not the Tawe or Tamar (for this was a third copper-smelting family originating from Cornwall).

Economic forces, to some extent, determined that copper-manufacturing entrepreneurs would live near their main markets rather than, or as well as, at the centre of their copper-smelting activities; the Morrises had also had London houses and helped found the first purpose-built art gallery in Britain, at Dulwich rather than in Swansea. The Grenfells were merely following *Twm Chwarae Teg*, Thomas Williams, the outstanding 'Copper King' of Anglesey to political and economic eminence at Marlow, alongside the mills that rolled Swansea copper for the London market and which had earlier been used by the Swansea and London entrepreneur, Chauncey Townsend.

The Industrialists and their Mansions

The small scale of coal-mining in the Swansea area prior to the Industrial Revolution did not result in any large change in the population. However, the scale and depth of mining progressively increased and by the seventeenth century several local gentry families, such as the Mansels, the Mackworths, the Seys of Swansea, the Prices of Penlle'rgaer and the Popkins of Forest were actively involved in mining; they were joined by several of the more well-to-do yeomen. In the early eighteenth century, such local enterprise was probably producing in some years a figure of between 45,000 and 50,000 tons of coal annually, some two-thirds of this being exported

through Swansea and the rest through Briton Ferry and Neath.[175]

These Swansea industrialists of the late seventeenth and early eighteenth centuries lived in fairly plain, large, gabled blocks such as those at Penlle'r-gaer and Forest, the homes of the Price and Popkin families respectively. The latter stood directly alongside where the water-powered Upper Forest Iron Forge and cornmill were founded at the end of the seventeenth century, while the Prices' ancestral seat was some three kilometres north-west of the area where coal could most easily be exploited on their land. The brick Queen Anne-style house at Gwernllwynchwith, home of the coal-owning Vernons, may represent much greater aesthetic aspirations. However, that too seems to have had a coal-winding engine added to its various outbuildings at the end of the eighteenth century, during its ownership by the coal-mining Smith family.

The absentee Duke of Beaufort, owner of many manorial and mineral rights in the western lower Swansea Valley and at Sketty, and with his great house at Badminton in Gloucestershire, ensured that a great deal of the profits from coal-mining and copper-smelting went to create elaborate architecture away from the Swansea area.

Of the eighteenth- and nineteenth-century copperworks, it was only the relatively short-lived Cambrian works that was founded by a partnership of local people in 1720. Generally the works were founded by incoming interests with capital raised from mercantile, copper-smelting, copper-mining or manufacturing activities elsewhere. Only in the case of two out of the 11 copper-smelting works founded between 1717 and 1828 did the managing partner live locally for over half the productive life of the associated copper-smelter. This was the situation with the Morrises and the Forest Copperworks in the later eighteenth century and with the Vivians and the Hafod Copperworks in the nineteenth century.

The Morrises

The success of the first non-ferrous smeltery at Llangyfelach (1717) produced profits that initially went largely to swell the London mercantile interests of the Lockwood and Gibbon families, whose capital had largely rescued the works from initial bankruptcy under the Bristol entrepreneur, Dr John Lane. Only gradually was Robert Morris, a resident manager under Lane and then initially a minor partner in the succeeding firm, able to accumulate capital. At first he seems to have lived in the Strand, but growing prosperity allowed him to enlarge what had been a single-block minor gentry house, much like that of Forest and Penlle'r-gaer, at Tredegar Fawr to the north of Swansea (now just north of the M4).

When Robert Morris acquired it, Tredegar Fawr was a small seventeenth-century hall-house, three

Fig. 232. Portrait of the coal-mining entrepreneur John Morris II. (George & Henry Dare, Morris Estate Trust)

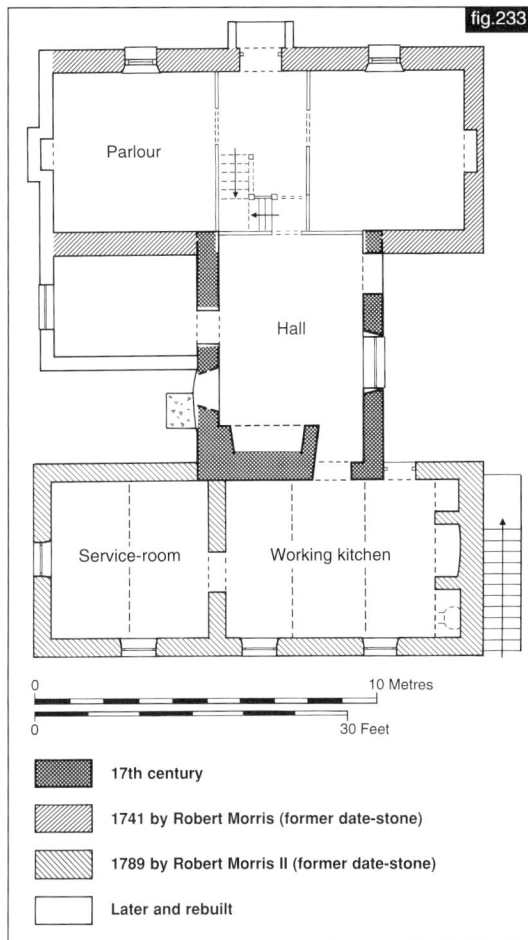

fig.233

Parlour	
Hall	
Service-room	Working kitchen

0   10 Metres

0   30 Feet

17th century

1741 by Robert Morris (former date-stone)

1789 by Robert Morris II (former date-stone)

Later and rebuilt

miles to the north-west of the Llangyfelach Copperworks. In 1741 he seems to have had a fairly large cross-wing added to one end of the hall, thereby producing a substantial house some 2,055 sq ft in size. This was about four times bigger than each of the two dwellings provided for his head colliers at Tirdeunaw coal shaft (or about five times the amount of full-height accommodation if the low-height sleeping loft is discounted). Robert Morris I lived at Tredegar Fawr until his death in 1768.[176] This period may have marked the zenith of the copper-generated wealth of this locally-based family, for a much bolder statement was made by his younger son John (later Sir John) Morris I, who in effect created a complete 'picturesque' landscape of mixed copper industry and architectural elements in the lower Swansea Valley. Sir Herbert Mackworth was doing similar things in the lower Vale of Neath to the east, emulating the similar 'mixed' landscape that William Champion had just completed at the earlier copper-smelting and rolling complex at Warmley, near Bristol. Morris's landscape was dominated by his new, large and pretentious Palladian mansion of *c*.1775, only one mile from, and overlooking, the circular and octagonal pavilions (the smelting-houses) of his Forest Copperworks.

Despite this lavish expenditure, the Morris family (in the person of the older brother Robert Morris II) was able in 1789 to flank its seventeenth-century hall at Tredegar Fawr with a second, matching, cross-wing that transformed the fairly humble and homely

fig.233a

Figs. 233, 233a. Robert Morris, the managing partner of Lockwood, Morris and Company, bought a small gentry house at Tredegar Fawr to the north-west of Swansea, which was gradually aggrandized by the family.

fig.234

Fig. 234. Painting of the Morris mansion of Clasemont. Both the earliest resident eighteenth-century copper-smelting dynasties of the Neath and Swansea Valleys had impressive mansions built in the 1770s overlooking picturesque mixed rural and industrial landscapes. The architect of both houses was John Johnson. (National Museums and Galleries of Wales)

pile into something more like a baronial mansion.[177] The symmetrical mansion was 3,225 sq ft, equivalent to about five late eighteenth-century Morriston houses, or eight typical colliers' flats in Morris Castle.

In the eighteenth century all other copper-smelting entrepreneurs lived away from the area, the works generally being run by fairly large, and generally absent, consortia. However, by 1774 the locally based Morris family had been entrepreneurs in the copper industry for almost 50 years. John Morris I was also the first of the Swansea-based copper magnates to marry into the aristocracy, in 1774. Some 30 years after the construction of his grand Palladian House, John Morris I was made a baronet in 1806 for helping to raise volunteer forces to fight in the Napoleonic Wars. In 1809 he retired from the copper-smelting business and was made portreeve of Swansea; he also saw his son married to Lucy Juliana Byng, daughter of Viscount Torrington. He built a fine Palladian house on a bluff dominating the lower Swansea Valley, reminiscent of the great mansion which the copper magnate, Herbert Mackworth, was building alongside his aristocratic wife's old family home on the Gnoll at Neath. Both employed the Leicester born architect John Johnson to design striking white houses dominating a mixed picturesque landscape of industrial pavilions, structures and water features.[178]

Clasemont, probably stuccoed and shining white, was laid out along the skyline for maximum effect. A large double-pile block had its roof neatly screened by a horizontal parapet. Flanking single-storey glazed passages curved gracefully forward along the crest of the northern end of Graig Trewyddfa to connect with two fairly ornate two-storey pavilions covering the service wings.[179]

Clasemont housed the Morris family within sight of their works for some 45 years. However, during that period the rural vista from the crest of Graig Trewyddfa changed drastically. In the 1770s John Morris extended his underground canal and mining of thick coal seams under the mansion itself. In the 1780s he also expanded the industrial activity at the foot of the Graig by extending the Lockwood, Morris & Co. surface canal south to Landore and attracting concerns from Birmingham and Leeds to build their own copper-smelting works along it. Then, in 1790, he moved his own copper-smelting operations from Forest, at the foot of the hill where Clasemont stood, south to Landore, but his old works was leased to the principal Bristol-based copper-smelting concern. The ever-increasing, noxious copper-smoke poisoned the upper slopes of the hill and made both the view from Clasemont and its healthy breezes things of the past. Subsidence from the extensive mining underneath must also have been a factor in the decision to abandon this site as a residence.

The grand house was situated between Morriston and Llangyfelach on the high ground occupied by what are now the 1960s housing estate roads of Mount Crescent and Pen-yr-Yrfa (off Pentre-poeth Road), with much of the former parkland sloping down to the east surviving as Morriston public park on the hillside above Morris's former township. A home farm with a substantial symmetrical quadrangle of buildings was constructed as

'Pengwern' on Banwy Road and Clase Farm on Clasemont Road; the site of the former now stands immediately to the east of the tower block of the Driver and Vehicle Licensing Agency. Clase Farm formed a lodge at the south park gate and survives, a single range with, to its west, the matching truncated coachhouse, both flanking the former south drive to Clasemont. The drive gates to the north of the buildings and the former coach house have cast copper-slag quoins.

The huge capital accumulated by John Morris by the 1770s encouraged him to invest in west Swansea, where the seams of the south Wales coalfield stretched to the sea. Here, in 1771, he purchased a very large area of land at Sketty, just west of Swansea, from Francis Greville, earl of Warwick; it included the site of what became Sketty Park.[180] There is no sign at this time that any of the coppermasters were thinking of an eventual move away from the copper-smoke to some sort of industry-less idyll. In fact, the opposite was the truth and with continuing commercial zeal the Morrises made every effort to exploit the underlying coal deposits at Sketty by means of their Oystermouth Railway schemes. It may only have been the comparative failure of these that encouraged them to develop the rolling terrain as parkland. The other result of this failure was the adaptation of the connecting industrial railway as the world's first passenger-carrying line.

The main mansion in the area at the time was Sketty Hall which, with its central pediment and double-piled hipped roof, had more pretension than did the seventeenth-century Tredegar Fawr, but it appeared homely compared to the stately Clasemont. The late seventeenth or early eighteenth-century mansion forms less than a quarter (the lower south-east part) of the present mansion facade. It had been built as 'New Hall' either by Alexander Trotter, collector of customs in Cardiff between the 1690s and 1716, or subsequently by Rawleigh Mansel who came from a long-established Gower gentry family.[181] It continued to be let to tenants in Sir John Morris I's ownership but he enhanced the value of his investment with the addition of a first-floor timber balcony, carried on double, pink-brick bay-windows, which commanded the fine sweep of Swansea Bay. In 1785 he drafted particulars for the letting of this house which emphasised its 'marine villa' aspect: 'the house on the sea-side near Swansea: The house is called Sketty Hall, and well known as a beautiful situation.'

The house had three principal rooms on each of the two main floors, five service chambers in the cellar and four attic chambers with a cupola-capped brew and wash house, stabling for six horses with an upper hay loft, a coach house for two carriages, a cold bath in the garden and 'pig-sties at a convenient distance'.

It was also noted that: 'the front of the house has been newly [improved with] two new sash

fig.235

Fig. 235. 'Marino', built in 1784 for Jane Morris and her husband Edward King, was the first of a substantial number of industrialists' marine villas built to view Swansea Bay. William Jernegan was responsible for the octagonal house and for the octagonal Mumbles Lighthouse visible in the background. (Bernard Morris)

[windows] with bows to the two front parlours, which are the drawing-room and breakfast parlour. [There is] A balcony over the bows, and it is this first which commands a view of the sea-bay.'

Enquirers had to apply in person to John Morris I at his London home in Mayfair, situated on the corner of Davies and Grosvenor Streets, mid-way between Grosvenor and Berkeley Squares.

By 1822-23 Sir John Morris II was newly ensconced in his fine new house of Sketty Park, with sweep windows commanding the view of Swansea Bay; he therefore sold the freehold of the smaller Sketty Hall to one of his tenants.

The age of what were consciously considered 'marine villas' began with the appropriately named 'Marino' down the slope and near the Swansea Bay foreshore, designed by the Swansea architect William Jernegan. This small octagonal three-storey picturesque house, with the same plan-form as Jernegan's Mumbles Lighthouse, but 38 ft across, was designed in the tradition of the playful waterside circular villa of Belle Isle on Lake Windermere, designed in 1774 (with the design subsequently published) by the architect John Plaw.[182] The delightful Marino was built with a central chimney in the form of a classical urn at the peak of its conical roof, in 1784. The facades were characterised by the large blind-arch recesses that were a prominent feature of both the work of John Johnson and William Jernegan and which are common to much of Swansea's surviving Georgian architecture. It was built for Edward King, whom the ageing Jane Morris (sister of the copper-magnate John Morris I) had brought to Swansea as her husband. A job was subsequently procured for King as deputy comptroller of customs in the Port of Swansea. Marino helped to establish a fashion for such multi-sided rooms - and hence with multiple-vistas - and it was followed by such buildings as John Nash's triangular 'Castle House' in Aberystwyth of 1791-94, with its three octagonal towers, enclosing one multi-vista room each on every floor[183]; and 'A la Ronde' of 1796 at Exmouth in Devon, with a conical roof covering a 16-sided villa with seven rooms branching-out from a central octagon on the main floor.[184] Marino was small with three rooms on each floor and a small back staircase. Nash, who was living at Carmarthen and is likely to have seen Marino, gave his Aberystwyth villa three-four main rooms on each floor.

Forever the astute businessman, John Morris I next built for rent (in 1799) a 'Marine Villa' in the Cottage Style of architecture on the slopes of Swansea Bay to the east of his Sketty Hall above his sister's house at Marino. A verandah, on the two seaward sides of the

fig.236

Fig. 236. The present south front of Singleton Abbey showing how, in modified form, the octagon of Marino (right) is incorporated. (9500348/2)

house, allowed a protected contemplation of the sweep of Swansea Bay from the harbour pier on the east (soon to be piers, as Morris and his associates completed the copper-trade harbour with William Jernegan's octagonal cast-iron light at its end) on the east to Jernegan's octagonal stone lighthouse on Mumbles Head to the west. It is probable that this third Swansea Bay villa, appropriately named 'Verandah', was also designed by William Jernegan, who was much admired by John Morris I for his innovative work.[185]

The seven-bay house had no less than 33 rooms, excluding outbuildings, and stood in a park of some 18 acres. Above its central doorway was a large circular window with glazing-bars similar to those found on Morris's old Forest Copperworks, and a central circle surrounded by radiating bars. This lit the 'print gallery' and the 'geometrical' or circular stair enclosed in a central semi-circular apse. This gentry 'cottage' also came to have two 83 ft long hot houses in its walled 'pleasure ground' and covered accommodation for nine horses, two carriages and a cart. The house was rented between 1800 and 1818 by Calvert Richard Jones whose father, in the 1770s, had built one of Swansea's first railways from his Cwm Level Colliery at Landore (and whose son of the same name was an important pioneering photographer). His brother George rented the adjoining Sketty Hall from 1811 to 1815 and then moved to nearby 'Glanmor'. In the first half of the nineteenth century, a large group of industrialists, merchants and gentry associated with the industrial and commercial enterprises of the growing copper metropolis came to be resident on, and around, the Sketty foreshore of Swansea Bay.

The period of critical development of the area as a rural idyll came with the laying out of two large parkland estates and attendant mansions at Sketty, in the single decade 1813-23 by the two main dynasties whose wealth was anchored in copper-smelting at Swansea.

John Morris's parkland at Sketty, on the eastern side of the colliery developments in the Clyne Valley, was already enclosed by 1813, but John Morris II was still resident in his house at The Bryn and his ageing father, Sir John Morris I, at his mansion and parkland at Clasemont. Then, in 1816, John Henry Vivian bought Marino from Morris's brother-in-law and in 1818-21 enlarged it.

Sir John Morris I died in 1819 in his London house, Hans Place.[186] Since the 1790s he had moved from his previous London residence in Mayfair to a desirable address in Knightsbridge, off Sloane Street. He had yet wider concerns. In 1810-13 he was responsible for improving the pleasure grounds in front of Cambrian Place, now under the Swansea Marina, and in 1812 he became a trustee of the Swansea Assembly Rooms (built by William Jernegan) which still remain in Cambrian Place. His social concerns, reflected in the founding of Morriston, were shown in 1816 when he became vice-president of the Morriston Savings Bank along with Pascoe Grenfell, Thomas Lockwood and John Vivian. He also chaired a meeting that year to relieve unemployment in Llangyfelach, Llansamlet and St John's.

Starting two years after his father's death, Sir John Morris II had the mansion of Sketty Park built in 1821-23. An ageing William Jernegan was responsible for this and Vivian's new mansion, both being simple rendered blocks designed without the flair with which the younger architect had conceived the octagonal Marino, although pavilion-mansions, such as Sketty Park, with asymmetrical service blocks were a common feature of large Regency houses. Interestingly, pre-eminence in the copper industry was reflected in major ownership of land and large houses in the Sketty area. The large Morris estate had been acquired in 1771 and members of the Morris family variously occupied, altered or built Sketty Hall, Marino, Verandah, Bryn, Sketty Park and Hafod - a total of six villas and mansions.

John Morris II had lived in the fashionable area of west Swansea at least since his marriage in 1809. His home was at The Bryn, while his father, John Morris I, continued to live at Clasemont, perhaps out of affection for the house despite its increasing disadvantages. He headed his letters 'Clasemont' up to his death in 1819. Following the death of John

Morris I, his son seems to have lost no time in abandoning Clasemont, and by the time the Ordnance Survey mapped the site in 1826 the house had been largely demolished.

In the early nineteenth century the Morrises moved their residence (in part, literally - certainly re-using the Portland stone ashlar work of Clasemont's facade) to the sylvan sweeps of Swansea Bay, immediately to the east of their failed industrial developments in the Clyne Valley. There they contributed to the development of an area of large houses surrounded by parks and gardens, built by the local capitalists and entrepreneurs. With the exception of the Clyne Valley Arsenic and Copperworks, the area was well away from, and generally upwind, of the toxic fumes of the smelting houses in the lower Swansea Valley. The large-scale Ordnance Survey maps of the area, published in 1826, show that by 1825 the long ranges of Sketty Park had been completed. Close inspection of photographs shows that it included architectural elements seen at Clasemont, which may have led to the, patently erroneous, story that Clasemont was taken down and removed in one night to Sketty Park in about the year 1805.[187]

Clasemont was certainly still occupied until 1819 and it is known that the furniture was sold *in situ* in 1821. Possibly the architectural fittings as well as ashlar masonry were transported in hundreds of carts after the demolition of Clasemont in 1821-22.[188] There is cartographic evidence of substantial ruins remaining on the site throughout the nineteenth century: presumably the Pennant sandstone core of

Fig. 237. The coppermasters and other Swansea entrepreneurs moved their residences to take advantage of the beautiful sweep of Swansea Bay, resulting in the creation of a large area of parkland that is still recognisable. (955062/52)

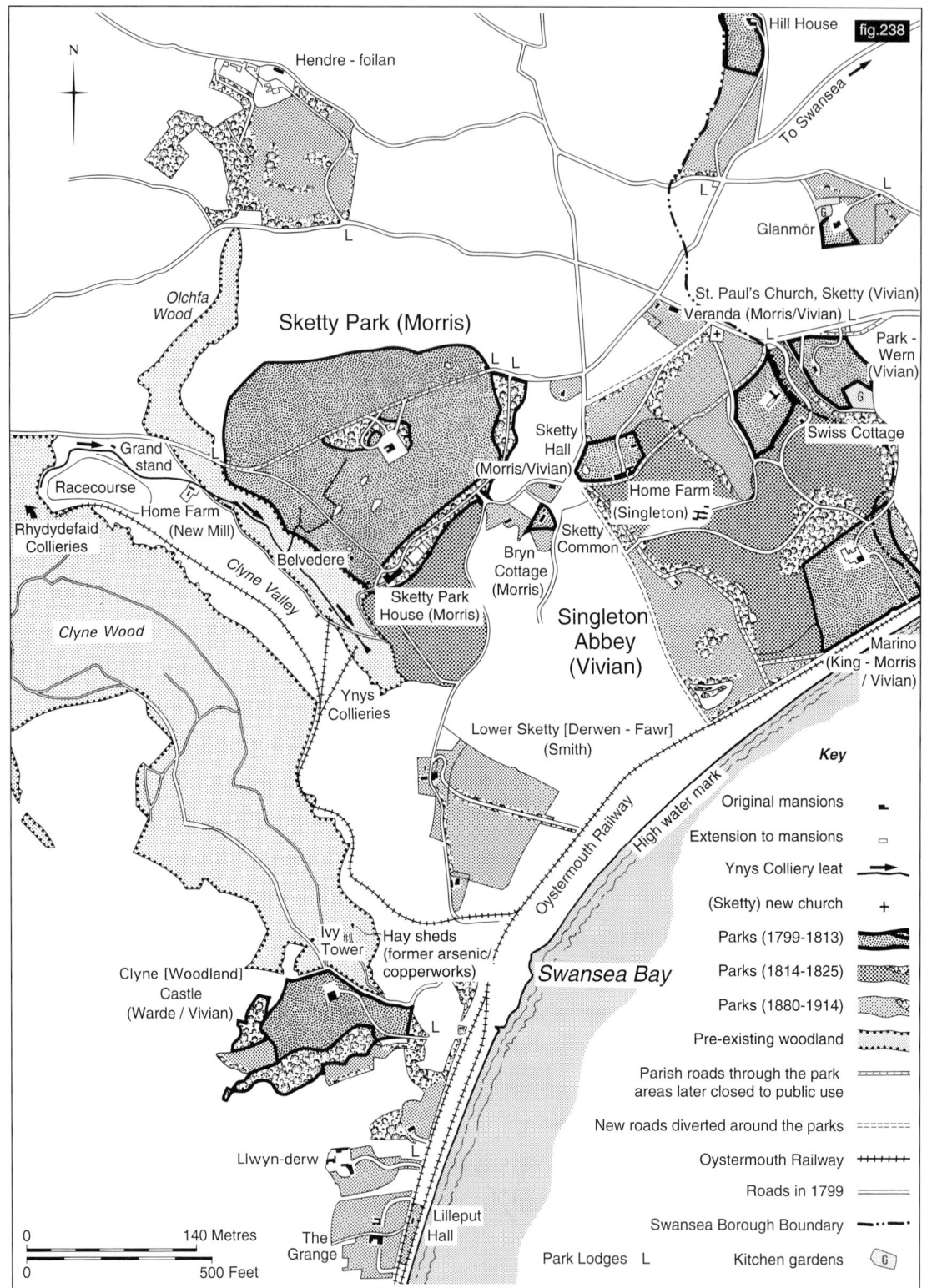

Fig. 238. Map showing the concentration of coppermasters' and other entrepreneurs' housing in the Sketty area of west Swansea. The entrepreneurs of Swansea may initially have been attracted to the area by the underlying coal reserves and the superb views of Swansea Bay.

the main external walls. Only the foundations of the house are not visible.[189] Examination of photographs taken of Sketty Park mansion before demolition in 1975 reveal that it was a cube, similar in dimensions to Clasemont but given a reformed Regency aspect.

Like Clasemont, the northern front had five bays with ground floor windows contained in arched recesses, and with the ground-floor capped by a broad projecting string-course immediately below the first-floor windows.

fig.239

Fig. 239. John Morris II's house in Sketty Park, west Swansea, where the main industrialists' residences were grouped together overlooking Swansea Bay. The landscape was still a mixed one of parkland and collieries. (Randall Davies; Swansea Archives D39)

At 158 acres, the newly laid-out Sketty parkland was of a similar size to Clasemont but was bisected by the public road leading westwards from Sketty.[190] The eastern section of this road survives as Sketty Park Drive. The south-western boundary of Sketty Park was formed by Mill Wood on the eastern scarp of the Clyne Valley; at this point the coal seams outcropped on the southern edge of the south Wales coalfield.

Sketty Park house is attributed to the elderly William Jernegan.[191] He was certainly capable of producing a finely balanced Georgian design, as can be seen in his Heathfield Lodge which, with its three pavilions similar to Clasemont, was built (before 1792; demolished *c*.1870) for Sir Gabriel Powell whose father had been the Duke of Beaufort's Swansea agent.[192] However, the main house at Sketty Park was not balanced by symmetrical wings as Clasemont had been, but instead had two asymmetrical service and stable wings to the west, each of (unequal) two storeys with hipped roofs.

What could have been taken from Clasemont to be re-used at Sketty Park? Re-used dressed stone may have formed the blind arcading of the ground floor and the string-course banding around the building. It is possible that the main staircase, fireplaces, panelling, floor joists and boards may also have been re-used. This would suggest that Sir John Morris II was keen to conserve his resources carefully. Contemporaries were quick to praise the design of the house:

'...Sketty-park, the splendid mansion of Sir John Morris, Bart., built after a design of Mr. Jernegan's, in a delightful situation, commanding the whole of Swansea Bay, and fine views of the surrounding country'.[193] The house was situated on a prominent knoll at the southern apex of the park as it existed in

1813. A drive through the wooded scarps on the south-eastern and south-western fringes of the park already existed and this determined the site of the house. By 1825 the park had been extended south to Derwen Fawr Road so that the sea vista from the house was already emparked. The public road was diverted away from the mansion, in the same as the Vivians would later do at Singleton Park.[194]

The Ynys Colliery, which Sir John Morris was developing in 1804-7 in connection with the Oystermouth Railway scheme, lay only 660ft south-west of the service court of the mansion. It would have even been visible from the garden in front of the mansion had it not been for a steep scarp into the Clyne Valley. It is arguable that the Morrises, with their move to Sketty Park, were not retreating from the scene of their industrial involvement at all. They had withdrawn from copper-smelting in 1808-9 and were then primarily involved in the development of local coal-mining activity. In 1840, a second Ynys Colliery was developed, also 660ft south-west of the mansion, but the shaft was located further north so that it was also out of sight. A copper and arsenic works was built in Clyne Wood in *c*.1837 and functioned until 1860. The fumes would have been seen and smelt from Sketty Park, 3,000ft to the north. It seems that even the third generation of entrepreneurial Morrises was not averse to living in the neighbourhood of industrial activity.

The house was occupied by Sir John Morris II's grandson, Sir Robert Armine Tankerville Morris, until 1930 and was then bought by Swansea Corporation and was used for such varying purposes, as hostel accommodation and offices, until it was demolished in 1975; by then it was surrounded by modern houses.[195]

Two of the most noticeable survivals of Sketty Park can be found on the eastern edge of the modern Clyne Valley Country Park. Much of the Pennant sandstone park wall survives on the upper slopes of Mill Wood on the eastern side of the valley. Overlooking the eastern scarp is the remarkable landscape feature called the 'Belvedere', a gothic

Fig. 240. The castellated, octagonal prospect-tower of Sketty Park House ('Belvedere') overlooking the Clyne Valley. The fine gothic tower, now derelict and decaying, was probably designed by William Jernegan. (950051/24)

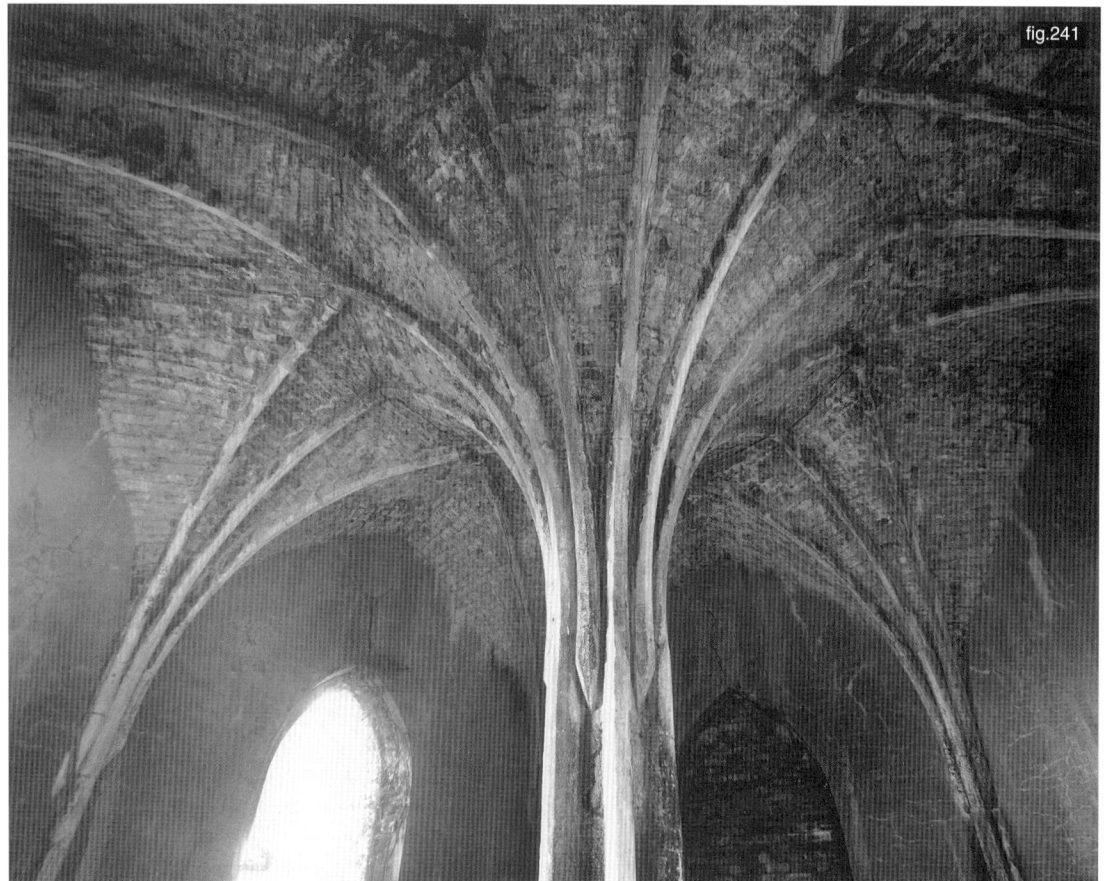

'In the park, which is extensive and well stocked, is a gothic building, the roof of which is supported by a central pillar, branching at the top, similar to the late Chapter-house, at Margam.'[196]

The building stands on a mound close to the junction of Saunders Way and Valley View. It may have been designed by William Jernegan. It has ashlar stone dressings and a taller polygonal stair turret to the north-west angle. There are diagonal corner buttresses with pointed, chamfered, gothic windows between them, containing Y-tracery and capped by drip moulds. The structure is crowned by panelled parapets over a projecting eaves string course. Below the parapet in raised stone lettering is the inscription 'ESTO PERPETUO' (indeed, the conveyance to Swansea City Council contained a covenant not to demolish the building). The entrance doorways are

Fig. 241. The interior of the 'Belvedere' with its vault supported by a central pillar is based on the design of the Margam Abbey Chapter House. (9500357/4)

prospect tower of octagonal plan, with rubble-stone walls of Pennant sandstone. The flair and expense shown in this structure suggests that it was built before the somewhat parsimonious Sketty Park mansion. In 1823 it was described in the following words:

now blocked; there is a tall chamfered archway on the western side and also a pointed doorway to the stair turret which has a corbelled top and cross loops. Now heavily overgrown, the interior retains its central ribbed pier and wall shafts supporting the slender ribs holding large vaults in place.

The third Morris house in this area, was Hafod, which lay to the north-west of Sketty Park. It was built by John Armine Morris, the eldest son of John Morris II. After his father's death he let Sketty Park and lived in France for many years. Like Lady Charlotte Guest, wife of the Dowlais ironmaster, he immersed himself in Celtic studies. Hafod was built c.1860 towards the centre of the park, taking full advantage of the potential offered by the landscaping. It was a simple, two-storeyed classical villa with a square ground plan which would not have looked out of place in any suburban area. It was demolished between 1879 and 1900, and Parkland School now stands on the site. [197]

The Morrises fortune was based on the family's copper-smelting activities from which they had withdrawn, after some 90 years' involvement, in 1808-9, the same decade in which Sir John Morris I had been given his baronetcy and retired from business. The family concentrated on their coal-mining activities after this, while new coppers-melting works began to be constructed by others on a much larger scale. However, they were to be involved later with other types of metal industries in Swansea: with the Millbrook Ironworks at Landore from 1830 and the Landore Tinplate Works from 1851. The family still own some of the huge area of land that was acquired at Swansea in the mid- and late- eighteenth century.

The Vivians

The Vivians built their new Hafod Works at Swansea in 1808-9. John Vivian, the son of a Cornish vicar, had become very active in copper-mining and smelting at Hayle, in his native county; the latter in order to challenge what he and the other Cornish

mining captains saw as the exploitation of themselves by the low ore-prices imposed by the 'Swansea smelters'.[198] In order to break this perceived cartel, he initially came from Cornwall to work the Penclawdd Copperworks, to the west of Swansea, on behalf of the Cornish copper-mining interests. The family continued to live in Cornwall, where they moved in polite society. John Vivian died in a hunting accident at Truro in July 1826 and was buried in the family vault in St Mary's Church (part of which is now incorporated in Truro Cathedral). His second son, John Henry (1785-1855), succeeded his father in his industrial activities and chose to live largely in Swansea to more effectively operate the family smelting works. To do this, the family chose to adapt a residence some 2½ miles to the south-west of the works rather than build a mansion overlooking the smelting-works and workers' village, as John Morris I had done some forty years earlier.

John Henry Vivian was the first member of his Cornish family to settle in Swansea after he married Sarah Jones, the daughter of a Flintshire gentleman, in 1816. His father-in-law moved to Hill House and then Bryn Newydd on Sketty Green, while he supervised the rebuilding of Marino as the couple's new home in 1818-21.[199]

The ubiquitous Swansea architect, William Jernegan, added a rather conventional two-storey rectangular block on the landward side of Jane Morris's three-storey octagon (to the same roof height with consequent ill-matched staggered floors), for the use of John Henry Vivian and his wife. This contained a dining-room, drawing-room, staircase in-between and bedrooms over. Among the 250 books left in the house by Jane's widower, Edward King, for John Henry Vivian were, ironically, volumes from John Morris I's library and

Fig. 242. In 1817-20, John Henry Vivian commissioned William Jernegan to extend 'Marino' into a much larger mansion. The architect's original design was much more elaborate than the plainer and utilitarian final form of the first transformation. (Bernard Morris)

bearing his stamp. Between 1817 and 1830 Vivian acquired virtually all the individual properties that were to form the extensive Singleton estate.

The Oystermouth Tramroad, which Morris had promoted with others, ran along the seashore close to the house and Vivian considered that 'The Tram Road passing close to it is a great Convenience', he indeed used a passenger vehicle on it to attend Oystermouth Church on Sundays. He could also have used it to travel to town and most of the way to the Hafod Copperworks.

In 1825 Vivian also bought the Morris-built mansion 'Veranda' the freehold of which the Morrises had sold in 1818. However, at that time John Henry Vivian was only interested in extending Singleton Park and he quickly resold the house. In 1827 land passed directly from Sir John Morris II to the emergent coppermaster when John Henry Vivian bought the 17 acres of Elm Farm (Tir Sirie or Issa) from him.

The gothic style had already reached the Sketty estates with Jernegan's early-nineteenth-century 'Belvedere' on the Morris's Sketty Park estate. In fact, the building of a fine new house at Sketty Park by Sir John Morris II may well have prompted John Henry Vivian, the pre-eminent copper-smelter of his generation, to improve his own house. Hence in 1823 he decided to extend and modernise Marino in the latest fashion. He approached the nationally known architect Peter Frederick Robinson, who reclothed and enlarged the mansion in gothic revival style between 1827 and 1836.

Medieval idioms were attractive to contemporary industrialists who were seeking to acquire a status comparable with that of the aristocracy, so a wave of gothic recloaked the mansions of the Sketty coppermasters and other industrialists in the 1820s. In fact, this trend had late-eighteenth-century origins in the Swansea area and 'Sketty industrialists' gothic' continued to be built until quite late, at least the 1860s and beyond (Parkwern had gothic wings added in 1924).

The first use of the medieval idiom may have been by John Morris I who had built his Morris Castle workers' flats in 1773. Its designer, John Johnson, went on to build the castellated mansion crowning the Gnoll in Neath for Herbert Mackworth. The associated fort and crenelled Ivy Tower were also built for Mackworth. Much of this medievalism may have arisen from the business connections that Swansea area had with the earlier Bristol copper and brass industry, to which much Swansea copper still went for manufacture into bowl and other goods and to be alloyed into brass. One Bristol coppermaster in the 1760s had built a complete sham castle of copperslag blocks to complement a gothic copper-

smelter's mansion at Arno's Court; a similar and smaller structure with copperslag crenellations ornamented another copper-smelter's lake and grounds at nearby Warmley.

In c.1803 the Swansea entrepreneur, George Haynes, built his 'model village' at Clydach, with a rendered gothic mansion, workers' houses, stables and banqueting house on the scarp above, but with the mansion in a position directly alongside the Swansea Canal and tramroad wharves that formed part of his diverse industrial empire. Parkwern, at Sketty, which became a 'gentleman's residence' c.1799, was certainly embellished with 'toy fort' crenellations and arrow-looped facades by 1817-18.[200]

On the opposite, western, edge of Clyne Valley, 'Woodlands' was bought by the industrialist General Warde and rebuilt as a substantial castellated mansion, in 1824, with William Jernegan again as architect. As with Morris Castle, it was limewashed to attract attention from a distance.[201] More serious medievalism was in vogue by this period and General Warde had rebuilt Woodlands as 'Woodlands Castle' with corbelled-out battlements

fig.243

and towers. The surviving lodge on Oystermouth Road is also castellated. In the 1860s, William Graham Vivian, then owner of Woodlands Castle (which he renamed 'Clyne Castle'), acquired the woods on the western side of the Clyne Valley, including the Clyne Valley Arsenic and Copperworks, and adapted the upper condensing chamber and chimney as the castellated Ivy Tower (the main works building was converted to hay

Fig. 243. The early nineteenth-century lodge, Woodlands (Clyne) Castle, on the Oystermouth Road is castellated in the medieval style beloved of Swansea industrialists in the 1820s. (9500347/2)

EAST FRONT.

sheds); this stood on the opposite, western, rim of the Clyne Valley from the Morris's Belvedere. Both the Ivy Tower and Woodlands Castle (now serving as a university hall of residence) can still be seen.

In the context of all this architectural improvement, John Henry Vivian felt obliged to contact Peter Frederick Robinson in 1823. Instead of demolishing Marino and its later side-wings, Vivian ordered the encasing of the house in rich ornament and heraldic mouldings. Work started in 1827 and its completion was marked by a grand ball which was held on 20 January 1832. Vivian showed a sense of economy in adapting an existing house. The original peaked roof of the octagon (now largely submerged by the later additions) was incorporated into a new longitudinal roof running the length of flanking wings and terminated by hipped roofs broken by conventional upstanding chimney-stacks.[202] However, the use of local rubble-stone covered in a cement-wash, compounded by the structural change from classical to gothic, has created problems of maintenance in a building that now serves as the administration building for the University of Wales, Swansea. Gables, enclosing gothic hood-moulded lancet windows, were added to every surviving plane of the original facade. A large, overhanging bay

window replaced the central first-floor balcony of Jernegan's octagon and a mere string-course below the multiple gables shows where the elegant, spired roof once commenced. The recased, more coherent south front was almost four times as wide as the original house of Marino, 121ft compared to the Marino octagon's 38ft. Vivian's grandiose new house was also almost as deep as it was wide, a substantial new east wing was built, extending back 125ft. This had a slender, 63ft high, octagonal, crenellated tower with spirelets at its corners which gave the house some claim to its new pseudo-medieval title of Singleton Abbey. This turret, reminiscent of Beckton's Fonthill, also held a water cistern. The surviving building has lacked this feature since its was demolished in the 1980s.[203]

The west of Swansea was still dominated by the Morris's elegantly landscaped Sketty Park. Not to be outclassed, John Henry Vivian acquired 42 acres of land and the farm of Singleton in 1829. Vivian rebuilt the old Singleton farmhouse as his home farm, incorporating window-mullions and other stone dressings from the medieval Plas House in Swansea, which he obtained when the house was being demolished in 1840. The name of the farm also provided the medieval-sounding title for his reconstructed mansion.

Fig. 244. In 1823, John Henry Vivian approached the London architect Peter Frederick Robinson to prepare plans to transform his plain Regency-style house into a large Tudor gothic mansion. This was completed in stages during 1827-36. (University of Wales Swansea, Library & Information Services, Archives)

223

Fig. 245. Sarah Vivian (born Sarah Jones) and John Henry Vivian pictured at Singleton Abbey in the mid nineteenth century. (City & County of Swansea: Swansea Museum Collection)

Fig. 246. Prominent among a bay of Vivian family tablets in Truro Cathedral is that of John Henry Vivian who built up the Hafod Copperworks with profits generated from the family's extensive Cornish copper-mining interests. (990236/19)

In 1843 John Henry Vivian's eldest son, Henry Hussey Vivian, married and his father repurchased Veranda as the matrimonial home. Henry's wife died within a year, shortly after giving birth to a son. Most of veranda was subsequently demolished (between 1851 and 1853) and the remnant of the house was renamed Wern Eynon. Sketty Church was built nearby by the Guildford architect, Henry Woodyer, as a memorial to Henry Hussey's wife. In 1853, John Henry Vivian bought the castellated villa of Parkwern and had Woodyer double it in size to form a suitable residence for Henry Hussey and his new wife. It remained one of the Vivians' principal residences for nearly 70 years. The only other residence in the area with castellated bay projections

similar to Parkwern, was William Jernegan's 'Woodlands Castle'. Parkwern had been built in 1799-1800, (at the same time as John Morris I was building his marine villa further along the coastal ridge to the west) either by James Coke, or by John Minshul who was the occupier between 1800 and 1809.[204] Later tenants included William Henry Smith in 1842 and then Lewis Llewelyn Dillwyn, son of the former owner of the Cambrian Pottery, who left in 1853 for the neo-Tudor gothic mansion of Hendrefoilan which had been built for him a mile to the west.

The walls of the early mansion of Parkwern were incorporated within Woodyer's house. During construction Woodyer's plans seem to have been altered and the house was built as a three-storey, rather than a two-storey, building. The multi-gabled house was fenestrated with short, grouped lancets and in one corner had a keep-like crenellated tower with projecting bartizans on its top corners. A water-wheel pumped water from the nearby Cockett Stream. The house was occupied until 1886, when Henry Hussey Vivian and his wife moved to Singleton Abbey on the death of his mother (John Henry Vivian had died 31 years previously, in 1855).

Singleton Abbey was then considerably enlarged and Henry Hussey Vivian became a baronet in 1882 and subsequently the first Baron Swansea in 1893.[205] He died the following year but his widow and

Fig. 247. After 1836, Marino assumed the form of a large gothic mansion which was already known as Singleton Abbey by 1830. In 1920, the house was sold to become the centre of University College, Swansea. (Illustrated London News, 1881)

daughters continued to live there. The Vivians' copper-smelting at Swansea ceased in 1924 after 115 years, and the second Lord Swansea spent his time in London, or travelling, or at the family's Tudor-gabled country mansion of 'Bryn Howell' near Builth Wells (now a hotel).

Under the terms of John Henry Vivian's will, Parkwern was to revert to the second of his four sons, William Graham Vivian, but by 1886 he had already been living at nearby 'Woodlands Castle' for some 26 years. He had bought and enlarged it with

higher castellated blocks in 1860, later adding his own church-sized chapel; he saw no reason to move to Parkwern. Medieval French fireplaces and other exotic features were brought in to adorn the mansion, just as Glynn Vivian was to do at nearby Sketty Hall and Randolph Hearst at St. Donat's Castle in the Vale of Glamorgan. Parkwern stood empty for 26 years until Graham's death in 1912 and subsequently it became a military hospital. In 1920 it was bought by the local businessman, Roger Beck, for the Swansea hospital authorities, and when it became a nurse's

Fig. 248. Henry Hussey Vivian (centre) ran Hafod Copperworks in succession to his father from 1855, but he only became resident at Singleton Abbey after the death of his mother, Sarah in 1886. A widower twice, he finally acquired a family of six with his third wife, Averil Beaumont. They are pictured in the orangery at Singleton in 1893; Henry had added a new orchid and hothouse in 1889. (Colonel Douglas of Mains)

225

training school in 1922 it was renamed 'Parc Beck'. A large wing was added in 1924, also with lancet windows, and in 1989 a large part of the grounds, long used as allotments, was sold and built on.

In 1898 the Vivians came into possession of the seventh of their mansions in the fashionable Sketty area, surpassing the number once held by the earlier Morrises. Indeed, this last house, Sketty Hall, was one of no less than three that had previously been held by members of the Morris family.[206] Richard Glynn Vivian, John Henry's youngest son, was far more interested in art collecting than in his family's copper-smelting activities and only returned to Swansea when he bought Sketty Hall in 1898. He added a roof-top gazebo (surviving), a parapet balustrade (gone) and marble balustraded balconies (restored). Exotic additions included the Italian garden, which incorporated the contents of 197 crates of interior fittings of a demolished Genoan church which he had shipped into Swansea Docks in 1901.

He endowed and laid the foundation stone for the Glynn Vivian Art Gallery in Swansea in 1909 and in 1910 died at his London home. After his death, the large mansion was occupied by Averil, Dowager Baroness Swansea (Henry Hussey Vivian's widow) and her daughter Violet. The hall was sold to Swansea Borough Council in 1936 and later became the research laboratories for the British Iron & Steel Research Association; it has now been refurbished as Swansea College's Conference Centre and Catering School.

In and around Singleton Park, the Vivians had eight lodges and park buildings built in a variety of gothic, cottage and more exotic styles.[207] Of these, the architect Peter Frederick Robinson built three of the Singleton lodges or cottages: Front Lodge (Mumbles Road), Sketty Lane Lodge (thatched) and Swiss Cottage. The notable gothic tower on the foreshore road beside Swansea Bay, at the corner of Brynmill Lane and Mumbles Road, was built by Henry Woodyer c. 1854.

It is interesting to speculate why Sketty, on the foreshore of west Swansea, developed into a landscape of industrialists' mansions and parks. It may have been no more than a combination of the superb view of Swansea Bay and the comparative failure of the Clyne Valley collieries, that induced John Morris I to extend the sea-viewing facilities of Sketty Hall and to build a second scenically-placed 'Verandah' to complement the marine villa that his brother-in-law had already built at 'Marino'. It was far easier and more convenient for the frugal John Henry Vivian to buy an existing mansion in this developing genteel quasi-suburb, than to build a mansion and park overlooking the Hafod Works

where there was no convenient elevated bluff as there was at Clasemont. Marino was $2\frac{1}{2}$ miles south-west of the Hafod Copperworks.

The industrial wealth of Swansea created an extensive landscape of leisure around Swansea Bay for its main entrepreneurs. Much of the Vivians' wooded and landscaped parkland can now be visited as a public park, whilst all of Morris's Sketty Park has disappeared under suburbia except for a slender strip of its western edge now in the Clyne Valley Country Park. Parts of the grounds of Graham Vivian's Clyne Castle form the western part of the Clyne Valley Country Park and public gardens. Outside this area, besides Tredegar Fawr, there were two important coppermasters' mansions at Gwernllwynchwith and Maesteg House to the east of the River Tawe.

Chauncey Townsend and the Smiths

Chauncey Townsend's son-in-law, John Smith, bought a mansion at Llansamlet 15 years after Townsend had founded the Middle Bank Copperworks in 1755 (with a lead and a zinc smelting works at adjacent Upper Bank immediately afterwards). Townsend was also responsible for the post-1750 development of the Llansamlet coalfield, and industrial development on a substantial scale around Llanelli and, more modestly, at Neath and Landore. All this enterprise was financed by the considerable wealth Townsend had accumulated by providing British forces in Nova Scotia with supplies. He may have largely controlled his Swansea interests from his merchant premises in the parish of St Peter le Poor in London. It is possible that he also used the existing Gwernllwynchwith Mansion that stood on the north of the Llansamlet coalfield, beyond the parish church.

Townsend died in 1770, leaving one-fifth of his property to each of his two sons, James and Joseph, and to his three daughters Elizabeth, Charlotte and Sarah. John Smith, Elizabeth's husband, had already been involved in Townsend's business affairs and took over management of his collieries.[208] The split ownership inevitably caused problems and Smith took steps to obtain a further three-fifths of Townsend's original coal leases from James, Joseph and Sarah Townsend, through recourse to the law in 1786. The remaining one-fifth share (Charlotte's) was to be the subject of legal arbitration after his death.[209] On John Smith's death in 1797, his four-fifths' interest in Townsend's leases passed to his two sons, Charles and Henry Smith.[210]

In 1770 John Smith bought Gwernllwynchwith House,[211] which survived until about 1960. It was a fine cubic house of two full storeys with a recessed attic storey, which had something of the grace and

Fig. 249. The now-demolished mansion of the coal-mining Smith family at Gwernllwynchwith; the colliery engine house that stood among the walled gardens still survives. (GL6201)

elegance of Sir John Morris I's Clasemont across the valley. It was built of brick, similar both to the dressings of the Gwernllwynchwith engine house, that still stands next to the mansion's walled garden, and to the slightly later brick arches of the over-bridges and tunnel on John Smith's Llansamlet Canal of 1783-5. However, its Queen Anne styling suggests that the brick mansion may pre-date John Smith's occupancy. The hipped roof had two hipped dormer-windows on each face, with a recessed central pediment containing a distinctive half-round or lunette window. The garden walls also have castellation of cast copper-slag blocks.

Charles Henry Smith left Gwernllwynchwith in about 1840, and followed most of the other industrial proprietors to west Swansea. The family had sold their three non-ferrous smelting works in the later eighteenth century and concentrated all their efforts on coal-mining. Charles Henry bought Waun Kysha, otherwise known as Ty'r Tom Hopkin, at Lower Sketty and enlarged the house, renaming it 'Lower Sketty'; the house was later renamed 'Derwen Fawr' after a massive oak tree in the grounds. In 1852, his daughter, Emily Matilda, was married to George Byng Morris, son of Sir John Morris II.[212] Smith left Derwen Fawr in 1866 and settled in Tenby, where he died in 1878. The house survives and is now part of the Bible College of Wales.

The Grenfells

The Grenfells were the fourth dynasty of managing partners of a copper concern to live in the Swansea area. In terms of the age of the works they ran, they were late in acquiring a large mansion near the town. The Grenfells came originally from the small, strongly Methodist mining town of St Just in Penwith, in western Cornwall, and by the eighteenth century were merchants with an active interest in Cornish mining.[213] Pascoe Grenfell (1761-1838) was at least the third of that name and was a well educated much-travelled merchant who spoke both Dutch and French as well as English.

In 1794 Thomas Williams funded Grenfell and his own son, Owen Williams, in a speculative venture with John Vivian to buy copper ore in Cornwall (to offset the falling-off of the formerly immense Anglesey production) and to sell it to Williams for use in the Middle and Upper Bank Copperworks. By 1801 Pascoe Grenfell held a half-share in the partnership, his brother William one-sixth, and Owen Williams one-third.[214] This partnership bought the Upper Bank Copperworks from the Parys Company in 1803-4 and took over the Middle Bank Copperworks from the Stanley Company at the same time. The Grenfells were the more commercially astute partners (Pascoe was responsible for the publication of the Bank of England accounts) and the

Williams family retired from the copper business in 1825. In 1825 the firm of Pascoe Grenfell and Sons took over the works completely and dominated the east side of the lower Swansea Valley until the 1890s.

Pascoe Grenfell succeeded Thomas Williams as MP for Marlow in Buckinghamshire; the family had a residence at Taplow on the River Thames. Two sons were involved in the acquisition of property in Swansea: firstly, Riversdale William Grenfell took a sub-lease of the Maesteg estate on the Earl of Jersey's land from Mrs M.E. Tennant. His elder brother, Pascoe St Leger Grenfell, built Maesteg House in the 1840s (he died in 1879); to what extent he also resided in the large family mansion at Taplow is not clear. Maesteg House was about one mile south of the Middle Bank Copperworks on the high south-western slopes of Kilvey Hill, looking seaward over the Swansea harbour mouth at Fabian's Bay. Foxhole, the main copperworkers' settlement for those employed at the Freemans' White Rock Copperworks and at the Grenfells' Upper and Middle Banks Copperworks, was half a mile to the north-west of his new residence; the copperworkers' settlement built by the family at Grenfelltown was a mile to the north.

Maesteg House was situated on the southern slopes of Kilvey Hill, overlooking Fabian's Bay, and compared to Clasemont or Singleton Abbey it was a modest residence, but of course the Grenfells had an equally grand mansion near their coppermills, at Taplow. Charles Grenfell lived at Taplow and among the family's philanthropic acts based on their copper industry profits, was a gift of land for the large Anglican church and schools of All Saints, Boyne Hill in 1857; the only parish church in the borough of Maidenhead.[215]

Their Swansea house consisted of an irregularly shaped block, 70ft wide by 60ft deep, with a hipped roof. On the front of the house there were two full-height bay windows with peaked slate roofs. A conservatory formed an extension on the eastern side. Overall it occupied about four times the area of a typical dwelling built for an employee of the firm. The main lodge, with the stables, greenhouses and Maesteg Cottage, stood at what is now the junction of Morris Lane and Grenfell Park Road. Grenfell Park Road follows the line of the drive and the house itself stood at what is now the junction of this road with St Leger Crescent. There was a second lodge near the junction of Kinley Street and Port Tennant Road. Maesteg House accommodated Belgian refugees during World War I and was pulled down soon afterwards.

The grid-iron layout of the new Swansea suburb of St Thomas was immediately in front of the garden of the mansion in the 1850s and 1860s.

The involvement of the Grenfell family with Swansea, as managing partners and then owners of the Middle and Upper Bank Copperworks, was in the period 1804-92 (the Grenfells were also partners in the Rose Copperworks in the period 1823-29). Pascoe Grenfell was fully involved in the economic life of the town, being a member of the Corporation and Chairman of both the Swansea Harbour Trust and the Swansea Vale Railway.

All this was in addition to the host of civic, military and business activities and offices that one might expect of a major industrialist. He was Deputy Lieutenant of Glamorgan and a town councillor, but he refused to be Mayor of Swansea. He was also the first chairman of the Released Prisoners' Aid Society in 1868. The regard in which Pascoe St Leger Grenfell was held locally was reflected in a vote of condolence which was inserted in *The Cambrian*, following his death, by '... the whole body of the officials, men and boys, working at the Middle Bank ...'. In it they declared that they were '... thankful for a master who took such an unfailing interest in us - an interest not confined to our work, but extending to our spiritual and mental welfare, our homes and our children. We feel that we have lost our best friend ....'

The level of Grenfell family involvement in the Swansea business started to decline when Pascoe St Leger Grenfell died in 1879. After his death there may have been an unwillingness to introduce new and expensive technology, or to incur the costs of importing copper ore or semi-refined copper from the available reserves overseas. In 1892 the family firm was liquidated, the London offices at 27 Upper Thames Street sold, and the new firm of Williams, Foster and Pascoe Grenfell was formed. In this inevitable rationalisation of a declining industry, the Grenfells retained one seat on the board of a company which now owned the Middle Bank and Upper Bank Copperworks as well as the Morfa Copperworks. Copper smelting ceased at the works of the company in 1924. Their earlier nineteenth-century house at Taplow is now a hotel and their much grander late nineteenth-century mansion is a Buddhist study centre with Pascoe St Leger Grenfell's remains entombed in the family vault alongside. The philanthropic work of the Grenfell family in Swansea largely ended with the death of Mary Grenfell in 1894.

George Haynes and Clydach

The model settlements examined so far were connected with the Swansea copper industry. However, not all Swansea industry involved copper-refining. Coal was needed in vast quantities to fuel the smelters, and iron-castings and machinery were needed to run the works and rolling-mills. Coal and iron were also produced for markets beyond the Swansea works. Coal fuelled the local pottery industry and this large centre of industry in an old mercantile and maritime town produced the market for the first newspaper produced in Wales.

George Haynes, born in Henley-in-Arden, Warwickshire, has been judged one of the most under-rated influences on the development of Swansea in the late eighteenth century.[216] Haynes was an industrial entrepreneur who had been active in business in the Quaker colony of Philadelphia, Pennsylvania. His obituary in *The Cambrian* in January 1830 recorded that 'the earlier part of his life was a scene of active industry and utility' and it claimed that he had 'retired' to Swansea.[217] However, he was soon involved in running the Cambrian Pottery and in other business affairs, such as founding *The Cambrian* newspaper, the Cambrian Brewery, the Glamorgan Pottery and the Swansea Bank. The last-named activity resulted in his bank's foreclosing on Richard Parsons and taking Clydach Upper Forge and Ynysgedwyn Ironworks into administration. His philanthropic side was revealed in his substantial interest in the paving and lighting of Swansea town, in the Swansea Tontine, the Swansea Harbour Trust and the Theatre.

Haynes's initial interest was in the Swansea Pottery. He later said that he had become involved as an 'interest' but then discovered that the enterprise needed total reorganisation. The Swansea Pottery had been founded by William Coles in 1764. George Haynes became a partner with Coles's three sons in 1786-7 and the managing partner from 1790, when he re-named it the Cambrian Pottery. During 1801-2 the Quaker William Dillwyn purchased the lease of the pottery and the controlling interest in the business from George Haynes on behalf of his son, Lewis Weston Dillwyn.[218] The capital released from the sale of the pottery enabled Haynes to pursue his building schemes at Clydach.

Haynes had done much to improve the running of the pottery business. He had experimented with the production of porcelain in *c.* 1796, but production was only successfully carried out by the subsidiary Cambrian China Works in the short period 1814-17, before the expert potter Samuel Walker returned to Nantgarw. Under the management of Haynes, the Cambrian Pottery produced a far wider range of wares than it did under Dillwyn or his successors and was a more consistent commercial success. Haynes's aim was to produce a high quality earthenware suited to a wide range of fashionable taste. All the available sources make it clear that George Haynes was the driving force in the recruitment of new talent to the Cambrian Pottery, and it was his judgement which was

behind the reorganisation of the pottery in 1788-9 in imitation of Josiah Wedgwood's methods.

After the sale of 1801-2, the pottery company was known as Messrs Haynes, Dillwyn & Co, with George Haynes remaining as manager until 1810. In 1811 he left as the result of friction with his partner and in 1813 he founded the neighbouring Glamorgan Pottery with his son-in-law, William Baker. The potteries were alongside each other, and had docks on the Swansea Canal for the reception of coal from further up the valley. Haynes also founded the evil-smelling soapworks upwind of the Cambrian Pottery. At their height the two potteries employed 400 people. The 1816 *Swansea Directory* of William Mathews noted 'George Haynes [to] whose public spirit the town of Swansea is much indebted'. In 1838-9 the Glamorgan Pottery was acquired and closed down by the Dillwyns, much of the plant going to Llanelli; in 1870 the Cambrian Pottery itself closed.

In about 1803 George Haynes founded what *The Swansea Guide* of 1813 described as:

> 'a sweet romantic spot, where is the tasty retirement of Mr Haynes who has built some cottages in a style that adds peculiar beauties to the spot. Here is a very large walled enclosure with hot and cold greenhouses.'

This complex, including Haynes's mansion, stood alongside a wharfage on the Swansea Canal at Clydach. This was unique among the settlements so far considered in having the owner's mansion balanced by a similar block housing the workers. His workers' housing and the stables, as well as his mansion, were erected in Georgian 'gothic' style; the only parallels for such 'polite' workers' architecture in the lower Swansea Valley were John Morris I's castellated workers' flats and pedimented Poor House.

The settlement was alongside the Swansea Canal, in an area 270ft in breadth between the canal and the main valley road; George Haynes was a member of the Swansea Canal Committee. The origin of the settlement lay 450ft to the north-east, where the new Clydach Railroad terminated in wharves which were occupied by the owners of new mines in the lower Clydach valley. The railroad ran north-westwards from the canal and passed immediately west of the Upper Forge. A canny businessman, George Haynes proposed to build his own house and the houses of his workers on the spoil dumps resulting from the excavation of the neighbouring canal. There are interesting later parallels in the Upper Swansea Valley of the use of such waste land: George Crane (of the Ynysgedwyn Ironworks) built his workers' housing on a riverside reclamation embankment,

fig.250

Key to Clydach Landscape

Pre-existing roads

Cwm Clydach Railroad 1798

Watercourses

George Haynes's improvements after 1803

Parkland in Haynes's ownership

Woodland in Haynes's ownership

Orchard in Haynes's ownership

Fields in Haynes's ownership

Arable fields in Haynes's ownership

Paths & trackways on Hayne's Estate

Boundary of part of Estate leased by Thomas S. Strick in 1838

John Strick's Foundry

Heol Dywyll

Fields

To Swansea

Swansea Canal

Barn & Cottage

Cwm-Dŵr

Smithy

Houses

Summer house

Park

Dry dock

Stables

Ynys-Tanglwys

Ice house

Gardens

Orchard, walled-garden and walks

Fruit walls

Fields

Cwm Clydach Railway

Canal feeder

Canal lengthman's cottage

Arable

River Tawe

Canal feeder

0   300 Metres

0   1000 Feet

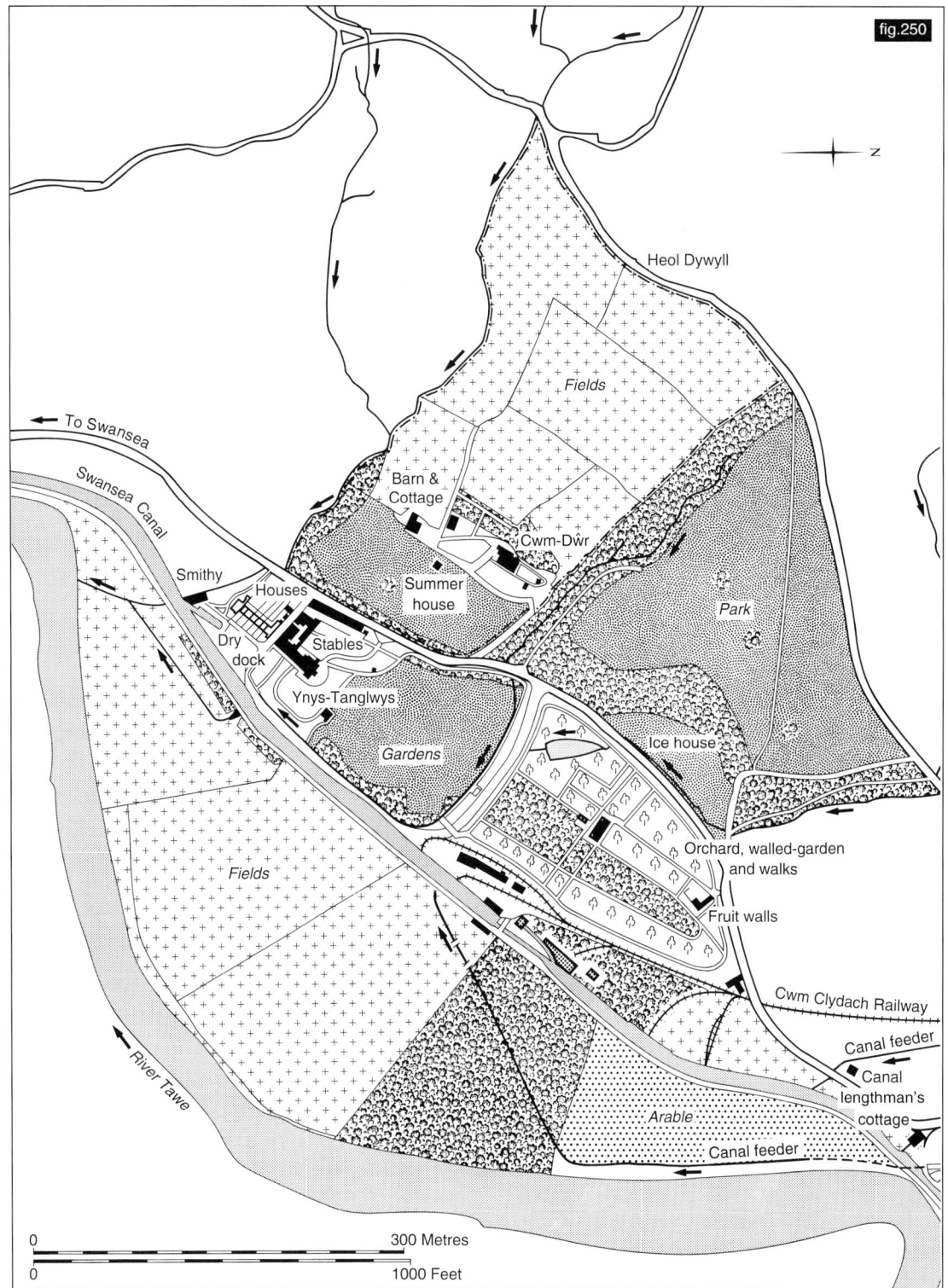

Fig. 250. In 1803, George Haynes bought land laid waste by the construction of the Swansea Canal at Clydach, and with adjoining areas turned it into a mixed industrial and rural landscape. He erected two blocks in gothic style alongside the canal, one formed of seven workers' houses and the other of his new mansion of Ynys-Tanglwys.

while John Dickson (contractor, Neath & Brecon Railway) constructed housing on a narrow strip between two old tramroad formations.[219]

Thus, on 1 February 1803, the Swansea Canal Company minutes record: Mr. Haynes - proposed - purchase those parts of the:

'Spoil banks and wastes on Clidach Farm not used by the Canal Company ... the company allowing Mr. Haynes the expense of making three gates on or near the Bridge and for fencing the road to join the Waggonway.'[220]

The road between the houses and the Clydach

CLYDACH-ON-TAWE DOCKYARD

Fig. 251. 'Clydach-on-Tawe Dockyard', originally the working area of George Haynes's estate. Middle background: smithy and dry-dock for canal-boat maintenance; extreme right: later boat-house where the steam-tug used to tow canal-boats was based; beyond the boat-house: piles of pit-timber imported from Swansea Docks. (John Hutchings)

Railroad's wharves is the key to understanding why Haynes' buildings were constructed. Haynes needed coal to fuel his potteries, brewery and soapworks, which were sited alongside the southern end of the Swansea Canal. He may have owned one of the several collieries which were being opened in Cwm Clydach alongside the new railroad: in 1802 'Hayes [sic] & Co' are one of six proprietors noted as owning collieries alongside the canal.[221] In any event, throughout the nineteenth and early twentieth centuries boats built, repaired and manned from Haynes's wharf and buildings at Clydach helped to supply the vast amounts of bituminous coal which were needed by the works of Swansea.

Haynes' Buildings[222] consisted of three long blocks aligned south-west/north-east alongside the Swansea Canal. The two blocks, nearest to the canal, which were almost symmetrically balanced, consisted of housing for seven families in the south-western terrace; while the north-eastern block consisted of Haynes' mansion, which was set in a garden about 300ft square with summer-houses, and with a service court and buildings to its rear. Behind this, aligned with the main valley road, was a large block of stabling with hay lofts above. Orchards lay to the north and pleasure grounds for the mansion were on the hillside above, including what appears to have been a banqueting house or prospect tower which has now been converted to a dwelling-house.

Haynes's settlement may have been wholly, or partly, occupied by Quakers. Samuel Hill, a Quaker,

lived in Morriston in the 1790s but moved to Clydach to build canal boats. His sons Samuel (born 1810) and John (born 1813) also became boatbuilders, while Benjamin (born 1801) became the Swansea Canal manager/engineer.[223]

The end houses were larger and presumably intended for the foremen. The north-eastern house extended back along the entrance to the site for 48ft and was mirrored by a similar rearward extension from the mansion block. The house at the south-western end extended forward of the frontage of the main block, forming a hipped-roof pavilion and terminating the block in a way that was also mirrored at the north-eastern (or mansion) end of the north-eastern block. Overall both blocks formed a fairly ornate frontage (265ft long) to the Swansea Canal, with an originality of composition that might suggest the hand of William Jernegan.

The main three blocks (which were demolished in the 1960s) were subsequently much altered, but enough cartographic and photographic evidence remains to establish their original form. The mass of walling was in Pennant sandstone rubble with brick used for the heads of openings and for chimney-stacks. The workers' housing had two, fairly small, rectangular openings on the upper front of each house with flat, triangular heads that were a simplification of the four-centred arches that crowned the lower doorways. These doorways were originally crowned with flowing gothic tracery, executed in cast iron and flanked by a single window. Both upper and lower windows had six-light casements but the lower casements had sharply

231

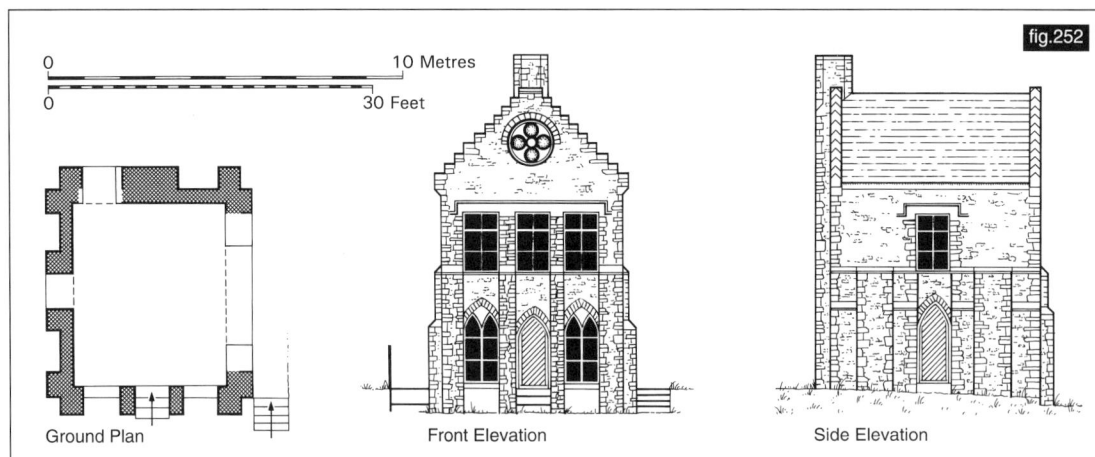

Figs. 252, 252a. The surviving summer-house on the valleyside above the site of Ynys-Tanglwys Mansion. (960365/1)

Ground Plan | Front Elevation | Side Elevation

fig.252a

pointed heads infilled with 'Y' tracery. The end pavilions probably had circular windows looking back along the façades of the two blocks. This was certainly the case with the north-eastern end which retained elaborate quatrefoil cast-iron tracery intersected by a delicate diagonal cast-iron cross. All roofs had characteristic, Regency-style, hipped ends, rather than gables. On the south-eastern façade, the mansion pavilion had a small upper window and a high-arched 'Y' tracery opening on the ground floor. On the adjoining face of the main block there was a deep, triangular-headed staircase window and a similar small bedroom window alongside it to the south-west, with a doorway into the garden below. The mansion block was later converted into five terraced dwellings and the rest of the early fenestration to the south-west was replaced by large Victorian openings. Two other blocked gothic doorways remained in the façade of this block until the 1960s.

A service court lay behind the south-western half of the mansion block, with a large, two-storey range of stables to the north, backing onto the main road. Its dimensions were 148ft in length and 18ft in breadth. Large openings in two storeys alternated with circular openings on the upper floor and high-arched gothic arches on the floor below. The second, eighth and tenth bays from the north-eastern end had large loading and access doors on both the ground and first floors, for horses and hay respectively. Both sets of openings had depressed triangular tops, reminiscent of those found in the upper floors of the houses. The fourth and sixth bays from the north-eastern end of the stables had blind, two-storey, gothic-arched recesses with upper loading doors formed in the sharp points of the arches and lower openings under wide depressed triangular heads. All openings and walls seem to have been in Pennant sandstone. The block was later extended to the south-west in similar style

fig.253

but completely executed in brick. There was a robust dentiling at eaves level.

A secondary, rather plain white-rendered mansion survives on the hillside overlooking the cleared site of Haynes's 'tasteful retirement'. This is 'Cwmdwr', originally the home of Samuel Fox Parsons; in 1817 George Haynes (as the mortgagee) took over the failed business of Richard and Samuel Parsons, owners of the Clydach Upper Forge and the Ynysgedwyn Ironworks. The house was built after 1813 and came to form part of the Ynystanglwys land-holding even if this was not the original situation. In 1819 Samuel Fox Parsons married Felicia Haynes (George's second child) and he became a partner in the Glamorgan Pottery. Two of George's grandchildren were born probably at Cwmdwr in 1820 and 1825, and George Haynes died in 1830.[224]

Capitalists and Workers Houses: Conclusions and Contrasts.

Three distinct and widely differing types of copperworks' housing have been examined in detail: those for resident managing partners, for agents or managers, and for the mass of workers. Yet there were some common traits: all the copper-smelting concerns were primarily concerned to see that their activities were a commercial success and that all available capital should initially reserved for the productive process. This explains why three out of four of the resident dynasties of copper-smelters first moved into existing small gentry houses and either enlarged them as the money became available, as in the cases of the Morrises at Tredegar Fawr and the Vivians at Marino; or left them much as they were, as did the Townsend/Smiths at Gwernllwynchwith. Only when the works were successful and producing enough profit were these relatively modest mansions enlarged or, more rarely, completely new houses constructed, for example Clasemont.

A similar picture emerges with the management housing. Two of the four copperworks' agents' houses were conversions of former local gentry houses, for example at the White Rock Copperworks and the Upper Forest Coppermills. Only later, when the works were more successful, were these replaced by newly-built large houses. With the large works, such as the Hafod Copperworks, a couple of large houses might be built for under-managers when the works were at the height of their prosperity, 50 years after the construction of the original works.

Fig. 253. Gothic-style stable-block at Ynys-Tanglwys mansion facing south-eastwards towards the house and the Swansea Canal. Horses from here may have drawn George Haynes's canal-boats as well as coaches for his mansion. (Gareth Mills)

233

fig.254

Fig. 254. Gothic-style window-openings on the Haynes (Ynys-Tanglwys) workers' housing facing the Swansea Canal at Clydach (1803). The house on the left retained its original 'Y' wooden tracery in its ground-floor window beneath pointed twin-casement windows to a fairly low upper floor. (University of Wales Swansea, Library & Information Services, Archives)

The situation with regard to the general mass of the workers was rather similar. There was an initial concern to house only those workers essential to the productive process, while all available capital was used in the building up of the core business. However, the remoteness of the works from Swansea town meant that substantial accommodation needed to be provided for a greater number of workers as the works expanded. Sometimes this could be provided in accommodation adapted from existing buildings, as in the 24 dwellings made available at White Rock or in the first collier's house provided by the Morrises at Tirdeunaw Colliery. New houses on a large scale might be provided some 30 years after the construction of a works. They were generally made with sufficient living space to attract workers but did not usually have what the works' owners might consider unnecessary ornamentation. There is evidence that houses at both Trevivian and Grenfelltown were enlarged as general standards of housing improved and profits became available.

Conclusion: Houses, Settlements and Enlightened Self Interest.

What does this material evidence, and the documentary sources tell us about relationships between masters and men in eighteenth-and nineteenth-century Swansea? How was this relationship expressed in spatial and landscape terms in the lower Swansea Valley? It would be inappropriate to judge standards of living in terms of those at the beginning of the twentieth-first century, but how did they compare to those in the same industry elsewhere in Britain? How do they compare with other industries and industrial centres within Wales and the British Isles and with the wider European experience as industrialisation spread.

The basic question concerning the social impact of the copper industry is whether the industry raised living standards for the majority of the local inhabitants in the short-term. The answer is that of course it did, not only in the strikingly higher wages that were available in the copper industry compared to the subsistence wages available to workers on the surrounding estates,[225] but also in the switch from small cramped, single-storey cottages centred around a single cooking and heating hearth. Living developed from a necessarily communal activity by the construction of two-storey houses which allowed a degree of privacy and specialised space for individuals of all social classes. Workers' settlements upwind of the smelting-works might also avoid some of the worst pollution spewing from the chimneys and flues, for at that time the measures for controlling the sulphurous fumes were seriously deficient. The houses built downwind of prevailing winds, such as those constructed by the workers themselves at Foxhole, were not so well placed and even window glass turned opaque.

It has been said that the other world centre of metals smelting in Glamorgan, Merthyr Tydfil, exhibits many of the features of a colonial town; such factors are much less in evidence in Swansea. It was already one of the largest towns in Wales prior to the boom in the copper industry, and much of the early development in the coal industry was undertaken by local gentry families who also encouraged the foundation of copper-smelters and water-powered forges. Merthyr Tydfil, by comparison, prior to industrialisation, was a small sparsely populated farming community surrounded by vast tracts of upland waste.

Both of the main copper-smelting resident dynasties in the lower Swansea Valley - the Morrises from the west Shropshire hills (at Cleobury Mortimer and then Bishop's Castle, where many families have retained surnames of Welsh origin) and the Vivians from west Cornwall - quickly intermarried with local families, the latter for two generations. The Morrises accumulated wealth through managing the capital of moneyed partners based in the south-east of England. They did not have large estates or interests in England and invested their capital in considerable tracts of land in and around Swansea, where their economic, social and political interests became centred. The Vivians already had considerable economic, political and social interests and contacts in west Cornwall, so for the first half of the nineteenth century the senior family members remained resident in Cornwall with the junior line of the family in Swansea. The financial interests of the residual family in Cornwall

were bought out by that part of the family resident in Swansea in the mid-nineteenth century. As with the Morrises in the eighteenth and nineteenth centuries, the political, economic and social focus of the Vivians for the majority of the nineteenth century, and for the first quarter of the twentieth, became centred at Swansea. This strong local interest and involvement largely explains why the only two large copperworkers' settlements, laid-out on grid-plans and given churches, were those planned by these resident dynasties.

The third resident family, the Grenfells, was the only other coppermasters' family to build a church in a copperworkers' settlement. Unlike the other two dynasties, the main managing family partner was only resident in Swansea for one generation and did not intermarry with the local gentry, though others in his family did. This helps to explain why they did not construct a large grid-plan of streets for their workers, although the pre-existing settlement at Foxhole and the steeply-graded local topography are other factors that militated against such a course of action. The main residence of the family remained around their coppermills and large mansions in the Thames Valley; it was there that they intermarried with the local gentry and exercised their financial powers of philanthropy in equal measure to their benevolence at Swansea.

The Cornish family who owned the biggest copperworks, the Williamses, remained resident in Cornwall, close to their copper-mines. Some housing was built at Swansea but not a large expanding community equipped with churches or schools as in the examples above. Michael Williams, as MP for north Devon, did practice philanthropy but in the area where his political interests lay.

There was always an element of self-interest in building workers' housing to ensure that there was a sufficiently skilled workforce, attracted and kept in order to operate a copperworks. Anything more than that had a strong element of philanthropy. There is little doubt that the neighbouring gentry had a strong sense of duty and paternalism towards their surrounding communities. The resident coppermasters, who mixed socially and intermarried with this gentry, also clearly felt a sense of duty to their dependent workforce. Their political standing in the community was also partly dependent on their local social responsibility and 'enlightened self-interest' may be the best way to describe this undoubtedly complex amalgam of motives.

References

1. D. Bick and P.W. Davies, *Lewis Morris and the Cardiganshire Mines* (Aberystwyth, 1994), 8; S.J. Hughes, 'The Cwmystwyth Mines', *British Mining,* 17 (1981), 7-8.

2. S.D. Coates, *The Water Powered Industries of the Lower Wye Valley* (Monmouth, 1992), 11.

3. J. Davies, *Hanes Cymru* (London, 1992), 252.

4. W. Rees, *Industry before the Industrial Revolution* (Cardiff, 1968), 313; P. Riden, *A Gazetteer of Charcoal-fired Blast Furnaces in Great Britain in use since 1660* (Cardiff, 2nd ed., 1993), 32.

5. W. Rees, *op.cit.,* 275.

6. J. van Laun, *The Clydach Gorge: Industrial archaeology trials in the south-east of the Brecon Beacons National Park* (Brecon, 3rd ed., 1989), 26-7.

7. 1841 census of Llansamlet Higher and *c.*1793 map of the *Case of the Duke of Beaufort* (Swansea Ref. Library).

8. 1841 census of Llansamlet Higher.

9. B.L. Ordnance Survey Drawings 179, 'Swansea Bay', Serial No. 135, Index 1" No. 37, 1813 .

10. J.B. Lowe, *Welsh Industrial Workers Housing, 1775-1875* (Cardiff, 1977), 8-9, 48.

11. S.R. Hughes, *The Archaeology of an Early Railway System: The Brecon Forest Tramroads* (Aberystwyth, 1990), 83-7.

12. T. Rees and R. Thomas, *Hanes Eglwysi Annibynol Cymru,* Cyf. II, (Liverpool, 1872) p.55.

13. Copy of a 1761 map entitled 'Land under present day Landore, Plasmarl & Morriston' in N.L. Thomas, *The Story of Swansea's Districts & Villages*, Abridged Vol. 1 (1969) opposite p.109.

14. For the alum works, see chapter one.

15. Jersey Archives (destroyed) quoted in A.H. John, 'Introduction: Glamorgan, 1700-1750', In *Glamorgan County History.* 5, 1-46, 28.

16. Ibid.

17. UWS, Morris MS., 'History of the Copper Concern' (1774), 14.

18. R.O. Roberts (ed.), 'The Copper Industry of Neath and Swansea; Record of a Suit in the Court of Exchequer, 1723', *South Wales and Monmouth Record Society, Publications No. 4* (1957), 123-64 (135).

19. A.H. John, 'Introduction: Glamorgan, 1700-1750', In *Glamorgan County History,* Vol. *5,* 40.

20. Ibid.

21. Copy of a 1761 map entitled 'Land under present day Landore, Plasmarl & Morriston' in N.L. Thomas, *The Story of Swansea's Districts & Villages*, Abridged Vol. 1 (1969) opposite p.109.

22. E.g. at Pontypool, see ref. *Arch. in Wales*, 37 (1997), 51.

23. A.H. John, *op.cit.,* 24.

24. This and other dates for the ownership of Swansea copper smelting-works are taken from: R.O. Roberts, 'The Smelting of Non-ferrous Metals since 1750', *Glamorgan County History,* Vol.. V: *Industrial Glamorgan from 1700 to 1970* (Cardiff, 1980), 47-96.

25. 1871 Census, Parish of Llangyfelach.

26. A.H. John, *op.cit.,* 28.

27. 'Case of the Duke of Beaufort'.

28. 1851 census, St. John's Swansea.

29. T. Sheasby, *Plan of an Intended Navigable Canal from Swansea to Pentrecrybarth in the Counties of Glamorgan and Brecon,1793* (WGRO).

30. Information given for 1771 in this section is taken from Jones, 'Map of the River Swansey' (Swansea Museum 304/463A).

31. J.R. Harris, *The Copper King: A biography of Thomas Williams of Llanidan,* (Liverpool, 1964), 181 and L.A. Cook, *An Examination of the Social Impact of the Vivians on Swansea, 1809-1894,* unpublished Ph. D thesis (Univ. of the W. of England, 1997), 45.

32. A.H. John, *op.cit.,* 24.

33. D. Boorman, *The Brighton of Wales: Swansea as a fashionable seaside resort, c.1730 - c.1830* (Swansea, 1986), 1.

34. B. Morris, 'Swansea Houses - Working Class Houses, 1800-1850', *Gower*, XXVI (1975), 53-61.

35. J. Davies, *A History of Wales* (London, 1994), 328-29.

36. L.A. Cook, *op.cit.,* 45.

37. P.R. Reynolds, 'Industrial Development' in *Swansea: An Illustrated History*, ed. G. Williams (Swansea,1990), (29-56, 36).

38. Ibid.

39. B. Dean, *Slums: Living Conditions in 19th Century Swansea, Part 1* (Swansea History Project, *c.* 1980), 56.

40. 'Report to the General Board of Health, on a Preliminary Inquiry into the Sewerage, Drainage, and Supply of Water, and the Sanitary Condition of the Inhabitants of the Town and Borough of Swansea. By George Thomas Clark, Superintending Inspector,' London, 12 July 1849, 7.

41. OSD, Swansea Sketch H, Jan. and Feb. 1826.

42. 'Waun Gwn' is an area of land between what are now the Llangyfelach Road, Heol Gerrig, Moriah Road and Heol Fach.

43. 1841 census.

44. On the land of Pwll-yr-Oer farm to the north of Heol y Glo (now Bryn Terrace).

45. The references to the two dates 1826 and 1838 in this section are drawn from two maps: B.L., OSD Swansea Sketch 'H', 1826 and N.L.W., Tithe Map of Llangyfelach.

46. *Hanes Eglwysi Annibynol Cymru*, entry for Seilo.

47. N.L. Thomas, *Swansea's Districts and Villages*, abridged Vol.1, 54.

48. O.S. 1/2,500, Sh. Glam. XV.13, 1899

49. R.O. Roberts, 'Banking and Financial Organization, 1770-1914' in A.H. John *op.cit.,*

50. ibid,. 368-69, 401 and 408-11.

51. P.R. Reynolds, *op.cit.,* 36.

236

52. The total of the population with occupations included a copperman's widow, a woman receiving parish relief and a person of unsound intellect.

53. WGRO, 'Schedule of Deeds of Whiterock Copper Works', D/DXhr 30/1-6.

54. 'Report to the General Board of Health. By George Thomas Clark, Superintending Inspector,' London, 12 July 1849, 17.

55. R.O. Roberts, 'Banking and Financial Organization' In *Glamorgan County History,* 5 , 401.

56. B. Dean, *Slums: Living Conditions in 19th Century Swansea, Part 3* (Swansea History Project, *c.*1980) 39-58. Swansea was the only town in Wales to make use of the 1875 'Artizans & Labourers Dwellings Improvement Act'.

57. 'Report to the General Board of Health. By George Thomas Clark, Superintending Inspector,' London, 12 July 1849, 8.

58. B. Morris, *op.cit.,* 53-61.

59. 'Report to the General Board of Health. By George Thomas Clark, Superintending Inspector,' London, 12 July 1849, 8.

60. Jones, 'Map of the river Swansey' (Swansea Museum 304/463A); R.O. Roberts, 'The White Rock Copper and Brass Works, near Swansea, 1736-1806' In Stewart Williams, *Glamorgan Historian*, Vol. 12 (Cowbridge, 1980), 136-51.

61. NLW Badminton Collection, Group II MSS., 1454.

62. Ibid.

63. (NLW) Badminton Collection, Group II MSS., 1283, d.1768.

64. Ibid.

65. v.i. Bernard Morris and B.Morris, 'More Evidence for the Date of Morris Castle', *South West Wales Industrial Archaeology Society Bulletin*, (November 1994) 4-5, and 'Another Early Reference to Morris Castle', *SWWIAS Newsletter*, (September 1983) 9.

66. v.i. Bernard Morris.

67. B. Morris, 'More Evidence for the Date of Morris Castle', citing *The Cambrian*.

68. v.i. Bernard Morris.

69. S.R. Hughes, *The Industrial Archaeology of Water and Associated Rail Transport in the Swansea Valley*, unpublished Univ. of Birmingham M.Phil. Thesis (1984), 156-59.

70. B. Morris, 'Swansea Houses' *op.cit.,* 53-61.

71. Ibid., 55.

72. Swansea Canal Committee Minutes, 3 March 1812.

73. 'Hill Sketches', 1" Sh. 41, 'Ystradgynlais 2'; 1841 census and field observation.

74. J.B. Lowe, *Industrial Housing* (1977), 33.

75. Ibid., 28.

76. J.B. Lowe, 'Industrial Housing Stock' in *Buildings Archaeology: Applications in Practice,* ed. J. Wood (Oxford, 1994), 59-88, 64.

77. Ibid., 65; and J.B. Lowe, *Welsh Industrial Workers' Housing*, 28.

78. B. Morris, 'Swansea Houses', 55.

79. J.B. Lowe, *Welsh Industrial Workers' Housing.*, 54.

80. Ibid., 3.

81. v.i. Jeremy Lowe.

82. M. Palmer, 'The Richest in all Wales!: The Welsh Potosi or Esgair Hir and Esgair Fraith Lead and Copper Mines of Cardiganshire', *British Mining,* No. 22 (1983), 9.

83. N. Briggs, *John Johnson, 1732-1814, Georgian Architect and County Surveyor of Essex* (Chelmsford, 1991), 36-8, 45-6.

84. v.i. the late Harry Green, citing manuscript maps in private possession.

85. Lowe, *Welsh Industrial Workers Housing,*

86. 'Report to the General Board of Health. By George Thomas Clark, Superintending Inspector,' London, 12 July 1849, 17.

87. Case of the Duke of Beaufort.

88. All 1844 references in this section refer to the Llangyfelach Tithe Map of 1844.

89. UWS , Vivian MSS, C9, Plan of the White-Rock Copperworks Leat.

90. Cook, 'An Examination of the Social Impact of the Vivians on Swansea', 78.

91. G. Gabb, *The Morris Family*, Lower Swansea Valley Factsheet 10 (Swansea, 1980s), 32, quoting from a letter to Lord Liverpool concerning Morris's opposition to coal duties which would disadvantage Swansea.

92. N.L. Thomas, *The Story of Swansea's Villages,* Vol. 2 and Abridged Vol. 1*,* (Swansea, 1969), 72; quoting from Walter Davies, *General View of the Agriculture and Domestic Economy of South Wales.*

93. S.R. Hughes, *The Archaeology of the Montgomeryshire Canal* (Aberystwyth, 4th edn 1988), 102.

94. I. Donnachie and G. Hewitt, *Historic New Lanark: The Dale and Owen Industrial Community since 1785* (Edinburgh, 1993), 40, 158).

95. G.T. Clark, 'Report ... on the Sanitary Condition of Swansea,' (1849), 19.

96. Ibid.

97. G. Gabb, *Later Copper Works,* Lower Swansea Valley Factsheet 6 (Swansea Museum, n.d.).

98. R.R. Toomey, *Vivian and Sons, 1809-1924* (London, 1985), 161, 174.

99. J.B. Lowe, *Industrial Housing Stock*, 56-7.

100. Ibid.

101. Ibid., 70.

102. p.c. Jeremy Lowe.

103. Lowe, *Welsh Industrial Workers' Housing,* 8-9, 48.

104. J. Day, *Bristol Brass: The History of the Industry* (Newton Abbot, 1973), 78-90.

105. WGRO, Schedule of deeds of Whiterock Copper Works, D/DXhr 30/1-6.

106. Harris, *The Copper King*, 172-73.

107. K. Davies and C.J. Williams, *The Greenfield Valley* (Holywell, 1977), 26.

108. Toomey, *Vivian and Sons*, 152.

109. P. Jenkins, *Twenty by Fourteen: A History of the South Wales Tinplate Industry, 1700-1961* (Llandysul, 1995), 173.

110. B.R. Mitchell and P. Deane, *Abstract of British Historical Statistics* (Cambridge, 1962), 472-73.

111. L. Donnachie and G. Hewitt, *Historic New Lanark: the Dale and Owen Industrial Community since 1785,* (Edinburgh, 1993), 25-27

112. C. Haslam (ed.), *The Landmark Handbook* (Maidenhead, 1989), 134-5.

113. v.i. Jeremy Lowe.

114. Toomey, *Vivian*s, p.153.

115. M. Fisk, *Rhondda Housing*, 47.

116. 'W.G.W.' , *A Brief History of the Hafod Copperworks' School From its Foundation* (Swansea, 1905), p.4 (supplement to *St. John's Parish Tidings*, copy in NLW, Vivian Papers, L46).

117. N.L. Thomas, I,. 64-5

118. Toomey, *Vivians*, 48 and 238.

119. 'Memorandum Book of Mr. William Jones, for many years manager of the Hafod Works', (NLW MSS 15113B)

120. W.S. Williams, *Hanes Llansamlet*, p.154.

121. Toomey, *Vivians,* 153.

122. p.c. Yngvar Ramstad, Sorsenja Museum, Norway.

123. GGAT, Swansea Chapels survey reports, (NMR)

124. Toomey, *Vivians* , 173.

125. Ibid., 259.

126. Ibid., 154.

127. L.A. Cook, *An Examination of the Social Impact of the Vivians on Swansea, 1809-1894,* (unpublished Ph.D. University of the West of England, Bristol, 1997), 78.

128. Lowe, *Industrial Housing Stock*, p. 66.

129. Fisk, *Rhondda Workers Housing*, 47.

130. Toomey, *Vivians,* 161, 174.

131. Ibid..

132. Fisk, *Rhondda Workers Houses*, 49.

133. Lowe, *Welsh Industrial Workers Housing,* 46.

134. M.H. Yeadall, 'Building societies in the West Riding of Yorkshire and their contribution to housing provision in the nineteenth century', In M. Doughty (ed.), *Building the Industrial City* (Leicester, 1986), 59.

135. Jevons, *Coal Trade*, 646.

136. Lowe, *Welsh Industrial Workers Housing*, 47.

137. M.J. Daunton, 'Miners' houses: South Wales and the Great Northern Coalfield, 1880-1914', *International Review of Social History*, 25 (1980), 143-75.

138. Fisk, *Rhondda Housing*, 48-9.

139. Ibid, 48.

140. Ibid, 51-3.

141. M.J. Daunton, *Coal Metropolis, Cardiff, 1870-1914* (Leicester, 1977), 96.

142. Fisk, *Rhondda Housing* , 57-9.

143. Ibid.,.58.

144. Ibid.,.47.

145. M.J. Fisk, *Housing in the Rhondda,* 24; S.R. Hughes, *The Archaeology of an Early Railway System: The Brecon Forest Tramroads* (Aberystwyth, 1990), 204-5.)

146. Lowe, *Industrial Housing Stock,*. 63.

147. J.E. Ross, *Radical Adventurer: the Diairies of Robert Morris, 1772-1744* (Bath, 1971), 183.

148. J. Childs, 'The Growth of the Morris Estate in Llangyfelach Parish, 1740-1850', *Gower*, 42, (1991), 50-69.

149. J.M. Davies, 'The Morris Family and Swansea', *Gower,* 5, (1951), 26-30, cites John Morris I as providing houses for his poorer workers at both Morriston and Sketty, and the foundation of a sick scheme for his workers, in glowing, eulogistic terms but without giving references. One wonders if the Morriston reference is to the dwellings provided in the split-level 'Wychtree House' and the Sketty one to estate-workers' cottages. He certainly was involved with William Bevan and others in schemes to help workers save.

150. p.c. Bernard Morris

151. G. Gabb, *The Morris Family*, Lower Swansea Valley Factsheet 10 (Swansea, *c*. 1985), 32, quoting from a letter to Lord Liverpool concerning Morris's opposition to coal duties which would disadvantage Swansea.

152. Walter Davies, *General View of the Agriculture and Domestic Economy of South Wales* (London, 1814).

153. *Case of the Duke of Beaufort against the Building of the Swansea Canal* (*c*. 1793, Swansea Reference Library).

154. T. Sheasby, Swansea Canal Map, 1793.

155. *Case of the Duke of Beaufort, op.cit.*

156. British Library, O.S. Drawings, Hill Sketches, 'Morriston', 1828.

157. 1849 Public Health Report.

158. British Library, Ordnance Survey (Hill) Drawings 179, serial no. 135, no. 37, 1813. 2" to the mile and 1826 OSD.

159. *Plan of the Town of Swansea and Adjoining District, 1852*, and Ordnance Survey 1: 2,500, Glamorganshire (Gower) Sh. XXIV.1. Surveyed 1879.

160. *A Map of the County of Glamorgan from an Actual Survey made by George Yates of Liverpool on which are delineated the course of the rivers, and navigable canals; with the roads, parks, gentlemens seats, castles, woods, 1799.*

161. Much of the information on the Grenfells in this section comes from M. Chamberlain, 'The Grenfells of Kilvey', *Glamorgan Historian*, 9 (1973), 123-42, and G. Gabb, *The Grenfells, Lower Swansea Valley Factsheet 9* (Swansea, n.d.), produced by Swansea Museum.

162. Yates 1799 Map, and British Library, Ordnance Survey (Hill) Drawings 179, serial no. 135, no. 37, 1813. 2" to the mile.

163. v.i. one of the present occupants.

164. N.L. Thomas, *The Story of Swansea's Villages, abridged Vol.1* (Swansea, 1969),.53.

165. 'Report to the General Board of Health, on a Preliminary Inquiry into the Sewerage, Drainage, and Supply of Water, and the Sanitary Condition of the Inhabitants of the Town and Borough of Swansea. By George Thomas Clark, Superintending Inspector,' London, 12 July 1849, 9-10.

166. O.S. 1/2,500 GL. XXIV.I., 1879.

167. Information on this section on Saltaire and Ackroyden is drawn from J. Reynolds, *Saltaire: An introduction to the village of Sir Titus Salt*, (2nd ed Bradford, 1985), and W.J. Thompson (ed.), *A Brief Guide to the Industrial Heritage of West Yorkshire* (Ironbridge, 1989), 16.

168. W.J. Thompson , *A Brief Guide to West Yorkshire*, pp.14-6.

169. J.M. Davies, 'The Morris Family and Swansea', *Gower* , 5 (1951), 26-30.

170. Much of the information in this section on the Belgian workers' settlements is drawn from J. Miller and L.-F. Genicot, 'Les cités ouvrières', in L.-F. Genicot and J-P Hendrickx (eds.), *Wallonie-Bruxelles: Berceau de L'Industrie sur Le Continent Européen* (Louvain-La-Neuve, 1990), 241-54.

171. Most information on Bois-du-Luc comes from K. Simonis-Boon, 'Bosquetville ou les carrés de Bois-du-Luc: architecture sociale au 19ᵉ siècle' in J. Liebin (ed.), *Bois-du-Luc: 1685-1985* (La Louvière, 1985), 59-76.

172. Information on La Colònia Güell comes from B. Costillas i Pérez, I Dal Maschio Eisele and M.T. Montero Gómez, *La Colònia Guell. Modernisme i Indústria* (Barcelona, 1992).

173. Roberts, *Copperworks Chronology.*

174. *Inside Cornwall*, 1997.

175. A.H. John, 'Introduction: Glamorgan, 1700-1750', in A.H. John and G. Williams (eds) *Glamorgan County History,* Vol. V (Cardiff, 1980), 1-46.

176. P. Morgan, 'Art and Architecture', in G. Williams (ed.), *Swansea: an illustrated history* (Swansea, 1990), 195.

177. Information given on the development of Tredegar Fawr by the owner, and copies of the date plaques held by Henry Dare of the Morris Estate.

178. Briggs, *Johnson*, p. 2.

179. Four views of the handsome, spectacularly sited house at Clasemont seem to have survived. The clearest is the view of 'Clasmont' in the National Museum of Wales (published in T. Lloyd, *The Lost Houses of Wales* (2nd. ed London, 1989), p. 74. There is also a painting by Sir Paul F. Bourgeois which shows the park in front of the house fairly clearly (now Morriston Park). The house appears to have been modified by the addition of a central doorway and porch giving onto a large gravelled area above the park itself. This is illustrated in W.C. Rogers, *A Pictorial History of Swansea* (Llandysul, 1981), p. 128. A third illustration is Thomas Rothwell's 1792 engraving, copies of which are in Swansea Museum and NLW and which is reproduced in M. Gibbs and B. Morris, *Thomas Rothwell: Views of Swansea in the 1790s* (Cardiff, 1991), pp. 50 and 56. A fourth illustration is that made by Booth in 1783.

180. J. Childs, 'The Growth of the Morris Estate in Llangyfelach Parish, 1740-1850', *Gower* 42 (1991), 50-69, and N.L. Thomas, *The Story of Swansea's Districts & Villages*, Vol. II, Parts IV-VIII (Swansea, 1969), pp. 142-3.

181. This section on Sketty Hall is drawn from B. Morris, *Singleton Houses*, pp. 86-117.

182. Ibid., pp. 9-13.

183. R. Suggett, *John Nash: Architect in Wales* (Aberystwyth, 1995), 66.

184. L. Greeves, *The Good Country House Guide* (London, 1994), 24.

185. B. Morris, *Singleton Houses*, 53-69, for Verandah.

186. G. Gabb, *The Morris Family: Lower Swansea Valley Factsheet: 10* (Swansea Museum, *c*. 1980s) 44.

187. W.C. Rogers, *A Pictorial History of Swansea* (Llandysul, 1981), 128.

188. G. Gabb, *The Morris Family*, 27.

189. J. Childs, 'The Growth of the Morris Estate in Llangyfelach Parish', *Gower,* 42 (1991), 50-69.

190. *George Yate's Map of Glamorgan (1799)*, published under that title by the South Wales Record Society, Vol.2. (1984), with introduction by G. Walters and B. James.

191. T. Lloyd, *The Lost Houses of Wales* (London, 2nd. ed., 1989), 83.

192. Ibid.

193. N.L. Thomas, *Swansea's Villages*, Vol. II (Swansea, 1969), 138.

194. v.i. Bernard Morris citing W.C. Rogers's tracings of Morris MS maps (after Warwick estate maps).

195. W.C. Rogers, *A Pictorial History of Swansea* (Swansea, 1981), 128, and v.i. Henry Dare of the Morris Estate.

196. Ibid., 139.

197. Ibid., 129-30.

198. B. Bradford Brown, *A History of Coppermining in Cornwall* (Truro, 1966), 55.

199. B. Morris, *Singleton Houses*, 2-18, for a history of Marino alias Singleton Abbey.

200. Ibid.., 72-5.

200. Ibid., 72-84.

201. A drawing of Woodlands exists in Swansea Museum and is illustrated in Rogers, *Swansea*, 132.

202. A painting of Marino with side-wing extensions also exists in the possession of the present Lord Swansea and is illustrated in Rogers, *Swansea*, 124.

203. R.A. Griffiths, *Singleton Abbey and the Vivians of Swansea* (Llandysul, 1988), 24-8.

204. For Parkwern, see B. Morris, *Singleton Houses,* pp.72-84.

205. Ibid., 14-26.

206. Ibid., 110-17.

207. Ibid., 118-44.

208. M.V. Symons, *Coal Mining in the Llanelli Area, Vol. 1, 16th Century to 1829* (Llanelli, 1979), 79.

209. Ibid., quoting Cardiff Central Library, Glamorganshire deed 35.32 (Dec. 1799).

210. Ibid., quoting from the *Report from the Committee on the Petition of the Owners of Collieries in South Wales 1810. Evidence of Henry Smith, M.P.*

211. T. Lloyd, *The Lost Houses of Wales*, 80.

212. G.Gabb, *The Morris Family,* 48.

213. Much of the information on the Grenfells comes from M. Chamberlain, 'The Grenfells of Kilvey', *Glamorgan Historian*, 9 (1973), 123-42, and G. Gabb, *The Grenfells, Lower Swansea Valley Factsheet 9* (Swansea, n.d.), produced by Swansea Museum.

214. Harris, *Copper King*, 154-55.

215. M.J. Best, *The Parish Church of All Saints, Boyne Hill, Maidenhead* (Maidenhead, 1989), 5.

216. M. Gibbs and B. Morris, *Thomas Rothwell: Views of Swansea in the 1790s* (Cardiff, 1991), 8-10, is the source for this evaluation of George Haynes's career.

217. Philadelphia University Library, Joshua Gilpin's Tour, 1796.

218. N.L. Thomas, *The Story of Swansea's Districts & Villages*, (Swansea, 1969), xi-xiii, summarises much of the involvement of Haynes with the Swansea potteries.

219. S.R. Hughes and P.R. Reynolds, *The Industrial Archaeology of the Swansea Region* (Aberystwyth, 2nd ed, 1989), 51.

220. P.R.O., Swansea Canal Committee Minutes, 1 February 1803.

221. *Oldisworth's Guide to Swansea* (Swansea, 1802).

222. 'Haynes Buildings' are named on the O.S. drawings 179, 'Swansea Bay', Serial no. 135, index 1" No. 37, 1813, but had been renamed 'Ynys Tanglwys Buildings' by the time of the 1879 O.S. 1/2,500 map, Glamorganshire Sheet XV.2.

223. J.H. Davies, *History of Pontardawe and District* (Llandybie, 1967), 92; J.M. Davies's M.A. thesis on 'The history of the Swansea valley', 83.

224. H.L. Hallesy, *The Glamorgan Pottery, Swansea* (Llandysul, 1995), 75-7.

225. A.H. John, 'Introduction: Glamorgan, 1700-1750' in A.H. John and G. Williams (eds), *Glamorgan County History,* Vol. V (Cardiff, 1980), 1-46, 38-9.

CHAPTER 5

THE INSTITUTIONS OF THE COPPER TOWNSHIPS

Introduction

The industrial settlements of Swansea had rows of workers' housing interrupted by three types of larger institutional buildings: schools, churches and chapels. All three were generally present in the townships built by the workers themselves and also in those completely, or mostly, constructed by the coppermaster employers. The most numerous class of communal structure comprised the nonconformist chapels. It might be argued that the Swansea industrial settlements have a distinctively Welsh character, with copious provision of what were large and elaborate nonconformist places of worship, including, of course, Tabernacl at Morriston. It is worth asking how central these were to the workers'; how they were funded and what factors determined which architects built them.

There is also the question of the influences determining the architectural vocabulary used in these substantial chapels and whether these design features were used in a form distinctive to Swansea and to Wales. It may be that this large and reasonably affluent population, attracted and sustained by an international centre of industry, was in a unique position to determine the form of a specific class of building - the nonconformist chapel - that is a focal point of rural and urban landscapes throughout Wales.

At a more complex level the building of chapels, churches and schools may provide information concerning human motivation and trends in economic and social history. These buildings, themselves physical manifestations of the historical process, may tell us about the aspirations of the workforce as well as of the business dynasties. It may be possible to begin to understand the issues that concerned these early industrial communities, and their influential employers, by examining the bricks and mortar of institutional buildings. There may also have been a need to ensure social control and to ensure the docility of a large workforce by providing educational and religious buildings; but, with the provision of workers' housing settlements, it

could be that there were more altruistic motives of philanthropy and paternalism at work. All of these factors may have manifested themselves in the provision of institutional buildings that were, in effect, material displays of enlightened self-interest.

In the following section the multiple and widely diverse providers of educational and religious buildings in the industrial communities and the surviving institutional and religious fabric are examined with these important questions in mind. The more significant conclusions are summarised at the end of each section and in the general discussion at the end of the chapter.

Schools for Industrial Workers

Introduction

The copperworks and collieries of the lower Swansea Valley were not located within the old borough boundaries of Swansea. The communities that grew up around each industrial concern were a complex result of the initiatives of cottager-workers and employers. Each concern not only provided employment but also, depending on the partners' consciences, a degree of involvement in the local community, a sense of duty and social idealism and, more selfishly, education as an aid to the efficiency of the business: those concerns catered for the social betterment of the workmen and their families, particularly by the provision of schools for their children.[1]

The Growth of Popular Education in Wales

There had been grammar schools mostly for the sons of the more affluent families of Welsh market towns and rural centres since Elizabethan times. A later foundation of this type was the grammar school founded by Bishop Hugh Gore in his native town of Swansea in 1682.[2] At the end of the seventeenth century there were also 16 'illegal' nonconformist schools in Wales, including one in Swansea.

More significant was the way in which the education of the poor was pioneered in Wales at the end of the seventeenth and in the eighteenth centuries. This innovatory work was firstly carried out by the 'Welsh Trust', an organisation of Anglicans and nonconformists drawn from the gentry and other benefactors and activists in Wales and England, acting under the inspiration of the Carmarthenshire ex-cleric and translator Stephen Hughes.

The Welsh Trust were involved in two types of activity during the years *c.*1672-81: the handing-out of Welsh bibles and religious literature to adults and the setting up of English medium schools. The Trust disappeared under the impact of the religious and political turmoil of the 1680s that preceded the 1688 Revolution. Two schools were founded in the lower Swansea Valley, one at Swansea (having 20 pupils) and one at Llangyfelach to the north. The Welsh Trust was the inspirational parent of all the voluntary societies for the education of the poor in Britain in the eighteenth century and particularly of the Society for the Promotion of Christian Knowledge (the SPCK).[3]

The SPCK continued the work of the Welsh Trust from 1699 and was active in Wales until 1737 (i.e. 20 years after the establishment of the first Swansea copperworks, and the year in which the third copperworks was founded). Welsh gentry such as John Vaughan of Derllys (father of the influential Bridget Bevan), the lord of the manor of Marros, in south-western Carmarthenshire, and the gentlewomen of Brecon threw their influence and money into the scheme to found schools. The Neath coppermaster, Sir Humphrey Mackworth, one of the original five members of the SPCK, urged industry to follow the lead of the landed gentry. He persuaded the company of the Mines Adventurers, of which he was deputy governor, to support schools in Neath and at the Esgair Hir metal mine in Cardiganshire, to which the children of smelting-works, battery mills and mine employees could go.[4] These two schools were the first charity works' schools, although an earlier works' school had been founded at Tintern in Monmouthshire.

William Waller, steward of the mines in Cardiganshire, described Mackworth, who was a keen Churchman 'as taking delight in bringing advantage to others, especially to his miners and labourers' and with 'giving encouragement to his workmen to be careful and industrious in their own interest as well as in that of their masters'. The published objectives of the Company of Mine Adventurers stated that one-twelfth of its profits was to be appropriated to charitable uses under Mackworth's direction, including the relief of 'poor Miners and Labourers at the works, their wives and children, and the provision of schooling for the

latter'. These schools differed from the later works' schools in being maintained for a few years at the entire expense of the company.

The Esgair Hir school was set up in 1700, an annual grant of £20 became the norm and the schoolmaster was paid £15 per annum. However, the mines proved disappointing and the partners quarrelled; by 1709 the miners had the sum of 30d per quarter deducted from their wages towards the education of their children 'since the miners were eager to have their children educated'. These contributions were a forerunner of the deductions that became fairly universal in the works' schools of the nineteenth century. The Esgair Hir school had closed by 1721. The second school was established near the Melincryddan Copperworks at Neath before 1705; this also received a £20 annual grant from the company. Mr Williams, the schoolmaster, was paid £30 annually for the 'education of the children of the poor workmen'. The Neath copperworks school closed at some date after 1718. Forty children of the workmen of the Mine Adventurers Company were educated at the two schools, mainly in religious instruction and reading, conducted through the medium of English.

Circulating schools consisted of itinerant teachers spending some time (three to six months or so) in a particular locality; they taught the basic skills needed to read the Bible and other pious literature and they had been suggested by the coppermaster Sir Humphrey Mackworth in 1719. The Pembrokeshire organiser of the SPCK, and brother-in-law of one of its main activists and financiers (Sir John Phillips of Picton Castle), was the Revd. Griffith Jones, rector of Llanddowror in south-western Carmarthenshire. He was also the administrator of the small college there that provided a brief training for many of the itinerant teachers (who were only paid £3 or £4 a year). He and his treasurer and successor, Bridget Bevan of Laugharne (wife of the MP for Carmarthen), set up, between *c.*1730 and 1779, over 3,000 short-lived schools which taught in Welsh or English according to the language of the locality. This promoted popular education in Wales which reached its zenith under Bridget Bevan between 1763 and 1773.[5] It has been estimated that some 200,000 children and adults were taught to read Welsh out of a total population of some 400,000; in its scale this was unprecedented in the British Isles. These temporary, but very successful, schools did not have permanent buildings but used parish churches or other available buildings. The movement spread dramatically over south Wales in 1738-39. The school held in Swansea in 1745-46 had 201 pupils, but schools were also held in all the industrial settlements around Swansea, at Cwm (Llansamlet), Cnap-llwyd (near Morris Castle), Graig Trewyddfa, Llansamlet, Olchfa, Pantmawr (Hafod), Pentre-estyll (at the top of Cwm Burlais), Penylan, St. John and other places. The miners of Trewyddfa even petitioned the vicar of Llangyfelach to arrange a circulating school for their children.[6] What was known as 'Mrs Bevan's circulating school', stayed in Llansamlet for two years in the early nineteenth century (a revisitation in the second year being common to correct any deficiencies in the first year's visit).[7]

Like the schools provided by the earlier 'Welsh Trust' and the SPCK, these schools were financed in substantial part by subscribers in England. Griffith Jones believed his schools, which he saw as a continuation of the work of the SPCK, were 'pious [spiritual] nurseries'. His orders to the itinerant teachers was that the primary role of the schools was to be 'the poor men's guide to Heaven' but also 'to make...good Men in this World and happy in the next.' Interesting, in the light of the theory that later employer-providers of schools were promoting social control, were the comments received from clergymen in 1752 that there was 'visible change for the better in the lives of the people' and that the children 'now prefer praying to playing'. Some have criticised the role of these impermanent schools however, for not teaching writing and arithmetic and for discouraging any local initiatives to establish permanent schools.[8]

A large blow to the provision of circulating-schools came with Bridget Bevan's death in 1779, when her will leaving £10,000 to trustees to continue the work was contested by relatives and her funds were frozen in the Court of Chancery for no less than 31 years. Others persisted with organising the circulating school system but it was largely superseded by the Sunday school system. Thomas Charles, who started out as an Anglican clergyman and who by 1784 was a Calvinistic Methodist minister in Bala, was one of those who carried on the circulating school system for a while. However, the inadequate finance available in Wales was unable to support a large number of salaried schoolmasters, whilst the new Sunday school movement promised the use of voluntary teachers on Sundays and for one or two nights during the week. The Sunday school movement had been started in Gloucestershire by Robert Raikes, and in 1798 Charles asked for and received financial help from the English Sunday School Society, whose correspondent he became. Charles noted that the 'Grand principle of subordination', such a prominent part of teaching in English Sunday schools, and the teaching of manners were not the primary ethic of the Calvinistic schools.

He was sure that by teaching the pupils Welsh first 'we prove to them that we are principally concerned about their souls ... that most important point is totally out of sight by teaching them English, for the acquisition of English is connected only with their temporal concerns.'[9]

The Sunday schools in Wales, the weekday, and particularly the Sunday and night schools which had preceded them as part of the circulating schools, included a significant number of adults. In the Sunday schools the small groups for intensive study, the discussion class and the training in self-expression, had no equivalent in the early English Sunday schools. As with the circulating school earlier, the Sunday school established a place for widespread adult education in Wales that had no equivalent in the rest of the British Isles.

The Sunday school system expanded greatly in the nineteenth century both in Swansea and in the rest of Wales. By mid-century there were 54 day schools in Swansea and 20 Sunday schools, but the latter were generally larger and had 1,987 pupils in total, while the smaller day schools had an only marginally greater total of some 2,122 pupils.[10]

There had been a gap in the provision of works schools throughout much of the eighteenth century. There is no evidence that even the fairly enlightened and wealthy resident coppermaster, Sir John Morris I, thought of setting up a works' school during the period of the circulating schools. They were only being set up again towards the end of the period when circulating schools were such a large influence on the lives of ordinary Welsh workers. The next recorded works' school in south Wales was founded in 1784 to serve some 25-30 children of the employees of the Beaufort and Sirhowy Ironworks in the Heads of the Valleys area of Monmouthshire. This was the Capel Waun y Pound school built by the employers, Atkinson and Barrow, and their neighbours the Kendalls.

Then followed schools in the copper-smelting region centred on Swansea. In 1804 R.J. Nevill and Company built the Llanelli Copperworks and opened the Free School for their workers. In 1818 59 pupils attended whose parents worked at the copperworks and the associated Caemaen and Box Collieries. Unusually for early nineteenth-century works' schools, it was non-contributory and in 1823 £31 was subscribed by local industrialists and gentry. It grew into the respected Llanelli Copperworks Schools, opened in 1847. By the mid nineteenth century, these schools ranked in size with the Hafod Copperworks Schools as the largest and most important in the western coalfield, being almost as large as the famous Dowlais Schools founded for ironworkers' children at Merthyr Tydfil.

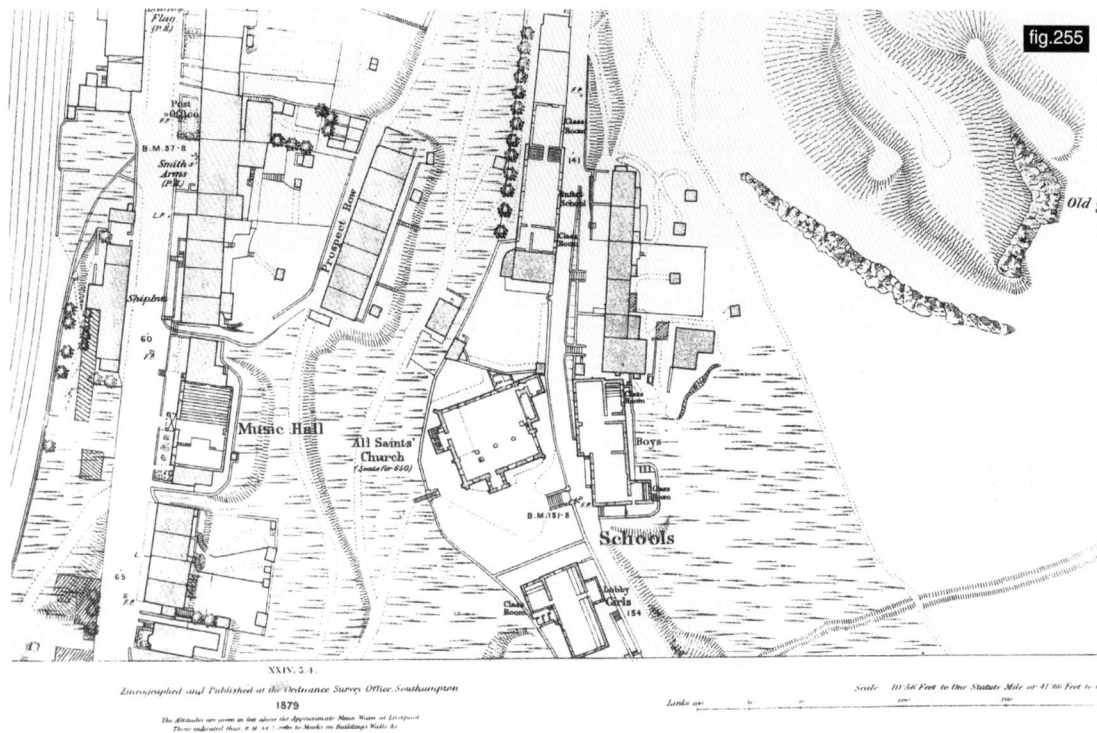

Fig. 255. Large-scale Ordnance Survey map of the 1870s showing Kilvey Schools, church and Music Hall. All communal buildings at the centre of the copperworkers' settlement of Foxhole were largely provided by the Grenfell family of coppermasters. (Reproduced from 1879 Ordnance Survey Map)

Fig. 256. Kilvey Copperworks School buildings. The remains of the schools of 1806 stand on the right of Kilvey Road, the southerly of three rooms remaining roofed as the Gwyn Mission. In the background, beyond Kilvey Church, is the building of the Kilvey Infant School of 1839, which is now a private house. (9500349/4)

The Kilvey Copperworks Schools, Foxhole, 1806

The smelting-works of the lower east side of the Swansea valley were all founded in the mid-eighteenth century, but it was not until 1806 that the first copperworks' school in Swansea was started. The site at Foxhole was given by John Freeman and Company, owners of the White Rock Copperworks, and the cost of building was met by Pascoe Grenfell and Company, owners of the Middle and Upper Bank Copperworks. Like other later industrial schools, it was maintained by a stoppage of 1d. a week from the workers employed in the three works. The girls paid separately at the works' offices. The original school was later converted to the boys' school which stood on the eastern side of Kilvey Road, on the hillside above All Saints' Church. The most southerly of three large rooms, which had two classrooms annexed, survives in use as the Elim Gospel Hall, with the ruins of the rest of the school, built of local Pennant sandstone rubble, to the north.

In 1847 the Commission into Education in Wales reported on the Foxhole schools. The Assistant Commissioner, William Morris, noted that the 'Kilvey Infant and Juvenile Schools' were:

'... supported by stoppages upon the wages of the men employed in the copperworks of Messrs. Grenfell and Sons and Messrs. Freeman and Co. The cost of accommodation is borne by the employers.

Mr. Grenfell told me that he had it in contemplation, after a time (he had but recently resided in the neighbourhood), to admit no young people to his works who could not read and write, or at least to make such, at extra hours, attend school. Just before my visit, five young men in his employ, aged from 18 to 22 years, had all signed an agreement, respecting their work, with marks.

Infant School. - I visited this school on the 17th of February. The building is commodious and furnished in the usual manner. The site was given by Messrs. Freeman and Company, and the cost of erection defrayed by Messrs. Grenfell and Sons.

I heard a gallery-lesson given. There was a stand with a frame into which prints could be fixed. Each scriptural lesson was illustrated by a scriptural print ... [The lesson consisted of a discussion of the print.] The master's manner was good and animated. The children appeared pretty well interested and attentive.

The girls were much older than the boys; the latter were drafted off into the juvenile school. There is no similar school for the girls. They sew with the mistress in the afternoons; and members of Mr. Grenfell's family attend twice a week to teach them writing. Beyond this, their daily instruction is confined to the routine of the infant-school. Almost all the answers came from the girls. Among others, they repeated the number of miles which the earth is distant from the sun, its circumference, its diameter, and ... . They performed the following additions very readily ... without slates ...

This school appeared to be efficiently conducted.

Juvenile School. - I visited this school on the 17th of February. It is held in a dingy, dilapidated building. I found the old master (a mason disabled 41 years ago) sitting stick in hand. The 12 senior boys present were reading the Epistle of St. James. The class to which they belonged had been reading straight through the Testament ....

The writing was middling. The books were not very clean.

The master complained that the children could not come early in the morning, because they had to take their parents' breakfast to the works; and that they were removed at a very early age from school ....

A list of attendance is sent in weekly to the Companies whose workmen support the school.'

Even if the condition of the juvenile school revealed deficiencies (it was, after all, the original 41-year-old building of 1806, still with the original teacher) it is hard to imagine the family of today's managing directors making the time and effort to teach the children of workers. Frances Madan, Pascoe St Leger's second wife, took her religion and philanthropy seriously and she and her step-daughters, Gertrude, Mary, Kate and Eleanor, divided up the Pentre-chwyth and Foxhole areas for the purposes of visiting and other charitable works. Mary Grenfell was a trained nurse and out of her own money, she set up the Golden Griffin coffee house (named from the family coat of arms) next to the Swansea Vale Railway station in St. Thomas as an alternative to the local public houses. The eldest daughter, Madelina, married Griffith Llewelyn of Baglan Hall but supported Mary's work and also endowed the Eye Hospital attached to Swansea General Hospital.

Pascoe St Leger Grenfell commented that his workers at Foxhole in 1847 were part of a largely monoglot Welsh-speaking community. In answer to a question from the government education inspectors, Pascoe St Leger Grenfell commented on the extent to which English was understood by the copperworkers: 'Very limited. Many of the workmen speak none at all, and those that do, scarcely understand anything beyond the common routine phrases applying to their own peculiar station.' The bridging of the mouth of the River Tawe and the suburbanisation of the east side, with the building of St Thomas and the provision of education in English, changed that, and by 1908 it was noted that the majority 'neglected their Welsh'.[11] Even enlightened employers regarded monoglot Welsh-speakers as having severely limited horizons. Moreover, a great value was put on education (in English) and on teaching the scriptures so as to raise the level of morality of the workers. This is especially clear from the evidence given by Pascoe St Leger Grenfell to the 1847 Inquiry:

'I believe an improved system of education, especially to give the rising generation a good knowledge of the English language, by which their views may be extended beyond the narrow circle to which their own language confines them, to be the most effective means of raising the exceedingly low and defective tone of principle, morality, and truth of this people. From their great want of order and system, I should conceive a sound Scriptural education, conducted on moderate Church principles, the best adapted for that end. The present means are, or have been hitherto, very defective in this neighbourhood, but the public attention has been roused lately, and many schools are in progress and in contemplation.'

Because of very strong nonconformist feelings in the district, the teaching in the schools was strictly non-sectarian, but still subject to fierce religious criticism. The schools were put up for sale when most of the Grenfells wanted to sell the family firm in 1892, but Mary Grenfell ran them for a further two years with financial assistance from her half-sister Madelina. In 1894 the Swansea School Board took over the schools and in 1897 they were replaced by the large building which is the present St Thomas Primary School.

In 1839 a new infants' school was built to the north of All Saints' Church at Foxhole, on the western side of Kilvey Road. John Freeman and Company again provided the site and the Grenfells the building. The infants' school had one large room with eight rows of raked seating at its northern end and a classroom at each end of the building. By 1846 over 200 pupils attended the school. The building survives, now converted to housing.

The original school was converted into the boys' school in 1842 and a new girls' school, capable of accommodating 500, was built to the south of the church on the western side of Kilvey Road. This consisted of a large hall with two rakes of seating down one side and a classroom. This large gothic building, important as being the first building recorded to have been built by the local chapel and school architect John Humphrey; it was later converted into the parish hall and survived to the 1970s.[12] These Kilvey Copperworks Schools were considerably enlarged in 1850 to accommodate children from the new Swansea suburb of St Thomas to the south. Pupils were admitted whose parents were not employed by the copperworks companies provided there was room for them; they also had to pay a higher rate of fees.

The Birmingham Copperworks School, Morriston, 1815

There is no fully comprehensive list of works' schools. However, the next foundation in Swansea, after that of the Kilvey Copperworks School in 1806, was probably the school that the Birmingham Copperworks Company built on John Morris II's land at Morriston in 1815. The new school was built on the western side of Martin Street, at the corner of Banwell Street. It was rebuilt in 1845 and called the Lancastrian School after Joseph Lancaster's monitorial style of teaching. The Birmingham Copperworks closed in 1833 and by 1848 it was noted that this Morriston infants' school was 'partly supported by the proprietors of the different copper-works and collieries in the neighbourhood, for the instruction at a moderate charge of the children of

the persons employed by them'. A third, larger 'Neath Road School' that could accommodate 1,000-1,200 children, was built further to the south in June 1868 with £1,000 of the £3,300 building costs provided by H.H. Vivian MP; in that year Vivian & Sons converted their part-owned Forest Copperworks nearby into the Morriston Spelter (i.e. zinc) Works. It was commented upon in the architectural press that it was completed in utilitarian style with 'No attempt at architectural display in plain native stone with Gothic windows'.[13] This school was the fourth building known to have been designed by the Morriston chapels' and schools' architect John Humphrey[14] and this continued in use until July 1988.

In addition, Sir John Morris II had built the old market in Morriston in 1827 with a large room over the market arcade. This was used as a schoolroom in much the same way as the space over the stables was used at the Dowlais Ironworks in Merthyr Tydfil, but he may not have directly funded its operation. It was later known as a 'twopenny dole' or 'charity school'.

Y Coleg, Landore, 1824, and the Brynhyfryd School.

A few of the many nonconformist schools in the Swansea industrial communities grew into day schools. Landore was one of the few substantial industrial communities near the large copperworks in the lower Swansea Valley not to have a school provided by the nearest copperworks. Landore (1790), Nant Rhyd-y-filais (1814), Hafod (1809-10) and Morfa Works (1828 and 1835) stood nearby. The Nant Rhyd-y-filais Air Furnace Company of Messrs. Bevan (Quaker industrialists) developed into the Landore Iron Forges. This was almost certainly the 'Gwaithbach' ('small works') where the Landore members of the Welsh Independent congregation at Mynyddbach were allowed to hold their school and meetings in the works office; Robert Monger, the manager of the nearby Hafod Copperworks was one of the deacons who started the cause. In 1824 they were given a convenient piece of land (on the eastern part of what is now Siloh Road) and built a new school there. Unlike the works' schools, but like several of the other nonconformist buildings, this seems to have been built by the congregation themselves (at a cost of £90), with the female members of the congregation collecting stones from the adjoining Cwm for their menfolk to build into walls.

The building was called 'Y Coleg' ('the College') as a day school was kept there and the congregation worshipped every Sunday night. In 1829 the religious services were transferred to the adjoining new chapel (which cost £560). In the 1840s a substantial number of scholars, whose parents worked at the Hafod Copperworks, were transferred to the new Hafod Copperworks schools. Finally, in 1862 the congregation extended the chapel and its gallery over the old Y Coleg site and a new schoolhouse and house were built at a cost of £900 by the architect-minister Thomas Thomas.[15]

The adjoining industrial suburb of Brynhyfryd to the west of Landore, home to many colliers, also benefited from the school-building activities of the congregation. The large day school established here (at a cost of £600) also housed a night school to satisfy local workers' quest for education. The night school was led by the industrialist Sir Richard Martin, who was mayor of Swansea in 1898-99 and an important founder of the University College of Swansea.[16] The Sunday school here was allowed to continue when the building was sold to the new Swansea school board in the 1880s and part of the old school was incorporated in the enlarged building. The Seilo congregation in co-operation with other Landore chapels and Libanus at Morriston, also paid £160 to jointly fund a school in the new adjoining industrial suburb at Plasmarl to the north.[17]

Ysgoldy Gapel-y-Gwm, 1850

Another day school attached to a nonconformist chapel was formed towards the mid-nineteenth century on the lower east side of the lower Swansea Valley, at Capel-y-Cwm, in the Llansamlet collieries area to the north-west of Bon-y-maen. This school was held at the Ysgoldy Gapel-y-Gwm (i.e. the Capel-y-Cwm Schoolhouse) by the disabled William Hopkin, who was probably an ex-miner.[18] At an inspection in 1851-52 the books, discipline and furniture were said to be as 'sub standard' and it was described as 'a village-school of poor accommodation & defective character'.[19]

Other Nonconformist Chapel Day Schools

The Landore, Brynhyfryd and Cwm day schools, attached to chapels in the industrial satellite communities around Swansea, were paralleled by other schools in the north Swansea area. The Welsh Baptist Chapel of Bethesda, in Bethesda Street in the northernmost suburbs, had an English-medium school attached to it with 202 pupils by 1851. Likewise, the English Wesleyan Chapel towards the centre in Goat Street, with copperworks management participation in its deaconate, had a day school with 180 pupils by 1851 (it was originally founded in 1805).[20]

Private and Dame Schools

In addition to the works and chapels schools, which were usually part of the 'British Schools' system (and therefore eligible for government grant), there were a number of schools in the industrial satellite communities run by private individuals. One early nineteenth-century dame school held in the room over Morriston Market has already been mentioned. Another was run by a male teacher in the old Barracks building near the Hafod Works gates. A third was founded by the schoolmaster Robert James in the old Scott's Tramroad stables at Round Pit, near the old Smith's Canal.[21]

Most of these small private schools were superseded by the much larger works schools. In fact, in the case of the Round Pit School a process of direct transformation into a works school can be traced. Mrs Charles Henry Smith, wife of the proprietor of the Llansamlet Collieries in 1827-69, continued the Round Pit School as a works school and re-opened it in a schoolhouse at the newly rebuilt Charles Pit, with Robert James (of the local Welsh Independent congregation) remaining as teacher in this 'Ysgol yr Yard' ('Yard School').[22] Charles Pit was the important industrial centre of the Smith Collieries on the east of the lower valley and had its own foundry from 1834.

There were some other initiatives in the area of the industrial settlements. Mysydd Farm was on the south side of Brynhyfryd and Landore (around the modern Mysydd Road) and to the north of Hafod; by 1851 it was reported that there were 'the Wickcliffe Schools, established by the Revd. Thomas Dodd, in the populous neighbourhood of the Mysidd Fields'.[23] There were Church schools other than those provided by John Henry Vivian at Sketty, for instance in 1857 the colliery village of Cockett to the north-west of Swansea was given a new school designed by the well-known architect R.K. Penson, at a cost of £700 'to correspond in character with the recently erected church'.[24]

The Works Schools Legislation of 1833

It is difficult to assess how the Factory Act of 1833 influenced the provision of education for the children of workers in the Swansea copperworks. Two of the schools, those at Kilvey (1807) and Morriston (1815), seem to have catered for all the children of parents employed at each of the three nearest copperworks (not just the minority employed in the works), both schools were founded well before the legislation was passed. The Act obliged works' owners to make sure that their child labourers were receiving some education before being given employment. What clearly happened at Swansea, and elsewhere, was that many more than the minority of workers who were children were given an education. The willingness to establish schools on a voluntary basis was crucial, for even the Lord Chancellor made known his views that he was against establishing a school system because of the cost involved. An extensive system of works schools was established at Swansea, as elsewhere in south Wales, on the voluntary principle.[25]

The Hafod Copperworks Schools, 1847-48

After Kilvey, Morriston and Landore, the fourth major concentration of copper-smelting was at Hafod, where the large Hafod Works, with 84 smelting furnaces at its peak, was founded by John Vivian in 1809. To the Vivians, education was clearly the path to success; John Vivian and his eldest son, Richard Hussey, had both been partly educated in France. As a result of the French revolutionary wars, however, the younger son, the 16-year-old John Henry Vivian, went to Germany. In 1803 he enrolled at the famous mining institute of the University of Freiburg, where he studied mineralogy, geology, chemistry, metallurgy and mathematics. He returned home in 1804 via a tour of the mining regions of Austria, Germany and Hungary. It has been claimed that this combination of academic and practical training was the secret of John Henry's extraordinary success as an industrialist and his considerable reputation as a scientist.[26]

J.H. Vivian regarded education as the key to industrial efficiency and commented that he was 'constantly deeply conscious of the almost complete lack of educational facilities in the Swansea district'. In the 1820s he subscribed 'most liberally' to the National School in Swansea. His wife was 'a tireless supporter' in his educational efforts and Mrs Vivian started a small school for 40 girls in the parish of St John (i.e. the Hafod area) in 1825. Sarah Vivian also organised the Swiss Cottage Dame School in Singleton Park for 25 boys and girls who received 'individual instruction from an excellent matron in reading, writing and arithmetic, with knitting for the boys and both knitting and sewing for the girls'.

The foundation of the company town of Trevivian c.1837 provided a focus for the Vivians' educational activities. At first there was a temporary school there, and the imminent government enquiry into the state of education in Wales may have prompted the building of the permanent Hafod Schools. In February 1847 the large, purpose-built Hafod Copperworks School was opened in what was intended to be the centre of Trevivian. These schools at Hafod cost between £2,000 and £3,000, which was about the average cost of the many ironworks' schools built at the heads of the eastern valleys of south Wales. The larger and more elaborate Dowlais Schools, serving a population of some 12,000, cost Sir John Guest around £20,000 for the erection of just one phase.

The Hafod schools were built in three stages. It was at first thought that one building would be sufficient for a mixed school (the present northerly wing of the building attached to the former school-teachers' houses). A wooden partition was erected along the length of the schoolroom with boys on one side and girls on the other; however, the room soon became overcrowded. The original schoolroom was used by the girls and a new boys' department was added. A new infants' schoolroom completed the scheme in 1848. In 1849 they were described as 'three large schools erected by Mr. Vivian for the use of his workpeople, which rival in size those of Sir John Guest at Dowlais'. In 1859 the Government

Fig. 257. The Swiss Cottage in the grounds of Singleton Abbey, designed by Peter Robinson, was one of the buildings used by Sarah Vivian for her educational initiatives in the 1820s. (9500348-3)

Fig. 258. The Hafod Copperworks Schools (1846-8) were one of the largest works schools in Wales and still continue in community use. The design of the south wing of 1848 is strictly utilitarian but has gothic detail. The east side was the main boys schoolroom, two smaller classrooms were inside the central gabled block, and the main infants schoolroom lay beyond. (9500353/3)

Inspector's report referred to them as 'excellent and efficient schools, and at the head of the list of the best schools in Wales'. They began with 350 pupils in 1847; this had increased to 521 in 1865, and had reached 1,114 by 1893 with an average attendance of 889. The schools were taken over by the Swansea School Board in 1898.

Henry Hussey Vivian was active in Trevivian and its school from an early period it was especially important after his father entered Parliament in 1832. Other schools promoted by the family were the Vivians' Court Herbert Colliery school at Skewen,

the Sketty National School and a site for, and part payment of the construction of, the new school replacing the old Birmingham Copperworks British School at Morriston. It was noted that Vivian's wife visited the Hafod Schools weekly and listened to reading and dictation lessons, often with one of her sons in attendance. This was similar to the intimate relationship with the copperworks schools shared by the womenfolk of the other resident coppermaster's family, the Grenfells, in Swansea: an interesting matriarchal concern.

Pentre-chwyth (1854) and St. Thomas Schools

Later, other smaller schools were founded by the Grenfells at Pentre-chwyth and St Thomas on the eastern side of the river. At the former, children from the adjacent terraces of Grenfelltown were started in their education by their schoolmistress Miss Jones and then moved to the schools at Foxhole to be educated by her father David Jones.[27] All these schools were later taken over by the Swansea School Board and the former Pentre-chwyth school building is the present Anglican churchroom.

This provision of works schools was fairly comprehensive and was certainly the equal of that in any other town in south Wales. The one missing link in the coppermasters' provision for schooling was at Landore, where the non-resident proprietors of the

large Morfa Copperworks may have considered the pre-existing day schools attached to the chapels sufficient. Otherwise, the three main areas of copper-smelting in and around the old borough of Swansea were well served by large works' schools.

The Voluntary Societies and Nineteenth-century Education in Swansea

It was noted at the Kilvey schools in 1847 that English books were used but that Welsh was used to explain them. This facility would not have been available at the Hafod Schools, where the Vivians had sought qualified teachers from London schooled in the latest continental European ideas in education (notably Heinrich Pestalozzi's work at Stanz and Verdun). The successive census returns record the Trevivian population's general progress from Welsh monolingualism to bilingualism to English monolingualism in three generations. In 1847 it was noted that the Hafod workforce 'were mostly Welsh, with few English or Irish among them'. Ironically, Henry Hussey Vivian lamented the decline of the Welsh and Cornish languages because of the introduction of English into Anglican church services.[28]

At this time the most contentious issue was that of Christian denominational teaching rather than the alternative use of the English or Welsh language. Controversy was avoided on the issue which the workforce felt most strongly: that the teaching of the Anglican church should not be imposed upon their children. In England, some 45% of religious attendance in the early nineteenth century were Anglican, although the figure was considerably lower in the newly industrialised areas where the control of the traditional aristocracy was noticeably weaker. In Wales it has been estimated that only between 9% and 20% of religious attendance were Anglican at this period.[29] Indeed, in industrial areas the proportion of nonconformists is likely to have been much larger: at one stage only 14 out of 779 children in the Hafod Copperworks Schools belonged to the Anglican Church (or 0.02% of the total).[30] This was commonly attributed to the Sunday schools introduced to the Hafod area in c. 1820 (presumably for both adults and children) which facilitated a study of the scriptures through the medium of Welsh. To this influence in turn was attributed a general improvement in moral standards. One of the clerks at the Hafod Works commented in 1847 that 'At that time an Independent minister came to look after us - if he had been Baptist, Churchman, or anyone else to have drawn the net, he would have had us all'.

Swansea employers, and indeed employers elsewhere in south Wales, were wise enough to

Fig. 259. Sketty National School, near the mansions and church built by the Vivians, was one of their several educational foundations.

consult their workforce on the type of religious education which they favoured, despite the fact that most of the employers were church-building Anglicans. Pascoe Grenfell, builder of the Kilvey Copperworks Schools, is on record as having done this and to have agreed to a request that the schools be conducted on a non-sectarian basis. One government inspector in 1868 commented that 'It is much to the credit of a great majority of employers that they consulted the wishes of the workers in the class of schools which they established, and wherever that was the case, the system produced admirable results... All the schools give entire satisfaction to the workmen....' Provision of school accommodation by employers in south Wales provided a non-sectarian educational structure that would otherwise have been unavailable.

The two British voluntary societies to aid the provision of elementary education were set up early in the nineteenth century. The British and Foreign School Society of 1814 developed from the Royal Lancastrian Association set up in London in 1810, and the rival National Society for Promoting the Education of the Poor in the Principles of the Established Church was founded in 1811. The former laid down that 'no formulary peculiar to any religious denomination, shall be ... taught' and initially it consisted of both Anglicans and Nonconformists. The National Society totally alienated the mass of nonconformist working-class people in Wales around 1835 when it made the teaching of the Church catechism a condition of financial help. Both societies helped day schools that taught through the medium of English but there was some flexibility in the amount of Welsh that was used. The British Society had neither the resources nor the diocesan organisation that enabled the National Society to make rapid progress. In industrial areas of Britain, such as south Wales and the textile areas of Lancashire, little progress was made in the establishment of nonconformist British schools during the first 40 years of the nineteenth century. The requisite school sites were also difficult to obtain from, (sometimes unsympathetic) landowners. The collection of building-grant plans for Wales includes 110 British Schools and 318 National schools.

A viable, non-sectarian alternative, however, was provided by the works' proprietors and their schools. By 1846, in the four industrial counties of south Wales (Brecon, Carmarthen, Glamorgan and Monmouth) there were recorded 11,724 pupils in 199 National Schools, 2,809 in 20 British Schools and 5,532 in 37 recorded separate works' schools. The recorded works' schools had an average of 150 pupils each, the National Schools were only a little over a third as large, while the British Schools had an average of 140 pupils each. In addition, it is estimated that a further 21 smaller works had schools, probably taking the total in works' schools to some 7,050 pupils, or 50% of the total attributed to ordinary British and National Schools.

The number of works' schools in Wales expanded during the nineteenth century:

> 1847 - 58 schools
> 1853 - 67 schools
> 1860 - 79 schools
> 1870 - 80 schools.

The dispersed collier population which produced fuel for the works was more difficult to cater for; however, the involvement of the Smiths has already been noted and the following indicates a continuing use of circulating schools in such areas. The Government Education Inspector said in the 1847 report that:

'When Mr. Morgan [the Revd. Rice Morgan, the vicar] came into the parish [of Llansamlet] a school-room was fitted up (lent by C.H.Smith, Esq., the coal proprietor of the parish), by subscription of the parishioners, for the accommodation of one of Mrs. Bevan's schools which was in the parish for two years [i.e. a 'circulating school' on Griffith Jones's model]. This was in 1841. After Mrs. Bevan's master had left, the present master entered upon the school, and by an arrangement with Mr. Smith, 1d in the pound is stopped upon his workmen's wages to maintain it. The school, though the master is an Independent, has been under the control of the clergyman only, and not visited by any one else. The Church Catechism has been taught in it, but not to all.'

The differences from the copperworks' schools which served large concentrations of working families are evident. With such a dispersed rural population, the Anglican clergy (and elsewhere the gentry) could place their own perceptions before those of the dispersed and smaller workforce. The inspector continued:

'A site had recently been granted by the Earl of Jersey for a National School close to the parish church. Subscriptions had not yet been collected; they had been waiting to see what Government would do.

It was contemplated that the new National School would supersede the present school.

Mr. Morgan expected that the 1d. would still be contributed to a national school.'

Elsewhere in the lower Swansea valley it is possible to see how the large metal-smelting works created working-class settlements that gave their workers an opportunity to influence the kind of education they wanted for their children. They had the co-operation of a new class of employer who was in turn dependent on his employees for the effective running of his business. The dense concentration of population around the smelting-works in Swansea provided a strong influence on the overall balance of education in south Wales. Works' schools were so numerous, and the numbers attending them so large, that it has been claimed that these schools constituted a system equal in importance to the voluntary societies in this area in the early nineteenth century.

Fifty per cent of the schools in the voluntary sector in south Wales were works' schools. Up until 1853, when the first full-time agent for the British and Foreign School Society was appointed in south Wales, works' schools completely eclipsed traditional British Schools in the region and were the main sources for elementary and further education in the industrial districts. In 1860, 50% of the south Wales British Schools were works' schools and by 1870 the number of pupils in attendance at works' schools almost equalled the total at those of the voluntary societies.

However, even in the lower Swansea Valley the more remote north-eastern part of the copper-smelting and coal-mining district around Grenfelltown and Bon-y-maen lacked school facilities until 1854. The Assistant Commissioner, William Morris, conducting the 1847 State of Education survey, wrote:

> 'the Revd. Thos. Harries, Perpetual Curate of the district, informed me that he lately went into every house in the village of Pentre-with. He found 130 children under 15; above 100 of whom were old enough to go to school. Not very many (he could not say the exact number) were going to school. There is a village above Pentre-with called Ponymaen, scattered, but with a good many people living about; too far from the present schools. It was contemplated to raise schools which should serve the population there and at Pentre-with. Mr. Grenfell was the only person in the district anxious about a school. The population was mainly employed by him.
>
> There were 15 or 16 public houses between Pentre-with and the ferry, about a mile and half. Drunkenness was the prevailing sin of the district. Wages were good, and there was little suffering except by the people's own fault.'[31]

Pascoe Grenfell soon provided a school at Grenfelltown thereafter (1854). The local Welsh Independent Minister commented that he might have been a Churchman but still exhibited true Christian generosity.[32] One of his daughters continued with more altruistic philanthropy in the Swansea suburbs which were expanding eastwards across the new river bridge. Mary Grenfell asked her father to give her land on which to build a school for the children of the inhabitants of St. Thomas's and then raised over £300 in three weeks with which to build it. The school, on the north side of Morris Lane, opposite its junction with Mackworth Street, was completed in 1862 and used as a day and Sunday school until the new Board School was completed in 1897.

The 1870 Education Act and the establishment of the Swansea School Board eventually caused the demise of the Swansea works' schools. By 1897, only the Hafod Copperworks Schools in the Swansea Inspection District were still independent. In their time, these schools were a very important element in the social landscape of nineteenth-century Swansea and reflect a variety of motives for their construction, to which the term enlightened self-interest may be appropriately applied.

School Buildings

The development of the Swansea works' schools was similar, in general terms, to that known elsewhere in industrial south Wales. Most of the earlier schools in the metal-smelting centres of the coalfield had more than one phase of development, whilst the largest, that at Dowlais, went through at least three or four phases of expansion. In Swansea, the Birmingham Copperworks School at Morriston had three recorded phases of development; the Kilvey Copperworks Schools four phases; and the Hafod Copperworks School five phases. The first phase of development of the south Wales works' schools was usually humble; the school would be held in any spare room attached to the works or in disused buildings near them - even in the lofts of stables. The second phase, after about 1840, saw the provision of proper and substantial accommodation.

How appropriate this general model is to Swansea can be seen by the 1847 Government Inspector's comments on the Kilvey Copperworks Schools. He noted that the Juvenile School built in 1806 was then a 'dingy, dilapidated building' while the Infants' School, built later in 1839, was a 'commodious' and 'good building'. At Trevivian a temporary school building was first used in 1846, perhaps the institutional looking building still standing on the east side of Neath Road, between the entrance to Vivian Street and Bowen Street.

The works' proprietors usually built and paid for the school buildings but exacted a levy on workmen's wages to maintain the schools. At Kilvey

1d. a week was stopped from each workman at the White Rock, Middle Bank and Upper Bank Works, so that their children could attend school. At the Hafod Copperworks, 1d. a week was also stopped from the wages of the workmen and ½d from the children employed. In 1847 the Hafod Works employed 500, a total which included some 100 children over 12 and 30 women, with the average weekly wage for furnacemen of 21-25 shillings, free coal and a cottage at 1s. 4d. a week.

From 1833, the Committee of Council on Education gave building grants to approved schools with no doctrinal strings attached, including those works' schools that were 'neutral' (non-affiliated) and British Schools. The Hafod, Kilvey and Birmingham Company (Morriston) Schools were all British Schools. In 1858-9, for example, the Kilvey Copperworks Schools received grants of £73 for the augmentation of the salaries and pensions of teachers with certificates; there was also a grant of £38 2s. for pupil teachers and a £10 grant for books and apparatus. The annual grant increased with the number of pupils: it was £212 1s. 8d for 326 day and 42 evening pupils in 1865-9; £477 6s. 9d. for 537 day pupils and £32 8s. for 65 evening pupils in 1889; there were 842 pupils in 1894, when it was taken over by the Swansea School Board, receiving an annual grant of £536 10s. 6d. The Swansea

coppermasters were enthusiasts for education, in their works they used processes that required considerable technical knowledge and members of both the Grenfell and Vivian families were personally involved in giving lessons in their schools. The school buildings at Hafod and Kilvey were singled out for special comment by Government Inspectors and Commissioners alike after 1850. Words used by them to describe the works' schools included 'magnificent', 'extensive' and 'good design of artistic beauty'.[33]

Most works' proprietors also provided houses for the teachers. The Hafod Schools had three houses (all still remain with very much their original appearance), one for each department of the school, a 'Master's House', a 'Mistress's House' and an 'Infant Mistress's House'. The two ends of the Infant School at the Kilvey Schools also had houses attached ; all of these dwellings remain in use. The houses were considerably grander than those for the ordinary workers: the master at Hafod had two large reception rooms flanking a central hall, with service rooms at one end. The attached Mistress's House had one large reception room entered via an internal porch. The slightly later Infant Mistress's House was similar to the Master's House, with two smaller rooms flanking a central stairwell and with a large front garden (which the earlier houses lacked). All

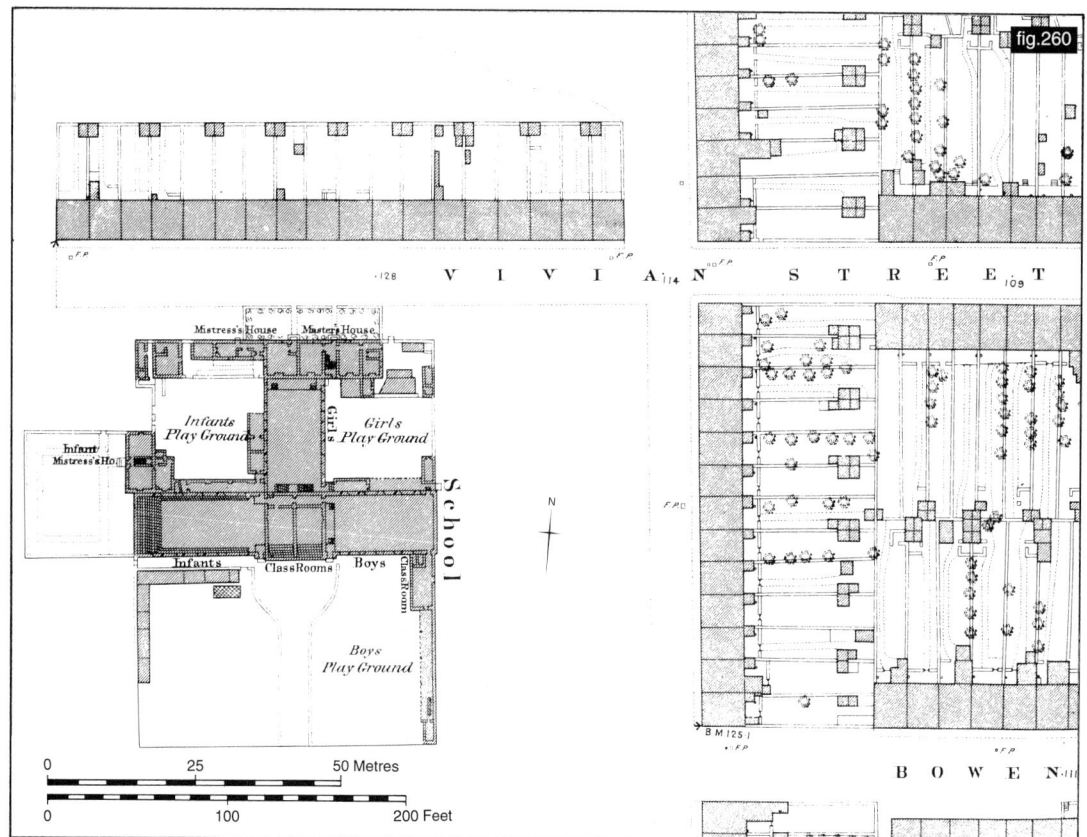

Fig. 260. The position of the original Hafod Copperworks School of 1846 is on the upper part of the plan with the cross-wing of 1848 below. Eventually, the workers' housing surrounded the schools on all sides; the large gardens with pig-sties and toilets of the earlier 1840s housing between Vivian and Bowen Streets are to the right. (Reproduced from 1879 Ordnance Survey map)

three houses backed onto wings of the T-shaped school buildings.

The monumental school building at Trevivian was erected in 1846-7; by 1849 Evans, the copperworks' surgeon, was able to report that there was '...a large pile of buildings, containing immense school accommodation, and residences for master and matron attached'. At Trevivian, in the same year, G.T. Clark 'found 410 children, boys, girls, and infants, at school, in three distinct rooms, airy and clean';[34] by 1893 1,114 children were attending school here. The fee of 1d. a week was deducted from the wages of the employees whose children attended, 'Strangers' (non-employees) were charged 2d. a week. The school closed in 1905 and the building became a clothing factory; it has now been refurbished for community purposes and as accommodation for the Hafod Brotherhood.

The main classroom ranges form a T-shape with a larger block 170ft long and 42ft wide, aligned east-west on an axis that bisects the first residential housing block built between Vivian Street and Bowen Street. This later massive and mildly gothic range may have been designed by the Vivian's architect, Henry Woodyer. The earlier, plainer and smaller lower block (of about half the length of the main block) extends at right-angles northwards towards the back of the master's and mistress's houses on Vivian Street and has a hipped roof integral with the houses. The walls were built in the

local Pennant sandstone rubble with corner-stones formed from moulded rectangular copper-slag blocks. The teachers' housing forms a range parallel to the south side of Vivian Street, and with its hipped roof and overhanging eaves, is reminiscent of the earlier Regency style. The teaching block at its rear - presumably that built in 1846 - is integrated with the houses, while the large main teaching block has high gables towering over Odo Street (and the unmade alley that gives onto the rectangle of undeveloped ground that lies between the school and

Aberdyberthi Street). The dressed-stone, gothic detailing of this massive block has been somewhat obscured by later rebuilding and the southerly playgrounds have been built over. Like the rest of the settlement, the schools originally lacked what might be considered basic necessities; G.T. Clark noted that: 'There was, however, though a well-constructed cesspool, no water supply'.[35]

Fig. 261. The original main elevation of the Hafod Copperworks School of 1846-7 to Vivian Street. The main hipped-roof block (right) fronted the first large schoolroom. The schoolmistress occupied the dark, rendered house to the right of the door, whilst the schoolmaster occupied the larger house with the addition nearer the camera. (9500353/4)

The primary determinant of the size of the school buildings was the likely size of the intake. The Kilvey Copperworks School of 1806 stood opposite the site which was later selected for All Saints' Church, on the east side of Kilvey Road. Its ruins still remain (the southernmost part is now the 'Elim' church mission). The school had a small single room 15ft high, 26ft by 17ft: that is, 441sq ft, or 12.3sq ft for each of the 36 pupils attending in 1847, when they were said to be using 'insufficient furniture and apparatus in bad repair', presumably on a simple, flat floor. The Birmingham Copper Company's School of 1815 at Morriston was in a similarly small building, but both were not much smaller than the genteel Swansea Grammar School whose measurements were 41ft by 23ft.

This sort of school was fairly common, as the educational reformer, Joseph Lancaster, noted in his *Improvements in Education* (3rd. ed., 1805), which record the 'bad accommodation common school-rooms afford to the poor children who attend them; many of whom suffer materially in health, by the confinement at their seats, winter and summer, without variation'. He also commented that these schools rarely accommodated more than 30 children each, and that 'disorder, noise, etc., seem more the characteristic of these schools, than the improvement of the little ones who attend them'.[36] This implied the need for larger and better regulated schools.

Lancaster had much to say about the monitorial system of teaching. His ideas about the physical arrangements of schools received publicity with his publication of *The British system of education*

Fig. 262. The western gable-end of the former large Infants Classroom at the Hafod Works Schools block of 1848. The doorway is a later addition into the base of large tiered staging that once supported infants receiving instruction. The Infant Mistress's garden was formerly in the background with her house remaining to the left. (920057/12)

253

(1810) and *Hints and directions for building, fitting up and arranging school-rooms on the British system of education* (1811). The first of these books recommended placing pupils' desks facing the master, instead of being arranged along the sides of the room; this was a revolutionary layout for the time. The desks were to be used for writing, but sufficient space was to be left at each side of the room for the children to congregate in small groups to practise reading, spelling and arithmetic under the supervision of monitors.

Lancaster's 1811 book was much more detailed and included 'a technical description of a plan for a school-room, intended for the guidance of a builder'. There is no doubt that Lancaster's ideas influenced the form of the schools built in Swansea's industrial settlements. The Birmingham Copper Company's School was rebuilt as 'The Lancastrian School' in 1845 and the three schools at Hafod were built to the basic form of the large Lancaster schoolroom. The Lancastrian model plan of 1811 was for a schoolroom to accommodate 320 children. The room measured 70ft by 32ft and contained 20 rows of desks and forms arranged to face the master's platform and so spaced as to allow the monitors to move between the rows. This plan allowed 0.7sq ft for each child and provided for 16 children to each 22ft bench. Lancaster considered it essential to leave aisles five foot wide on each side so that the children could stand in semi-circles facing a lesson board hanging on the side walls, below fairly highly placed windows. The level of the floor was to rise gradually towards the back of the room, so that the master could see every child clearly when at their desks. The building of a ceiling was not recommended, since this would act as a sounding-board, so increasing the level of noise which would have been considerable when all the monitors were busy questioning the children in their groups. Floors of wood, or flagstones, were also considered to be too noisy.

The first copperworks schools pre-date any legislation concerning the education of children and seem to reflect an acceptance by employers of responsibility for the education of the children of the workforce; these children who would later be expected to succeed their parents in the works. The coppermasters' concern for the education of the children of their workforce may have been encouraged by the 1833 non-binding legislation for the education of factory children and the 1847 enquiry into education in Wales. Part of their motive may have been to provide an educated workforce better suited to performing skilled tasks and able to understand and converse in English. Another motive was undoubtedly the paternalistic concern which the entrepreneurial families felt towards the communities they had created and which were to a large extent dependent on them for employment. As a class they may have felt obliged to promote philanthropic activity, although it was the families of the *locally resident and politically active* coppermasters, such as the Vivians and Grenfells, who built several particularly large and well-constructed educational institutions.

Religious Buildings

Introduction

The type, size, number and shape of religious buildings in the industrial townships around Swansea depended on the wealth and beliefs of the worker congregations and their employers. It is often assumed, when examining the eighteenth and nineteenth century copper and iron industries of Wales, that the industrialists were English Anglicans in Swansea and that the workforce were Welsh-speaking nonconformists.

An examination of this centre of the copper industry allows this assumption to be reassessed; additional factors which help to determine the architectural focus of the workers' settlements. For example, as has already been noted in relation to workers' settlements, there is the question of whether the local residency of the managing partner determined whether a particular copper-smelting concern practised philanthropy: in this instance in building churches or occasionally contributing to the building of nonconformist chapels for the workforce. It may be that if the partners in an industrial concern were absent their managers assumed responsibility in the local community. If so, were the managers likely to support the established Anglican Church or a nonconformist denomination?

A prime motive of the coppermasters may have been to produce a quiescent workforce through the social control of the established Anglican Church. Presumably their access to varying degrees of readily available resources would affect how the buildings were constructed and this may have changed over a period of time. It may be that the buildings they produced were always of a similar or distinctive type.

If the workers were largely nonconformist they presumably constructed buildings that were distinct from the religious buildings provided by their employers and these may have formed a distinctive class or type. It might be argued that it is this category of buildings that defines the Welsh character of towns and settlements, both in the number of nonconformist places of worship and in their contribution to the townscape. The numerous nonconformist chapels built at Swansea may provide information on whether the designers of these buildings had similar backgrounds, how they were trained and where their architectural ideas came from. There is the question of how relatively poor worker congregations could afford to employ able architects. An examination of the buildings in the industrial communities may show whether there are characteristics shared with others constructed in the area. It must also be asked to what extent Swansea contributed to the development of the nonconformist Welsh chapel and how important the town's industrial settlements, were in defining the character of Welsh nonconformist buildings.

If attendance at places of worship is taken as an indicator of piety, then Swansea in the Industrial Revolution was a deeply religious town. One denomination - the Welsh Independents or Congregationalists - predominated, especially in the new industrial satellite communities like Morriston. The established Anglican church had an unhelpful church organisation that regarded pluralist livings, with one vicar serving several churches and communities, as tolerable and indeed normal.

At the time of the Religious Census of 1851, the totals of morning and evening religious attendance (copperworks' shiftwork was one of the factors producing considerable disparity between attendance at these services) out of a population of 46,907 was 29,898 64%. By far the most numerous were the Welsh Independents, with a total of 11,912 (40% of all worshippers) at the two Sunday services. The copperworkers' communities were much more strongly nonconformist, and much more widely Welsh-speaking than was the population of Swansea town. In Morriston, the largest of the copperworkers' settlements, in 1851 just 108 of the 2,798 seats available in religious buildings were Anglican, or 4% of the total. This huge difference did not greatly vary when the number of actual worshippers is counted: 117 of the 1,839 worshippers at Morriston were Anglican, or 6% of the total. There were seven nonconformist congregations meeting in this settlement at this date, of which five were worshipping in Welsh and two in English. St John's, the Anglican Church, had services in English and Welsh but the latter congregation was much smaller. Overall there were 2,402 people worshipping in Welsh and only 396 in English, 16% of the total.

There was not a great difference in the balance between the various religious buildings by the time the 1905 Royal Commission on Religious Observance reported; then the centre of Morriston, in and just around the boundaries of the eighteenth-century township, had no less than 16 nonconformist chapels and only two Anglican churches. This great preponderance of nonconformist congregations still existed in 1960, when Anglican worshippers formed only 10% of the congregations of the town, despite the decline in the use of the Welsh language.

There was a distinct division of religious opinion between the largely Anglican coppermasters and their mostly nonconformist workforces. Perhaps this is visually most apparent in Morriston, where the unfinished Anglican church occupies a focal point in the street layout but the 'Cathedral of Nonconformity', complete with towering spire, dominates the main street. All of the most influential locally resident coppermasters' dynasties in Swansea were Anglican: the Morrises, Vivians and Grenfells. In fact, this was not a great change from the domination which the established Church had had amongst influential local families. In the generation before the Morrises, the local early eighteenth-century landowner-industrialists were, and remained, Anglican.

However, as in other contemporary industries, there was a strong Quaker influence in the Swansea copper-smelting industry of the eighteenth century. John Lane of the Llangyfelach Copperworks (1717) had a father-in-law and business partner at the Neath Abbey Copperworks, John Pollard, who was a Quaker. Pollard stayed at Landore after the bankruptcy of the Llangyfelach Copperworks and Joseph Pollard became a local Welsh-speaking Quaker minister. James Griffiths (who had worked at the Llangyfelach Copperworks), Silvanus Bevan, Michael Bevan and their partners at the Cambrian Copperworks (1720) were Swansea Quakers.[37]

Welsh Quaker influence had also extended in a reverse direction to the earlier copper mercantile metropolis of Bristol. This was then the second richest city in Britain and was responsible for much investment at Swansea. William Bevan of Swansea (d. 1700; father of Silvanus and grandfather of a later William) is known to have corresponded with Edward Lloyd, a merchant of Corn Street in Bristol.[38] The Lloyds, Welsh Quakers associated with the Dolobrân Forge in Merioneth (and some of whose members were later responsible for the founding of Lloyds Bank), were involved in the Bristol copper trade with Abraham Darby (from 1702-3) and were related to the major Bristol coppermaster William Champion. In turn, this Bristol Quaker family was also heavily involved in non-ferrous metal manufacture in Wales, a nephew John was one of the partners in the large brass battery mills at Holywell in Flintshire and his son, also named John, had proposed the reopening of the Upper Bank zinc smelter at Swansea. The elder John Champion had also been one of the partners in the

Mines Royal Copperworks at Neath Abbey between 1694 and c. 1716.[39]

Partly because of Bristol's strong Quaker tradition, it has been claimed that the third Swansea copperworks at White Rock (1737) had Quaker connections,[40] but in fact the main owners belonged to a rival Anglican group of Bristol industrialists, who were the second largest Bristol copper-smelting concern by the mid eighteenth century.[41]

The biggest Bristol concern, which did have Quaker partners, acquired the Forest copper-smelting works at Swansea in c.1790 upon new smelting orders being placed by Thomas Williams of Anglesey (an Anglican), who himself was managing partner at the Middle and Upper Bank Works. This large Bristol concern was at that time known as Harfords & Bristol Brass (Wire) Copper Company (and earlier the Brass Warehouse (Wire) Company). Six of the 10- member board were drawn from the Bristol Quaker banking family the Harfords, themselves descendants of Edward Lloyd and also ironmasters at Ebbw Vale, Nantyglo and proprietors of the Melingriffith Tinplate Works.[42] This pre-eminent Bristol company transferred all of its copper-smelting activity to Swansea until it had ceased such activity by 1820.[43]

In 1814 the Swansea Quaker William Bevan, son of Silvanus Bevan of the Cambrian Copperworks, started the Nant Rhyd-y-Filais Works to extract copper and slag from old copper-slag. The larger and adjacent Landore Copperworks were also being run by the locally resident Quakers, Henry Bath & Co. (with R.J. Nevill) between 1827 and 1837.[44]

There was also Quaker entrepreneurial participation in other significant Swansea industries and non-ferrous smelting activity. George Haynes, the founder of the first Welsh based newspaper, *The Cambrian* involved in the Cambrian and Glamorganshire Potteries, came to Swansea from the Quaker colony of Pennsylvania. The Quaker William Dillwyn, who was born in Pennsylvania (in 1699 the Deulwyns had emigrated from Llangorse to Pennsylvania), bought the Cambrian Pottery from George Haynes in 1807 and placed his son Lewis Weston Llewellyn in charge, where he remained until the pottery was sold in 1817. In turn, his son Lewis Llewellyn Dillwyn (1814-92) MP set up and ran the Landore Silver Works (*c*. 1853-67) and the Dillwyns later operated the Llansamlet Zinc Works between *c*.1858 and 1896.[45]

The company that came to employ the most copperworkers by the mid nineteenth century had three main partners from one Cornish copper and tin-mining family which respectively followed the Quaker, Wesleyan Methodist and Anglican traditions! The Methodist head, of the Williams family of Scorrier House, was John Williams, the so-called 'King of Gwennap', who had become a Methodist, built Wheal Rose Chapel in west Cornwall and was principal subscriber to the remodelling of John Wesley's preaching pit at Gwennap in 1806-7.[46] In the period before 1841, under the leadership of this influential Wesleyan Methodist, the family partnership, known as Messrs. Williams, Foster & Co., became a very significant copper-smelting and manufacturing concern at Swansea. The Williamses firstly became involved with their Cornish compatriots the Grenfells and Foxes, at the Rose Copperworks from 1823-1829, at which latter date the significant partnership of Williams, Foster & Co. assumed responsibility for the works; by this time the expansion on the huge Morfa site was beginning and continued after 1835. From *c.* 1839 they also became involved in the running of the Crown Copperworks at Skewen in the adjacent Vale of Neath.[47] The company was active until 1924, but in the critical period after the death of John Williams senior in 1841, the managing partner was the Anglican Michael Williams rather than his elder Quaker brother, John Williams junior.

A major factor in predisposing the many nonconformist works' operators to act as benefactors to local nonconformist working populations, was whether the key managing partners had been locally resident. The second major factor in persuading the coppermasters to help with the cost of building a chapel was if the denomination they belonged to coincided with that of their workforce. This did not often happen, nonconformist industrialists in south Wales tended to be Quakers or Wesleyan Methodists and the Swansea copperworkers were nearly all Welsh Independents (Congregationalists). It was Congregationalist industrialists from Bristol and west Yorkshire who helped fund English-language Congregationalist chapels in Swansea, and later tinplate industrialists in Morriston who supported the congregations they belonged to in order to build the huge Welsh Independent chapels. Having said that, it is evident that the Swansea coppermasters were willing to give or lease cheaply land for chapel construction, to lay chapel foundation stones, and to speak at the opening services of chapels. Hussey Vivian is even recorded as donating money to the construction of a Welsh Independent Chapel (Seilo Newydd) near the Hafod Copperworks and building another for his colliers at Pant-lasau. Many of the various copperworks owners were predisposed to let their management play a leading role in the local communities and many of these influential and locally powerful figures were nonconformists.

It has been estimated that a third of this influential and fairly affluent tier of local society were Welsh (and Welsh-speaking) local men who were likely to be members of the Welsh Independent denomination. [48] However, it was far from the case that it was only managers with surnames of Welsh origin who were members of nonconformist congregations. For example, Robert Monger, the manager of the largest copperworks at Hafod, was a founding deacon of the important Welsh Independent congregation at Seilo, Landore. Several other eighteenth-century managers were drawn from the earlier Bristol copper industry and were likely to be Quakers. The Cornish mining industry's nineteenth-century investment in Swansea smelting brought in managers who had been involved in some of Cornwall's 1,500 or so Wesleyan Methodist chapels.

Quaker managers possibly included the Pollards at Llangyfelach from 1717 and there were probably Quaker managers at the Cambrian Copperworks (1720). It is likely that Quaker management (the Phillips family) persisted for at least 25 years at the large White Rock Works (31 furnaces and 15 calciners by 1760; with brass-making from 1762). The remote Committee of Management met weekly in the Three Tuns public house in Corn Street, Bristol, until 1762, when twice weekly meetings were switched to a specialist counting house in Small Street, Bristol. [49]

On the evening of 28th June 1778, Catherine Phillips, the Quaker Minister, was noted as preaching at the White Rock Works, amongst the workmen and others in the copper-furnace hall. Her husband William, of this Swansea Quaker family, was also a manager or agent in the copper trade at that date, travelling between the Swansea smelting-works and the Cornish mines. [50]

The Quakers generally showed much greater understanding and support for their fellow nonconformists than did members of the established Church, but by the time of the Religious Census in 1851, the number of Quaker workers was tiny compared to members of the other nonconformist denominations in Swansea town: there were only 35 Quakers compared to 1,232 Wesleyan Methodists, 2,013 Calvinistic Methodists, 2,936 Baptists and no less than 3,958 Independents (Congregationalists). Cornwall was a stronghold of Wesleyan Methodism and the early nineteenth century brought Cornish mining interests to Swansea with the founding of the much larger Hafod (1810) and Morfa (1828) Works. The Cornish managers included William Morgan, manager of the Hafod Copperworks Rolling mills, who with Thomas Evans took over building operations at the large and ornate Wesleyan Methodist chapel in Goat Street, Swansea town, in 1847. This elaborate three-towered Italianate structure, the 'cathedral of Methodism in west Wales' was at that time the finest chapel that had been erected in Swansea. [51] The general manager of the Morfa Copperworks, Captain Rundle, was also instrumental in building the Wesleyan Methodist chapel in the nearby industrial settlement of Landore (Wern Road) in 1860-61. [52] Many Cornish Wesleyan families had been attracted by the employment opportunities available at the copperworks.

By the same period, the influence of the many local copperworks' managers, who had acquired their experience in the long established local trade, became apparent and the great majority followed the Welsh Independent (Congregational) denomination. Robert Monger, the main manager of the largest copperworks at Hafod, was a founder deacon of the very influential Independent congregation of Seilo (Siloh) at Landore, whose minister/architect was the most influential chapel designer in south Wales. [53] Another founder deacon, William Williams, may have been the individual who was manager of the Hafod Iron Foundry.

At the time of the great post-revival building in the mid 1860s, Elias Evans, the manager of the large White Rock Copperworks was a member of the town chapel in Castle Street, but considerably assisted his workers financially in the rebuilding of Canaan chapel at Foxhole, which he also attended. The £702 6s. debt was paid off within six years with the help of Elias Evans.

The 1851 Religious Buildings Census reveals that the other nonconformist chapel at Foxhole also had a copperworks' manager, John Mathews as a prominent member of the congregation. The Baptists were a smaller, but still influential, denomination in the industrial settlements and this small Baptist Chapel, built on the hillside next to the White Rock Copperworks, could only seat 50.

Anglican Churches of the Coppermasters

The late eighteenth-and nineteenth-century Anglican churches built in the copperworkers settlements were partly or entirely funded by the resident coppermasters. Relations between the industrialists and the Anglican Church varied. John Morris I said he had been persuaded to build the first Anglican church in Morriston to counterbalance the influence of the nonconformists and that the bishop of St. David's had congratulated him. At the time of the 1851 census his son, Sir John Morris II, was patron of St. Mary's Church in Swansea (capacity 1,371), St. John's juxta Swansea in the mainly Welsh-speaking northern suburbs (capacity 1,500), and St. John's, Morriston (capacity 108). The Vivians, however, delayed funding a church at Trevivian for some 40 years because of a dispute with the Bishop

Fig. 263. The copperworkers' township of Morriston was initially dominated by the tower of the coppermaster John Morris's Anglican Church. Later it was overshadowed by the steeple of Tabernacl Chapel the 'Cathedral of Welsh Nonconformity'. (9500356/1)

of St. David's over plurality of Church livings and having to share a prospective vicar with St John's juxta Swansea. The Grenfells at Foxhole were persuaded to fund the building of Kilvey Church (capacity 346) by a particularly enthusiastic vicar of Llansamlet.[54]

These places of worship did not grow from within the working-class communities as house-churches as was usually the case with nonconformist chapels; instead they were provided largely from the funds of the patriarchal coppermasters. An exception was the growth of the Anglican congregation at St. Thomas's before 1864 in two houses purchased by Mary Grenfell. Sometimes there was a short interim stage, as with the nonconformists, when the nucleus of an Anglican congregation worshipped at the works' schools. There were several examples of this in the Heads of the Valleys ironworks and it happened at Trevivian from 1874 until the new church was ready in 1880.[55] Even when the coppermasters did not meet the whole cost of a church, they still determined the style, ornamentation and, hence, the price of building a structure. Of the £5,000 needed to build St. John's Church at Trevivian in 1878-80 in the Cornish Perpendicular style favoured by the Vivians, only £1,500 was given in cash by the firm of Vivian and Sons, with £500 from Henry Hussey Vivian personally. That left a large debt which took

the small congregation until the early years of the twentieth century to pay off. However, such large and long-continuing debts were not confined to Anglican traditions, the enthusiastic Welsh Baptists at the neighbouring Philadelphia Chapel had spent £1,850 on their elaborately façaded chapel 14 years earlier in 1866 and the last of the £1,200 debt was not cleared until 1906, whilst the huge cost of Tabernacl, Morriston (£10-15,000 in 1873) was not cleared until 1914.

The religious revivals and growth of nonconformist congregations, and the consequent enlargement and elaboration of their chapels, evoked a response in terms of the enlargement and elaboration of the Anglican churches of the copperworkers' communities. For example, in 1839-40 the members of the Welsh Independent cause at Foxhole dug a shelf on the hillside and laid foundations for Canaan Chapel after working their shifts at the copperworks (self-building was common in nonconformist chapels but was probably unknown in contemporary Anglican churches),[56] the 342-seat chapel was erected at a cost of £702 6s 0d.; three years later the Grenfells were persuaded to fund the first Anglican church in the community with a capacity of 346 for the sum of £1,350.[57]

At Morriston there seems to have been a similar process at work. Libanus Chapel, which it had been

Fig. 264. The coppermaster's church at Kilvey was built in 1843 above the copperworkers' settlement of Foxhole. The south aisle and apsidal chancel added in 1858-9 can be seen behind a copper-slag capped churchyard wall. (9500349/1)

necessary to rebuild and enlarge twice since William Edwards first designed it in 1782, and which already had a capacity of 602, was completely rebuilt in fashionable style in 1857 by the rising chapel architect, Thomas Thomas of Landore, to provide a capacity of 750. The 1851 Religious Census makes it clear that St. John's Church, built by William Edwards in 1789 and with a capacity of 108, was more than adequate for the size of the Anglican congregation; even so, work was started on rebuilding the church in a more acceptable full-blown Decorated Gothic style in 1859, two years after the new Libanus was completed, by the well-known architect R.K. Penson. Not content with providing a building for a projected congregation of 350, the grandiose church was left in 1862 with its west (liturgical north) aisle unbuilt and so it remains. It was not alone in being an ambitious church built for a copperworkers' community on a scale that was too large for its congregation and with a grand plan that was never realised. St. John's built at Trevivian in 1878-80 was very similar, it was completed one bay short of the original plan, without its tall and fine perpendicular tower and with its north aisle missing, as can be seen today. The expansion of this copperworkers' settlement church may have inspired

the Grenfells to do the same at their church at Foxhole where the fine apsidal chancel and south aisle were added, again by R.K. Penson, in 1858-59.

As the two Anglican churches were being rebuilt at two of the largest copperworkers' settlements in the 1850s, a major religious revival erupted in the nonconformist chapels which resulted in much grander rebuilding on a scale, and with architectural ornamentation, that provided as much of an architectural focus to the workers' settlements as those churches funded by the coppermasters.

It is easy to be cynical about the coppermasters' motives for promoting the Anglican Church, but there is no doubting the strength of the religious convictions of most of the coppermasters even if Henry Hussey Vivian said he found it impossible to drive thought of business out of his head during church services! The endowment of churches, schools or workers' houses for each of the copperworkers' settlements was not advertised as a stratagem to produce a more quiescent workforce, although that was what was achieved. The evidence for a genuine interest in religion and Christian charity and philanthropy is clearly there. John Morris I, for example, had translations of William Edwards's sermons specially made for him. Equally,

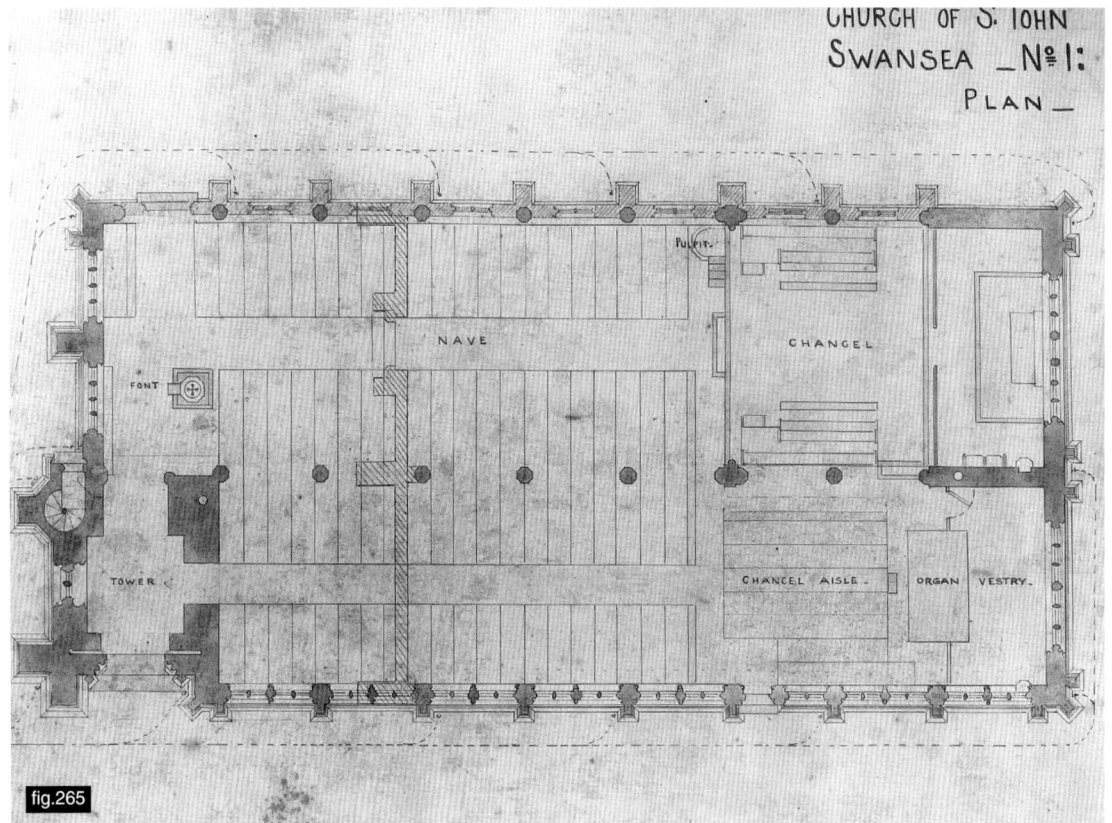

CHURCH OF S: IOHN
SWANSEA _ Nº I:
PLAN _

Fig. 265. Henry Woodyer's plan of the proposed Anglican church of St John at Trevivian. Henry Hussey Vivian insisted it should be a copy of St Mary's, Truro. Among the three alternative designs, the plan shows free-standing octagonal columns built into the north wall of St John's to allow for a possible future expansion to more closely resemble the Cornish prototype. (Held by the West Glamorgan Archive Service; P/106/CW/62/1)

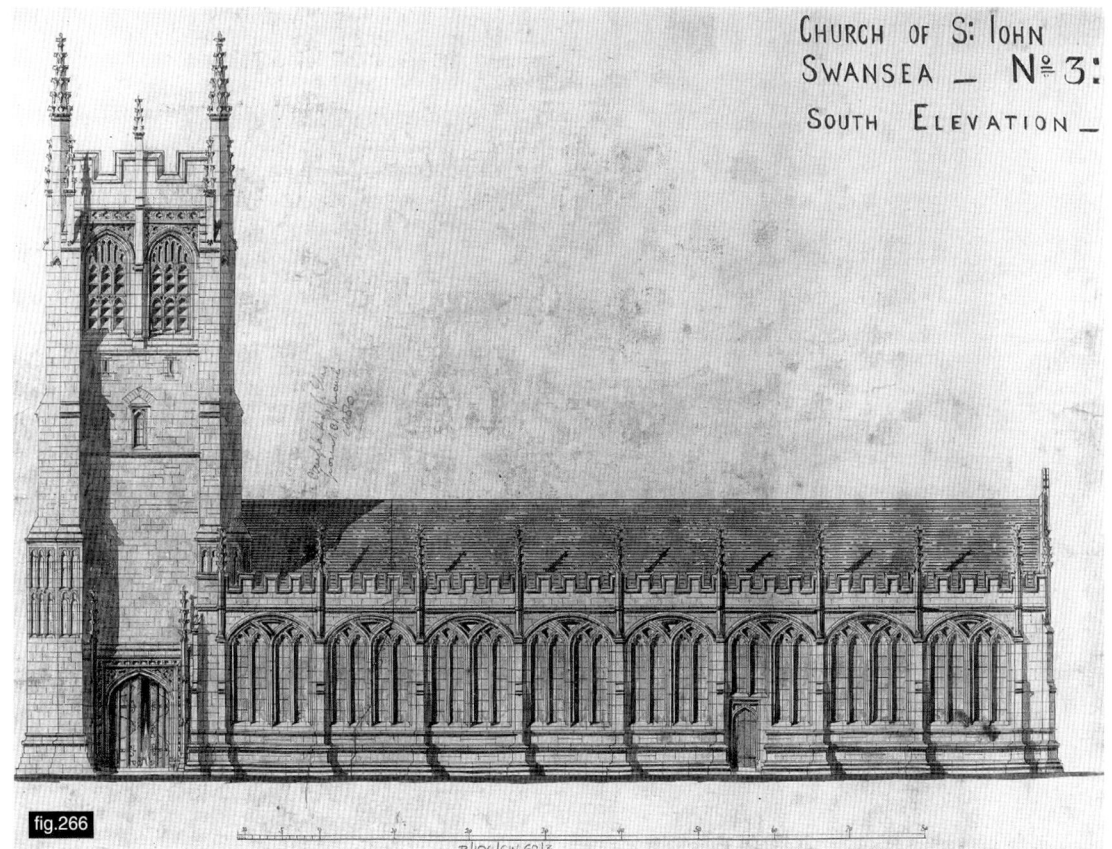

CHURCH OF S: IOHN
SWANSEA _ Nº 3:
SOUTH ELEVATION _

Fig. 266. The long facade of the south aisle of St Mary's, Truro was the most elaborate piece of church architecture in Cornwall, and Henry Hussey Vivian had Woodfield produce an exact copy of the form of its tracery and a simplified version of its parapets for the south aisle of the new Trevivian church (completed in simplified form). (Held by the West Glamorgan Archive Service; P/106/CW/62/3)

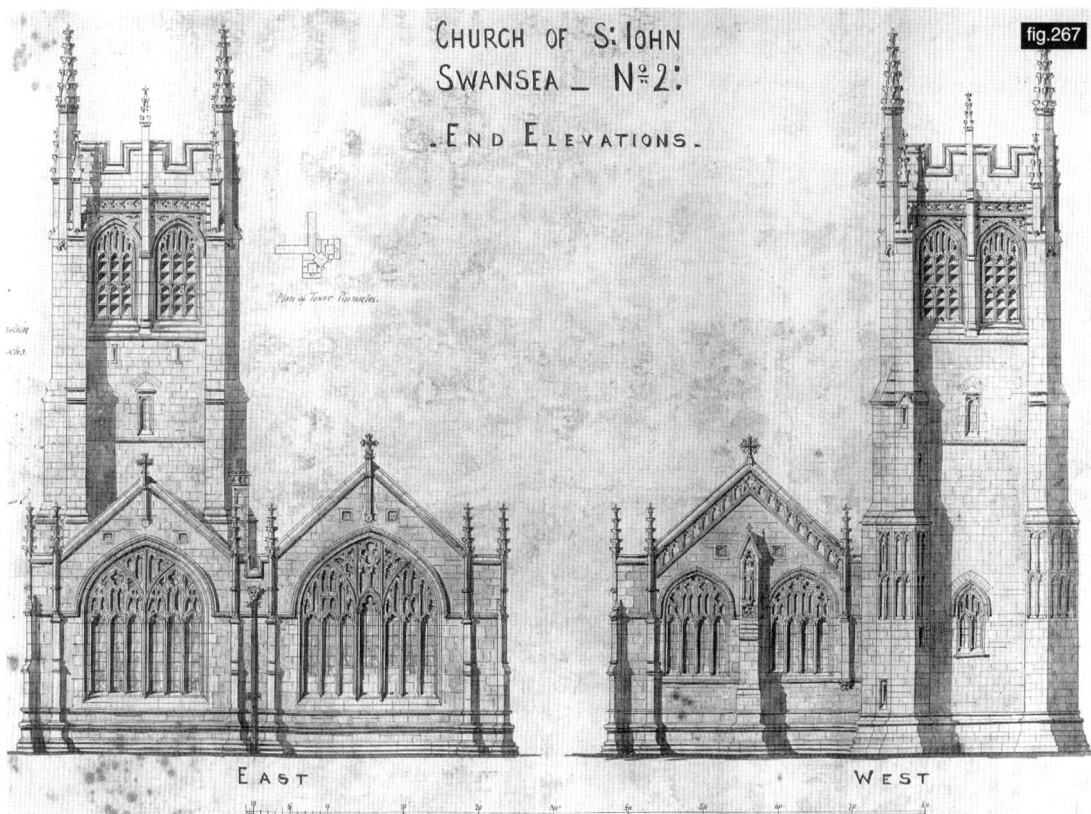

CHURCH OF S. JOHN
SWANSEA — No. 2.

.END ELEVATIONS.

fig.267

EAST   WEST

Fig. 267. The east end of St John's, Trevivian was modelled on St Mary's, Truro but the grand be-towered west end was Henry Woodyer's own creation; it was never built except for the form of the windows incorporated in the shortened nave. (Held by the West Glamorgan Archive Service; P/106/CW/62/2)

Pascoe Grenfell's obituaries in *The Cambrian* (4 and 11 April 1879) show the strength of his religious convictions:

> 'For thirty years he taught a Bible class in the Sunday school, and it is said that no less than seven of his scholars have since gone into the church... .
>
> The Swansea Savings Bank experienced the benefit of his careful direction, and the British and Foreign Bible Society, the London Missionary Society and other kindred associations received his warm and personal support; for though a Churchman, he was most tolerant and kindly in his religious views...'.

The first copperworkers' settlement at Morris Town, founded in 1779-82, may have been planned uniquely to have both nonconformist and Anglican places of worship. The founder, John Morris I, was an Anglican but he was influenced by his architect and town planner, William Edwards, a Calvinistic Methodist (and subsequently Independent) minister. Edwards apparently designed both a church and a chapel each with a tower. The church was constructed at the centre of the square to be built in 1789 in the second grid of the township, the first Morriston chapel (Libanus) having been built seven years earlier at the centre of the first phase of the township. It was designed by William Edwards, a

simple, barn-like structure (with or without the supposedly intended tower) on a site with a 1,000-year lease at one shilling a year. Morris later provided a site for a second nonconformist chapel, in the same area, at a peppercorn rent. Symbolically, the huge Tabernacl chapel of 1872 took the Welsh-speaking congregation from Libanus to a central position on the later main street of the town, Woodfield Street, with the largest tower ever built on a Welsh nonconformist chapel dominating the surrounding townscape. John Henry Vivian, made plans in the years 1837-41 to build an Anglican church at Trevivian based on the church in which his family were buried in Cornwall. The unwillingness of the Church of England authorities to allow the creation of a new parish, together with the early death of Vivian's daughter-in-law, the first wife of his son, Henry Hussey Vivian, resulted instead in the decision to build a new church in her memory at Sketty, on the edge of their Singleton estate. This situation was compounded by John Henry Vivian having to mortgage the Hafod Copperworks in order to be able to purchase his late brother's share of the family firm. Part of the centre of the new workers' town may then have been re-designated as the site for the extension of the Hafod Copperworks Schools.

John Henry Vivian had wanted to build a church at his own expense, having told the bishop of St. David's that he would no longer tolerate a pluralist vicar who officiated at both Llangyfelach Parish Church and St. John's, Swansea. However, the medieval church of St. John's (now St. Matthew's) had already been rebuilt by William Jernegan in 1824 (interestingly with a 'preaching-box' extension to the side holding 1,500, thus making the internal design not dissimilar to that of a nonconformist chapel) and the bishop insisted that any vicar of the new church at Trevivian should also officiate at St. John's church.[58]

John Henry Vivian approached the architect Henry Woodyer in 1849 for designs for his proposed churches at both Sketty and Trevivian.[59] At this stage, the Trevivian schools had just been completed, leaving a space to their immediate east at the centre of the development which may well have been the second site intended for the new church. However, 'the donor changed his mind' for the reasons given above. Hence, unlike Morriston, a new church was not built as the centrepiece of the town. When eventually built, as one of the last elements in the township, it was built on its western edge. It did not even align with Odo Street or with the new east-west cross street, Gerald Street, down which its eastern facade can be partially glimpsed. It did however dominate a slope looking south-westwards over Cwm Burlais where Swansea's church for its large Roman Catholic worker community was built in 1875. The present large, seven-bayed Roman Catholic Cathedral (also lacking its intended tower and spire) was rebuilt by Pugin and Pugin in 1886-88. The Trevivian Anglican church was built a full 35 years after the Vivians had funded schools for the settlement and 13 years after the first nonconformist chapel, and when eventually realised, it was only part-funded by the Vivians.

The church authorities had been given the site at Trevivian which, in the third quarter of the nineteenth century, was the centre of population growth for the parish. Consequently, from 1874 the new vicar, the Revd. J. Stephen Davys, began building an Anglican community by holding services in the Hafod Copperworks' School. There was a depression in the copper industry and the Vivians' plans to build a church in the Cornish style were only partially brought to fruition. Henry Woodyer, the architect, wrote in 1879 that Henry Hussey Vivian was resolute at meetings with the him in his intention to build a church:

> 'which then and now shall be a copy of St. Mary's Truro, but with greatly enlarged width - this explains the peculiar treatment and length. You will see that we are building only a portion of the intended church - at the last interview I had with the Donor he pointed out to his Son how he should in his time carry out the other

portions of the church [i.e. the tall tower and north aisle].'[60]

The Vivians contributed over £2,000, which was half of the £4,000 which had been raised at the time of the church's opening in March 1880, when £1,000 was still owing. A debt of £440 was still outstanding on the church in the early twentieth century.[61] Consequently, what was built was only part of the planned church. A typical Cornish church had a long, low nave with multiple aisles of equal width and height. The division of the nave and chancel was not

marked by any change in roof height and a high tower stood at the west end. At Trevivian a long, low nave and aisle were executed but the tower was replaced by a bellcote at the west end of the south aisle and the north aisle was never constructed. A copy of Woodyer's original design is engraved on a brass plate on the inside of the west front.

St. John's is an elegant, seven-bayed building in local Pennant sandstone rubble with ashlar dressings. The surviving south aisle of St. Mary's, Truro, now part of the cathedral, was built in the first decade of the sixteenth century and has a richly ornamented crenellated parapet, with quatrefoils for its three-bay chancel and it also has flat four-centred arches with ornamentation in their spandrels - the inspiration seems fairly clear. After the opening of St. John's, the old parish church was renamed St. Matthew's and the Trevivian church became the new parish church of St. John-juxta-Swansea. St Matthew's (William Jernegan's church of 1823-4) was then rebuilt at the considerably smaller sum of £1,587 for the local Welsh-speaking congregation of Swansea.

John Henry Vivian also financed the building of a new Anglican church, the richly ornamented St. Paul's, Sketty, which was completed in 1850. Unlike some contemporary industrialists, he did not build his first church for the use of his workforce. The industrialists' move to west Swansea had separated owner and workforce and his differences with the

Fig. 268. The interior of St John's Church, Trevivian looking east to the pulpit and altar in 1999. The south arcade is blocked to reduce the size of the church in use, and that first intended for the north was never constructed. (990270)

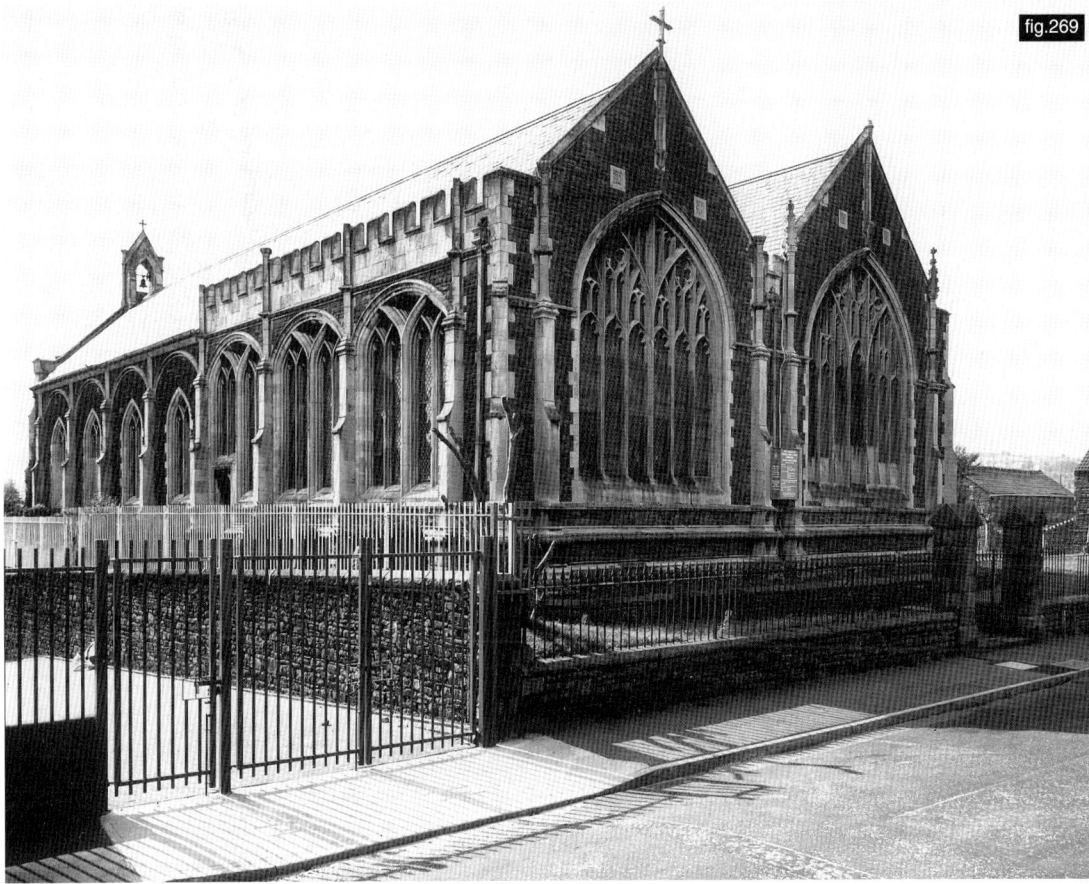

fig.269

Fig. 269. The photograph of St John's as built illustrates how the Vivians' expectations were not matched in construction, with only the three bays of the chancel being finished with elaborate Bath stone tracery and crenellations in close imitation of the south aisle of St Mary's, Truro. (9500353/1)

Fig. 270. The south aisle of St Mary's, Truro, now incorporated into the side of Truro Cathedral, showing how closely the Trevivian church was modelled on it. Note the 'Y' tracery of the sides, the castellated parapet and the perpendicular tracery of the east end. (990236/17)

Fig. 271. St Paul's church, Sketty, erected in 1850 in memory of Henry Hussey Vivian's first wife and designed by Henry Woodyer in the Early English style. The Vivian Chapel is to the right of the larger window.

fig.270

fig.271

hierarchy of the Anglican Church precluded the building of a church at Trevivian. John Henry Vivian succeeded at Sketty where he failed at Trevivian in that the church became the centre of an independent parish in 1851. It was very much the family church on its new estate, being conceived initially as a family memorial, one of two coppermasters' churches which originated in this way. In 1849 the first wife of Henry Hussey Vivian died and the bereaved husband and his father constructed the church in her memory.[62]

The Guildford architect, Henry Woodyer, was commissioned to build a 'village church' in Early English style, though using the Cornish stone of the Vivians' homeland. This design was preferred to that of the Cornish town church, which they later insisted

263

fig.272

Fig. 272. The interior of the Vivian Chapel at St. Paul's Church in Sketty.

should form the basis for St John's at Trevivian. The rural landscape setting of Singleton Abbey was improved with the new tower and spire of the church dominating the northern part of the park.

The 1851 *Guide to Swansea* described the new church in glowing terms:

'The Church is 97 feet long by 37 feet, and ... affords accommodation for 350 persons. It consists of a well-developed chancel, south chapel to chancel; south aisle to nave, south porch, and baptistry... The tower and spire, which is covered with shingle, is 100 feet high, and forms a prominent and pretty object from the Mumbles and Swansea roads.

The Communion Table covering was embroidered by the Ladies of Mr. Vivian's

264

Fig. 273. Statue by Tenerani in the Vivian Chapel flanked by memorial plaques to members of the Vivian family. (9300681)

Fig. 274. Doorway into the Vivian family vault under the Vivian Chapel at St. Paul's Sketty.

family, aided by several intimate friends. The east window, by Warrington, was the gift of Mrs. Vivian... The tower of the Church contains a capital new peal of eight bells in F, cast by Messrs. Mears, of London. The tenor weighs 15 cwt.; and the bells are, without exception, one of the finest in the kingdom for their weight. They are composed of the finest copper and tin, especially prepared at the Hafod Works, to which is probably due their fine tone. The Church has an endowment, and a district attached to it.'[63]

At this date Henry Hussey Vivian lived in 'a pleasant country seat, called Wern Eynon ... a short distance from the church'. Other magnates from the rural seats of west Swansea contributed to the church, such as L.L. Dillwyn (of the Cambrian Pottery), who lived in the castellated Parkwern and contributed the organ to the new church.

Grenfelltown originally had no place of worship; later the two nonconformist chapels were complemented by the conversion of an infants' school into an Anglican church in the early years of the twentieth century. The Grenfells' Swansea home of Maesteg House was originally built in grand isolation on the southern, seaward, slopes of Kilvey Hill. The bridging of the Tawe and the foundation of the Swansea suburb of St. Thomas near their home induced the family to build a family memorial church there. The congregation was started when Pascoe Grenfell's daughter, Mary, had purchased first one and then two houses in Benson (later

Pinkney) Street and held religious meetings there until St. Thomas Infant School was built in 1862. A curate was employed and an organ was presented by the workmen at Middle Bank Copperworks, some of whom presumably lived in the new suburb.

In about 1879 a corrugated-iron church was built at St Thomas. It was removed to serve as St Stephen's Church in Port Tennant when the new stone church at St Thomas was built at the Grenfells' expense. The first iron church was built on the main axis of St Thomas, just as St John's, Morriston, was at the centre of that other grid-iron settlement, but the larger stone church was asymmetrical to the settlement. It was designed in gothic style, with seating for 500, by Nicholson & Son of Hereford and cost £4,000. The foundation stone was laid in July 1886 by General Sir Francis Wallace Grenfell,

Fig. 275. The coppermaster's infants' school at Grenfelltown (right) was later extended to form the small Anglican Church that remains. Note the front wall (left) with its capping of copper-slag blocks from the Grenfells' Middle and Upper Banks Copperworks. (9500351/4)

fig.276

Fig. 276. The Grenfells provided three successively larger Anglican places of worship for the new Swansea suburb of St Thomas's, which was constructed next to their Swansea residence. In 1886, General Sir Francis Wallace Grenfell laid the foundation stone for the substantial stone church which remains today. (Held by the West Glamorgan Archive Service; P/326/7)

commander of the Egyptian army and Mary Grenfell's brother. Her sister, Mrs Llewellyn of Baglan Hall, paid for the tower and chancel to be added.[64]

Landore was the last of the substantial copperworkers' settlements to be given an Anglican church, in 1890. By that time, the community was (and is) dominated by the very large nonconformist chapels of Dinas Noddfa and Seilo Newydd, both representing the tertiary chapel building phase of their respective congregations by elaborate and sophisticated buildings. By contrast the mission church built in 1890 at Landore was completely of roughcast local Pennant sandstone, and built to house a congregation of 300 at a cost of £600. It sat with lowly presence on Dinas Street, near the soaring façade of the Dinas Noddfa Baptist chapel with its great rose window; J. Buckley Wilson and Glendinning Moxham of Swansea were the architects.[65] The first design for a larger permanent structure was abandoned in 1899, but then the high dominating bulk of St. Paul's Church was built on the main Neath Road in 1902-3 to the designs of E. M. Bruce Vaughan executed in flowing decorated gothic style.

Other daughter churches and missions expanded into late nineteenth-century Swansea under the patronage of the resident coppermasters. One such example was the building of St. Michael's Church at Manselton with the financial support of Glynn Vivian, while others in the Vivian family funded the restoration of the medieval Oystermouth Church.[66]

The Anglican churches of the industrialists at Swansea fit a much wider tradition of entrepreneurs constructing places of worship. The first chapel built by an industrialist for his workers in Wales may have been the Anglican chapel of ease built by Sir Hugh Myddleton of Chirk Castle, at Cwmsymlog Lead Mine in the seventeenth century. St. John's at Morriston, built by John Morris I in 1789, was probably the first built in industrial south Wales after local clergy had appealed to his sense of social obligation to the established Church. Kilvey was the second coppermaster's church at Swansea, built in 1842 by the Grenfells, again after local clergy had appealed for the dominant resident local employer to do so.

The coppermasters of Swansea were not the only industrialists to endow Anglican churches for the use of their workers. At the end of the eighteenth century, Thomas Williams, the 'Copper King' of Anglesey, and his partners provided £600 for a new church for their copper-miners and smelter workers at Amlwch. Indeed, the 1851 Religious Buildings Census makes it clear that the ironmasters of the Heads of the Valleys area were building, substantially contributing to, or making available a number of buildings for the use of Anglican worship: Blaenavon Church (1805); Dowlais Church (1827), Nantyglo School (1837), Clydach 'Lecture Room' (1842), Ebbw Vale School (1842), Rhymney Church (1843), 'Llynvi Iron Co.' School, and the Beaufort School. The Tai-bach chapel of ease had four main sources of funding, including the Quaker ironmaster John Reynolds and the English Copper Company, while 'The Spelter School', owned by the 'Llynvi Iron Co.', was also used for worship.

The industrialists thus fulfilled the role that fell to the moneyed local gentry in the rich agricultural lowland areas. The practice continued elsewhere after 1851, as at Pontardawe in the mid-Swansea Valley where the wealth generated by the Swansea Canal powered Primrose Forge (1843), its successive tinplate and steelworks providing the motor for the town's growth. The old medieval church of the parish, Llangiwg, was in the remote rural area to the north. The tall 197ft 9in spire of St. Peter's Anglican church at Pontardawe dominates the valley, for which the ironmaster, William Parsons paid £5,607 in 1858-60 to the architect J.H. Bayliss of Swansea. This was substantially more than the coppermasters were prepared to pay for churches further down the valley, but even at Pontardawe relations between the Church and industry were not as smooth as they might have been. Bishop Connop Thirwall of St David's refused to consecrate this church for two years because he considered the endowment inadequate, despite Parsons offering £1,000 and his daughter contributing £70 a year for a curate. Parsons left in 1861 and the church was consecrated during 1862 in the presence of his successor at the tinplateworks.[67] Both Welsh and English services were held in St. Peter's.

Twenty-four years later the different aspirations of industrialists and the established Church again became apparent. In 1886 Arthur Gilbertson built a private church, All Saints, and many of his workmen and managers transferred to the new place of worship. He intended this building, originally in the parish of St. John's, Clydach, as a memorial chapel to his father. The church was endowed and transferred to the Anglican Church after his death, the services being in English.

There was no Welsh Independent chapel in the town until 1880 when the foundation stone was laid by the Morriston tinplate industry magnate Sir John Jones Jenkins for a building designed by William George of Ystalyfera and John Griffiths of Pontardawe.[68]

Swansea industrialists also built churches in other parts of Wales where they had their country residences, such as the originally Quaker Dillwyn-Llewellyn family which had built the fine spired Anglican village church at Newbridge-on-Wye. Ironically, the Crawshays, ironmasters of the Cyfarthfa Ironworks at Merthyr Tydfil, had a seaside home at Langland Bay near Swansea and partly funded the restoration of Oystermouth Church with the Vivians.

The evidence shows that there were four types of Anglican churches built in the lower Swansea Valley in the era when Swansea was the copper-smelting capital of the world: churches built by coppermasters as memorials to members of their own families; churches built by coppermasters as their local places of worship; churches built by coppermasters as places of worship for their workers; and the town and suburban churches of Swansea built for the general populace. There was obviously some overlap between these types, particularly between the first two. The building operations were dependent on there being a resident Anglican managing-partner of a copper-making concern at Swansea.

Nonconformist Chapels of the Copperworkers

The nonconformist congregations generally started as house churches that is where small numbers of a new congregation or 'cause' met in the house of one of its members. The seventeenth-century congregation that built the Welsh Independent chapel on the edge of Mynydd-bach Common (one mile west of the site of Morriston) started in this way and came to have 12 daughter congregations, all of which built their own chapels in the course of the late eighteenth and nineteenth centuries, dispersed throughout the growing industrial satellite communities of Swansea on both sides of the River Tawe. Sometimes these churches failed, as with the Baptists in early nineteenth-century Morriston, where the Baptist cause was revived by deliberately planting parts of several surrounding congregations.

Sometimes congregations outgrew individual houses and, as an interim stage, worshipped in available communal buildings until they had enough cash for purpose-built places of worship. This was the case at Foxhole, where the Welsh Independent congregation used the Kilvey Works' Schools from 1806 to 1840; at Landore where the Welsh Independent cause used an industrial works office in the 1820s, and at Morriston where the Wesleyan Methodists used John Morris II's market building from c.1829.

The nonconformist congregations regularly outgrew their, initially modest, buildings as their causes expanded in waves of intermittent fervour inspired by successive charismatic leaders and the perceived manifestations of the Spirit. Some of these religious 'revivals' originated within Wales, some in the North American Welsh diaspora, and some had effects that were apparent in nonconformist congregations throughout Britain. The 1851 census reveals a situation in which the capacity of a chapel was measured not only by its available seating but by standing room, which some revealing entries explain as 'standing room in the aisles'. Several of these crowded structures were rebuilt soon after the census when the new waves of converts overflowed the chapels in the 1858 religious revival.

The only nonconformist place of worship mentioned in the 1851 census as being in the ownership of industrialists (and 'rented' rather than provided) was the Salem Baptist chapel at the Llynfi Spelter Works. This was started before 1846 in 'an old Store belonging to the Llynfi Co.'. Other similar congregations are known to have started in premises loaned by industrialists, the congregation at Penuel Calvinistic Methodist at the Ebbw Vale Ironworks of the Quaker Harfords, for example.

One of two similar examples at Swansea has been mentioned already - the main nonconformist congregation at the copperworkers' settlement of Foxhole. This began as a 'house-church'; it soon outgrew these premises and permission was asked from, and given by, the proprietors of the White Rock, Middle Bank and Upper Bank Copperworks for the worshippers of this Welsh Independent congregation to use the Kilvey Schools as a place of worship. This situation seems to have lasted from about 1806 to 1840 when the copperworkers excavated a shelf on the hillside on which to build Canaan chapel at a cost of £702 6s. The new chapel would seat 342, but when the copperworkers returned from their shifts and could attend on Sunday night, 389 were packed into the seats and aisles. Both chapels at Foxhole were demolished after the houses were taken down in the 1930s.

A second example at Swansea was the Landore Welsh Independent cause that became the Siloh chapel. In the 1820s it is recorded that the female members of the congregation carried stones (presumably water-worn rocks from the stream bed, or possibly the remains of the old Llangyfelach Copperworks) from the Cwm Nant Rhyd-y-Filais valley to construct their first building at Landore. The men then dressed the stones and erected the walls of 'Y Coleg'. This self-building philosophy allowed a poor worker congregation to provide itself with a modest multi-purpose building for just £90.[69]

Another recorded example of the worker congregation carrying out their own building work, is that of the rebuilding of the chapel of Bethel Welsh Independent chapel at Llansamlet in 1849-51. The old building was pulled down and the population of colliers and other workers both quarried and carried stone for the new 47 x 37ft chapel, which was completed at the low cost of £508.

The great national revival had started in North America in 1857 and produced perceived manifestations as far as Aberdeen, but in 1858 congregations started to throng to the existing chapels of south Wales and everywhere the need was felt to expand these buildings. Up to the chapel-building boom of the 1860s when congregations were well-established and numerous, it was common for the committed worker builders of the industrial communities to contribute physically to the building of their own chapel. Cash resources above subsistence level were not plentiful, even if more available than in the surrounding rural communities.

Five years after the extension of the Foxhole church on the hill, a new Canaan, the main chapel of the community on the hillside below, was opened. This chapel, built by John Hacche of Swansea, was considered 'one of the most beautiful in the Principality' and cost £1,100.[70]

As elsewhere in the Swansea townships, the Welsh Independent congregation at Canaan chapel soon outgrew its building and a new one was built in 1864, by the denomination's prominent Swansea architect, the Revd. Thomas Thomas of Landore. The early 1860s seem to have been an era when there was an explosion of chapel building inspired by Thomas's architectural signature of the giant arch within a pediment, as seen at the Philadelphia and Siloam chapels built on either side of Trevivian in these years.

English-language Chapels and Gothic Nonconformist Architecture in the Industrial Townships

In the later nineteenth century Welsh-language congregations from the predominant Welsh Independent (or Congregationalist) denomination sought to improve the provision of buildings for their English-speaking fellow workers and their children. The Revd. Thomas Rees was prominent in this movement, concerned at the 'Godlessness of the English language'. They were aided in this by funds and help from the Chapel Building Society of the Congregational Union of England and Wales and various wealthy Congregational industrialists. Four of the most prominent of these included the tobacco magnate John Wills and Samuel Morley MP, (both from the so-called 'Welsh [trade] Metropolis' of Bristol), and two prominent founders of model workers' settlements in west Yorkshire - Sir Titus Salt (the alpaca-wool magnate of Bradford and

'Saltaire') and John Crossley (the carpet tycoon of the largest textile mills complex in Britain at Dean Clough, Halifax). These sources of external finance were applied to buildings, often in the gothic style. Halifax, for instance, was dominated by the soaring spire of a new Congregational church to replace the old Georgian classical style chapel.

These philanthropists were not the first to use the gothic style in Swansea, where its early use has been little understood.[71] 'The Gothic was favoured in Wales after 1860, especially by the English-speaking Wesleyan Methodists'. In Swansea town the evidence is that the Calvinists, Unitarians, Presbyterians and Congregationalists, in admittedly English-speaking congregations, were building in full-blown gothic well before this date. However, the watered-down basic 'poor-man's' gothic of the exiled Cornishmen's Bible Christian chapel at Hafod was the only gothic to reach the industrial communities in the later nineteenth century.

Even so, large and innovatory classical chapels were (and are) a feature of Swansea town as well. The Swansea Wesleyan chapel (Goat Street) of 1845 was a very large and influential early essay in the 'Italian' classical style with 'a fine lofty steeple'.[72] In Swansea gothic was introduced by Selina, countess of Huntingdon, one of several aristocratic English patrons of Calvinism. She commissioned William Jernegan to design a gothic chapel in Swansea in 1787-89, not dissimilar in style from the fine but richer chapel of the sect that still remains in Bath or the college building she erected near Howell Harris's college at Trefecca. The Unitarian chapel (1689) in High Street was also rebuilt in 1840 in 'the pure Tudor [Perpendicular] Gothic style' by 'an eminent London architect', Joseph Gwillt (1784-1863), who was a consultant to the Office of Woods and Forests in London.[73] He had designed Tudor gothic churches from 1813 and had published and edited a new edition of Sir William Chambers's *Treatise on the Decorative Part of Civil Architecture* in 1825.[74] Following this two gothic style places of worship were constructed in north-west Swansea, almost as large as cathedrals - the twin-towered gothic Presbyterian building of St. Andrew's on St. Helen's Road in 1862-64 (financed by Scottish drapers and at present being converted to a mosque) and the Walter Road Congregational chapel in Walter Road, designed by the Manchester architect H.J. Paull and built for the substantial cost of £5,000 (a fine building whose tall spire was demolished in the 1960s).[75]

This latter chapel was built at the initiative of the Revd. Thomas Rees, the influential minister of Ebenezer (1861-85) in the industrial suburbs of north Swansea and a significant author on the history both

of Welsh Nonconformity and of the Welsh Independent denomination. When the chapel opened in early June 1863, there was, first, a day of English sermons and then two days of sermons in Welsh. It was clearly very common in the industrial suburbs and settlements in and around Swansea for the Welsh-language congregations to initiate and host equivalent English-language causes and subsequently to promote the use or building of separate Sunday schools and chapels. Thus, to build the new north Swansea chapel, the Revd. Rees Thomas and invitees preached in the large auditorium of the Swansea 'Music Hall' (now the 'Albert Hall') on Sundays with a sponsorship of £50 from the Bristol MP Samuel Morley and £10 from a Mr C. Jupe.[76] The cost of the new building was

Fig. 277. St Andrew's Presbyterian chapel in Swansea town was one of several large gothic structures built for English-language nonconformist congregations in the urban centre. (9300633/4)

£5,850, of which £2,000 was collected from benefactors all over England; for example Samuel Morley gave £700, John Crossley of Halifax £350, C. Jupe £200, H.O. Wills £100 and Sir Titus Salt £100. The foundation stone was laid by the Bristol industrialist and Congregationalist, H.O. Wills, on 27 May 1866 in front of a large crowd, as a fund-raising event for the large gothic chapel that was completed in the summer of 1868. Subsequently, in 1872 Rees collected another £800 to pay for an 'immense' schoolhouse for his English-speaking brethren.

The same happened in the new industrial settlements near the Port Tennant Copperworks. In about 1856 Morgan Hussey of the copperworks moved a rented house to make way for a school and chapel founded from the Independent cause in the older copperworks settlement at Foxhole. Then, further east, in 1860 several Independent ministers began preaching in English at Fabian's Bay. A schoolhouse was built, at a cost of £400, in 1862 to serve the new congregation and then the Bristol tobacco magnate H.O. Wills laid the foundation stone for a new chapel on 13 October, 1870.[77]

In 1849 the northernmost Welsh Independent congregation in Swansea town (at 'Soar' on the Carmarthen Road) decided to try to build a new chapel on the other side of the road and to leave its old chapel to the English congregation. However, the duke of Beaufort (or his local agent) would not let the cause have the use of the required land and consequently it had to knock the old chapel down and build a new one on the site in 1869-70. Samuel Morley, laid the foundation stone for a new chapel that continued to serve congregations in the two languages.[78]

At Landore the Welsh Independent congregation of Siloh succeeded where Soar had failed; on moving out of their old chapel on 3 February 1878, they established separate Welsh and English Sunday schools in the old building until the new Sunday school building was available on the new site at Brynhyfryd. On the 18 November preaching in English was established at the old Siloh chapel and on the 10 February 1884 an English cause was established there.[79]

The dual support of local Welsh-speaking congregations and nonconformist businessmen for the building of chapels for English-speaking congregations was taking place all over industrial south Wales by the later nineteenth century. In 1859 the Welsh Independent cause at Capel Ivor in Dowlais, together with the other three congregations of the denomination in the town, decided to lease land from the iron company and build a chapel for an English-language congregation. The foundation stone was again laid by the Bristol entrepreneur H.O. Wills, this time on 22 May, 1860.

The Wesleyan Methodist Guests of Dowlais were nonconformist entrepreneurs in south Wales, and yet do not seem to have endowed the Independent or Congregational chapels built by workers in their company township. However, denominational loyalty was a key factor in determining benevolence, Sir Titus Salt built a lavish chapel (where he is buried) for his Congregational co-religionists in his own company township near Bradford, but he left his other workers to build their own Wesleyan Methodist place of worship nearby.

Chapel Architects in the later nineteenth century

Outside the town of Swansea, the architecture of the eighteenth and early nineteenth-century industrial townships was generally that of simple meeting houses with an element of self-building by relatively poor working-class communities and few sources of funds to provide ornamented facades. The situation in the more affluent town was rather different.

By the mid nineteenth century in Swansea several influences had come together to produce two of the most influential architects in Welsh chapel architecture. The nonconformist congregations who formed the great majority of the population of the townships were becoming more self-confident. Greater concentrations of employment led to bigger congregations with employment at larger coppers-melting works such as Morfa, Hafod and White Rock. Congregations, or 'causes', were themselves often led by nonconformist copperworks managers from these works or from the 1870s by a new group of local nonconformist industrialists who were able initially to re-use the old water-powered coppermills as tinplate-works, with relatively modest injections of capital. The great religious revival of 1857-58, with its various manifestations, attracted a larger proportion of the industrial workers. It is estimated that over 75,000 new worshippers overfilled the existing meeting houses of Wales.

The Swansea industrial townships, produced new architectural ideas that dominated the great boom in gable-façaded, urban-type chapels of the 1860 and 1870s that changed in the urban settlements and landscapes of industrial Swansea and indeed elsewhere in Wales. Earlier rural chapels, evolved from barn meeting houses, had generally been simple, long-wall façade structures, often erected by local congregations under the supervision of local masons, builders and ministers. In the Swansea industrial communities, examples of the latter are recorded in the construction of 'Y Coleg' at Landore (by what became the Siloh congregation) in the 1820s and Canaan chapel at Foxhole in the 1840s.

A new class of ministers and deacons specialising in chapel and school design (the schoolhouse or 'ysgoldy' was an integral part of every later nineteenth-century chapel complex and was sometimes used as a British day school as well as a Sunday school) arose on the crest of this 'great national revival'. The work of two of its main exponents had its origin, early growth and most notable work in the industrial settlements of the lower Swansea Valley; between them these two are known to have been responsible for the design, or substantial rebuilding, of no less than 175 chapels.[80] The two principal architects based in the Swansea were Thomas Thomas, generally known to contemporaries as 'Thomas, Glandwr' (Landore), who was designing from 1848-51 until 1885, and John Humphrey of Morriston, who was designing from 1865 until 1888.[81]

Thomas Thomas was a minister and John Humphreys a deacon with the Welsh Independents (Congregationalists), by far the largest sect in nineteenth-century Swansea. Swansea became the headquarters of this denomination, with a theological college in the former Sketty house of the Smiths, the coalmasters of Llansamlet. The standard five-volume history of the denomination, *Hanes Eglwysi Annibynol Cymru* (Liverpool, 1872-75), was co-written by Thomas Rees, minister at Thomas Thomas's designed chapel of Ebenezer in the northern suburb of Swansea town.

Thomas Thomas and the Giant Arch

Thomas Thomas (1817-88) started building chapels in the late 1840s, when he was about 30 and almost 20 years before John Humphrey. As a minister, he also had the distinction of giving one of the first sermons in each of the buildings he designed, during the two-or three-day preaching festivals that marked the opening of every new or substantially rebuilt chapel. He was one of the two most prolific chapel architects in Wales and seems to have built, or rebuilt, at least 119 chapels as well as one of the main Welsh Independent theological colleges, at Brecon costing £11,000 in 1867-69.[82] His standing was such that his influence in the Independent denomination spread eastwards across the breadth of England. He built for the Welsh Independent denomination in Shrewsbury (1862)[83] and Liverpool, eastwards to a cluster of three buildings serving miners in the Durham coalfield (1871-75),[84] and also what seems to have been the main London chapel of the denomination (1872-73: £4,352).[85] Very late in his architectural career (1880-81) with a reputation spreading into the English-speaking Congregationalist community, he also built at least one large chapel in the border counties of England, at the woollen-industry centre of Stroud in Gloucestershire.[86]

He was born on 24 June 1817, in Llandeilo Fawr, Carmarthenshire where his father was a deacon and where he was also received into the Welsh Independent denomination.[87] He was trained as a carpenter but was chosen to start preaching to his

Fig. 278. Thomas Thomas. (The National Library of Wales)

fig.279

Fig. 279. The main gable-end elevation of Libanus Independent Chapel at Morriston (1857) seems to have been one of the first examples of a design that Thomas Thomas was to use from the late 1850s all over Wales and for Welsh congregations as far as London. A three-bay front is divided by tall pilasters; tall round-headed windows illuminate the gallery and stairs; a triple round-headed version of the Venetian or Palladian window fills the upper central bay above the central door. (9500352/2)

congregation by the minister, Mr W. W. Williams. He married and moved to Swansea but was soon called upon to begin his career in the industrial settlement at Clydach-on-Tawe on the northern edge of the lower Swansea Valley. He had not attended formal training at an Independent college but was ordained on 16 June 1846. He started his career as a minister in the Welsh Independent chapel of Hebron at Clydach, which had a small chapel holding 200-300 originally built in 1821. By contrast the local Anglican church of St John the Baptist was built in early gothic style at Clydach in 1845-47 funded by the resident (Ynys-pen-llwch) tinplate entrepreneurs - the Miers.[88] Under Thomas, a 'beautiful and extensive church' was then built, in 1848 for the Independents, on their site further to the south along Clydach High Street.[89] Thomas was also responsible for the ministry at the chapel at Cadle and at the beginning of 1848 he was called to be minister at Siloh, Landore, but continued also to be minister at Clydach until 1853.

The industrial settlement at Landore was populated by workers from the Landore, Morfa and Hafod Copperworks and prominent in its congregation were the deacons from the Hafod Copperworks management. After his involvement in the Clydach rebuilding, Thomas's preaching and chapel contacts rapidly led him to become involved in building for other congregations. The first occasion seems to have been a chapel built for the cause at the iron-making centre of Abersychan straight after his involvement in rebuilding his own Clydach chapel. Spiritually, the earlier 1850s were quiet before the frenzied storm of the great revival at the end of the decade, which unleashed a force demanding large new buildings for substantially increased congregations. Materially, the early 1850s were a time when Thomas was building a few chapels in the industrial settlements around Swansea with an outlier at the second coppersmelting, coal-mining, tinplating centre to the west at Llanelli (1852-53). In 1851, he had built a Welsh Independent chapel for the colliers of Llansamlet and in 1853 another for the neighbouring congregation of copperworkers at Landore's Baptist congregation of Dinas Noddfa. The latter is significant, for in the great chapel building boom of the 1860s and '70s, both Thomas and John Humphrey, although designing largely for their own Independent denomination, were also supplying plans to a range of nonconformist denominations without discrimination, this was to satisfy a huge demand for architectural guidance without paying exorbitant architects' fees.

In his early and local chapels Thomas had the time and inclination to work as a surveyor/building supervisor. This is recorded to have been the case at the Bethel Welsh Independent chapel at Llansamlet - the old chapel was of 1818, a gallery was added in 1839, demolished in 1849 and a new chapel opened on 13-14 February 1851 at the low cost of £508.[90] He also worked on Canaan, Foxhole (1864) and at Saron in Tredegar, and presumably also fulfilled this role with the chapels, houses and schoolrooms built and rebuilt under his ministry at Clydach, Siloh, Landore (1862), Brynhyfryd schoolhouse (1862) and the schoolroom at Plasmarl (probably in the 1860s).

In all, least 12 of the chapels and schoolhouses designed by Thomas Thomas are known to have been built in the industrial settlements and suburbs to the north and east of Swansea, whence his influence spread by way of to which more than 110 chapels his name is linked throughout most of Wales. His later chapels in the Swansea area included Tanygraig Welsh Independent, the cause was started in 1856 to

Fig. 280. The facade of the Welsh Independent Chapel in the poor coal-mining community of Pentre-estyll (1864) was one of the earliest to use the giant arch in the pediment as a design feature. (970075/35A)

serve the population of the new Port Tennant Copperworks and the chapel was opened on 9-10 September 1860 at a low cost of £600;[91] Ebenezer Welsh Independent and English Congregational (the old chapel was demolished in 1862 and the new one finished at the end of May 1863 at a cost of £2,000);[92] Seion Welsh Independent in Glais (1861-62); Cwmbwrla Welsh Independent, [93] and 'Siloam' Welsh Independent at Pentre-estyll on the west side of Trevivian.[94]

In Glamorgan as a whole Thomas was responsible for the design of at least 33 buildings largely serving the growing populations of the industrial settlements. Of the firmly dated buildings, one was built in the 1840s; three in the 1850s and after the national religious revival, a substantial 14 in the 1860s and eight in the 1870s. In 1875 he resigned from the ministry of Siloh Landore and (like the coppermasters) moved to the rural west of Swansea Bay, but he still designed a chapel in Glamorgan from his new home in Mumbles in the 1880s. His influence was carried far beyond the Swansea area into the rest of the coalfield and its metal-smelting areas - some three-quarters of his total chapel design work seems to have been carried out in industrial south Wales, especially the west, for while he is recorded as designing 34 chapels in Carmarthenshire he designed five in Monmouthshire.

By the mid-1860s, the publicity attending Thomas's architectural work in the Welsh Independent denomination's journal *Y Diwygiwr* was enough to stimulate commissions to design, or substantially modify, 10 chapels in the mid-Welsh counties of Cardigan, Montgomery and Radnor. In addition in these and those in other areas, his innovative and striking later designs prompted commissions and imitation by congregations and denominations. The latter were partly carried out by the local mason/builders in each area who had worked with Thomas in fulfilling his commissions. An example was David Davies, who had worked on a Thomas design in Cardiganshire and had then built another 20 similar chapels in the same county.[95]

Similar patterns in the spread of chapel designs developed in Swansea are discernible in the more distant, but more heavily industrialised and populated, north of Wales. Thomas was commissioned to build a Welsh Independent chapel in the slate quarrymen's village of Llanddeiniolen; this was opened in 1859 and seven other chapels were built in north Wales during the next five years, some featuring Thomas's striking use of the giant arch in the pediment of the gable-end façade, which had its early use in Salem Welsh Independent chapel at Porthmadog in 1860. By contrast, Thomas's first chapels were simple and seem to have been based on fairly unremarkable round-headed architecture. His rebuilding, in 1858, of William Edwards' simple Libanus chapel in the copper workers' settlement at Morriston, produced a very ordinary gable-end structure that still survives.

The characteristic feature that Thomas Thomas helped to spread throughout Wales was the use of a giant arch, capping a large blind recess and breaking through into the triangular pediment framed by the top of the gable. Giant blind recessed arches were

273

also a common feature of Georgian architecture (developed by the influential Scottish architect Robert Adam) and can still be seen on the façades of Thomas Jernegan's terraces in the Swansea Maritime Quarter and on other surviving eighteenth-century buildings in Swansea.

The use of a giant arch breaking into the pediment, as a central feature of a Classical Renaissance façade, would have been known to anyone undertaking the Grand Tour. A multiple use of the arched entablature certainly occurs in one of the main sites - around the pool of the Emperor Hadrian's villa at Tivoli, near Rome. The roots of this architectural device lay in classical antiquity, but had reappeared by the end of the twelfth century in Umbrian Romanesque churches. The main west-end gable façade of St. Rufino's Cathedral, in the pilgrimage centre of Assisi has a façade divided vertically into three by projecting pilaster bands; the wider central bay, enclosing a large doorway with big wheel-window above, breaks upwards through the corbel table at the base of the triangular gable and forms a large, slightly-pointed, arched central recess almost reaching into the apex of the façade.[96] This type of façade, with a giant arch breaking through a more sharply defined triangular pediment, was also being copied locally in 1854 in the church of St. Mary of Rivotorto.

Semi-circular giant arches reappeared in the early Italian Renaissance for next to the famous church of Santa Croce in Florence is the Cappella de' Pazzi, built by Brunelleschi in 1430; this has just such an arch springing from the two-bay entablature that flanks it on either side. A complete façade that almost exactly mirrors the use of the giant arch in British Nonconformist architecture is the narrow version of the arch that breaks through the large horizontal band of the classical entablature at the base of the gable-ended pediment of Alberti's San Sebastiano church in Mantua (1460). Similar devices were designed in Rome in the following century, following earlier antique classical models which Robert Adam would also have observed. One arch that Adam did see occurred in a pediment in the façade of the Emperor Diocletian's palace at Split in Croatia. Indeed, the fully-developed architectural device, commonly used in Welsh chapel architecture by, and after, Thomas Thomas had a classical entablature capping an arcade bent upwards into a semicircular arch; such can also be seen in examples of late Roman architecture in Syria and Asia Minor, at Termessus, Palmyra and Baalbek.[97]

In seventeenth-century Britain, Sir Christopher Wren was among the first to use this device on the transept ends of St. Paul's Cathedral. Later, his former assistant Nicolas Hawksmoor, used such an arch in his six baroque London churches with their tall dominating towers. The various stages of the west front and steeple of Hawksmoor's Christchurch, at Spitalfields in London, as examples of giant arches with, and without, frames of arched entablatures (1723). Both the Swansea architects would have seen these prominent London landmarks, for Thomas seems to have designed and then preached at Southwark Bridge chapel, whilst Humphrey is recorded as visiting the 1862 International Exhibition at Kensington with his wife.[98] Other British architects working in the neo-classical style were using the device, for example, as the architect Joseph Pickford on the front of his house in Derby (1768). By the 1830s and 1840s, prominent architects such as C.R. Cockerell were making common use of giant arches breaking through an entablature, as seen on the monumental façades of Cambridge University Library (1835-36: with several designed as central features of gable-ends), the Ashmolean Museum and Taylorian Institute at Oxford (1841-45) and framed, in a pediment, at the Bank of England branch in Manchester (1845-46), while the contemporary Manchester Theatre Royal adopts the same device.

Nonconformist chapel architects may first have copied this design idea from one of its London uses or from its fairly common use in late eighteenth-and early nineteenth-century polite architecture in Britain. Perhaps a more likely source for chapel architects in Wales may have been the contemporary illustrated books on buildings and architecture, such as R.N. Shaw's *Architectural Sketches from the Continent* (1858) or G.E. Street's *Brick & Marble Architecture in North Italy* (1855). Gable-end facade urban chapels became quite common in eighteenth- and early nineteenth-century England, generally earlier than in Wales. Certainly the idea of a tall and fairly narrow giant arch (but without an entablature frame) over a central recess, with its head breaking through the banded lower part of a pediment in a Georgian-style façade, was already being used in the nonconformist architecture of the west Midlands in 1811 when the Uttoxeter Wesleyan High Street chapel in Staffordshire was constructed.[99] A further elaboration of its use with flanking pilasters dividing into three a front chapel gable-end, and supporting an elaborate classical entablature through which the central narrow arch rises, could already be seen in the early nineteenth century Baptist chapels of Hartley Wintney in Hampshire and Truro in Cornwall (1850) and then, in 1858, in the border county of Shropshire on Aenon Baptist chapel, High Street, Madeley.[100] However, by the time *The Congregational Year Book* (of the Congregational Union of England and Wales) featured a printed drawing of a wide giant arch, flanked by double columns, almost filling the gable of a large, winged Birmingham chapel ('Lozells Chapel') in 1862, Thomas Thomas had been using the idea of the giant arch in the pediment for at least two years.[101]

One of the first secular uses of a giant arch set within a pediment in Wales may have been on the Penrhyn Castle Estate at Bangor, where the land agent, Benjamin Wyatt, built double-gabled, inclined-plane winding-drum houses with recessed giant arches breaking into the pediment. Wyatt, the brother of the eminent British architects Samuel and James Wyatt, built the lower incline at Marchogian near Port Penrhyn in 1798-1801;[102] the two pedimented pavilions sited on an eminence are still prominently visible from the old coast road. The 1830 designs for a new slaughterhouse in Swansea also had a giant arch set in a gabled pediment on the street facade.

An early use of the giant arch in Welsh chapel design seems to have been Daniel Evans of Cardigan, at Harmon (1832) in Fishguard and also at Bethania, Cardigan (1847).[103] Another early example is in the full-blown Norman Romanesque of Commercial Road Wesleyan chapel in Pillgwenlly, Newport, of 1847, closely based on the twelfth-century west front of Tewkesbury Abbey in Gloucestershire. However, the giant arch breaking though an entablature, is in essence just a larger version of the almost universal motif of Georgian (and indeed later chapel) architecture, namely, the Venetian, or Serlian, window with a semicircular central arched window emerging from a flat entablature covering the smaller side lights. By 1844-47 the grand new Goat Street Wesleyan chapel in Swansea had a large prominent version of this Venetian window in its central western tower, its head actually breaking into the lower part of the pediment entablature.

Thomas Thomas made this architectural motif, when applied to Welsh chapel architecture, very much his own. He had perfected his use of the giant arch soaring upwards into the chapel front pediment by 1858 at Saron in Tredegar. At Porthmadog in 1860 the facade was divided into three unequal bays by tall Tuscan-style pilasters supporting a thick horizontal banding (or entablature) flanking the wide central arch at impost level. The banding rose in a semi-circular arch over a recess accommodating a large central window over the three entrance doors. The side panels accommodate tall windows framed by ashlar bands. At Swansea, an early example of this design is in Thomas's own Ebenezer chapel of 1862-63 in north Swansea with its substantial population of copperworkers. If Thomas did not plan other contemporary chapels in the Swansea industrial suburbs and elsewhere, he certainly had a large influence on their main design features. Immediately after Ebenezer a new and elaborate chapel, was built in 1864 by Thomas costing £1,500, for the Welsh Independent congregation to the north-west at the colliers' settlement of Pentre-estyll on the west side of Trevivian (Hafod); this had a very similar scheme on its main facade.[104] The idea of the giant arch in the pediment spread rapidly in the Swansea area in such buildings as Soar Maesyrhaf Welsh Independent chapel of 1864-66 at Neath.[105]

Fig. 281. Ebenezer Welsh Independent Chapel in north Swansea, built by Thomas Thomas for a prosperous and influential urban congregation in 1862. An elegant and elaborate version of a design first used four years previously for Saron Welsh Independent Chapel in Tredegar, executed using fine Bath stone dressings on mass-walling of Pennant sandstone (960074/36)

Fig. 282. Portrait of the architect/deacon John Humphrey. (The National Library of Wales).

Fig. 283. Mynydd-bach chapel to the west of Morriston was the mother chapel to most of the Welsh Independents in the Swansea area. John Humphrey's chapel (modified in the 1930s) has many of the characteristics of Thomas Thomas's earlier chapels but his great arches are narrower and are surmounted by a roundel window. (960070/31)

Fig. 284. Interior of Mynydd-bach chapel showing how John Humphrey followed a stylistic innovation pioneered by Thomas Thomas in dipping the gallery at the rear of the preacher's pulpit. (960071/31)

John Humphrey: architect/deacon

The other designer of local chapels in the later 1860s, whose designs featured the giant arch in the pediment on the facade was John Humphrey. He was the second important architect to emerge from the Independent denomination based in Swansea's industrial suburbs. His period of activity coincided in part with that of Thomas Thomas. He was a collier's son, born in 1819, and he trained and practised as a carpenter and joiner from his home in Morriston.[106] He is not known to have designed a building before the coppermaster Pascoe Grenfell commissioned him to build the Kilvey School for girls in 1865, although The Cambrian's comment on its opening that 'Mr J. Humphreys ... really does excel in school architecture' implies that he had previously been building chapel schoolhouses.[107] He built Mynydd-bach, his first chapel, in 1867 when he was 47. The reason why he, and other active members of denominations were asked to design buildings for

their causes was probably a willingness to work at cost or without fees. The newspaper report on the opening of Mynydd-bach chapel noted that:[108]

'It was built from a design provided by Mr J. Humphreys of Morriston, one of the deacons. The gentleman not only generously presented the plan of the building, but supervised its erection in the most economical manner without any professional fees, an act which claims the gratitude of the building committee, and all who assemble within its walls.'

Both Humphrey and Thomas, also paid regular supervisory visits to sites, to see that local builders were executing their designs in a satisfactory manner. Humphrey, for example, in 1877 was engaged in three chapel-building projects between 12 and 18 miles from home and he visited each project on average 30 times. These local chapels providing firm foundations for a wider practice.[109]

In the 1860s Humphrey's design skills were not so well developed as were those of Thomas Thomas. His giant facade arches of the later 1860s tended to be rather narrow (i.e. approaching the one-quarter façade widths of earlier Georgian arched recesses, rather than the wider measurements of later Victorian arches) and were not framed by any elaborate arched entablatures or banding. They were executed 'in the most economical manner' from local cut and uncut green Pennant sandstone. A feature that distinguishes them from Thomas's designs is the small oculus (or circular opening) sqeezed into the tip of the roof apex above the central giant arch. In Humphrey's later, and grander triple arch facades this circular opening had space to descend to the centre of the gabled pediment where it produced a far more balanced and mature design.

Humphrey's first known chapel building work in 1866 used a simplified version of Thomas's use of a giant arch in the pediment design, Humphrey's first work using this motif was in rebuilding the influential mother chapel of the Swansea Independent causes at

Mynydd-bach, and his second was at New Dock chapel, Llanelli (1868). Some of his less elaborate façaded chapels, such as Ffald-y-brenin (Carmarthenshire, 1873) and Carmel (Gwaun-cae-Gurwen, 1877), also used the same design.[111]

It was in the 1870s that Humphreys chapels entered a more sophisticated design phase, utilising elaborate Bath stone dressings imported through Swansea docks; the Chapels were funded by the growing resources of the workers and foremen from the copperworks and the new tinplate entrepreneurs. He very often used Bath stone dressings for giant columns on the façades and flush banding. However, the stylistic device for which he is best remembered is the triple giant arch gable façade, set on tall attached columns reaching up to the chapel pediment and guaranteed to dominate the townscape of any industrial town; it did in north Swansea (Dyfatty), Llanelli, Llanidloes and, finally, at Morriston in 1873, where Tabernacl's triple arch was supported

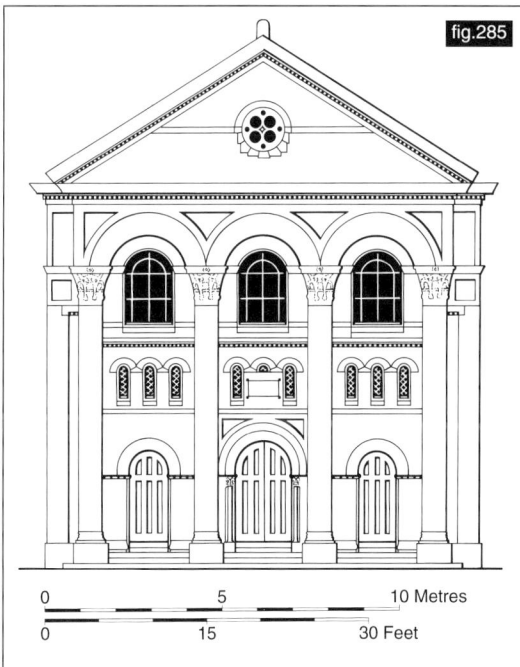

fig.285

on double columns and flanked by its unique multi-style steeple. Tabernacl was described in the standard history of the Welsh Independent denomination, (*Hanes Eglwysi Annibynol Cymru,* published in 1875), as 'the most extensive and excellent chapel in the whole Principality'. The costs are given as ranging from £8-10,000[112] to about £12,000 [113], and were later said to be £15,000 in Volume V of the *Hanes.* Tabernacl was a material statement, by locally-born tinplate entrepreneurs, that their distinctive culture and religious denomination were characterised in the Swansea industrial settlements.

The triple arch façade, with a minor enclosed upper arcade, would have been noted by any Italian Grand Tour visitor on the façades of the great octagonal Baptistry of St. John the Baptist in Florence (seventh to eighth century, restored *c.* 1200) and presumably Humphrey saw an engraving of this. *The Builder*, impressed by Humphrey's first use of giant columns, said that Soar was in the 'Grecian Style' - but the Greeks never used a high arched entablature topping the columns as did the Romans and medieval Italians. The first use of the triple arch façade in Welsh chapel architecture may be the monumental façade, designed by the local architect A.C. Watkins of the Victoria Road Congregational chapel at Newport, Monmouthshire in 1859.[114] It was subsequently used at Llangollen and then at Mold, in 1863, by the architect W.W. Gwyther of London.[115] Presumably Humphrey saw least the Newport chapel which belonged to the English Congregational denomination.

Prominent Victorian architects, such as C.R. Cockerell used triple groups of arches springing from giant columns. Cockerell did so in the designs for Cambridge University Library (1835-36) and the Ashmolean Museum and Taylorian Institute, in Oxford (1841-45).[116] A wide central façade arch supported by double flanking columns, a device later used at Humphrey's Morriston Tabernacl, was illustrated in the *Congregational Year Book* for 1862 ('Lozells Chapel', Birmingham, 1863, by Poulton and Woodman).[117] Indeed, it is recorded that John Humphrey and one of his from the tinplate industry, entrepreneurial sponsors toured England to collect ideas of chapel design. It seems clear, however, that the final version of the masterful design used at Morriston was drawn directly from published drawings of north Italian architecture. Based on the Florence Baptistry, in particular, all four versions of Humphrey's sophisticated three-arched design have an intermediate band of small blind arches above the chapel entrance doors and below the first floor windows; this is not found in the earlier Newport chapel but is found on the eight three-arched façades of the Florence Baptistry. It also occurs elsewhere, such as on the Romanesque west front of Assisi's cathedral of St. Rufino, which also provides a precedent for the arrangement of a main central door and two subsidiary doors dispersed between the three bays of the façade.

Unlike many contemporary chapels, Tabernacl has three elevations exposed to the gaze of neighbouring streets and there were sophisticated designs on the other elevations, besides the grand front façade. The full height, frontal Corinthian columns with their recessed intermediate arcade, however, were not reflected on the two simpler tiers of windows on the

fig.285a

Figs. 285, 285a. The elevation of Soar Chapel was the first by Humphrey erected with his monumental triple-arched facade. The building was for the use of a north Swansea congregation of workers significantly involved in the local copperworks and was demolished in 1978.

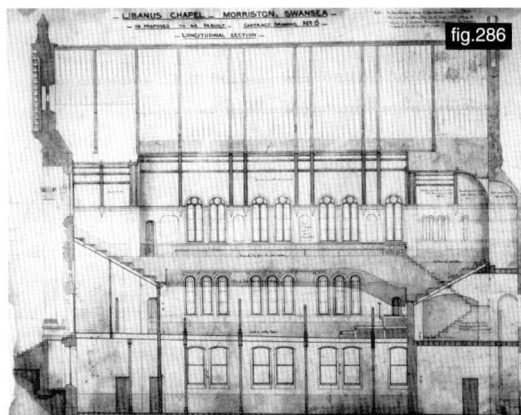

Fig. 286. Cross-section through John Humphrey's Tabernacl at Morriston, showing the large Sunday school room in a typical position under the auditorium of the main chapel. (The National Library of Wales).

Fig. 287. 'The Cathedral of Welsh Nonconformity': Tabernacl, Morriston. (9300650/2)

side façades. It was left to other architects working locally to produce fully integrated temple designs.

Humphrey produced a distinctive steeple - the highest on a classical nonconformist place of worship in Wales and possibly in the whole of Britain. Classical steeples are fairly rare on chapels built by working-class congregations, but this was not the first to be put on a chapel in the Swansea area. In 1844-47, the Wesleyan Methodists in Swansea town, led by the manager of the Hafod Coppermills, William Morgan, along with a local chemist, Thomas Evans, built a large chapel of classical design on a prominent corner site at Goat Street.[118] The main gable-end facade had low towers flanking a pediment, through the centre of which a central square tower rose, with a clock in the tower stage set at the head of the pediment and placed in front of a large round-headed belfry stage opening. Above the pediment, the square tower became an unequal octagon with tall round-headed openings set in the alternate four wider faces of the tower (which in that case was capped by a cupola). A source closer to that of the multi-stage Tabernacl steeple is to be found at St. Martin-in-the-Fields whose seven alternative classical steeple designs are illustrated in the most influential eighteenth-century pattern book - *A Book of Architecture* by the London architect James Gibbs (1728). Two of the steeples had octagonal stages with tall round-headed openings capped by classical pediments on their four wider faces; one of these also was flanked by pilasters, as found on the Tabernacl design. The tower bore a clock, as on the Swansea Goat Street chapel, and the upper open stage of the Tabernacl steeple consisted of eight equal round-headed arches set in a regular octagonal stage, as found in two of Gibbs' published designs. No less than five of Gibbs' designs culminated in a short octagonal spire, as found in Humphrey's design. A main aesthetic difference between the multitude of classical designs produced for places of worship in London in the seventeenth and eighteenth centuries and those in Wales in the nineteenth century, was that the former were richly endowed designs executed in ashlar masonry, while the latter were produced by poorer congregations, who even at Tabernacl and Seilo Newydd used mass walls of squared local stone with a selective use of ashlar masonry. The Morriston steeple was a robust version of a type that had developed *via* such versions as the Presbyterian church of St. Andrew, Glasgow (1739-56); the Anglican church of St. George, Hardwicke Place, Dublin (1802-3); North Leith Presbyterian church, Edinburgh (1814-16) and Holy Trinity Anglican church at Leeds (1839). One development at Morriston was the incorporation of round-headed tracery in the steeple openings, although this distinctive feature of much nonconformist architecture was already used on the main tower of the Goat Street Wesleyan chapel in Swansea in 1844-47.

The interior of Humphrey's Morriston Tabernacl is also very imposing, with the gallery sweeping round on all sides to culminate in a sinuous line that descends towards the all important pulpit, itself set against the soaring rows of organ pipes. The size of this ensemble is impressive, but its form is firmly part of the dominating 'Swansea School' of chapel design. Thomas's north Swansea Ebenezer of 10 years earlier (1862-63) has just such a grouping of a four-sided upper gallery dropping under the organ and emphasising the place from which the Word was

proclaimed. This device was still in use in 1879-80 when Thomas designed Bethel chapel at Llansamlet.

Humphrey was not just involved in these displays of newly acquired nonconformist social, political and religious power. He is known to have been connected with the design of at least 54 buildings, that is, about half the number that seem to have been designed during the longer career of Thomas Thomas. Fifty of these were chapels and four were specific school buildings, a total probably sufficient to make him the third or fourth most prolific chapel architect in Wales. It is also difficult to distinguish between the totals of chapels and schools designed, as most chapel complexes included a schoolroom or rooms, used for Sunday schools and sometimes day schools as well. Even the great Tabernacl at Morriston had a substantial schoolroom in its basement. Two of Humphrey's first four commissions were for schools and his last recorded one was for the large (597 pupil) Terrace Road Schools in west Swansea (1888); he was also clerk of works to the Swansea School Board at this time.

As with Thomas, it was the industrial communities of the Swansea area that provided the concentration of religious enthusiasm, funding and building experience to launch an architectural career. Humphrey is known to have been responsible for 10

Fig. 288. The backdrop to the broad preaching platform at Ebenezer Welsh Independent Chapel (1862) has a striking sinuosity in its balcony front and seating. Thomas Thomas seems to have set a precedent in using a dipped balcony front behind the preacher creating a single focus to the chapel. (960074/30)

Fig. 289. The magnificent interior of John Humphrey's Tabernacl Chapel (1872); the sophisticated design earned it the epithet 'The Cathedral of Welsh Nonconformity'.

Fig. 290. Philadelphia Chapel on the east side of Trevivian shares many characteristics with Thomas's Siloam Chapel at Pentre-estyll on the west side of Trevivian, built two years earlier, including tall windows within recessed arches. (9500352/1).

Fig. 291. In the 1860s-70s, Henry Thomas designed a series of full temple-like chapels for the Baptist congregations around Swansea. The design style included three upper windows of equal size over one or more round-headed main doorways with simple mouldings, and a multi-foiled central window in the pediment. (9300642/1)

buildings in the industrial communities of the lower Swansea Valley, one more than Thomas. All five of Humphrey's buildings in Glamorgan in the later 1860s were built in the industrial satellite communities of Swansea - two day schools funded by coppermasters at Kilvey (1865) and Morriston (1868), and three chapels funded largely by their workers at Mynydd-bach near Morriston (1866), Philadelphia at Trevivian (1867) and Salem or Capel y Cwm at Llansamlet (*c*.1867).

One of Humphrey's first commissions outside the Swansea area was in the nearby metals and coal centre of Llanelli. The majority of his seven known commissions in Glamorgan in the next five years were in the Swansea industrial settlements, two (Soar, Dyfatty, in north Swansea, 1870, and Tabernacl, Morriston, 1873) exhibiting the first examples of his grand three-bay columned façades. Other Swansea area chapels such as Horeb, Morriston (1870, costing £3,000), and Tabernacl, Mumbles (1871), were also substantial buildings. Humphrey's work in south Wales seems to have included 44 chapels and four separate schools. He also seems to have built one chapel (in his grand Dyfatty/Morriston style) in a mid-Wales textile-industry centre (Zion, Llanidloes, 1879) and three in north Wales. He died in 1888 aged 68, the year he completed the large Terrace Road Schools in Swansea.

The 1860s, following the great revival of the late 1850s, were also the years when the newly augmented and more confident communities around the copper and tinplate works of the lower Swansea Valley produced some of the first recognisable copies of classical temples in Wales. These distinctive and impressive chapels had full-height Tuscan pilasters carrying a pediment. Unlike the enriched façade-centred designs that came to dominate Wales under the tutelage of Thomas and Humphrey, these buildings had an equal emphasis on sophisticated matching side elevations to complement the elaborate façades of the completed chapel buildings. In the copper-communities it was the Baptist denominations that formed the secondary causes and it seems appropriate that the minister-architect who used the industrial wealth of the lower Swansea and Neath Valleys to build these classical temples should have been based outside these centres. The Revd. Henry Thomas came from the iron-making and coal-mining community of Ystradgynlais, towards the head of the Swansea Valley.[119] He may have designed an early example of the style at Neath (1862-63) and he certainly designed the splendid examples at Clydach (1868), Morriston (1869-70) and Caersalem Newydd at Tre-hoeth (1873). This latter chapel in the 'Tuscan Style' was reported, in the contemporary national building press, as being 'one of the largest in Wales' and could hold no less than 1,280 (the old chapel of 1845 became its vestry).[120] It seems likely that the existing Welsh Independent congregations were then prompted to become ever more lavish in their rebuilding of Hebron and Libanus ('Tabernacl'). Indeed, it has been said that the powerful and

fig.290

fig.291

prospering tinplate industrialists, under the leadership of a young and dynamic minister, decided to build the biggest chapel in the country.[121]

Thomas Freeman

There is a third local architect of the dominant Welsh Independent denomination in the Swansea area, who also produced an impressive temple design. It seems plausible that the Landore Welsh Independent congregation felt they had to emulate the neighbouring Independent congregation at Morriston which had funded the completion of Humphrey's Tabernacl two years previously. After all, the Landore congregation seems to have included several managers of the large Hafod copperworks as deacons. It might have been expected that they would have employed their own architect-minister Thomas Thomas, the unchallenged master of chapel

architecture in Wales in the 1860s (when he rebuilt their own chapel). The circumstances of Thomas Thomas's resignation as minister at Landore in 1875 are unclear. His obituary states [122] that 'on the 26 of March, 1875, Mr. T. Thomas gave warning that he was resigning from the ministry, a king to his congregation to 1875, when he ended his connection with them, rather unexpectedly, and moved to Mumbles to live [when 58 years old] and there lived-out the rest of his days.'

The passing of the Artizans & Labourers Dwelling Improvement Act allowed the promotion of the Borough of Swansea Improvement Scheme of 1876 which proposed the clearing of 'slums' owned by four main owners. Those belonging to the Revd. Thomas Thomas were classed as some of the poorest workers' housing in north Swansea, though it is not clear who had designed them. In 1879 Thomas Thomas was awarded £2,350 in compensation for the loss of an annual income of £143.50 from the houses.[123] It may be that 1875 was the year in which Thomas Thomas's ownership of what was regarded as unsavoury housing became public knowledge. His obituary notes that his frenetic period of activity seemed to have ended by his last 13 years for 'He preached little after this but was noted for his faithfulness as a member and teacher in the Sunday School which he greatly liked. He died after a short illness on the 16 March 1888 and was buried at Sketty.'

His abrupt departure in these circumstances presumably explains why he did not design the large new chapel at Seilo. The powerful deacons ran the congregation without a minister for a considerable period and used a style for their grand new building that was somewhat 'newer' than that so successfully propagated in the 1860s by Thomas, but perhaps by the 1870s it seemed dated and insignificant compared to Humphrey's magnificent Morriston Tabernacl. However, even though nearly £7,000 was spent, it was less than the sum of £8-£12,000[124] reportedly expended initially by the Morriston congregation which included wealthy tinplate industrialists; even so it was still one of the largest sums expended on a contemporary chapel in Wales.

Alderman (and Siloh deacon) Thomas Freeman, J.P., 'architect and overseer' of Brynhyfryd, was selected to design the new chapel. This astute politician, later Mayor of Swansea, had gained his architectural experience as builder and surveyor with Thomas as designer. He had undertaken building work on the nearby Landore Baptist chapel of Dinas Noddfa, with Thomas as supervising architect. The pair also seem to have been involved in work at the Baptist Capel y Graig at Newcastle Emlyn in Carmarthenshire.

Fig. 292. The interior of Caersalem Newydd (1873) showing the balcony dip behind the pulpit in which Henry Davies seems to have imitated the design principles evolved by Thomas Thomas; the organ was added later. (960069/6)

Fig. 293. Thomas Freeman. (The National Library of Wales)

fig.294

Fig. 294. Thomas Freeman executed an elaborate design for this chapel at Seilo Newydd, Landore, which extended to all sides of the building. The building cost was second only to Tabernacl, Morriston. (960074/11)

fig.295

Fig. 295. Seilo Newydd Independent Chapel showing its spacious interior, second only in size to Tabernacl, Morriston. There is the characteristic Swansea area balcony dip under the organ and over the broad rostrum or platform-pulpit. (960074/9)

The new building at Seilo was moved from the site of Thomas's old building to a larger and more prominent corner site (rather like the Morriston Tabernacl), further up the hillside at Landore and overlooking the copperworks in the valley below. Freeman produced a complete four-façade temple design, with distinctive full height Tuscan pilasters (this time variegated stone pillars composed of Pennant sandstone framed in ashlar) reaching-up to an ashlar Bath stone entablature running all round, except for the front facade where a variation of Thomas's arched opening into the pediment was produced. However, the front facade became rather a fussy arrangement of disparate elements; a triple doorway with pointed north Italian style hoods was squeezed into the central bay, which resulted both in the two main facade pilasters being partly hidden by

them and in a squeezing of these vertical features far to the sides of the main façade. To compensate, the vertical pilasters were continued upwards through the main entablature to a gable emphasised with 'Lombardic' (that is, north Italian Romanesque) stepped corbelling - a clumsy division of a fine classical gable rarely attempted in antiquity or the Renaissance. The central large window breaking upwards through the base of the pediment was framed by a draped name banner of ashlar and capped by a small circular opening, thus echoing Humphrey's variation of the standard Thomas façade.

The possibility arises that Freeman was in close contact with his fellow builder-architect John Humphrey, for there is a clear sharing of design ideas between the latter's contemporary chapel at Capel-y-Crwys (Three Crosses), on the western fringe of modern Swansea, and of Seilo Newydd at Landore. The two buildings share the same façade details: triple semi-circular headed doorways, with pointed north-Italian style hoods, are set in a projecting lower wall enclosed between full height pilaster buttresses; these enclose, above the doors, a large semi-circular headed window projecting into the gable, with a head of radiating spoke-like stone enclosing intermediate circular openings set in recessed plate tracery; the head of each central bay has 'Lombardic' stepped corbelling in the gable. On Capel-y-Crwys there are four distinctive bull-nosed finials placed on the main gable-façade roof, which are echoed at Seion Newydd, as is the cross carried high on the gable apex of each of the main façade roofs. Both also followed the latest fashion of 'Bristol Byzantine' in the use of alternate blocks of different stone.

The actual pattern of cross-fertilisation of ideas (or copying) is not clear. It was decided to build Seilo Newydd in 1875, the foundation stone was laid in 1876 and the new chapel was opened on 19 February 1878.[125] The decision to build a new Capel-y-Crwys was taken in 1876 and the construction was completed by October 1877.[126] Differences between the two structures include the fact that Capel-y-Crwys is a 'façade chapel' with plain sides and rear while Seilo Newydd is a full 'temple chapel' as in the Revd. Henry Thomas's earlier designs; this is reflected in the respective costs of £2,500 and £7,000 for the two structures. Freeman also applied radiating tracery bars to the margin glazing of his subsidiary windows, while Humphrey used traditional 'Y' tracery for the smaller windows at Capel-y-Crwys.

The congregations were linked in other initiatives. Seilo, in conjunction with Libanus/Tabernacl (Morriston), had a daughter congregation in the newer industrial suburb at Plasmarl to the north-east, and in 1878, Freeman built Hermon chapel there in gothic style alongside the Sunday school already established. Freeman, however, never seems to have had the great design influence that both Thomas and Humphrey achieved from their home base in the industrial settlements of Swansea.

Within three years of the completion of the Independent's 'cathedral' at Morriston, the Swansea town congregations of the other denominations had replied by building two of the very few nonconformist chapels to have great free-standing columns in their façades. These were the Calvinistic Methodist's Argyle Chapel of 1873 (by Alfred Bucknall of Sketty) and the Baptists' 1874-76 recasing of the Mount Pleasant Chapel (by George Morgan of Carmarthen). Morgan was another of the productive and influential chapel architects, but from the Welsh Baptist denomination of the 1870s, 1880s and 1890s. In the late 1870s and early 1880s he was experimenting with variations of fully-blown Lombardic design, expanding, in this respect, way beyond what Freeman and Humphrey had touched upon. A reordering of standard elements of design was used at Abergavenny (1877), Haverfordwest (1878), Carmarthen (1880) and, finally in Dinas Noddfa, Landore (1884).[127] In this latter rebuilding of Thomas's earlier work, the Baptists achieved what no other denomination did - a splendid building on the high slopes below the ruins of Morris Castle which dominated most of the copperworks spread out on the valley below. The central element of the front facade towards the valley was a great wheel window, found in three of Morgan's Lombardic designs, but here set into a Lombardic version of Thomas's great arch in the

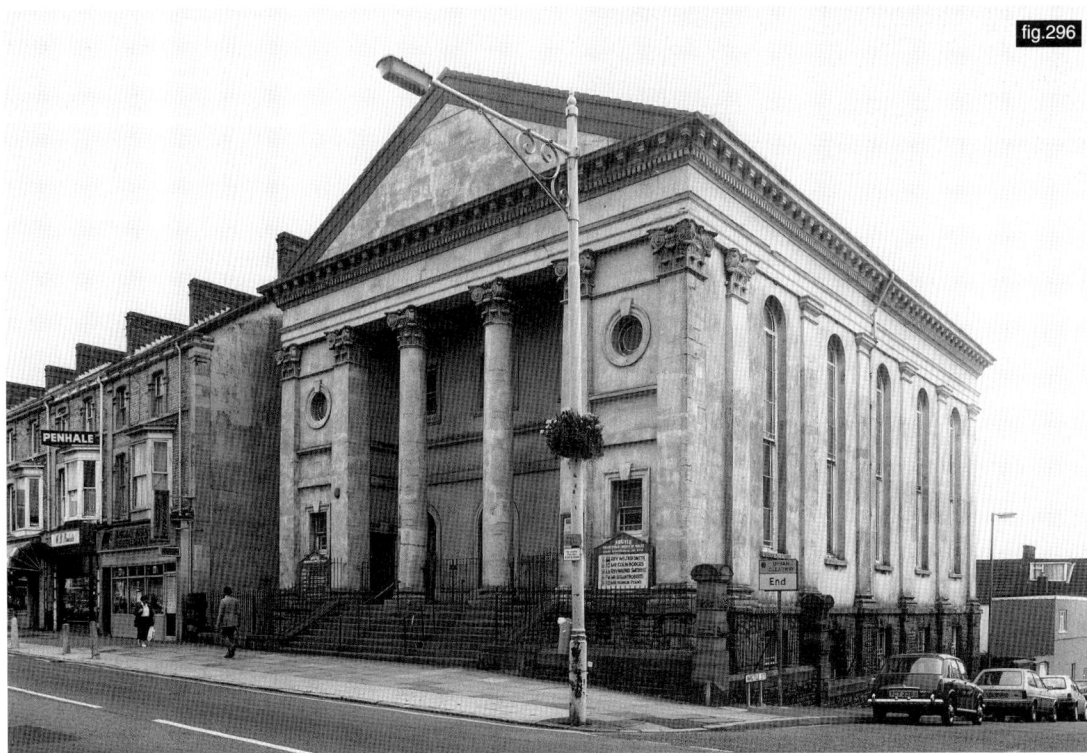

Fig. 296. Wealthy nonconformist congregations in Swansea town attempted to match the substantial columned façade of Tabernacl Morriston by building their own classical temples such as the Calvinistic Methodists' Argyle Chapel (1873). (871366/5)

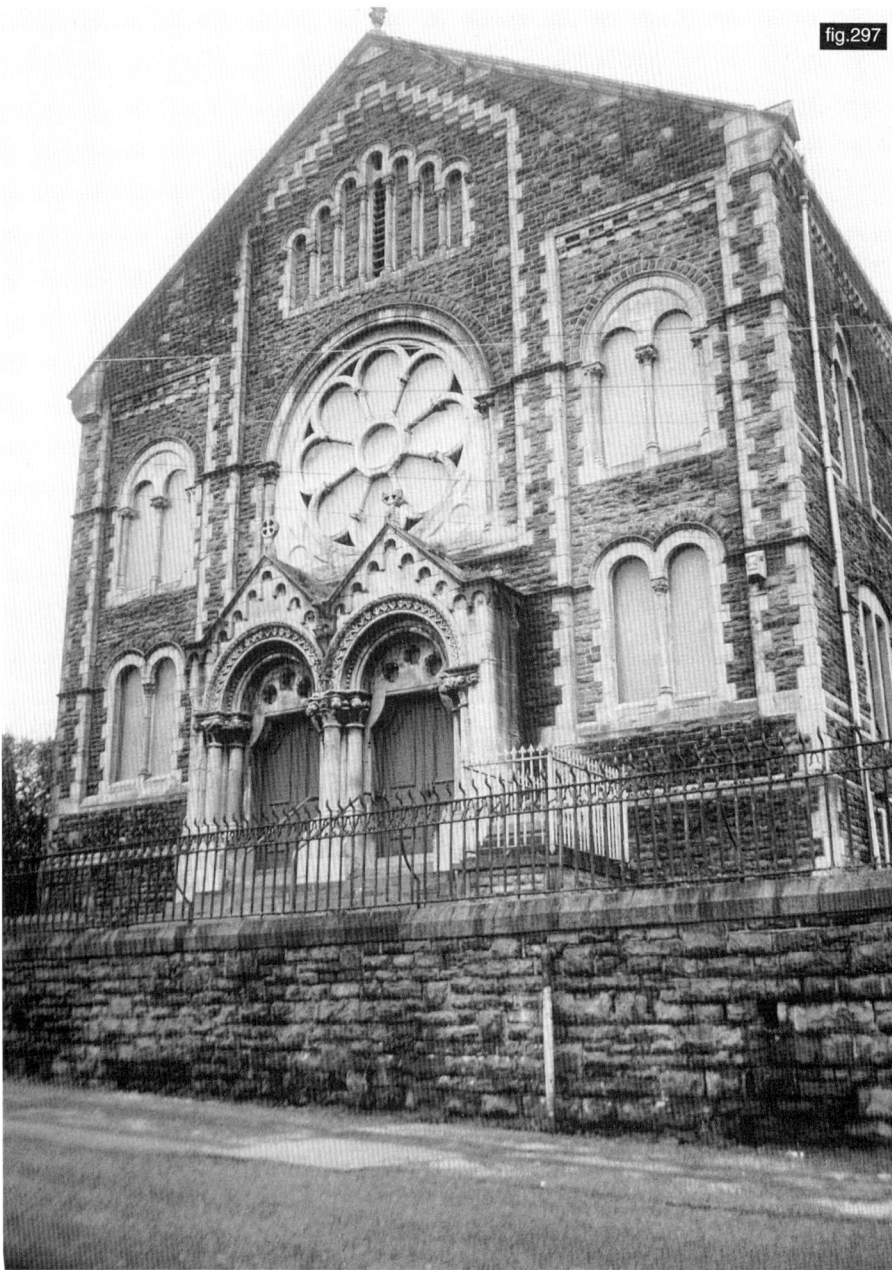

fig.297

Fig. 297. Dinas Noddfa Chapel at Landore.

Chapel Architects - Conclusions

The growth of a class of local yet influential artisan and minister-architects in the industrial settlements of the Swansea area in the mid and later nineteenth century is an interesting phenomenon. The buildings they produced became the focus of the social and spiritual lives of the workers and managers in the copperworks and the entrepreneurs of the new tinplateworks. The structures dominated in the townscapes surrounding the copper and tinplate works.

Thomas Thomas's work extended to the edge of the other international metal-smelting centre of Merthyr Tydfil. Merthyr's primacy in industry and indigenous chapel building was before the great religious revival of the 1850s had created a huge demand for chapel building all over Wales. No local Merthyr architects came to build chapels in the Swansea area. A hundred years earlier, William Edwards had been an artisan/engineer/minister/ architect of similar background, but one of a group of artisan-engineers creating structural and engineering innovations of importance to the world's first Industrial Revolution. One hundred years later Thomas Thomas, Humphrey, Henry Thomas, Freeman and Morgan were building in the Welsh townscapes created by that new order of economic activity and wealth.

What happened in the satellite industrial townships of Swansea, with initially their largely Welsh-speaking populations, makes a contrast with trends in the more anglicised town of Swansea. For example, the Welsh Independent minister of Ebenezer was concerned that the English-speaking population should be adequately served in order that they too could acquire the piety obvious in the diligent Christianity of the Welsh-speaking congregations. With preaching Sundays in Swansea's Albert Music Hall and support from the Bristol industrialist H.O. Wills, it became possible to employ a specialist architect from England for work in the town itself.

Religious Buildings at Morriston and Trevivian

Morriston

This section is of an examination of the comparative dates, sizes and forms of the places of worship of the several denomination. The seventeenth-eighteenth - and early nineteenth-century religious congregations ('causes') and buildings at this copperworkers' town are listed in chronological order up to the period when the tinplate industry became a dominant influence in the settlement. nonconformist places of worship are listed as 'chapels' and Anglican buildings as 'churches.'

pediment, supported by a fine double north Italian-style ashlar work porch.

There was a very large number of nonconformist chapels built to serve every group of houses in Swansea and its surrounding industrial townships in the later nineteenth century. Even in the Congregational/Independent denomination in the industrial suburbs of the 1870s, when Thomas, Humphrey and Freeman were at the height of their careers, it was possible for other architects to be active. For example, a Mr Williams 'architect' was designing a new chapel in 'Mixed Style' on the Carmarthen Road which cost the not insubstantial sum of £2,000.[128]

Chapel 1.1. 1682, Ty-coch House Church Congregationalist or Welsh Independent. Services began at Ty-coch, one of the squatter houses built on the common.

1779, Morris Town designed. The settlement plan was made by William Edwards, the architect-engineer and Independent minister. Two places of worship with towers were designed, the nonconformist one was probably never executed. Wide streets and two squares were planned, but only one of the latter was built up.

Chapel 1.2. 1782, Libanus 1. This replaced the 100-year-old Ty-coch chapel. It was one of 12 chapels dependent on Mynydd-bach (built by the Revd. Lewis Rees, 1762) 1.3 miles to the west. It was a simple, barn-like building with earth floor and benches and no building debts were left outstanding. It was built by William Edwards three years after the foundation of the town but probably never had a tower. John Morris granted a lease of the site for a thousand years at 1s. a year. It was very unusual for landowners or entrepreneurs to refuse land for their tenants' nonconformist chapels to be built, there was only one instance noted in the 1851 census for the whole of south Wales and that was by a non-industrialist landowner.

Church 1. 1789. St. John's. It was built with 108 free sittings only 10 years after the building of the town, a simple building with a tower. According to the 1851 census, a morning service in English attracted an adult congregation of 43 and 15 scholars and in the evening a congregation of 74 and 17 scholars; the afternoon Welsh-language congregation was much smaller. The church was nowhere near full even of its seated accommodation.

Chapel 1.2b. 1796, Libanus 1b. This was an extension of the first chapel by the Revd. David Davies, soon after the Anglican church was built and as a second Independent congregation was opening just along the same street. Seats replaced benches at this time.

Chapel 2.1. Between 1795-98, converted houses, pre-Philadelphia. A second Welsh Independent cause, possibly a reaction against the domination of the Mynydd-bach minister and/or a lack of adequate room at Libanus and/or because of theological differences. It was down the road from Libanus in two adapted small cottages on the later market site.

Chapel 2.2. 1802, Philadelphia. The congregation now had a purpose-built Philadelphia chapel, built only four to seven years after the completion of an adapted house-church. It was near the centre of the original south-eastern part of the grid. Land was leased from Lockwood, Morris & Co. and the congregation entered the Calvinistic Methodist denomination. In 1812, transepts (a 'croesdy') were added and the capacity was enlarged to 300-400 (that is three to four times the capacity of the Anglican church). Enlargement of the original Libanus had taken 14 years: at Philadelphia it took just 10 years before the congregation outgrew the original purpose-built chapel.

Chapel 3.1. 1809, failed Baptist cause. A Baptist cause was established in Morriston in 1809 but faded before it could achieve the momentum needed to build a purpose-built chapel.

1829, Expansion of chapels 1.3 and 2.3. In 1829 there was a revival which spurred expansion of the chapels in an attempt to accommodate and keep the converts attracted.

Chapel 2.3. 1829, Philadelphia. A new, larger, long-wall chapel was built with a gallery lit by upper round-headed windows. The architect was a local man, Daniel Roberts, with a design similar to Trinity Chapel in Swansea. Philadelphia was then the largest nonconformist chapel in Morriston with a capacity of 678. At times of the revival it may have accommodated 990, as in 1851: although then it had a usual congregation of 610, including standing accommodation (which could only be in the aisles) estimated at 262. This chapel alone had a capacity some 6.3 times that of the Anglican church built 40 years previously and not extended since then.

Fig. 298. 'Philadelphia' is the only chapel in Morriston to retain its earliest long or lateral-wall facade to the street. The original simple appearance of the facade was aggrandised in the 1930s with a central mock-gable. The chapel retains the characteristic window and end double-door arrangement of the type. (9500101/14)

Chapel 1.3. 1831, Libanus. During the revival of 1829 the chapel achieved independence from Mynydd-bach. The enlargement of nearby Philadelphia in 1829 may well have prompted the total rebuilding of the chapel in 1831, 49 years after its original building, so that it too could accommodate over 600. There were seats for 602, but by 1851 the congregation could often number at least 621.

Chapel 4. *c.* **1829, Market Hall.** A covered market was erected by John Morris II in 1827 down the street from Libanus, on the site of the Philadelphia congregation's original home. The Wesleyan Methodists may have been using the adapted houses that formed the original home for Philadelphia for by 1851 they were established in the large room, with gothic-traceried windows, that stood above the covered market (and was also used as a day school) and said that they had been using the site 'since 1800'. In 1851 this first recorded nonconformist English-language congregation had room for 88. Its afternoon school had 30 pupils and there was a congregation of 40 in the evenings.

Chapel 5. *c.* **1829 Market Hall?** The Welsh-language Wesleyan Methodist congregation may originally have shared the same home on Market Street. In 1851 they also used premises that were 'Not a separate and entire building; not used exclusively as a place of worship' but with a stated capacity of 50. They had smaller morning and evening congregations of 15.

Chapel 6. 1842, Horeb. This cause began as another house church of the Welsh Independents. For three months it was held in the house of John Evans in Edward Street and then the old Clyndu Colliery office was rented for £3 a year. It became the first nonconformist chapel to dominate the northern end of what was later the main street of Morriston. It was the first of a series of chapels built just beyond the perimeter road of the planned township, in this case on land held on a 999 year lease from the industrialist and landowner L.W. Dillwyn at 1s. a year. The foundation stone was laid in October 1842 and the building was in use from 7 April 1844. A gallery was added and opened on 10 November, 1845 and the total cost of the building had been £600-700.[129] The chapel had a capacity of 464; the maximum congregation in 1851 was 234 and a further 120 standing people could be accommodated.

In the 1840s and 1850s the land to the north of the planned town was being developed and a further two nonconformists chapels were built along the main roads of the area.

Chapel 7. 1845, Seion. This was situated on the Clase Road leading to the Wychtree Bridge, the first nonconformist cause in Morriston that did not have its origins as an indigenous house-church but rather was planted as a 'forward mission' by surrounding congregations to the south and south-west; over half of the original members came from Caersalem Newydd (Tre-boeth) and the rest from Dinas Noddfa (Plasmarl) and Bethesda (Swansea). There was seating for 500 (and 150 standing). The 1809 indigenous cause, and subsequent attempts at revival, had failed, but from 1845 the Welsh Baptist cause in Morriston was firmly established.

Chapel 8. 1840s, Zoar. This English-Language Baptist Church was built nearby with seating for 200 and in 1851 had a maximum congregation of 82.

In the third quarter of the nineteenth century, when the tinplate industry was a dominant industry in the town, the old centre of Morriston and its immediate environs had a total of 16 chapels with more in its surrounding suburbs. The Anglican response was to rebuild the central St. John's Church in 1857 on a larger scale, even though the size of the contemporary congregation did not warrant it. At the end of the nineteenth century, a second Anglican church for the English-language congregation was built at the northern end of Woodfield Street, in a position where it blocked the view of the hitherto dominant Horeb chapel. Then, the three most important elements of the main townscape competing for attention were St. John's, St. David's and, soaring over all, the elaborate triple-arched facade and tall spire of John Humphreys' Tabernacl.

Trevivian

The Anglican church which the Vivians originally intended to place at the centre of their workers' township has been discussed above. Unlike Morriston, the chapels at Trevivian lay on the fringes of its grid of houses and predated the Vivians' township, being situated in the earlier colliery settlement of Pentre-estyll. As has been noted, the Vivians' managers were prominent members of chapels to the north and south of the workers' township. Many of their workers lived in Landore to the north and it was from there, in the enthusiasm generated in the aftermath of the religious revival of 1859, that chapels with new elaborate and distinctive designs emerged.

The influence of Thomas Thomas's use of the giant arch can clearly be seen in the fine façades of the 1860s: to the west of Trevivian at Siloam chapel (1864, with seating for 900); in the coal-mining community of Pentre estyll in Cwm Burlais; and at the very similar Philadelphia chapel (Welsh Baptist, 1866; capacity 750) on the Neath Road at the eastern edge of Trevivian. The Philadelphia cause had originated in the Greenhill area of north Swansea in 1750, in a chapel that was recognised by the Glamorgan Baptist Association. However, most members lived nearer the Hafod Copperworks and so the minister, D.W. Morris, approached Henry Hussey Vivian who gave them the site for building and also laid the foundation stone. The chapel was opened in 1867 and cost £1,500 to build.[130]

A Vivian also laid the foundation stone of the Anglican coal-miners' mission building of 1905, which was erected in the year of the last great religious revival. It stands at the opposite end of a terrace of coalminers' houses from Siloam chapel. The Ruabon terracotta adornments were then matched on the neighbouring chapel in 1914, when a porch and balustrades were added to the handsome giant arch façade. At present, these three Trevivian buildings survive in various states of disuse and reuse (the mission is a hygienic goods wholesalers, Philadelphia is the Hamilton Dance Centre). The third nonconformist building on the edge of Trevivian was the mission or Sunday school planted by the Mount Pleasant English Baptist church in Swansea town, between 1851 and 1866 at a cost of about £1,300.[131] The cause still remains active in what is now a community in Hafod with a significant Moslem population; it is housed in a bungalow-like structure built to replace the old building in 1975.

The fourth nonconformist chapel at Trevivian was also the second English-medium place of worship; it was erected on the edge of the settlement and was the third such aided by earlier members in Swansea town. However, it was the first Trevivian chapel to have a very localised start as a house-church and to that extent was similar to several of the causes started in what was then the more remote township of Morriston. Its origins may lie with some of the Cornishmen who had followed the Vivians to Swansea. The Bible Christians had been a very active evangelical movement in Cornwall and a branch was built in Oxford Street in Swansea town; a nucleus of Cornish people at Hafod started meeting in the front room of 2, Maliphant Street and afterwards at 6, Earl Street. Henry Hussey Vivian was approached and he agreed to give as much land as was required for a nominal peppercorn rent.

The foundation stone was laid on 25 August 1873 and initially the members of the Oxford Street congregation bore 75% of the weekly costs of running the Hafod chapel. In 1883 a schoolroom was added at the rear and by 1905 this had been extended for a second time. The church became United Methodist and, in 1932, took part in the general Methodist Union.[132] Now on the east side of the Neath Road, the chapel is forlorn and disused, along with its manse. The structure is in the diluted simple, gothic beloved of chapel congregations (especially English-language and Wesleyan congregations) in the third decade of the nineteenth century, prior to their adopting full-blown gothic revival buildings from the Anglican tradition.

Fig. 299. The simple gothic facade of the Trevivian Bible Christian Chapel built in 1873 by Cornish workers at the Hafod Copperworks. (960213/4)

Chapels of the entrepreneurs

Coppermasters did not usually build nonconformist chapels in the way that they could be persuaded to build churches by Ministers of the Established Church. There were exceptions, however. John Vivian had been active in the copperworks at Hayle before he came to south Wales. There the Cornish Copper Company built an octagonal Wesleyan Methodist chapel from cast copper-slag blocks and John Wesley preached there on two occasions.[133] Vivian may thus have been influenced by Wesleyan Methodist partners for certainly no Swansea coppermaster built or owned a chapel intended specifically for his copperworkers.

This was to change significantly with the rise of self-made local entrepreneurs in the low-capital tinplate works (partly based in the old coppermills of the area). These entrepreneurs included John Jones Jenkins, the manager of the first Swansea and Morriston tinplateworks at Upper Forest (1845). He was born in Clydach in 1836 and entered the trade, undertaking the most menial work when he was first employed at the age of 14; he rose very quickly to be manager of the Upper Forest Tinplateworks at the age of 23. In that role he acquired books for a works library and introduced reading and music classes for the workers. He subsequently became a major shareholder in the Beaufort (1860, Morriston), Cwmfelin (Hafod) and Yspitty (Loughor) Tinplateworks. He was mayor of Swansea in 1889, MP for Carmarthen, knighted in 1885 and created Lord Glantawe in 1906. A large part of his life was spent in Bath Villa (on the site of the old alum works 'baths') within easy walking distance of Tabernacl Welsh Independent chapel.[134]

As a member of that congregation it was Jenkins, with fellow tinplate entrepreneur Richard Hughes (of the Landore Tinplateworks), the minister and a few others who resolved on building 'the cathedral of Welsh Nonconformity' on the main street of Morriston and who then persuaded their fellow tinplate industrialist Daniel Edwards to build it. Edwards was born in 1835 at Morriston, where his father was a stonemason and carpenter, he attended private dame and master schools there before commencing work at the Rose Copperworks. From there, he quickly moved to the Morriston Foundry and by the late 1840s, was working as a stonemason, by the 1850s he was working on the extension of the large Ystalyfera Tinplateworks which had been started by the Budds, formerly of the Hafod Copperworks. From there he moved on to the expanding Upper Forest Tinplateworks (then under Richard Hughes of the Landore Tinplate Works) and back to Upper Forest before the 1860s. Jenkins employed him at the Beaufort Tinplate Works until, together, they and William Williams (later mayor of Swansea) established the Worcester Tinplateworks adjacent to that at Upper Forest in 1868. This partnership was dissolved in 1870, the year when Edwards interrupted his promising tinplateworks career to build the grandest nonconformist chapel in Wales.[135]

The foundation stone for Tabernacl was laid in November 1870 and Edwards accompanied the architect, John Humphrey, and the minister on a tour of the most famous chapels in England in order to gather ideas for the new structure. Edwards supervised the building works and the team of masons, some of them from Bristol. The 1,450-seat chapel was opened on 22 December 1872,[136] the 'great redeeming feature in the whole of that huge manufacturing district - it is an oasis in a desert, an object worthy of admiration in the midst of unsightly works and manufactories of every size and description.' In the year of that report, Daniel Edwards & Company were already building the vast Dyffryn Tinplateworks at Morriston which employed 1,100 workers by the 1890s.[137]

The later coal-trade boom brought some other Congregationalist businessmen to Wales who actively aided the building of chapels for their own denomination. One such was John Cory of Cardiff who laid the foundation-stone of the chapel at Penrhiwceib and gave £25 of the £1,050 required for its construction.[138]

However, only a minority of the major entrepreneurs of south Wales were nonconformists, and often not of the same denomination as their workers; for example the Guests of Dowlais, were Wesleyan Methodists, and the Harfords and Darbys of Ebbw Vale were Quakers.[139] Yet, an evaluation of the Swansea evidence shows how the number and impact of nonconformist industrialists have been underestimated.

The iron-making towns at the Heads of Valleys had had the largest capacity chapels, but the growth of wealth in the copper-making and the new tinplate manufacturing communities in the later nineteenth century changed that. The Swansea area had the greatest concentration of large nonconformist centres of worship by the time of the 1905 Royal Commission: there were 33 nonconformist places of worship each of which could hold a congregation of 1,200 or more and almost a quarter (seven in fact) were in the greater Swansea area. Even Tabernacl at Morriston was eclipsed by other nonconformist places of worship that could hold larger congregations than its 1,450, even if not in such a magnificent building. In Morriston itself, the Calvinistic Methodist Forward Mission could hold

1,500, while the Albert Hall in Swansea held congregations on a Sunday of up to 2,800.

Religious Buildings - Conclusion

The 1851 Religious Census reveals that small Anglican congregations in the copperworkers' settlements were provided with large ambitiously designed places of worship, while the plain nonconformist chapels could barely contain their congregations, substantial numbers of whom resorted to the 'standing room' in the aisles.

The building of a new church or chapel seems to have roused a sense of obligation in neighbouring congregations to show that they had an equal commitment to their faith, so that there was almost a cyclical competition in building between the various denominations and causes in the industrial communities. The average cost of building a new chapel was between £1,000 and £2,000.[140] The new tinplate entrepreneurs of Morriston supported a congregation at Tabernacl that was able to afford a building cost of £8-£10,000 (1873),[141] while Seilo Newydd, with a congregation that included the Hafod Copperworks managers, was almost able to emulate this achievement two years later when nearly £7,000 was spent on the next Welsh Independent chapel down the valley at Landore.[142] This high level of spending on nonconformist places of worship was not possible elsewhere in Wales. The Swansea area was becoming the world centre of two metals industries; the only other international centre of a metals industry in south Wales was Merthyr Tydfil, where there seem to have been fewer Welsh nonconformist managers and industrialists for a lesser number of works. Between 1843 and 1890 the greatest sums recorded in *The Builder* for chapel building outside Swansea were £6,000 at both Colwyn Bay (1882) and Wrexham.

Communal Buildings

The chapels and their schools were undoubtedly the social focus for the industrial communities around Swansea. After the fervour of the 1857-58 religious revival, and with increasing affluence and numbers in the 1860s, many new large chapels and attached schools were the buildings in which social activities could take place. There was a strong tradition of adult participation in both Sunday and evening schools. There is also evidence that such works schools as Trevivian provided similar evening educational opportunities.

Beginning in the mid-nineteenth century, and increasingly towards the end of the century, employers in south Wales sometimes provided more general social and educational facilities for their workforce. In 1853 the Grenfells provided a community hall; this was the Foxhole Music Hall which stood at the south end of the long terrace in Foxhole, below All Saints' church on the hillside above. It was built on land leased from the Earl of Jersey and seems mainly to have been used for church purposes. When the Grenfells put their copperworks up for sale in 1892, the hall was sold to the parish; (together with the adjacent Foxhole Cottage), and became the parish hall. Both buildings were largely demolished in the 1960s although ruins remain alongside Foxhole Road. In the third quarter of the nineteenth century, Mary Grenfell built a reading-room and library for the use of railway workers it adjoined her 'Golden Griffin' coffee house next to St. Thomas railway station. During their dinner-break, she also taught the 'three Rs' to the boys employed at the neighbouring Patent Fuel Works.

Not all the coppermasters provided such facilities, it obviously made a difference if the managing partner's family was locally resident and active in the community. This was true elsewhere at other large metals works in south Wales, such as the institute provided in mid-century at Dowlais. The Guests were the only proprietors of the Merthyr Ironworks to provide such facilities. Later on, the Darbys and their (Quaker) partners at Ebbw Vale Iron and Steel Works did so and so did the Lysaughts in the Orb Sheet Metal Works at Newport.

The Swansea coppermasters do not seem to have provided almshouses or retirement homes for their workforce. This seems to have been rare elsewhere, Sir Titus Salt at his model settlement at Saltaire seems to have been an exception. It is worth mentioning that at the end of the eighteenth century John Morris I appears to have built the pedimented poor-house at Morriston which may have been one of the first structures with an 'almshouse' function to have been built by an industrialist.

The large sums generated by the copper industry facilitated the commissioning and collecting of fine paintings and works of art, by the entrepreneurial families. These paintings were generally hung in the large houses wherever the respective partners were resident. With the decline of the industry such collections were often dispersed, but two important public art galleries were directly financed by Swansea's copper-trade profit,: one each by the two main dynasties of Swansea coppermasters, the Morrises and the Vivians.

In 1786, Margaret (Jenkins), the wife of Robert Morris I, accumulator of the family's wealth, died at the London home (in Charlotte Street) of her daughter Margaret. The younger Margaret had been given a £5,000 dowry and. at least eight years earlier, had married Noel Desenfans.[143] He used the money to set himself up as a picture dealer and was commissioned to collect paintings for King Stanislas of Poland until the latter abdicated in 1795.[144] Left with the paintings, Desenfans tried to persuade the British Government to purchase them as the foundation for a national collection. When he died, he left them to Margaret and to his closest friend, Sir Francis Bourgeois, landscape painter to George III. Bourgeois in turn bequeathed the paintings to Dulwich College and Margaret gave the money to build the Dulwich Art Gallery, the first public art gallery to be built in Britain and the first constructed as a top-lit free-standing structure away from an aristocratic mansion. Margaret was one of the three founders whose ashes were buried in the mausoleum which Sir John Soane designed as part of the gallery complex (which itself subsequently became the model for Gilbert Scott's classic phonebox design!). The gallery was opened in 1817, 21 years before the National Gallery in London.

John Henry Vivian and two of his sons, Graham and Richard Glynn, were also avid collectors. The latter, resident in Sketty Hall, in 1905 offered Swansea town his paintings and also the money to endow a municipal art gallery. He was dead by 1909 when the present Glynn Vivian Art Gallery was opened.[145]

It is interesting that the Grenfells, resident coppermasters in Swansea only for one generation, did not endow a public gallery or museum but did endow the museum at the Royal Institution of South Wales with some of its best-known exhibits. One of the family members, Sir Francis Grenfell, while commander of the Egyptian army, was able to facilitate the delivery of various Egyptian mummies and exhibits to the museum.

General Conclusions:

It was the Bristol-based proprietors of the White Rock Copperworks who provided the site for the first lower Swansea Valley copperworks school at Foxhole and Grenfells from the Thomas Valley, who provided the building. The proprietors of the Birmingham copperworks provided the first works school at Morriston. Legislation and legal compulsion were peripheral to this activity, for these first schools were built before the 1833 Factory Act. That Act only applied to children employed in the works and there was no means of enforcement intended or applied.

The Hafod Copperworks Schools built by the resident Vivian family were among the largest in south Wales and came to educate over a thousand children of copperworkers at a time, while the number of boys over 12 employed at the Hafod Copperworks did not exceed more than a hundred. The Grenfells, resident in Swansea by the mid-nineteenth century, were also lavish providers of school buildings for girls and boys. There is evidence that some of the coppermasters felt both a social and a religious obligation to ensure that the children of the communities for which they felt a responsibility, could read the gospels. Equally, even the enlightened and religious Pascoe Grenfell said that he would not employ unschooled workmen. Such patronage undoubtedly aided the acquisition of skills and understanding of English, although the latter was not essential when many managing staff were bilingual. However, at least one Swansea industrialist, the Morriston colliery owner, John Glasbrook, saw education as a means of producing a disruptive rather than a quiescent workforce and implicitly criticised the provision of schools by the coppermasters. He successfully delayed the introduction of a free library in Swansea and in 1870 claimed that.[146] 'people have too much knowledge already; it was much easier to manage them twenty years ago; the more education people get the more difficult they are to manage.'

The works schools did not follow the rigid doctrines of the Established Anglican Church in a way that might be seen as enforcing social or political conformity. That was the issue that the local workforce felt most strongly about and which might have made them feel obliged to set up, on a much larger scale, independent schools attached to chapels, if they had not been satisfied that the works schools were non-denominational.

There is also little support for the theory that the coppermasters saw the provision of religious buildings as a means of direct social control, though they did undoubtedly share the opinions of many others that it improved moral behaviour and honesty. The great majority of their workforce were members of the nonconformist Welsh Independent (Congregationalist) denomination, although it was said that if other religious provision had been available they might have become members of the Established Anglican Church or another denomination. This latter opinion was shared by local clergy, who forcefully told locally resident Anglican coppermasters, such as John Morris I and Pascoe Grenfell, that they had a social responsibility to provide Anglican churches for their worker communities. The construction of these towered edifices made little difference to the workforce who had already started worship in their own independent way. The coppermasters tolerated, and gave some support to, their workforce in this but certainly did not actively support new causes choice of worship by the provision of complete new nonconformist chapel buildings (with the very occasional exception).

There is no doubt that the widespread provision of self-financed and built chapels helped to foster independence and self-reliance among the workforce. In this, the copperworks management, rather than the coppermasters, were heavily involved and a new breed of lower Swansea Valley industrialists evolved in that environment. There were a significant number of nonconformist coppermasters and industrialists, but often not of the same denominations as those of their workforce. There is no recorded instance of a Swansea coppermaster refusing a nonconformist chapel site to his workforce, or refusing the use of copperworks offices, schools or buildings in which a nascent copperworkers' congregation could begin and grow.

The institutional buildings were important to the spatial relationships and landscape of the lower Swansea Valley. Four of the works' owners provided (in part or in full) Anglican churches, but their function was not usually the same as those in some rural communities where the country squire's family church pew dominated those of his tenants by position and form. Further up the Swansea Valley, the spire of the ironmaster's Anglican church at Pontardawe (with a second successive ironmaster's church nearby) represents a rural location where an industrialist was at prayer surrounded by his managers and part of his workforce - his workers literally followed him when he decided to build a second church over which he had more direct control. The Grenfells may have worshipped with their workers on the more isolated eastern side of the valley, but there is little evidence of such a phenomenon with the Morrises or Vivians. Neither of these families rushed into building churches for their workers before sites were provided for their workers to build their chapels; the Vivians preferred instead to build a family memorial which also functioned as their own place of worship. It is interesting to note that the later nineteenth-century Anglican churches built at Morriston and Trevivian were only partially completed to their original plans and have remained viable places of worship to the present day, whilst the very large Anglican chapel-of-ease built by the Crawshays at their workers' settlement of Georgetown (Merthyr Tydfil) fairly quickly became disused as a church became a warehouse.

There are obviously differences of detail with what happened elsewhere and in other industries. If the denomination and language of an industrialist coincided with those of his workforce, then he might worship with them and would certainly substantially fund their building. This happened with important Congregationalist manufacturers from west Yorkshire and Bristol, who also partly funded the chapels of their co-religionists at Swansea and elsewhere. The tinplate manufacturers at Morriston, Port Talbot and other towns of south-west Wales largely funded the chapels in their communities including the spectacular, Tabernacl, built to dominate the main street of later nineteenth-century Morriston.

Two nonconformist chapels were built of cast copper-slag blocks in copperworkers' communities elsewhere in Britain, and John Vivian was presumably involved, with his Wesleyan Methodist business partners, in building such a structure at the copper-smelting works at Hayle in Cornwall. Of the Swansea coppermasters, only Henry Hussey Vivian is recorded as giving money (£50) to help build a chapel (Seilo Newydd at Landore which cost £7,000 to construct) and of actually building one for his colliers.[147]

At least one other Anglican operator of a copperworks elsewhere in south-west Wales (at Cwmavon) refused permission for a nonconformist chapel to be built on his land and the workers had to travel to worship, at Port Talbot where the local tinplateworks owner was a member of the congregation. John Guest (a Wesleyan Methodist) also refused permission for a local Welsh Independent congregation to build on, what he considered to be, Dowlais Ironworks land; although he had previously given the same denomination building land, and later found that he was not the owner of the land in question.

In northern Swansea, the Duke of Beaufort refused Soar Welsh Independent chapel permission to have a second piece of building land on the opposite side of the Carmarthen Road from their original building, though the motive for the refusal is not clear. The aristocracy were not universally hostile to nonconformists; the Wynns of Wynnstay may have put one preacher in their dog kennels, but female members of the aristocracy and gentry, such as the countess of Huntingdon, were active promoters of nonconformity in south Wales.

In these developments, the copper industry of the lower Swansea Valley is not greatly different from the rest of industrial south Wales, except perhaps in one respect. It has already been noted that although workers' housing had to be built by employers in the lower Swansea Valley because of its semi-rural character, it was not encumbrance on them to provide all the housing as it was on the ironmasters in the remote rural uplands of the Heads of the Valleys region. That same remoteness also encouraged the ironmasters to import provisions for their workforce. These could be provided in an enlightened way, as in the case of the Harfords in their non-profit-making shop at the rural Melingriffith Tinplate Works near Cardiff; but equally there was a great temptation to profit from what became the notorious truckshop system. Alternatively, covered market buildings could be provided by industrialists in their planned settlements, such as those provided by Sir John Morris II and John Guest at Morriston and Dowlais respectively.

The semi-urban environment of the lower Swansea Valley made it unnecessary to provide company shops and none is known to have been run by a coppermaster. Not all industrialists thought likewise, even in the later nineteenth-century. There was a 'company shop' near the first site of the Landore Steelworks and workmen had to purchase their goods there, queues forming after the monthly salary of grocery tokens and cash had been paid.[148]

Swansea coppermasters, in general did not actively promote nonconformity, but neither did they impede their managers and workforce in their allegiance to it. Sites were made available for building nonconformist places of worship and works' proprietors were prepared to lay the foundation stones of chapels, if invited to do so. At least one had Welsh nonconformist sermons translated for his own edification.[149] Another provided a secular meeting hall for the use of the employees of three local works and one coppermaster may have made provision for poorer workers. Indications from elsewhere, suggest that this level of provision in industrial communities may have accorded with a general practice of enlightened self-interest. [150]

References

1. L.W. Evans, *Education in Industrial Wales, 1700-1900; A Study of the Works Schools System* (Cardiff, 1971), i. Much of the background for the discussion is drawn from this publication. For the wider context of education in Swansea, see G.W. Roderick, 'Education in an Industrial Society', in *The City of Swansea: Challenges and Change*, ed. R. Griffiths (Stroud, 1990), 179-93; F.G. Cowley, 'Religion and Education', in *Swansea: An Illustrated History*, ed. G. Williams (Swansea, 1990), 145-76.

2. M. Seaborne, *Schools in Wales, 1500-1900; A Social and Architectural History* (Denbigh, 1992), 28. Works' schools are there placed in their historical perspective.

3. M.G, Jones, *The Charity School Movement: A Study of Eighteenth Century Puritanism in Action* (London, 1964), pp. 280-89.

4. ibid., 290.

5. ibid., 290-314.

6. Cowley, 'Religion and Education', in Williams *Swansea: An Illustrated History*, 145 176.

7. Seaborne, *Schools in Wales,* 46.

8. Jones, *The Charity School Movement*, 310.

9. Ibid.,318.

10. J. Lewis, *The Swansea Guide 1851* (Swansea, 1851), 46.

11. W.S. Williams, *Hanes a Hynafiaethau Llansamlet* (Dolgellau, 1908), 239.

12. D. Farmer, *The Remarkable Life of John Humphrey, God's Own Architect* (Swansea, 1997), 34.

13. *The Builder*, 20, (June 1868) 457.

14. Farmer, *The Remarkable Life of John Humphrey,* 37-38.

15. T. Rees & J. Thomas, *Hanes Eglwysi Annibynol Cymru,* 2, (1872)*,* 62.

16. N.L. Thomas, *The Story of Swansea's Districts & Villages*, abridged vol. I (Swansea, 1969), 54.

17. *Hanes Annibynol,* 2, 62.

18. S. Williams, *Hanes Llansamlet*, 153.

19. L.A. Cook, *A Critical Investigation into Nonconformity & Social Reform in Swansea, 1851-1914* (University of Wales, M. Phil., 1992 Swansea Institute of Higher Education), 84.

20. *Swansea Guide 1851*, 23 and 50.

21. S. Williams, *Hanes Llansamlet*, 149 and 188.

22. Ibid., 149.

23. *Swansea Guide 1851*, 50.

24. *The Builder*, 21, (November 1857) 679.

25. L. Cook, Ph.D., 202-3.

26. R.A. Griffiths, *Singleton Abbey and the Vivians of Swansea*, (Swansea, 1988), 32.

27. Williams, *Hanes Llansamlet,* 154.

28. Griffiths, *Singleton Abbey*, 43.

29. Seaborne, *Schools in Wales*, 115.

30. Evans, *Works Schools*, 312.

31. Commission of Inquiry into the State of Education in Wales (1847), quoted in G. Gabb, *The Grenfells, Lower Swansea Valley Factsheet 9* (Swansea Museum, n.d. [1980s]), 23.

32. Williams, *Hanes Llansamlet,* 239.

33. Evans, *Works Schools,* 312.

34. G.T. Clark, 'Report ... on the Sanitary Condition of Swansea,' 1849, 16.

35. Ibid.

36. J. Lancaster, *Improvements in Education* (London, 3rd. ed., 1805), 165, 188.

37. T.M. Rees, *A History of the Quakers in Wales and their Emigration to North America* (Carmarthen, 1925), 262.

38. A.N. Gamble, *A History of the Bevan Family* (London, 1923), 17.

39. J. Day, *Bristol Brass*, 35-7, 118.

40. Rees, *Quakers in Wales,* 262.

41. Day, *Bristol Brass*, 57, 67, 222-23.

42. Ibid., 35,110.

43. Ibid., 114, and Rees, *Quakers in Wales*, 255-56.

44. Ibid., 262, and Roberts, 'Copperworks Chronology', 89.

45. Ibid., 90.

46. T. Shaw, *Gwennap Pit: John Wesley's Amphitheatre, A Cornish Pardon* (Busveal, 1992), 17-18.

47. Roberts, *Chronology.*

48. R.O. Roberts ,'The Smelting of Non-ferrous Metals', *Glam. County History*, V, 63.

49. 'Committee Book of the Copper Company', MS. B.4771 (Bristol Reference Library).

50. Rees, *Quakers in Wales,* 233-35 and 262.

51. Cowley, 'Religion and Education' in *Swansea: an Illustrated History,* 145-76, 153.

52. This had started as a house church in 1859, run by the master tailor James Bennetto at his house in Siloh Road, N.L. Thomas, *The Story of Swansea's Villages,* Vol. 1, 58.

53. Ibid., 1, 11.

54. *Report of the Royal Commission on the Church of England and Other Religious Bodies in Wales and Monmouthshire* (London, 1911).

55. Ibid.

56. *Hanes Annibynnol* II, 83.

57. *Report of the Royal Commission on the Church of England and Other Religious Bodies in Wales and Monmouthshire* (London, 1911).

58. W.C. Rogers, *A Pictorial History of Swansea* (Swansea, 1981), 36-7.

59. Peter Howell, Note on 'Swansea - Church of St. John the Baptist', 15 April 1991.

60. Ibid., quoting from B.F.L. Clarke's work on Woodyer (at the Council for the Care of Churches), in a letter from Woodyer to the ICBS of 9 February 1879.

61. Thomas, *Swansea's Districts & Villages*, abridged Vol. I, 13-15.

62. *1851 Swansea Guide,* 15

63. Ibid.

64. Obituary of Miss Elizabeth Mary Grenfell, *The Cambrian*, 16 March 1894, reproduced in Gabb, *The Grenfells*, 31-3, and *The Builder*, 3 July 1886, 39.

65. *The Builder*, 4 October 1890, 272.

66. J. Newman, *The Buildings of Wales: Glamorgan*, (London, 1995), 610, & Thomas, *Swansea's Districts & Villages,* Vol.1, Parts I-III, 15.

67. J.H. Davies, *History of Pontardawe and District* (Llandybie, 1967), 136-7.

68. Ibid., 139-40.

69. Thomas, *Swansea's Districts & Villages*, abridged Vol. I, 52.

70. *Hanes Annibynol,* II, 83.

71. A. Jones, *Welsh Chapels* (Stroud, 1966), p.74.

72. *Swansea Guide, 1851*, 22.

73. Ibid.

74. G. Worsley, *Architectural Drawings of the Regency Period* (London, 1991), 122.

75. *The Builder*, 6 November 1869, 1894; demolished in the 1980s.

76. T. Rees, *Hanes Eglwysi Annibynol Cymru,* II (Liverpool, 1872), 31.

77. Ibid., 84-5.

78. Ibid., 78.

79. J. Thomas, *Hanes Annibynol,* V (Liverpool, 1891), 134-36.

80. The main sources of information on chapels in this section are on the database of chapels in Wales held at RCAHMW, Aberystwyth. Most of the information on architects of particular buildings is drawn from John Pritchard's research, funded by the Board of Celtic Studies, on nineteenth-century denominational literature.

81. Most of the information on Humphrey is drawn from David Farmer, *The Remarkable Life of John Humphrey, God's Own Architect* (Swansea, 1996) 13-21. citing Humphrey's obituary in *The Cambrian*, 10 January, 1888.

82. *Hanes Annibynol,* IV (1875), 494, and V. Hughes and M. Seaborne, 'Welsh Chapel Architects', *Capel Newsletter*, 27, (Winter 1993/94), 13.

83. *Hanes Annibynol,* IV (1875), 437.

84. Middlesborough, West Cornforth and Witton Park chapels are mentioned in *Hanes Annibynol,* IV (1875), 440, 443, 445); Liverpool (pers. communication John Pritchard).

85. Southwark Bridge Road chapel: *Hanes Annibynol,* IV, 392.

86. C.Stell, *An Inventory of Nonconformist Chapels and Meeting-houses in Central England* (London, 1986), 96.

87. There is a short biography of Thomas, from which details are drawn for this account in, J. Thomas, *Hanes Eglwysi Annibynol Cymru,* Vol. 5 (Liverpool, 1891), 133.

88. T. Rees and J. Thomas, *Hanes Eglwysi Annibynol Cymru*, Vol. 2, (Liverpool, 1872), 62.

89. Ibid.

90. Ibid., 65.

91. Ibid., 84.

92. Ibid., 43.

93. Ibid., 81.

94. Ibid., 78-79; Prys Morgan, 'Art and Architecture' in Williams, *Swansea: An Illustrated History,* 177-214, 201.

95. Julian Orbach, building historian, in a lecture at RCAHMW chapels seminar, 17 March 1997.

96. A. Cioci, *A Guide to Assisi* (Valdagno, 1998), 49-50.

97. David McLees and D. Watkin, *The Life and Work of C.R. Cockerell* (London, c.1980), 193.

98. Farmer, *John Humphrey*, 17.

99. Stell, *op.cit.,* 223.

100. Stell, *Central England*, 194 and *South West England*, .53, 140.

101. Stell, *Central England*, 231.

102. J.I.C. Boyd, *Narrow Gauge Railways in North Caernarvonshire*, Vol. 2:, *The Penrhyn Quarry Railways* (Oxford, 1985), 19.

103. v.i. Julian Orbach.

104. *Hanes Annibynol,* II, 78-79; Morgan, 'Art and Architecture', in Williams, *Swansea: An Illustrated History*, 177-214, 201.

105. J. Newman, *Buildings of Glamorgan*, 459.

106. References to the life of John Humphrey are largely drawn from Farmer, *The Remarkable life of John Humphrey, God's Own Architect.*

107. *The Cambrian,* 18 April 1865.

108. Ibid., 10 August, 1866.

109. Farmer, *John Humphrey*, p.28.

110. *Hanes Annibynol*, IV 440.

111. Farmer, *The Remarkable Life of John Humphrey, God's Own Architect*, 33-84.

112. *The Builder*, 26 April 1873, 333.

113. *Hanes Annibynol,* IV (Liverpool, 1875), 462.

114. Hughes and Seaborne, 'Welsh Chapel Architects', *Capel Newsletter*, 27 (Winter 1993/94), 13.

115. Ibid., 11.

116. McLees and Watkin, *The Life and Work of Cockerell.*

117. Stell, *Central England*, 231.

118. Cowley, 'Religion and Education', in Williams, *Swansea: an illustrated history* 145-76, 153.

119. Newman, *Buildings of Wales: Glamorgan*, 85, 458.

120. *The Builder,* 16 October 1869, 832. Swansea references from this source have been kindly supplied by V. Hughes and M. Seaborne.

121. Thomas, *Swansea Districts*, abridged Vol. I, 90.

122. Thomas, *Hanes Eglwysi Annibynol Cymru*, Vol. V (Liverpool, 1891), 133.

123. B. Dean, *Slums, Living Conditions in 19th Century Swansea, Part 3* (Swansea History Project, n.d. *c.*1980), 41.

124. *Hanes Annibynol,* IV (Liverpool,1875), 462.

125. Thomas, *Swansea Districts*, abridged Vol. 1, 52.

126. Farmer, *The Remarkable Life of John Humphrey*, 45.

127. A. Jones, *Welsh Chapels*, 67.

128. *The Builder*, 6 July 1878, 708.

129. *Annibynol Cymru,* II, 82-83.

130. Thomas, *Swansea Districts,* abridged Vol. I, 14-20.

131. *The Hafod: 1920's-1950's* (Swansea, 1996), 21-2.

132. Ibid., 15-20.

133. T. Shaw, *A Methodist Guide to Cornwall* (London, 1991), 18.

134. P. Jenkins, *Twenty by Fourteen: A History of the South Wales Tinplate Industry, 1700-1961* (Llandysul, 1995), 155-56.

135. Thomas, *Swansea Districts*, Vol. 1, 75-78.

136. *The Cambrian*, 3 January 1873.

137. Thomas, *Swansea Districts,* Vol. 1, pp. 79-80.

138. *Hanes Annibynol,* Vol. 1, 207.

139. E.T. Davies, *Religion in the Industrial Revolution in South Wales* (Cardiff, 1965), 18.

140. This and other statistics referring to chapels elsewhere in Wales in this section are drawn from Hughes and Seaborne, 'Welsh Chapel Architects', *Capel Newsletter*, 27 (Winter 1993/94).

141. Most references to building costs in this section are drawn from Vernon Hughes's index of *The Builder*.

142. Thomas, *Swansea Districts*, abridged Vol.1, 53.

143. J.E. Ross, *Radical Adventurer: the Diaries of Robert Morris, 1772-1744* (Bath, 1971), 42, 169.

144. A. Saunders, *The Art and Architecture of London* (London, 2nd ed., 1988), 418.

145. Morgan, 'Art and Architecture', in Williams , *Swansea: An Illustrated History,*. 177-214, 184.

146. J.A. Davies, 'The Cultural Tradition: Writing and Drama', in R.A. Griffiths (ed.) *The City of Swansea: Challenges & Change* (Stroud, 1990), 206-17, 208.

147. Thomas, *The Story of Swansea's Districts & Villages,* Vol I, Parts I-III (Swansea, 1969), 51.

148. Ibid.

149. John Morris I of the Forest Copperworks had William Edwards's sermons translated into English.

150. See J.B. Lowe, *Welsh Industrial Workers Housing, 1775-1875* (Cardiff, 1977): Evans, *Works Schools*.

LANDORE

Fig. 300. The deep valley of the Nant Rhyd-y-Filais in c. 1900, looking east from the spoil-tips of Pentre (Penfilia) Colliery (7). Visible are ruined engine houses of Cwm Pit on the left (12); the headframe and engine house of Townsend's Engine Pit (6) on the skyline to the right; on the right skyline, the casting house of the Millbrook Ironworks (17). The numbers in this and the following caption refer to the those given to water-power sites at Landore in the text. (University of Wales Swansea, Library and Information Services, Archives)

Fig. 301. Mouth of the Nant Rhyd-y-Filais Valley at Landore in 1935. Visible are: Landore Viaduct – on the right; Landore Steelworks (26) at the top right-hand corner; Millbrook Steelworks in the middle foreground (17); the colliery engine house of Townsend's (Callands) Coalpit (6) to the left of them; on the near side of the river, the ruins of the Landore Engine Pit (8) and Landore Copperworks (14); in the foreground, the workers' settlement of Landore. (National Museums & Galleries of Wales – Department of Industry)

The Nant Rhyd-y-Filais Valley at Landore

Introduction

This section examines a particularly densely developed piece of industrial landscape on the west side of the Swansea Valley, midway between Swansea town and the copperworkers township of Morriston.

The steeply-sided valley was formed by one of the tributary streams of the River Tawe. Flowing sharply down to the main river valley it provided the power source in the eighteenth and nineteenth centuries for a great variety of industries. A multitude of coal mines grew up to service them. The Nant Rhyd-y-Filais ravine was the next large stream north of Cwm Burlais; the old northern boundary of Swansea Borough. In the second decade of the eighteenth century a proposal was made to site Swansea's first non-ferrous smeltery in the latter valley. However, it was then considered to be too close to the existing town. So this first works - the Llangyfelach Copperworks, was founded instead at the mouth of the next stream north, the Nant Rhyd-y-Filais.

A large concentration of smelting works subsequently grew up in what are now the northern Swansea suburbs of Landore and Morriston. Several of these works and their ancillary mines, forges and ironworks were powered by this tiny stream with a fall of 575ft: it is with the intensive use and reuse of that stream that the following maps and gazetteer are concerned. The valley also acted as a natural funnel for transport ways both approaching the river and canal and for serving the water-powered installations in the valley and these are itemised in the second part of this section.

Fig. 302. The upper Nant Rhyd-y-Filais Valley in the 1950s prior to the valley being filled with copper-waste from the Hafod Copperworks tip. Visible are: a stone dam which dammed a large pool (Landore Map 3 above 19/21); another stone dam in the background (at 8 on Map 2); in the background, engine house and shaft-tower of Morris's Penfilia (Pentre) Colliery (7 on Map 1). (Fred Cowley)

Fig. 303. Four maps showing the development and changes in the water-economy at Landore in the two-hundred years during which copper-smelting was carried out at Swansea. The first copperworks was the 'Llangyvelach' Works of 1717 (map 1: site 3), which had a spectacular multi-arched aqueduct. In the late eighteenth century, some water resources were taken out of the valley via a long contour-leat (map 2: site 10). At Landore this powered the winders of two collieries and provided water to the Rose Copperworks. In the nineteenth-century, many watercourses fed water for use in large steam-powered installations.

Development of Water-power and Supply Sites at Landore

(numbers relate to maps nos. 1-4)

1. Landore Cornmill

(SS 6620 9593). Probable medieval manorial mill of the Fee of Trewyddfa to 1822; water feeds from Nant Rhyd-y-Filais, probably also from Morris' and Trewyddfa Canals.

2. Kergynnidd Cornmill

(SS 6561 9598). Possible medieval cornmill outside the Fee of Trewyddfa to 1786-1802. Feed from the upper Nant Rhyd-y-Filais.

3. Llangyfelach Copper, Lead and Silver Works

(SS 6613 9595). 1717 to c.1746-51. Feed from the tailrace of Kergynnidd Cornmill (see 2) and from the confluence of the Nant Rhyd-y-Filais and the Nant-y-Gwm-Gelli via a stone-arched aqueduct. Water from the tailrace was returned to the pond of Landore Cornmill (see 1) as required in a legal settlement of 1732.

4. Landore Water-wheel Coalpit

(SS 6665 9599). The coalpit was sunk in 1754 and a water-wheel provided power for winding and pumping until the installation of a Newcomen pumping engine in 1776 (see 9). The water-wheel was still used for winding until replaced by a steam winder in 1803. After 1814 the water-wheel was reused to drive the Glamorgan Pottery flint mill (see 18). The water to drive the wheel came from the adjoining pond of the Landore Cornmill (see 1) which must have had prior claim on the available water, as a horse-gin was also thought necessary for winding at this pit.

5. Reservoir for the Plas-y-Marl (Graig) Coal Pits Engine

(SS 6651 9552). In 1761 winding may have been by horse-gin but there was a Newcomen engine for pumping which was fed by a reservoir supplied by the small adjacent stream.

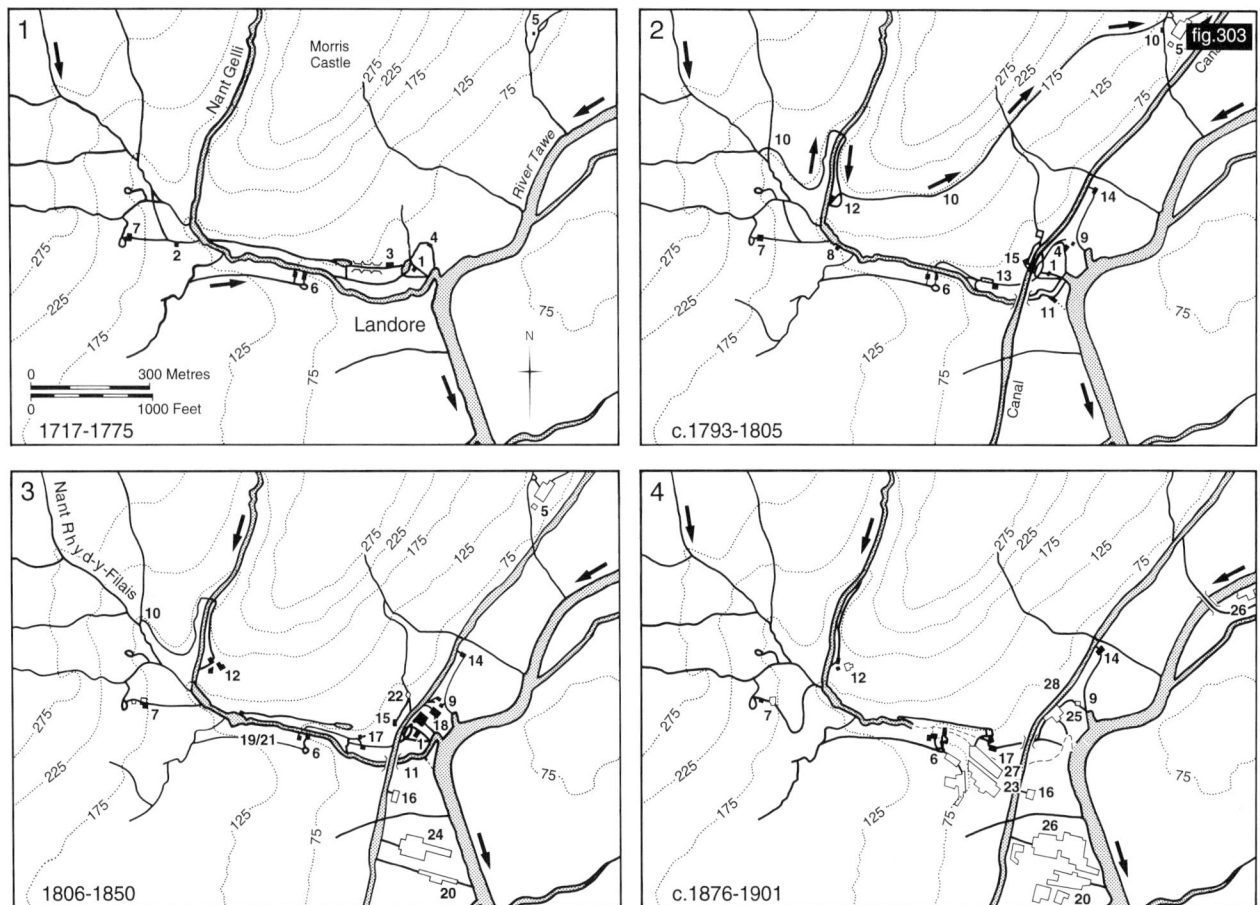

Fig. 304. Pentre (Penyfilia) Pit winding engine house (right) and stone tower supporting the pulley over the shaft (left), pictured in the first two decades of the twentieth century. (Fred Cowley)

Fig. 305. Sketch of Pentre Colliery from the south by the mining engineer T. Bryn Richard at the end of the pit's working life in 1920. (University of Wales Swansea, Library & Information Services, Archives; T. Bryn Richard Collection)

Fig. 306. The dual-pumping winding-engine at Pentre, Landore, had a cylinder 39 inches in diameter. Sketched by the late T. Bryn Richard who came from a prominent local family of mining engineers and surveyors. (University of Wales Swansea, Library & Information Services, Archives, T. Bryn Richard Collection)

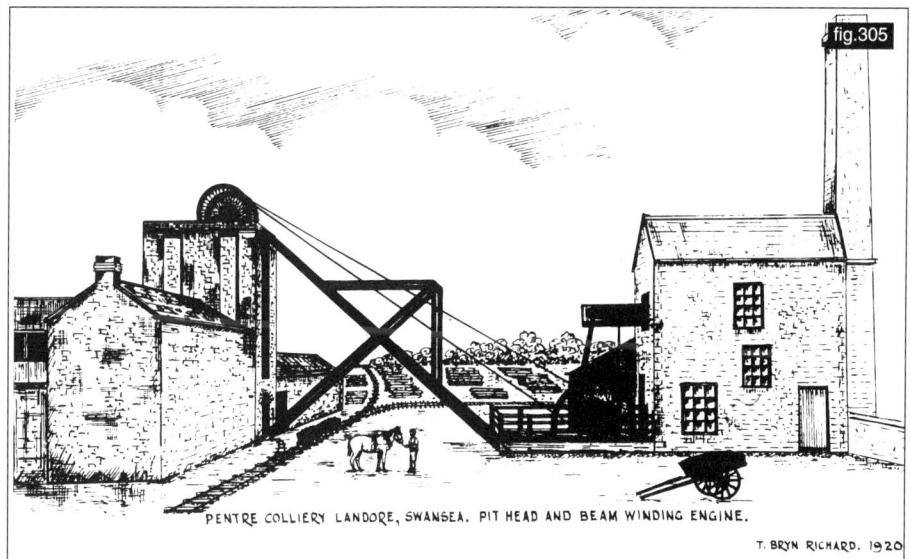

6. Townsend's Engine (Callands) Coal Pit Leat

(SS 6590 9588 to 6592 9588). Chauncey Townsend's Engine Pit was sunk in l762 [1], and by l768 it had a steam pumping-engine and probably a water-powered winder. The pit was abandoned c.l780, but re-opened soon after l800. In l858 a Cornish beam pumping-engine was installed in a new engine house at what was then only probably a pumping installation which closed in c.l926. Water came from a leat fed by a minor southern tributary of the upper Nant Rhyd-y-Filais. From perhaps c.l786 the drainage water from the Penyfilia Level was fed into this.

7. Penyfilia (Pentre) Pit Leat

(SS 6543 9605 to 6545 9600). This pit was probably working before 1748 and was certainly deepened in 1770. It may have had a water-engine. A steam

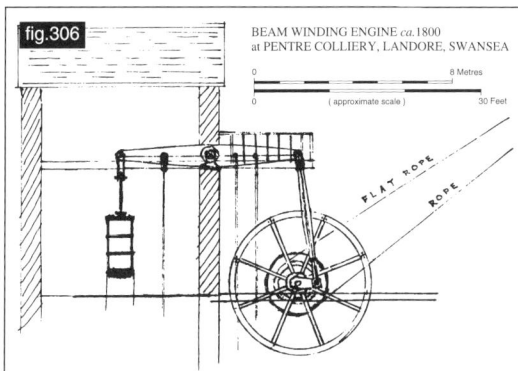

winding-engine was probably installed by Lockwood and Company c.l800-06 fed by water from a minor tributary of the upper Nant Rhyd-y-Filais. The pit closed in l926.

8. Chaff Mill

(SS 6567 9598). Possibly belonging to the adjacent Alltyscrach Farm. It was built between 1768 and l802, re-using the old Llangyfelach Works pond (see 3) below Kergynidd mill which had been redundant since c.l746.

297

9. Landore Engine Pit

(SS 6627 9599). A Newcomen engine was installed in 1776 but the pit was drowned in 1794. In 1800 a new large cylinder 25 h.p. Boulton & Watt engine was installed for pumping. In 1803 this was complemented by a Trevithick high-pressure steam winding-engine with an 8in diameter cylinder. It discharged exhaust steam directly into the atmosphere and so required no condensing water. Consequently water intake was only 5% of that required by a similar condensing engine. Water for the boilers (and the earlier condenser) was almost certainly drawn from the Landore Cornmill pond (see 1) and from the adjacent canals.

10. Plas-y-Marl Coal Pit Watercourse

(SS 6550 9618 - 6651 9552). A long feeder leat was built, possibly in 1775, from the Nant Rhyd-y-Filais more or less along the 175ft contour along the southern flank of Graig Trewyddfa to a new reservoir at Plas-y-Marl Colliery (see 5) to bring water to the new water-wheel winder. A winder was subsequently driven by the atmospheric engine pumping alongside and in 1785 a Boulton & Watt steam winder replaced it, fed by a local source. A probable water-wheel winder and pump at Cwm Pit (see 12) was fed from an intermediate point on the leat by c.1781. In 1803 a steam pump was built at Cwm Pit but winding was probably still carried out by water-power until later in the nineteenth century. The contour leat later ran no further than Cwm Pit after the complete conversion of Plas-y-Marl Pit to steam power c.1801 but the Rose Copperworks used the tailrace water of that pit from 1793 until possibly 1825 when water was drawn from the Trewyddfa Canal instead. Between 1852 and 1876 the leat was

truncated again, probably by the collapse of a timber aqueduct over the Nant-y-Gwm-Gelli, and a new supply channel was cut from that stream. This fell into disuse between 1876 and 1897.

11. Nant Rhyd-y-Filais Outlet Diversion

(SS 6623 9587-6626 9585). The outlet was probably diverted to the south of William Edward's New Coal Bank wharves during his construction of the first Landore Dock between 1772 and 1794. Most of the old course was left culverted as it now served as the scouring sluice for the New Dock.

12. Cwm Pit Water-wheel Winder

(SS 6567 9609). Situated on the Plas-y-marl coal pit watercourse (10), it was operational from about 1781 to between 1875 and 1897.

13. Landore Copperworks Stamping and Rolling Mill

(SS 6697 9588). This mill, in use from 1793 to c.1807, was supplied with water by a leat tapping the Nant Rhyd-y-Filais at a point well below the Chaff Mill (see 8). In alternative use by 1814 (see 17).

14. Landore Copperworks, water for granulating copper and crushing ore

(SS 6632 9612). A small feed led off the Trewyddfa Canal (formerly Morris' Canal) into a large building. The works was built in 1793 and closed in 1876.

15. Landore Forge

(SS 6617 9597). By 1793, possibly a blacksmiths forge or an adaption of the 'smiths shop' at the Llangyfelach Copperworks (3). It may have been absorbed by the later industrial-scale forge (17).

16. Tir Landore (Drew's Pit) Colliery Engine

(SS 6618 9584). A steam winding-engine was probably constructed by 1813. Chauncey Townend's other 'Engine Pit' (Callands) probably pumped this pit as well. The colliery was finally closed sometime between 1876 and 1897. Water for the boiler and condenser of the engine was probably drawn from and part returned to the neighbouring Swansea Canal.

17. The Water-wheel and Steam-engine of the Nantrhydyvilas Air Furnaces; later Landore Forge and Millbrook Ironworks

(SS 6697 9588). The history of the site is particularly long, the ironworks having been in production from 1814 until 1966 - the water-wheels in operation in 1846-8 are marked above 19/21 on the plan. The forge of the original works had been converted from that of the Landore Copperworks (13). It had two tilt hammers and a water-wheel of 18ft diameter and

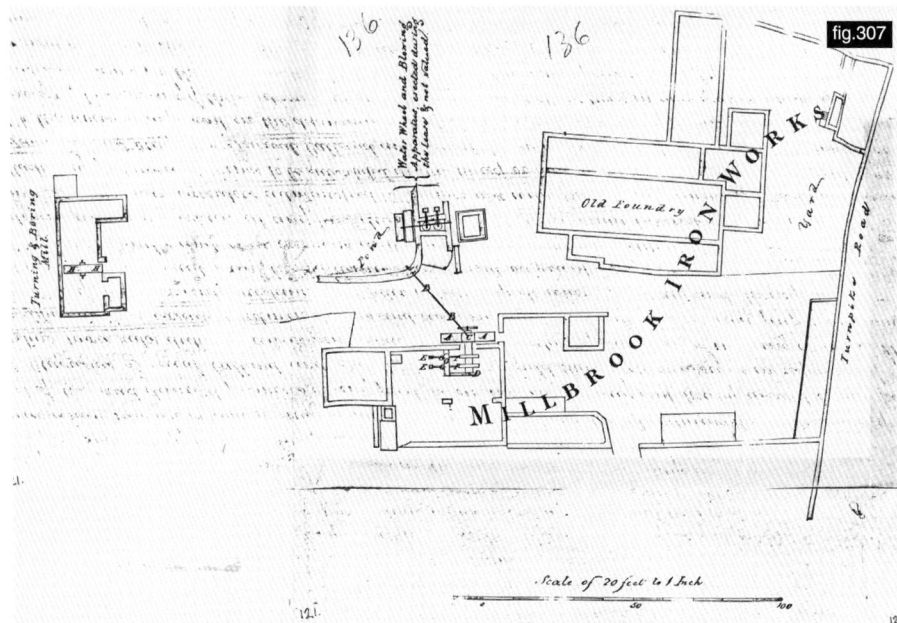

Fig. 307. Plan of Millbrook Ironworks (17 on Landore Maps 3-4) in 1846-48, showing at upper centre a water-wheel driving two blowing-cylinders for a cupola furnace. Also visible are a water-pipe (B); the second water-wheel (A); two trip-hammers (F-G); the Turning & Boring Mill, driven by a water-wheel (H). (The National Library of Wales)

continued in use until after 1897. This building became the nucleus of an integrated iron-making and iron-working concern established as the Air Furnace Company by the partnership of William Bevan and Sons in 1814. The first process was an attempt to extract the high iron and copper content of copper-slag through the use of reverbatory furnaces. The iron at least was then used in a foundry for casting both iron and brass, and the forge was used to produce the final products, edge tools and shovels. The iron obtained from the slag was reputed to be 'unweldable' and therefore a blast furnace was built with a 10ft. wide blast wheel. A new and higher leat was required to drive this wheel.

After the Bevans' bankruptcy during the deep depression in the iron trade of 1830 the works were bought by Sir John Morris II, Henry Habberley Price and William Price Struve, and its furnace was still operational until 1848-1854 and a second forge used the former blast-wheel to provide power from before 1876 until after 1897. The works also used steam-engines from at least 1830.

18. The Glamorganshire Pottery Flint Mill
(SS 6665 9599). Built in 1814 and probably in use until 1841, it may have had auxiliary steam power from 1819 onwards. It reused the wheel of the Landore Water-wheel Coal-pit (see 4) and thus acquired a dual feed from the Landore Cornmill pond and the Trewyddfa Canal. In 1817 and 1818 the use of the canal water seriously impaired navigation on the Swansea Canal and the canal committee ordered the pottery to stop taking its water.

19. Furze-Mill
(SS 6577 9597). Built between 1813 and 1826 to produce fodder from the prickly gorse on adjacent commonland, by 1848 the building had been reused as the Millbrook Ironworks turning and boring-mill (see 21) and its water-wheel possibly retained. The furze-mill was situated near an intake from the Nant Rhyd-y-Filais and had a long tailrace. It could have used all the available head of water between the chaff-mill (see 8) and the Landore Ironworks complex (see 17).

20. Landore Chemical (Acid) Works
(SS 6627 9557). Probably built c.1826-8 and certainly before 1838, it processed sulphates supplied by the owners' Rose Copperworks. Water for manufacturing sulphuric acid and for lixiviating (or dissolving) tanks was probably supplied by a pumped feed from the River Tawe. An illegal gravity feed was laid from the Swansea Canal but was stopped up in 1854. The works continued in use into the twentieth century.

21. Millbrook Ironworks Turning and Boring Mill
(SS 6577 9597). Possibly converted from the furze-mill (see 19) between 1842 and 1848, and demolished before 1876. It was powered by a cast-iron water-wheel 13ft in diameter and 30in wide, probably inherited from the furze-mill. Millbrook Iron Works was casting iron wheel segments by 1839 and probably supplied this particular wheel.

22. Calland's Landore Pottery
(SS 6622 9598). A pottery produced earthenware here between 1848 and 1856. In 1863 the disused premises were converted into the Little Landore Copper Works (see 25). The pug-mill (and perhaps the flint-mill) was probably driven by water from the old Landore Cornmill pond (see 1) and from the Trewyddfa Canal. The former supply may have been seriously depleted by the feed for a furnace at the Millbrook Ironworks (see 17). So much canal water was being used by the pottery by 1854 that navigation was seriously impaired. The Swansea Canal committee complained to the Duke of Beaufort who, as owner of the Trewyddfa Canal, may have considerably reduced the water-supply to the pottery so contributing to the shortness of its working life. There was still some illegal extraction taking place in 1856, immediately prior to closure.

23. Landore Tinplate Works
(SS 6603 9586). Opened in 1851 and closed in 1897, it was the first manufactory at Landore reputed to be totally dependent on steam-power. Ironically, it was situated over the stream that had given rise to a whole complex of earlier water-powered industry. It was also said to have been the first works in the British tinplate industry to be driven by steam. Seven stands of rolls were in operation by 1876. Condensing and boiler feed water was almost certainly extracted from the Nant Rhyd-y-Filais. A large pond was created just upstream of the works, which might suggest that some water-power was in fact used in the works, especially as the works seems to have reused a rolling-mill or cast-house building from the neighbouring water-powered ironworks which was at least partly under the same Morris family ownership as the tinplateworks.

24. Landore Silver and Spelter (Zinc) Works
(SS 6617 9564). It operated from 1853 to 1867 and was then converted into the first Siemens steel works in the area (see 26). It probably had at least one steam-engine, supplied with water from the canal.

25. Little Landore Arsenic and Copper Works

(SS 6622 9598). Operated from 1863 until 1896. The Landore Arsenic and Copper Company took over the then fairly modern buildings of the Landore Pottery (see 22) and the copper ore crushing mill may have used the wheel of the former pug-mill.

26. Landore Siemens Steel and Tube Works (Swansea Hematite Works, Mannesmann Tube Works)

Fig. 308. The Landore Steelworks in the late nineteenth century when it was one of the four largest in the world. The blast furnaces are visible standing against the South Wales Railway embankment on the right. (University of Wales Swansea, Library & Information Services, Archives)

(SS 6617 9564 & 6680 9615). This works was opened as an experimental plant in the converted buildings of the Landore Silver and Spelter Works in 1867-69 by a company chaired by L.Ll. Dillwyn, founder of the silver and spelter works. By 1870 it was producing steel rails. In 1871 a second works was opened to the east of the River Tawe, with two blast furnaces and extensive mills. By 1873 it was one of the four largest steelworks in the world, with a weekly output of a 1,000 tons of steel. In 1888 production of steel rails and plate ceased and the Mannesmann Tube Company (with the Siemens operating a franchise) converted large areas of the mills into production lines for seamless steel tubes. The blast furnaces and steel making plant were taken over by the Swansea Hematite Company in 1899. It operated as a foundry until 1980; tube production ceased in 1961. There were 64 steam-engines at work in the various mills. The large reservoirs on the north-east of the site with their steam pump house were fed by a long 15in diameter pipe from the Swansea Canal, and were probably also fed by gravity from the nearby Nant-y-Ffendrod that ran south from Llansamlet village. Later the feed pipes from the canal specifically fed cooling tanks on the roof of the Mannesmann Tube Works.

27. Copper House Chemical Works

(SS 6613 9587). This small works was built between 1838 and 1876 and converted into the boiler-makers' buildings of the Millbrook Ironworks by 1888. It was sited on the bank of the Trewyddfa Canal and almost certainly made use of its water.

28. Copperhouse Foundry

(SS 6621 9607). Built between 1838 and 1876 and still in production in 1897. It probably took water for the steam-engine of its machine shop from the adjacent Trewyddfa Canal.

Fig. 309. The blast engine house (left) and charging-bank (right) of the Landore Steelworks in the early 1980s after the closure of the last foundry in 1980 and before the clearance of the site.

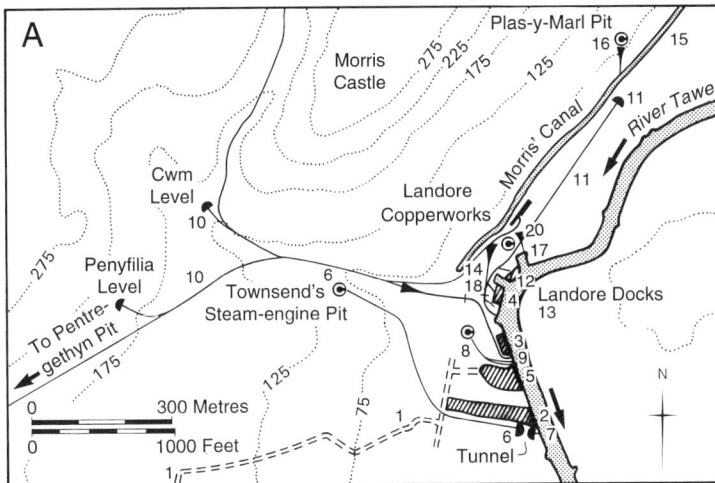

A

Plas-y-Marl Pit

Morris Castle

Cwm Level

10

Penyfilia Level

10

To Pentre-gethyn Pit

Townsend's Steam-engine Pit

Landore Copperworks

Morris Canal

River Tawe

Landore Docks

Tunnel

0   300 Metres
0   1000 Feet

1702 - 1793

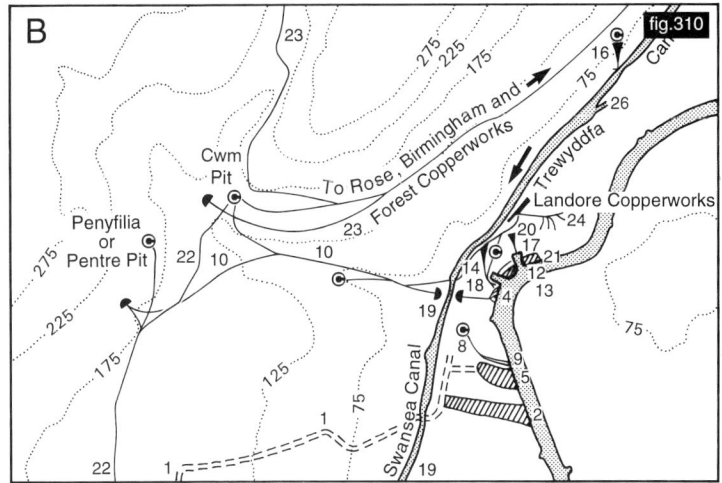

B   fig.310

Cwm Pit

Penyfilia or Pentre Pit

To Rose, Birmingham and Forest Copperworks

Trewyddfa

Landore Copperworks

Swansea Canal

1803 - 1829

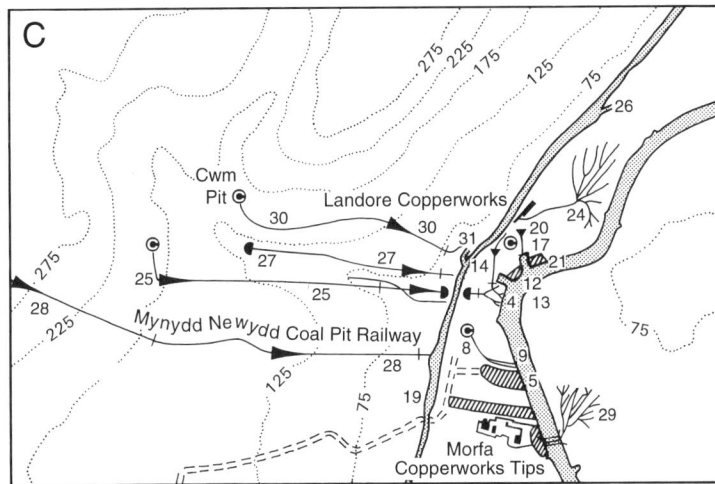

C

Cwm Pit

Landore Copperworks

Mynydd Newydd Coal Pit Railway

Morfa Copperworks Tips

1830 - 1846

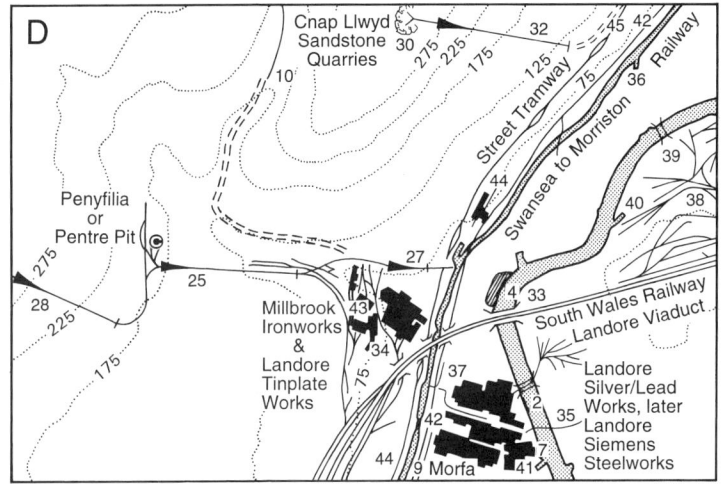

D

Cnap Llwyd Sandstone Quarries

Street Tramway

Swansea to Morriston

Railway

Penyfilia or Pentre Pit

Millbrook Ironworks & Landore Tinplate Works

South Wales Railway

Landore Viaduct

Landore Silver/Lead Works, later Landore Siemens Steelworks

Morfa

1867 - 1901

Key

Railway Incline
Top   Bottom

⊚   Coal Pit

Tunnel entrance: singly this represents a coal level

The Development of Transport at Landore

(numbers relate to maps A-D)

1. Heol y Glo (i.e. 'Coal Road')
(SS 6557 9548 to SS 6612 9558). This was possibly constructed in 1702 westwards from Thomas Price's wharf ('2') towards the Pentre Colliery. All three were certainly in Price's hands by 1745. This track may have become a public highway by 1768. By this time Price had built a graded track from Pentre Colliery through Cwm Burlais to his Swansea river wharf and Swansea town itself. This was later converted into a railway. The wharf which Heol y

Glo served ('2') probably still operated in 1838 but was built over in 1853 in order to construct the Landore silver and spelter works ('24'). Doubtless coal-carts still used this public right of way.

No other roads are described since none, whether turnpike or track, are known to have been constructed specifically to give access to linked industrial installations. However, routes to neighbouring villages and towns were obviously utilised to some degree by transport from small windlass pits and adits. The first turnpike road through Landore was adopted in 1764, but the only major turnpike improvement at Landore took place as an incidental consequence of major canal construction in 1794-96.

Fig. 310. Four maps illustrating how transport evolved for two hundred years around a side valley in the copper-smelting area. Map A: from the mid eighteenth century, four short wooden-tracked railways (6,8,10 & 11) were taking coal to be shipped on the river; the building of the Landore Copperworks (1793) resulted in one main valley railway (10). Map B: the break-up of the Lockwood, Morris & Co. smelting and mining partnership gave the impetus to construct railways north and south of the valley. Map C: the continued fragmentation of local mine ownership, and the building of new, large-scale copper smelters at Hafod and Morfa resulting in a multiplicity of short parallel lines. Map D: the later rationalisation of lines in the valley resulting from a new common ownership under the proprietors of the Hafod and Morfa Copperworks.

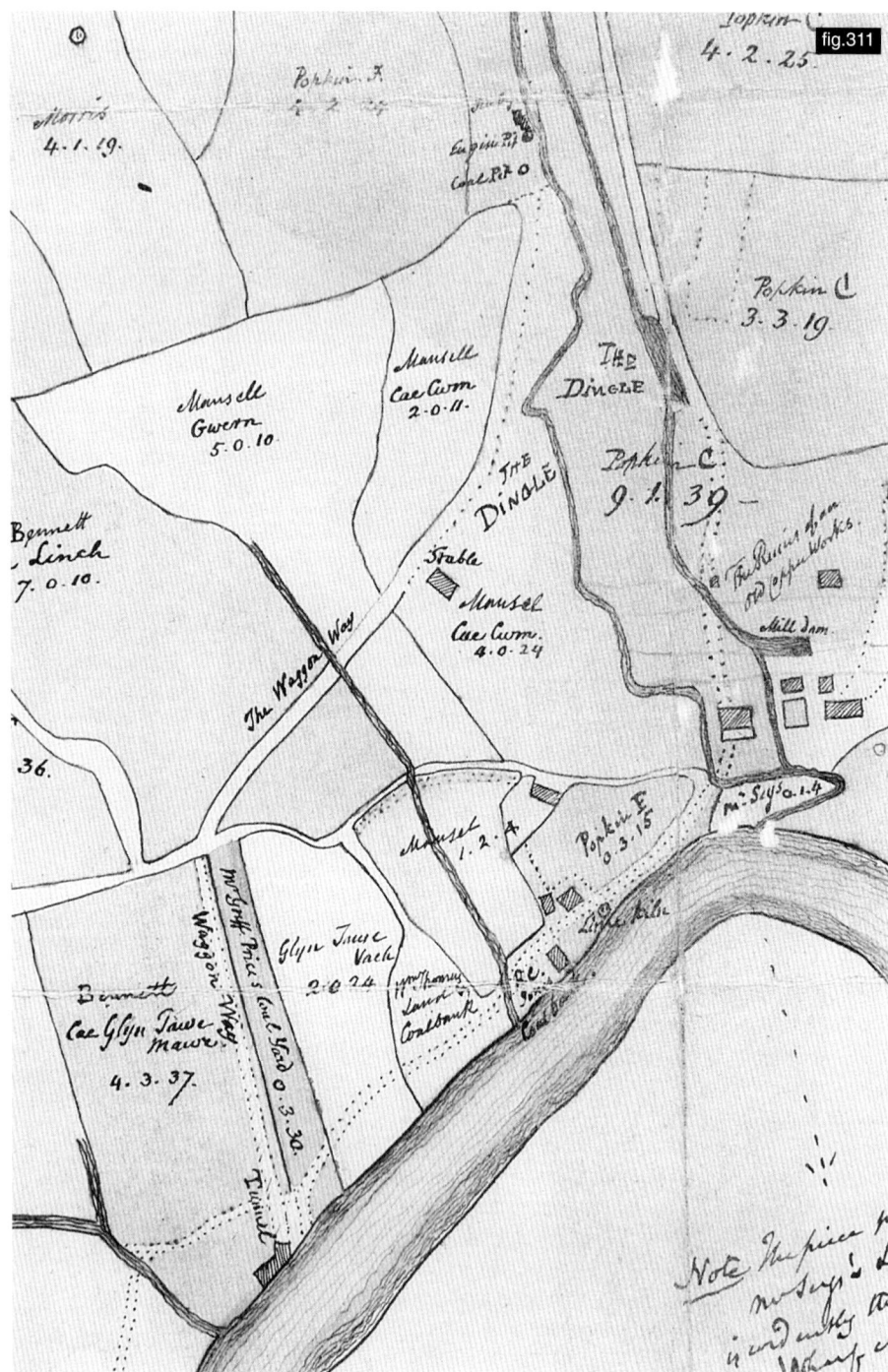

Fig. 311. Map of 1768 showing the course of Chauncey Townsend's wooden-railed Steam-Engine Pit Railway ('Waggonway' – 6 on Landore Map 1) with its tunnel and stable. Also, the ruins of Llangyfelach Copperworks (3 on Landore Map 1) to the right with one of its ponds above and the millpond of the Trewyddfa Cornmill (1 on Landore Map 1) below (The National Library of Wales)

2. Price's Wharf and Coal Yard

(SS 6633 9560). This 'Coal Place' was built or re-leased in 1702, later used by the Landore Silver and Spelter Works ('35') and the Siemens Steelworks ('37').

3. Copper House Wharf

(SS 6626 9579). The medieval collieries under the adjacent crag of Craig Trewyddfa may have shipped their coal from a quay here. Part of this wharfage (cottage and limekiln) was almost certainly re-let in 1710-11 to Thomas Price, squire and coal-owner of Penlle'r-gaer who had mined coal under the Fee of Trewyddfa since 1695 and under other nearby tenements since 1672-73. In 1713 a further part of the adjacent wharfage was let to Thomas Price: 'Two messuages [i.e. dwellings and attached land] and a parcel of lands called Place y Gloe ['Coal Place'], and a dock'. The dock was a longitudinal quay-wall, not an enclosed basin, for the 'landing of limestone and shipping of coal'. By 1717 building materials for the new Llangyfelach Copperworks were coming up to this wharf. Judging by the wharf's nineteenth-century name of 'Copper House Wharf', it must have become a trans-shipment point for copper and copper-ore. By 1768 the southern half of the wharf was again a coal bank, probably served by a railway by 1770 ('10'). The wharfage was probably in disrepair by 1876.

4. Seys Wharf

(SS 6627 9588). In 1672 and 1673 Richard Seys leased four veins of coal under large areas of adjacent land in partnership with Thomas Price. He probably shipped coal from this marshy bank.

By 1769-1771 John Morris' Plas-y-Marl railway ('11') terminated on a reasonably constructed quay-platform here. In 1772-74 the 'New Key' at Landore was constructed, probably immediately north of the earlier quay with its frontage at the low-water mark. It was built of rubble masonry and must have been erected behind a coffer dam. Work was directed by William Edwards, builder of Pontypridd Bridge, and John Morris I considered that it showed a considerable structural advance over the earlier river quays: possibly it was the first on the river built of stone bonded with the local Aberavon hydraulic mortar, rather than the earlier type of quay consisting of an earth platform revetted with timber-piling. The quay here may still have been in use in 1897.

5. William Thomas' Wharf

(SS 6631 9567). This coal-bank was in use by 1768 and probably continued to function until the Landore Silver and Spelter Works ('35') partly covered it in 1853. Coal must have been brought to the wharf by carts travelling on the public highways.

6. (Townsend's Steam-) Engine Pit Railway

(SS 6593 9593 to SS 6635 9555). Chauncey Townsend had built this short line soon after 1762 with a lineside stable (SS 6605 9579). The railway led to a low-level quay by the River Tawe through a shallow tunnel (SS 6631 9557) about 44 yds long.

The railway was probably used until the engine pit closed (see '6') in *c.*1780. John Morris bought the pit sometime in the early nineteenth century and connected it by a branch-line to the head of the Cwm Level Railway incline ('10') under the Trewyddfa Canal.

7. (Townsend's Steam-) Engine Pit Wharf

fig.312

(SS 6635 9555). Built for the above. A quay was built on its site to serve Landore Chemica (sic)/Alkali Works. It was probably in use from *c*.1825 until the twentieth century and may have incorporated remains of the earlier quay. This quay is still in existence.

8. Tir Landore/Glandwr (later Drew's) Pit Railway

(SS 6618 9584 to SS 6629 9569). This was built by Chauncey Townsend between 1768 and 1770. The railway was still in operation in 1827 but had probably closed by 1838.

9. Tir Landore (or Glandwr) Pit Railway Coal Banks

(SS 6629 9569). A slight quay carried the south arm of the Tir Landore Pit Railway and may have occupied a small northern piece of what was William Thomas' wharf ('5'). The site was empty by 1838.

10. Cwm Level (or Mr. Jones') Railway

(SS 6560 9610 to SS 6628 9572). This line had a particularly long and complicated history. The main line of the railway, which survived until the mid-nineteenth century, ran 800 yds along the valley at Landore from the outcropping of the Swansea Five-Foot seam to a wharf on the navigable river via a substantial incline (see Map A). The many connecting branches are shown on the successive maps of transport at Landore. The railway was originally built between 1768 and 1770 by Calvert Richard Jones. He was one of the last of a long tradition of lower Swansea Valley landowners to develop his own coal resources before increasing costs necessitated the injection of outside capital. Even so, Jones over-reached himself in mortgaging his property and in 1791 his son was unable to pay the interest on the loan. John Morris probably bought the railway then and built a short branch south of Jones' Cwm Level to his own level at Penyfilia. The branch may have used a line to Jones' Penyfilia Pit (see Map A) Jones' Cwm Level was disused by 1796. Most of the Cwm Level line was reused in a longer line which ran north

to the Tre-boeth Level up Cwm Gelli (see Map B). That longer branch was itself abandoned after 1805. The branch going south-west to Penyfilia level was considerably extended to a new pit on Wig Farm. Along the scarp in which the Swansea Five-Foot seam outcropped, the line also reached the original Pentre colliery, owned by Lockwood, Morris & Co., although it was probably already connected to navigable water at Swansea (see the entry on the Pentre-gethin to Swansea Railway). The new line, however, was almost certainly designed to bring coal from the Pentre Colliery both to the company's smelter at the new Landore copperworks and to the three smelteries to the north accessible via Lockwood, Morris & Co.'s Canal ('15' on the map). The self-contained industrial complex in the lower Swansea Valley was scarcely affected by the new trunk canal and the railway and canal intersected twice with no apparent connection (see Map B). One crossing was a canal aqueduct and the other was a swing bridge carrying the railway. The line south-west to Pentre colliery was abandoned when the lease on the colliery was surrendered in 1795, soon after Lockwood, Morris & Co. stopped copper-refining and channelled their individual enterprise solely towards coal production. By 1803 Cwm Pit, at the upper end of the Landore valley, was enlarged and a short branch line built to connect it to the railway which ran to the Landore wharves (Map B). After Townsend's steam-engine pit had been acquired by John Morris in the early nineteenth century, it was connected to the head of the incline by a short branch (see Map B). New patterns of trade were emerging as Map B clearly illustrates. New and separate lines were taking coal both north and south from the area of outcropping in the upper valley as the collieries were no longer subservient to the interest of the old firm of Lockwood, Morris & Co. By 1806 a line ('22') was taking coal to the deeper wharves at Swansea which were doubtless more convenient than the relatively shallow wharfage at Landore. The growing copper-smelters between Landore and Morriston also attracted a direct gently graded connection ('23') from the upper Landore valley. Previously coal had been transported to them by the Trewyddfa Canal ('15'). By the period of 1830-1846, represented by Map C, a further revolution in the pattern of industrial development had taken place. The great new copper-smelters, Hafod and Morfa, of the second and third decades of the nineteenth century had been developed on the marshlands of Hafod Farm just south of Landore. A spur was built to connect the head of Mr. Jones' old railway with the canal. This waterway was then funnelling the output from four railways in the Landore area to fuel the new large-scale smelting industry. The upper part of Mr.

Fig. 312. The remains on the site of Chauncey Townsend's Steam-engine Pit Wharf pictured during the creation of playing-fields over spoil-tips covering the railway tunnel of c. 1762. The brick wall on the nearside represents the original boundary with 'Mr. Griff. Price's Coal Yard' against which the railway ran. (960196/1)

Jones's railway, then serving Cwm pit, was partly or wholly replaced by the new Pentre Pit railway ('Y') in 1830. Certainly by 1848 only the incline down to the river wharfage was retained from Mr. Jones's original railway. The incline probably carried only a lesser and minor trade which had disappeared by c.1876, in the period of Map D.

11. Plas-y-Marl Railway
(SS 6653 9630 to SS 6627 9558). This was built for Sir John Morris (and probably his then partners in Lockwood and Co.) in 1769-1771 and abandoned before 1813. The detailed costing for the 567 yards surface length of this railway survive (see chap. 2).

12. Landore Upper Coal Bank
(SS 6630 9592). This was probably built by Lockwood, Morris & Co. before 1771 or 1796 under the probable superintendence of William Edwards. The 'Upper Coal Bank' was probably upstream of 'Seys Wharf' ('4') and between the two river-docks.

13. Landore Colliery Dock
(SS 6627 9593). This dock was constructed between the extensive 'New Key' building activity of 1772-74 and 1793, almost certainly by Lockwood, Morris & Co.

14. Landore Fire-engine Pit Railway
(SS 6627 9599 to SS 6625 9586) Map 3 only). This short spur was built by Lockwood, Morris & Co. before c.1793. It almost certainly terminated on the river wharf of the Plas-y-Marl Railway and may have been a branch of that railway from the beginning. After 1791 and certainly by 1793 it also connected to the Cwm Level Railway ('10'). In about 1800 the railway (probably an inclined-plane) was superseded by a spur of track at the head of the Trewyddfa Canal incline (see '18').

15. Morris' Canal
(SS 6628 9597 to SS 6698 9733), later the Trewyddfa Canal (SS 6613 9586 to SS 6716 9779): see a full history in chapter two. It was used until about 1930 and the 1980s line of the Plasmarl by-pass is built over its line.

16. Plas-y-Marl Coal Pits/Graig Coal Pit Railway
(SS 6652 9653 to SS 6652 9638 shortened to SS 6652 9648). A railway some 400 yds long had connected the early (c.1730) Plas-y-Marl pits to John Morris' wharves on the river. It was truncated in 1787-91 by Lockwood, Morris & Co. to communicate with Morris' Canal, their new waterway. The railway from the pit-head may have been replaced between 1793 and 1822 when the new

Graig coal-pit and its railway were built to the north of Plasmarl.

17. Landore Copperworks Dock
(SS 6631 9596) (Maps A-C). This long, narrow tidal dock may have been built by Lockwood, Morris & Co. in 1793 to receive incoming copper-ore at their new Landore Copperworks. Probably only the south side of the dock and the adjacent river quay were originally used for unloading. By 1826 the quay extended north of the dock entrance. The Landore Copperworks, and presumably its dock, became disused in 1876.

18. Trewyddfa Canal Inclined Plane
(SS 6624 9601 to SS 6623 9589). Built in 1794-96, this was part of the works associated with the Trewyddfa Canal. It only cost around £501 to construct, the same as a contemporary bridge on the Canal. It was a double-tracked self-acting inclined plane, the second or third to be built down to the main wharfage at Landore. It would have enabled Lockwood, Morris & Co. to avoid the tolls on the southernmost section of the Swansea Canal by shipping their coal at Landore directly into 200 ton sailing ships and smaller river barges. These could transfer their cargoes onto sailing vessels with a burden of up to 600 tons downstream. Its relatively early demise suggests that it was easier to take coal directly for loading onto larger vessels nearer the rivermouth.

19. Swansea Canal: southern section
(SS 6613 9586 to SS 6578 9339) (Maps B-D). The southern section was built in 1794-96 and led down to the river wharves at Swansea which were accessible to 600-ton sailing vessels. A double mitred gate at the junction of this part of the Swansea Canal with the Trewyddfa Canal presumably guarded each canal against bursts on the other's section and allowed independent draining. The gates did not function as a stop lock: the Swansea Canal's southern section had no independent supply of water. This led to much concern over the Trewyddfa Canal's sale of water to industrial concerns for driving machinery and condensing engine-steam.

20. Landore Copperworks Inclined Plane
(SS 6630 9598 to SS 6629 9601) Maps A-C. The main function of this, the third inclined tramroad at Landore was to convey copper ore up into the works from the Copperworks dock and would have been built in 1790-93. There was no steam-engine recorded at the works so the plane may have been water-powered. The plane was still working in 1876 when the Landore Copperworks closed down.

21. Copper-ore Wharf for the Landore Copperworks

(SS 6633 9595). A new wharf had been constructed northwards from the copper-ore dock ('17') by 1826.

22. Penyfilia Pit Railway

A line from the upper part of the Landore valley was constructed in 1806 to connect the Penyfilia Pit with the eighteenth-century railway that ran down Cwm Burlais to the Swansea shipping wharves and its head can be seen as '22' on Map 5. There were two reasons for preferring this route to the shorter west-east route down the Landore valley. Lockwood, Morris & Co. had sold the Landore copper smelter so did not need the Penyfilia pit to provide fuel for it. They, therefore, sold the pit to William Bevan & Son of Morriston, who built the Penyfilia pit line in order to carry their coal to the best market at the deep-water wharves at Swansea. John Morris promptly connected his Cwm Pit at the head of the Landore valley to the new line (branch 22, on Map B). By 1830 John Morris II had built a new line '25' (see Map C) from Penyfilia Pit (now renamed Pentre Pit) to connect with the Swansea Canal. This exploited

the demand for coal created by the very large new copper smelters, just south of Landore.

23. Cwm Pit, Tre-boeth and Cwm Level Railway, to Rose (SS 6683 9669), Birmingham (SS 6691 9681), and Forest (SS 6704 9715) Copper Works.

The railway, built after the Penyfilia Pit line, also ran from the head of Nant Rhyd-y-Filais valley but this time in a northerly direction. It again owed its origins to the demise of the Lockwood, Morris & Co. smelting concern. It was in the interests of the former partners, now solely involved in coal-mining, to supply coal as cheaply as possible to the smelters in the one and a quarter mile section of the Tawe valley from Landore to Morriston. The railway was built between 1813 and 1826 by Morris's Penvilia Vein Company. Interestingly, this link to the smelters was preferred to use of the Penvilia Vein Company's rights to toll-free navigation northwards along the Trewyddfa Canal ('15' on the maps). This is probably explained by the trouble and costs of trans-shipment that such use of the waterway would have implied. The Lockwood, Morris partnership lacked

Fig. 313. Cwm Coalpit (12 on map 2) and Morris Castle in c. 1900. The lower terrace on the left of the colliery buildings is the former watercourse (map 2, feature 10) powering the colliery water-wheels, and the higher is that of the railway formation from Tre-boeth Level to the Rose, Birmingham and Forest Copperworks (23 on map B). (University of Wales Swansea, Library & Information Services, Archives)

fig.313

the same rights to lay lines southwards as they did to the north. The Swansea Canal Company and local landowners would have defended their interests by preventing any such schemes by the partnership and this may explain the absence of similar lines along the southern section of the Swansea Canal to the later and much larger smelters of Hafod and Morfa. The new railway ran south from the Tre-boeth level (SS 6570 9685), north of the map area, skirted the northern flank of Nant Rhyd-y-Filais ravine, and headed north to the Rose Copperworks, the nearest smelter. Tre-boeth level ceased supplying coal between 1826 and 1838 and the railway had been replaced before 1832 by a new branch from the old Cwm Level. John Morris II, who still worked Cwm Pit, had a short branch constructed to link this pit to the railway. By 1832 the line had been extended northwards at the eastern end to the Birmingham Copperworks, and by *c*.1844 on to the Forest Copperworks. The line was joined at the copperworks by a similar railway running southwards and bringing coal from Clyn-du and Graig Pits in the Morriston area. The whole of the railway south of the Copperworks was out of use by c.1844. By *c*.1844 (see map D) the Tre-boeth level had been re-opened and a short length of railway relaid to connect it to a small wharf ('10'on Map D) equipped with a chute to facilitate the local sale of coal. After 1852, but before 1875, this line was disused, but most was again relaid and used in the period 1901-1914 as an outlet for the newly opened Cwmgelli Level.

24. **Landore Copperworks Tips Railway**
(SS 6629 9605 to SS 6645 9608) This may well have been in operation from the building of the copperworks in c.1790. The Landore Copperworks and, of course, its railway became disused in 1876.

25. **The Pentre Coal-Pit (previously the 'Penyfilia Pit') Railway**
(SS 6549 9593 to SS 6614 9588). The Pentre Coal-Pit railway was constructed by Sir John Morris II, William Price Struve and possibly Henry Habberley Price in 1830. It was a very long shallow incline that ran east down Nant Rhyd-y-Filais to connect with the canal. It also linked with the head of a line following Jones' old railway incline ('10') to the river wharves at Landore. The construction of the line was prompted by changing trade patterns. The new line superseded the earlier Penyfilia Pit railway (see 22) down to the deep-water shipping wharves at Swansea. Now coal was taken the short distance directly eastwards to the Swansea Canal at Landore in order to provide fuel for the copper smelters at Morfa and Hafod. After

34 years the trunk canal had directly affected the pattern of the transport infrastructure at Landore. The Swansea Canal Act had ensured that the Landore-Morriston area would be an independent transport enclave, but south of Landore the Canal Company was able to prevent railway connections over the main canal which linked coal-pits and smelters directly. The incline to the river ('10' on Map C) passed through Landore Ironworks and the connection to the Swansea Canal skirted the Works. This lease came up for renewal in the 1840s. Demands for a much larger ground rent almost certainly resulted in new railway owners, the Swansea Coal Company, diverting the line north-west to avoid the Works site. A large embankment of colliery waste from Calland's Pit (formerly Chauncey Townsend's steam-engine pit) connected the upper part of the railway with the Coal Company's line which ran eastwards to the Swansea Canal. By the period of Map D, from 1897, the railway from Mynydd-Newydd Colliery ('28' on the map) had also been linked to the top of the railway at Pentre Colliery. Three formerly parallel lines ('28', '25' and '27' on Map C) were rationalised into one route under the new common ownership of the Swansea Coal Company (the successors to the Penvilia Vein Co. jointly owned by the proprietors of the Hafod and Morfa Copperworks). There were now widely differing dates and origins of the formation used by these lines. The effects of the new arterial transport line in the Landore area is shown clearly on Map D. Some 38 years after the construction of the South Wales Railway a branch was built southwards from the Pentre Pit railway to connect with this new trunk locomotive railway main-line ('33' on Map D). The delay may have been partly due to differences in gauge. The line continued in use until the closure of Pentre Pit in 1926.

26. **Plas-y-Marl Dry Dock**
(SS 6648 9630) (Maps C-D). Built off the Trewyddfa Canal near the bridge to Plas-y-Marl Colliery, this may well have been constructed before 1838 by Lockwood, Morris & Co. to service their coal-boats. A yard could have been used for building new boats. By 1876 a second dry dock had been built to the north and the earlier dock became derelict.

27. **A Railway serving a colliery in Cwm Rhyd-y-Filais**
(SS 6570 9600 to SS 6616 9593) (Map C). This railway was built along the northern side of the Nant Rhyd-y-Filais between 1838 and 1848 by the Swansea Coal Company. The eastern half of its course was reused by the Pentre Pit railway ('25') as

shown on Map D. The original western half of the line had become disused by 1876.

28. Mynydd Newydd Coal Pit Railway
(SS 6387 9648 to SS 6611 9576) (Maps C-D). This railway of 1842-43 ran 1½ miles westwards from the canal. A 1 in 8 incline, 750 yds long, brought the railway down to the level of the upper Nant Rhyd-y-Filais at what is now Brynhyfryd Square. From here a gentle incline (of the 'Newcastle' type introduced at Pentre Pit in 1830 ('25')) took the line down to the Swansea Canal. Between 1876 and 1897 the lower incline was abandoned and a link line was made from the foot of the upper incline to the head of the upper Pentre Pit incline (see Map D). A significant shift in local trade patterns happened by 1897 when a spur was constructed to Landore station showing that both the Pentre and Mynydd Newydd railways were standard gauge by this date (the broad gauge used by the South Wales Railway from 1851 may explain why a connection was not made earlier). A disparity of gauge between the two colliery lines could also explain why the two systems were not rationalised by the Swansea Coal Company when it came into possession of both between 1854 and 1863. With the construction of the Landore station spur no colliery line remained totally dependent on the canal.

29. Railway to Landore Alkali Works Tips
(SS 6625 9557 to SS 6647 9565). This railway served a chemical works constructed between 1826 and 1838. At the latter date it was owned by Williams, Foster & Company who also owned the Rose and Morfa Copperworks and by 1876 was labelled on maps as an Alkali Works. The original tip was soon built over and a high-level (or movable) river bridge gave access to a fan of tracks on the east side of the river that was abandoned by 1876.

30. Cwm Pit Railway
(SS 6567 9609 to SS 6617 9598) (Map C). This railway replaced line '23' (Map B) in carrying coal from Cwm Pit from about 1845 onwards. Sir John Morris II was selling off his collieries, and by 1854 Cwm pit was owned by the Swansea Coal Company. Their interest lay in supplying the parent company's copper smelters to the south at Hafod and Morfa. It is obvious that line '23' (Map B) going north was not suited to carry this traffic south. Therefore line '30' (Map C) was constructed eastwards from Cwm Level along the northern flank of the Landore Valley to a new incline down to the Swansea Canal. Vivian & Sons, one of the co-owners of the Swansea Coal Company, still

possessed the colliery in 1864-65. The pit was later owned by John Glasbrook, probably by the 1870s or 1880s. This particular railway was disused well before 1876.

31. Cwm Pit Railway Dock
(SS 6618 9598) (Map C). Between c.1845 and 1876 a rectangular canal-dock was built as the loading point for coal boats going south from the Cwm Pit Railway.

32. Probable Railway from Cnap Llwyd Sandstone Quarries
(SS 6601 9649 to SS 6642 9642) (Map D). These quarries grew considerably between 1826 and 1876. A raised embankment running obliquely east down Graig Trewyddfa suggests that the most northerly quarry may once have had a self-acting railway incline down the hill slope. There was, however, no hint of this in 1838. Presumably access was to the main valley road.

33. South Wales Railway, Landore Viaduct
(SS 6628 9582) (Map D). The opening in 1850 of this broad-gauge railway, which later became part of the Great Western Railway, marked the advent of a new arterial transport system through the area. New and existing localised transport ventures now sought a junction with the public locomotive railway rather than the canal just as the latter had once usurped the role of the River Tawe.

34. Landore Tinplate Works Railway
(SS 6601 9582 to SS 6595 9554) (Map D). The Works was built on a canal-side site in 1851, but relied on a railway siding as its only transport connection. This clearly demonstrates the significance of the new South Wales Railway.

35. Landore Silver Works Tips Railway
(SS 6611 9564 to SS 6647 9581) (Map D). This silver-smelter and, later, possibly spelter works operated from 1853 until 1867 when the buildings were converted into the Landore Siemens Steelworks. Like many of the non-ferrous smelters in the Swansea Valley it seems to have outgrown its original tipping area. The owner, L. Ll. Dillwyn and the manager, E.M.Richards, threw a lifting-bridge over the River Tawe and carried the waste by railway to the marshy east bank of the river. With the conversion of the works to steel production and rolling in 1867-69 it is likely that the reduced production of waste made this railway, with a powered incline to the tip top, unnecessary and it was abandoned by 1876.

36. Plas-y-Marl Dry Dock No. 2

(SS 6652 9535) (Map D). By 1876 a second dry dock had been built immediately to the north of the first ('26'). It probably serviced the coal-boats of Plas-y-Marl Colliery.

37. Landore Siemens Steelworks (Landore) Railway

(SS 6611 9564 to SS 6625 9560) (Map D). The original works of 1867 had a short railway running from the Swansea Canal to at least three branch-lines in the works itself. It is likely that the previous non-ferrous works on the site from 1853 onwards also had such a railway but such lines had disappeared by 1897.

38. Landore Siemens Steelworks (Llansamlet) Railway

(SS 6649 9595 to SS 6688 9598) (Map D). The original works of 1871 had an extensive system of internal works railways linked to the adjacent Great Western Railway. The railway also served a river-dock (see '40') and by 1876 had a steam-engine house to draw trucks up to a series of tips on the eastern river bank. The engine house had disappeared by 1897.

39. Landore Siemens Steelworks (Llansamlet) Swing-bridge

Fig. 314. Looking downstream at a swing-bridge providing vehicular access from the Neath Road and carrying a pipeline conveying water from the Swansea Canal for the steam-engines powering rolling-mills and blast machinery in Siemens Landore Steelworks (Llansamlet). (West Glamorgan Archive Service – GWR Collection)

(SS 6660 9622) (Map D). The bridge over the river must have been built in 1871 to give access to the works when the latter opened. It may have been one of only two bridges on the river navigation that swung rather than lifted open.

40. Landore Siemens Steelworks (Llansamlet) River Dock

(SS 6653 9504 (Map D). This dock was formed by widening out the mouth of a stream filled in during the construction of the steelworks in 1871 but was disused by 1897.

41. Landore Alkali Works River Dock

(SS 6635 9551) (Map D). Between 1876 and 1897 the old marsh drain to the immediate south of the works was filled in, except for the mouth which was widened and given an adjacent mooring-post.

42. Great Western Railway, Swansea to Morriston Line

(SS 6602 9509 to SS 6717 9790) (Map D). The Great Western railway had bought the Swansea and Trewyddfa Canals in 1872 and in 1881 constructed a railway along their banks. It was designed to cream off the lucrative trade generated by the intensively developed industrial area between Swansea and Morriston the canal line. Elsewhere, the GWR was content to develop the Swansea Canal in its battle against the Midland Railway (owners of the parallel Swansea Vale Railway) for trade generated in the Swansea Valley.

43. Millbrook Ironworks Internal Works Railway

(SS 6594 9571 to SS 6596 9595) (Map D). The original site of the Millbrook Ironworks, just south of an incline (east '27' on Map C), was determined by its need for water-power. By 1876 its dependence on steam-power resulted in the works being re-sited (near '43' on Map D). It now lay by the future site of the line linking the Pentre and Mynydd Newydd collieries to the Great Western Railway ('33'). The works railways may have been built to connect with this line: they certainly existed when it was first mapped in 1888. The line into the works probably continued in use until 1966.

44. Copperhouse Foundry 'Internal Works Railway'

(SS 6619 9603 to SS 6625 9610) (Map D). The railway was probably a man or horse-powered line for internal movement of materials constructed between 1876 and 1897. In 1881 the foundry was physically separated from the adjacent canal by the building of the Swansea to Morriston Railway. It probably never had a direct connection with either.

45. Swansea to Morriston Street Tramway

(SS 6564 9323 to SS 6695 9733 (Map D). The first specialised transport way in the Landore area built primarily for the conveyance of the copperworks and tinplate workers. Opened between Morriston and the Hafod Inn, 12 April 1878, through running extended to Swansea High Street in 1880. Converted from horse power to electric power in 1900.

Fig. 315. The area around Cwm Pit
(maps 1-3: site 12) in the 1890s. The
lower terrace above Cwm Pit Road in
the left middle background is the
former formation of the Plas-y-Marl
Coal Pit Watercourse (map 2: feature
'10'). The upper terrace behind the pit
chimney and right across the picture is
an early railway formation. (University
of Wales Swansea, Library &
Information Services, Archives)

Fig. 316. A view of the Cwm Pit site in
1979 from the newly-reclaimed Pentre
(Penyfilia) Pit site, showing survival of
leats and early railway formations (at
top left) on the slopes below Morris
Castle. (1979/105/7)

Fig. 317. Nant Rhyd-y-Filais valley from the reclaimed tips of the Penyfilia or Pentre Colliery. The lower eastern valley of the Nant Rhyd-y-Filais was largely infilled with copper-slag from the Hafod Copperworks Tip in the 1960s and playing-fields constructed. (1979/105/12)

Landscape Dynamics at Landore

Introduction

The evolving industrial landscape of the side valley of the Nant Rhyd-y-Filais at Landore which flows into the River Tawe, was a microcosm of the development of transport and water-usage in the south Wales industrial area. Water for transport and power (to drive water-wheels and as condensing and boiler feed-water for steam-engines) was the predominant catalyst of industrial development in the area from the early eighteenth to the late nineteenth century. The watercourses built to service such needs formed a continuously changing web across this highly developed area. The successive railways of the valley formed a second network ancillary to that of the water economy of the defile. It constantly changed in response to varying sources of raw materials, sites of manufactories and varying patterns of entrepreneurial partnerships. It has fortunately been possible to reconstruct successive phases in this palimpsest of a landscape, thanks mainly to the scrupulous paperkeeping and extensive legal activities of the Dukes of Beaufort and their agents. This reconstruction may facilitate the planning of archaeological excavation, the understanding of the history of technology and landscape history, and aid the interpretation of visible industrial archaeological remains. The valley of the Nant Rhyd-y-Filais is now a buried historical landscape. It is submerged beneath the spreading suburbs of Swansea and the redistributed copper waste that has largely smoothed the folds out of the landscape and eliminated the gorge of the stream itself over its lower reaches. Beneath these sometimes superficial deposits lie significant early mechanical and structural artifacts, as is also the case in other south Wales valleys.[1]

Recent work on the Landore wharf area has already shown the extent of buried deposits here. With the use of optical enlargers it has been possible to work out the likely sites of early industrial features from traces surviving on the 1876 1:2500 Ordnance Survey map and thus plot detailed maps for the area illustrated. These maps are presented against a street pattern that is recognisable today to allow us to locate and relate present remains to the series of historical landscapes of which they formed a part. Most features did survive, albeit some in a modified form, from period to period. This was especially true of features in the periods 1770-*c*.1793 and *c*.1796-1806 which existed in a slightly developed landscape.

To gain a greater understanding of the factors governing survival it is worth examining each of the three categories of feature in detail: leats (artificial watercourses), railways and buildings.

Leats

Leats differed from the other features in distributing a natural resource of vital importance. Thus they were generally reused or modified by new undertakings. The only exception to this rule came with the abandoning of the Llangyfelach Copper Works leat in 1748 and the building of a new copperworks leat in the valley by the same company in *c*.1793. In the intervening period it would seem that Penyfilia chaff mill was built at the outflow of the pond to utilise directly the water formerly provided to the Llangyfelach Copperworks leat. The Landore Copperworks proprietors had, therefore, to construct a new weir and leat below this chaff mill to obtain sufficient head of water to drive their stamping and rolling mill. Similar incidents occured in the development of other industrial landscapes[2] but were not always an obstacle that caused reconstruction as in this example. One feature of special note is the existence of two leats specifically intended to supply water to steam-engines. They can be paralleled by at least three others in the lower Swansea Valley.

In Britain sites are known in Cornwall[3] where the corrosive properties of the copper-mine water necessitated alternative supplies for boiler and condenser use.[4] Although south Wales coalfield water is not corrosive, in this area canal water was preferred to pumped mine water for steam-engine use. Colliery proprietors were prepared to supply water to the canal in return for fresher water and to pay a rent for so doing.[5] However, there is a substantial difference between paying an annual rent of 20s. to a canal company and making a heavy capital investment to finance the construction of large watercourses. It seems so unlikely that long leats would have been constructed solely to supply water to steam-engines that it is reasonable to assume that the four-mile leat to Pwll Mawr Colliery of 1757,[6] and possibly also the shorter leats to Calland's Pit and the Plas-y-Marl Coal Pits, supplied water-wheels before the use of the succeeding steam-engines.[7] At this point it is worth considering differences between the development of the water-resources of this valley and those of similar valleys such as the Greenfield Valley in Flintshire, or the Anghidy and Whitebrook Valleys of Monmouthshire.

An obvious difference is in the character of the water-feed arrangements, which is not simply due to differences in the gradient of the valleys. The industrial development of the Greenfield Valley is characterised by a series of large dams holding back substantial reservoirs with works positioned under the dam tails. This is due to the requirements of large batteries or forges using several water-wheels in parallel from the same dam. These replaced a number of earlier installations of differing purposes operating in series along the main valley stream. The few large ponds of the water-supply system in the Anghidy Valley were constructed for similar purposes. The predominant type of feed in all four valleys tapped the main valley stream using a weir with or without a small pond, and leading into a simple lateral feed that terminated in a small pond adjacent to the works served.

The condensing-water leats of the Nant Rhyd-y-Filais Valley fall outside this category although parallels can be found in the Monmouthshire valleys. These can be described as contour leats to differentiate them from the linear leats described above which followed a single stream and did not cross minor watersheds and streams by means of aqueducts or weirs. The Plas-y-Marl and Calland's Pits had contour leats because their primary site was not governed by the need to tap a readily available water resource. The

coal pits obviously had to be near their available seams and the convenience of integrating wire-working processes meant that two sites in the Anghidy and Whitebrook Valleys were also served by contour leats. The leats of other minefields are similar.[8]

In contrast, manufacturing industry can generally site the works in a convenient place in a valley where a pond can be provided, with very short leats, and several wheels to be worked in parallel if desired. Examples of the contour leats are found in the Ceredigion lead mines, and of the latter in the Greenfield, Angidy and Whitebrook Valleys. It is interesting that in the Nant Rhyd-y-Filais both kinds occur.

Survival of leats

Not a great number of leat earthworks survive. Artificial watercourses are not intrusive on the landscape and did not have the same capacity as the broader railway formations for reuse as the new roads of suburban Swansea. They have largely disappeared below housing development or in the in-filling of the Nant Rhyd-y-Filais gorge.

Railway formations

Early railway formations, unlike leats, seem never to have been totally reused by successive owners although we do have one example where a track-bed was at least partially re-utilised: the Cwm Pit of Level Tramroad of pre-1826 used the disused alignment of the Pentre-gethin coal pits (Landore) railway of *c*.1790. Three main factors explain this failure to reuse old track formations: the short life of pits or levels worked with little capital, the competitiveness of local colliery owners and the changing markets served by the colliery tramroad outlets. This periodic revolution, rather than evolution, in the transport network is statistically very noticeable - there were nine railed-ways built from the area at the head of the Nant Rhyd-y-Filais Valley within the period c.1770-1843. Only the example previously noted made any attempt at reuse of a track formation. Less than 9% of the money invested was spent in opening up and arching a level.[9] The local terrain was of course restrictive and effectively prevented the reuse of the early Pwll-yr-Oer railway on the south side of the Nant Rhyd-y-Filais ravine by the Pentre-gethyn Coal Pits (Landore) railway on the north. The later railway may also have been partially built to supply coal to the newly constructed Landore Copperworks to the north of the valley, effectively preventing reuse of the earlier formation. The shift in markets was the prime factor in the building of two new railways in *c*.1806-1813. In 1800 John Morris I (who may possibly have been the driving force behind the company's copper-smelting activities) withdrew from Lockwood, Morris and Co. [10] and by *c*.1800 the Landore Copperworks had been bought by another company. In 1806 Lockwood, on his own behalf, bought a third share in the collieries owned by the firm of Lockwood & Company. He then formed a new company, known as the Penvilia Vein Company to exploit the collieries at Penyfilia, Tre-boeth and Clyn-du. It was this company that then built the Cwm Level, Tre-boeth Level and Lower Nant Cwm Gelli Levels tramroads to transport coal northwards to supply the Rose, Birmingham and Forest Copperworks directly by rail. Previously the coal had been supplied only to Lockwood, Morris and Co.'s Landore Copperworks by a line down the Nan Rhyd-y-Filais Valley. By 1807, a second tramroad (the Penyfilia or 'Pentre II' Coal Pits and Cwm Pit 'II' tramroad) had been built from the same coal-producing area southwards to an outlet at Swansea which could be reached by 600-ton sailing vessels. The theory that the remaining six tramroads or railways were constructed on quite separate alignments, for reasons of differing ownerships, is supported by the formation of the Mynydd Newydd Coal Pit railway in c.1843. This railway paralleled the Penyfilia or Pentre Coal Pit (1830) railway. These two lines came under the same ownership during the 1850s and between 1876 and 1897 the Mynydd Newydd incline was abandoned, so that both railways used the Pentre railway incline. The delay in rationalising the railway routes may have

been due to a difference in gauge. It may be that disparity of gauges and of track type reinforced the reluctance of colliery owners who may have felt unable to act together to develop railway systems. In five out of nine examples it seems that such factors ensured the separate development of new and isolated railways.

Railway survivals

It is perhaps surprising that so many topographical features of the late-eighteenth and early-nineteenth centuries survive in the landscape at Landore. The colliery railways became convenient local thoroughfares offering an alternative route to the local muddy tracks. They were easily adapted as tracks and then roads. It was possible to erect housing along such tracks as the lines followed the existing contours wherever possible. The levels of surrounding land surfaces did not differ significantly from those of the railway formations. The major early industrial archaeological survivors on the landscape are the slight earthworks of the formations of these early railways. The larger earthworks of the 1830 railway to Penyfilia (Pentre) Pit have not survived nor are there any remains of the more substantial canal or dock formations. The key to survival is adaptability and only those features that could fulfill new uses survive to remind us of the complex early industrial development at Landore. Many earthworks of early tramroads and leats have survived in the landscape above the urbanised

Fig. 318. The Pentre (Penyfilia) Pit Railway of 1830 (25 on the Landore transport maps) was largely operated as a long gently-graded inclined-plane. This early twentieth-century photograph shows people seated on a road under-bridge near the head of the incline at Pentre Pit, looking northwards towards Cwm Pit. (University of Wales Swansea, Library & Information Services, Archives)

valleys of south Wales in areas where there has been no activity by the Forestry Commission or the open-cast mining executive. At Landore the uplands have been urbanised by the growing suburbs of Swansea, so most such earthworks only survive on the steep scarps that are inconvenient sites for newer buildings. Such sites may have tramroad formations that have not become surfaced footpaths or roads and thus may yield sections of broken rail, wheels and stone sleepers *in situ*. These obviously can give us valuable information on rail types and gauge. At Landore sections of the branch tramroads to Tre-boeth Level (*c.*1806) survive on the north side of the Nant Rhyd-y-Filais ravine and Cwm Gelli.

Industrial Buildings

Most of the features noted at Landore were fairly typical of contemporary industrial landscapes. This area of West Glamorgan was the most intensively developed copper-smelting district of Great Britain and this is reflected in the three copper smelteries that existed at Landore. The industrial structures erected at Landore were largely non-intrusive. The local entrepreneurs had insufficient capital to ignore the natural restrictions of the landscape. The major exceptions to this rule were arterial transport features sanctioned by Act of Parliament and with a large shareholding, e.g. the Swansea/Trewyddfa Canal, the South Wales Railway and the Great Western Railway's Swansea to Morriston line. The locally financed railways and leats needed to follow the sinews of the landscape to avoid the large costs of embankments and cuttings. However, the influence of arterial route engineering was seen in the construction of at least one local railway. The long embanked inclines of the Penyfilia (Pentre) Pit (1830) tramroad reflect the age of locomotive railway civil engineering and its difference from predecessors was noted in the contemporary press. Buildings and their effect on the landscape were rather different. Morris Castle was designed to be an intrusive and eye-catching ornament to the local skyline, as well as housing 24 workers and their families. It was, however, hardly an industrial installation and so falls outside the scope of this section. To a lesser extent engine houses and copper-smelters were intrusive with their vertical emphasis and high stacks. Even here the earlier works' dependence on water-power to provide furnace blasts ensured that the Llangyfelach Copperworks was built onto a slope. Its stone aqueduct did increase the intrusion. A similar dependence on water-power determined the siting of the Landore Copperworks stamping and rolling mill and the Nantrhydyvilas Air Furnace Company's copper and iron works.

The survival of industrial buildings

Works buildings often survived from period to period. They were often merely sheds which housed machinery and could easily be adapted to other uses. Several shed-like structures survive in use as storage sheds, garages and warehouses. The late nineteenth-century Landore Tinplate Works office still performs a similar function for a couple of small local firms. The Landore Chemical Works of pre-1876 had become a building attached to the boiler-maker's Yard of Millbrook Iron and Steel Works by 1881-97 and is now a garage. One possibly early quarry face survives because of the relative difficulty of removing it in the days prior to the organised mass disposal of refuse. Some functional buildings that served one particular or related purposes survived through several of the periods covered by the chronological maps. The Landore Copperworks Stamping and Rolling Mill of *c.*1793 probably became the iron rolling mill and forge of the 'Nantrhydyvilas Air Furnace Company' of *c.*1814-17. By 1822 it was probably a similar installation in 'Landwr Forge', which later became Landore or Millbrook Ironworks. After 1881 the building was marked as the 'Old Forge' on a plan of 'The Millbrook Iron & Steel Works' and it survived into the early part of this century. Over half the stock of industrial buildings at Landore survived in their original or modified form between the periods noted except for the long period between *c.*1820 and *c.*1876. Over a longer period of time, of course, a majority of structures have not survived even in a modified or fragmentary form. If we discount two capped and refaced tops of mine-shafts (Drew's and Plas-y-Marl) and one possible quarry face none have survived from before *c.*1845. These structures are not buildings and survive as vestiges of the earlier industrial installations noted.

Landore survivals

Summary of Significant Remains at Landore

The following section lists survivals from the 73 transport and water-power features together with other structures of industrial significance.

1. The capped shafts of many collieries survive underground. Reclamation activity in the 1990s revealed the head of the Landore shafts (1754 onwards) prior to capping in concrete. The fine Boulton and Watt enginehouse had largely disappeared but two storeys of buried buildings remained in the deep infill around the site of the water-wheel coal pit.

2. Plas-y-Marl Coal Pits (Graig Coal Pit). There are traces of the shaft head; sunk by 1761.

Fig. 319. William Edward's 'New Key' of 1772-4 (4 on the Landore transport maps) still remains at Landore to the north of the Landore Viaduct (33 on the maps). To the north of it is the infilled entrance to the Landore Colliery Dock. (803752)

Fig. 320. The top of the Drew's or Tir Glandwr Pit shaft being exposed prior to filling and fitting with a lower-level cap in preparation for the new cross-valley link road in February 1999. (West Glamorgan County Council Engineers Dept.)

3. Calland's Pit, lower part of the rock-faced enginehouse of 1851 survives, but buried at present, on the site of Chauncey Townsend's Steam-engine Pit of 1762.

4. Morris Castle. Designed as an entrepreneurial 'eyecatcher': ruins of the multi-storey castellated workers' flats for 24 families built on the summit of Cnap Llwyd, on the north side of the Landore ravine, built between 1768 and 1775.

5. Townsend's Steam-engine Pit Railway. Formation now eastern two-thirds of Siloh Road. Operative by c.1768.

6. Landore Quay. William Edwards's masonry quay is largely intact: it was in use by 1771.

7. Drew's or Tir Glandwr Pit Shaft. Built c.1780. The base of the late nineteenth-century rock-faced engine house remained until 1991 when the shafthead was capped in concrete.

8. Plas-y-Marl Coal Pits engine pond 'feeder'. This survives on the western side of Cwm Nant-Gelli as a 2ft wide grassy ledge. On the east scarp of the valley it is more pronounced with a bank remaining on the valley side and is up to 5ft in width.

9. Pentre-gethin Coal Pits (Landore) Railway. Probably constructed c.1790. The line survives as Penyfilia Road and continues as a short track between Llangyfelach and Cwm Level Roads.

10. Nant Rhyd-y-Filais Culvert. Built before c.1793. Part survives south of Landore Quay.

11. Cwm Pit Railway. A tramroad built in c.1794. This survives as a surfaced footpath down the scarp of Cnap-Llwyd.

12a. The formation of the Tre-boeth Coal Levels Railway, built c.1806-1826, survives as Heol Nant Gelli north of Pont-y-Shoot.

12b. Pont-y-Shoot now carries a road (Heol Gerrig) over the Nant Cwm-Gelli and was almost certainly built originally as part of the railway from Cwm Pit to Tre-boeth Coal Level in c.1806-26. The north face of the bridge is now hidden by a large earth bank but the south face of the sandstone rubble structure is still visible. It is essentially a stone revetted causeway punctuated by a culvert similar to those at Pontardawe and Ynysmeudwy (Swansea Valley), Skewen and Aberdulais (Vale of Neath) and Hirwaun. Tre-boeth Level is on the western bank of Nant Cwm-Gelli and was worked by John Morris before 1775. Coal may have been carried on carts over a bridge on the Pont-y-Shoot site to supply Morris and associates's copper smeltery at Fforest. The south elevation of Pont-y-Shoot shows at least four periods in its construction (see drawing). The low stone abutments visible in the base of the present bridge may have supported a low timber deck to carry these carts or an earlier public road (a ford or possibly a bridge existed here in 1761) or alternatively they may have served as footings for a timber trestle bridge built perhaps in c.1806 to carry the tramroad to Tre-boeth Level. Such trestle bridges with stone abutments were a very common feature of late eighteenth- and early nineteenth-century railway construction in the Swansea Valley (e.g. the surviving abutment south of Pont-y-Yard, Aber-craf, and those that carried the Brecon Forest Tramroad at Cae'r-Lan, Aber-craf). This bridge may have been replaced after perhaps twenty years by one completely of stone. The battered coursed rubble walls that survive with their offsets in the flanking walls of the present arch are typical of

fig.319

fig.320

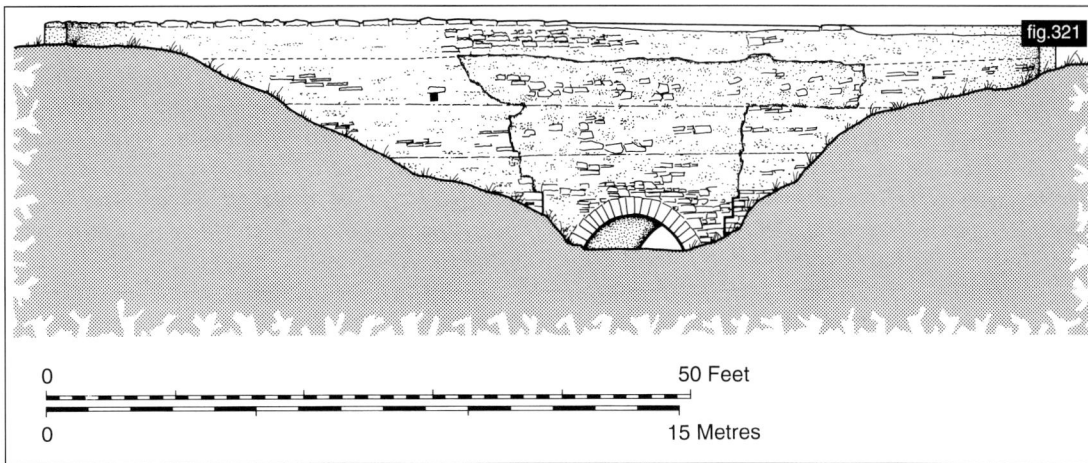

Fig. 321. South elevation of the causeway of Pont-y-shoot.

early nineteenth-century railway construction (e.g. the bridges at Quakers Yard on the Penydarren Tramroad, and at Abernant (Aberdare), on the Cefn Rhigos tramroad).The tooling of the arch does not look later than mid nineteenth century and could be either period '2' or '3'.

After a period of use the central portion of the stone bridge probably collapsed. This may have been due to either a combination of the weight of the earth and rubble fill of the causeway with water seepage leaching out the mortar of the side-walls or the original arch being undercut by the stream underneath (the aqueduct at the south end of the Brecknock and Abergavenny Canal was undermined soon after construction). The crude rebuilding of the central portion of the bridge with a mixture of stream-worn stones and uncoursed sandstone rubble (period '3') looks to have been a hasty repair after an unexpected collapse.

Finally, at about the end of the nineteenth century, the parapet was built or replaced and capped with a rusticated stone top typical of this period. The significance of this structure to the development of technology is minimal but it does help to quantify technological change: there were a number of technically interesting early iron bridges in south Wales but by far the larger number of early nineteenth-century railway structures were of wood or stone.

12c. The southerly course of the Tre-boeth Level railway to Cwm pit is clearly visible as a 6-12ft wide ledge on the eastern scarp of lower Cwm-Gelli above the smaller Plas-y-Marl engine leat. According to local inhabitants this ledge was reused in 1901-1914 to carry the Cwm Gelli Level railway to a loading chute for carts on the site of the disused Cwm Pit (now a church and car park).

12d. Cwm Pit, Lower Nant Cwm-Gelli and Tre-boeth Coal Levels Tramroad. A fourth remnant of this Tre-boeth Level branch survives as an overgrown footpath (originally a cart-track) running between Millbrook Street and Trewyddfa Road.

12e. Cwm Pit, Lower Nant Cwm-Gelli and Tre-boeth Coal Levels Tramroad. Adjoining the west end (Millbrook Road) of the above Tre-boeth branch is a hedge bank curving south-west behind some semi-derelict allotments. This bank is on the line of the *c.*1806 Cwm-Pit 'main-line'. A shallow ditch immediately to the south of the hedge bank is probably the old track formation.

12f. Cwm Pit, Lower Nant Cwm-Gelli and Tre-boeth Coal Levels Tramroad. A track crosses rough ground from the south-west end of Millbrook Street and descends in a shallow curved 8ft wide shelf to Cwm Level Road. This was the Lower Nant Cwm-Gelli branch tramroad that was constructed between 1826 and 1838. It joined the tramroad branches to Cwm Pit and Tre-boeth Level at the north-east end of Millbrook Street which is also built on its line.

12g. The long sloping sweep of Dinas Street, north of Landore, follows the main line of the Cwm Pit, Lower Nant Cwm-Gelli and Tre-boeth Coal Levels Tramroad which supplied coal to the Rose, Birmingham and Forest copper smelters and to the Trewyddfa Canal. The tramroad became disused between *c.*1844 and *c.*1855.

13a. The formation of the Penyfilia or 'Pentre II' Coal Pits and Cwm Burlais railway is represented by Eaton Road, Brynhyfryd. The exceptional width of Eaton Road (once known as Heol-y-Rail) is because it is built on the site of a marshalling yard where waggons from Lockwood and Company's Pentre Pit and John Morris II's Cwm Pit were stored - either full to go to the harbour, or empty to be returned to owner.

13b. Penyfilia or 'Pentre II' Coal Pits and Cwm Level 'II' Tramroad. Pentre Colliery retaining wall on the uphill (western) side of the reclaimed Pentre Coal Pit railways site.

315

14. The disused quarry faces survive of the Cnap Llwyd Sandstone Quarries. The earliest may be the quarry face to the immediate east of Morris Castle, the largest quarry to the north of the quarry worked until after 1897.

15. Penyfilia or Pentre Coal Pit (1830) Railway. At the southern end of the colliery retaining wall are the ruins of the (horizontal sheave) brake-drum housing of the upper incline, superseded by a second to the south by 1897.

16a. Mynydd Newydd Coal Pit Railway of *c*.1843. This is the most recent of no less than three ways converging on Brynhyfryd Square which owed their origins to colliery railway formations. A footpath (Penlan Fach) rises directly from Penyfilia Road up Pen-filia scarp to the Penlan housing estate at the top. This surfaced path with its shallow cutting and embankments is the old track formation of the former upper Mynydd Newydd Railway incline. It rises through 288ft in a horizontal distance of 2,684ft with several stone sleeper-blocks remaining alongside the path.

16b. Mynydd Newydd Coal Pit Railway. The south side of the former tramroad incline is marked by the rear property boundary of houses facing onto Siloh Road immediately west of Hosea Row. This boundary line is carried on eastwards by the south side of Pwll Road between Hosea Row and Byng Street. Both sides of Pwll Road lie above the Mynydd Newydd railway at this point.

17. South Wales Railway, Landore Viaduct. This was re-decked in 1978-79 but retains its original western abutment tipping the adjacent embankment and four of the five stone piers. This Brunel structure, originally mostly timber, dates from 1847-50. The four slender rock-faced 'Brunel piers' are pierced by twin semi-circular arches. The later piers and central span are of 1888-89 and the latter bears the inscription 'Edward D. Finch & Co., Steel Bridge Builders, Chepstow 1889'. It is one of only two bridges remaining on the Tawe River from those built to provide clearance for sailing vessels.

18. Landore Tinplate Works, 'Viaduct House'. This building on the main Neath Road was probably built as an office building for the tinplate works (founded 1851) during its expansion in the later nineteenth century.

19. Northern Cnap Llwyd Sandstone Quarry Railway Incline. About 40ft of a shallow embankment some 14ft wide remains in the scrub between Emlyn Terrace and Trewyddfa Road on the scarp of Craig Trewyddfa. This probably carried a railway in the mid-nineteenth century.

20c. Landore Siemens Steel Works Quay. These remains on the river side of the steelworks site survived until clearance work in the 1980s. The abutment-like structure dated from the period 1869-75.

21a. The Millbrook Iron & Steel Works Cast-House. This dates from before 1876. It is now a store on the Millbrook Industrial Estate and has slightly raised walls capped by a modern roof. The walls are of coursed rubble sandstone with fish-bellied cast-iron lintels spanning original windows and doors.

fig.322

Fig. 322. Mid nineteenth-century print of the Landore Viaduct from the south and its crossing of the Swansea Canal. The five masonry supports between the canal and river are visible. Morris Castle can be seen to the top left. (City & County of Swansea: Swansea Museum Collection)

Fig. 323. Mid nineteenth-century view of Landore Viaduct from the north. The River Tawe is on the left and the Swansea Canal on the right. At centre right are the wing-walls of the considerable embankment and aqueducts that carried the Trewyddfa Canal over the Nant Rhyd-y-Filais, the railway from Cwm Level and the headrace for the Trewyddfa Cornmill. (City & County of Swansea: Swansea Museum Collection)

Fig. 324. The surviving piers of Isambard Kingdom Brunel's high-level Landore Viaduct over the valley and navigable channel of the River Tawe, completed for the opening of the arterial South Wales Railway in 1850. (920058/10)

Fig. 325. The quays and docks at Landore seen from the south in the early part of the twentieth century and framed by the secondary main span of the Landore Viaduct (33 on the Landore transport maps). (British Railways Board)

Fig. 326. Photograph from the Landore Viaduct of the Landore Quay area in 1947, showing the quay and river dock, the 66" Boulton & Watt Engine house, a river warehouse and the mill tailrace. Above the quay are the ruins of the Little Landore Copper and Arsenic Works and Morris Castle. (University of Wales Swansea, Library & Information Services, Archives)

These were probably cast at the old foundry of the Millbrook Ironworks/Landore Forge. There are brick dressings to the openings.

21b. The Millbrook Iron & Steel Works, Steel Furnaces Building. This dates from the period 1876-1897 and is now a warehouse on the Millbrook Industrial Estate, largely reclad in corrugated asbestos. The coursed sandstone walls with brick dressings contain large blocked openings with cast-iron lintels and pillars.

22. Copperhouse Chemical Works. A semi-derelict building of red-brick dating from before 1876 that survived until the building of the new Plasmarl bypass road in the 1980s. It became the Millbrook Ironworks Boiler maker's yard and building sometime between 1881 and 1897.

23. 'Lisbon Llangyfelach Road Sandstone' Quarry. This quarry started work before 1876 and was probably still operative in 1897. As Hollett Road leaves Llangyfelach Road it ascends over the centre of the main working face.

24. Cwm 'Sandstone' Quarry. Also dates from before 1876 and was still at work in 1897. The quarry is now a gated yard with its working faces intact.

Landore Landscape Analysis - Results

Tabulated summary showing the approximate proportion of features reused or abandoned in each period

Period 1 - 1770 to *c.*1793 (23yrs) The Local Canal Age

Water *Economy* *%*	*Transport* *Development* *%*	*Other Industrial* *Installations* *%*	
0	0	0	DESTROYED
0	25	33	ABANDONED
0	25	33	TOTAL OUT OF USE
67	75	66	UTILISED
33	0	0	MODERNIZED
100	75	66	TOTAL IN USE

Periods 2-3 -*c.*1793 to *c.*1796-1806 (3 yrs)
The Arterial Canal Period

Water *Economy* *%*	*Transport* *Development* *%*	*Other industrial* *Installations* *%*	
0	0	0	DESTROYED
0	33	33	ABANDONED
0	33	33	TOTAL OUT OF USE
87	33	67	UTILIZED
13	33	0	MODERNIZED
100	66	67	TOTAL IN USE

Periods 3-4 - *c.*1796-1806 to *c.*1820 to *c.*1845 (14-24 yrs)
The Arterial Canal Period

Water *Economy* *%*	*Transport* *Development* *%*	*Other industrial* *Installations* *%*	
0	0	0	DESTROYED
14	38	0	ABANDONED
14	38	0	TOTAL OUT OF USE
71	63	10	UTILIZED
14	0	0	MODERNIZED
85	63	10	TOTAL IN USE

Periods 4-5 - *c.*1820 to *c.*1845-76 (31-56 yrs)
The Arterial Railway Age

Water *Economy* *%*	*Transport* *Development* *%*	*Other industrial* *Installations* *%*	
0	10	50	DESTROYED
40	40	17	ABANDONED
40	50	67	TOTAL OUT OF USE
53	40	33	UTILIZED
7	10	0	MODERNIZED
60	50	33	TOTAL IN USE

Period 6 - *c.*1876 - *c.*1897 (21 yrs)
The West Bank Valley Railway period

Water *Economy* *%*	*Transport* *Development* *%*	*Other industrial* *Installations* *%*	
0	0	0	DESTROYED
43	33	44	ABANDONED
43	33	44	TOTAL NOT IN USE
43	33	44	UTILIZED
14	25	11	MODERNIZED
57	58	55	TOTAL IN USE

Conclusions

The replacement rate of structures only exceeds half in the greatest period of time: inevitably more in 31 years than in 3. The periods considered have had to be chosen according to what sources are available and to the varying speed of change in the restricted valley of the Nant Rhyd-y-Filais.

Landore Survivals - Physical Remains in the 1980s 1

Water Economy, Transport Development and Industrial Structures

Period 1 - 1770

1. Recapped Plas-y-Marl Pit, Ruin [Morris Castle]

2. Most of Route of Pwll-yr-Oer Railway

3. Rebuilt Landore Quay

4. Rebuilt Drew Pit

5. Fragment of Plas-y-Marl Feeder

6. Recapped Landore Pit shafts

Period 2 - *c.*1793

1. Fragment Pentre Railway

2. Ruined Culvert

3. Fragment Cwm Pit Railway

Period 3 - *c.*1796-1806

NO SURVIVALS

Period 4 - c.1820-c.1845

1. Most of the Cwm Pit, Lower Nant Cwm-Gelli & Tre-boeth Coal Levels Railway

2. Largely the course of the Pentre II Railway

3. Cnap Llwyd Quarry

4. Small fragment of the Pentre II Railway (2nd)

5. One quarter of the Mynydd Newydd Railway

Period 5 - c.1876

1. Small fragment of the Landore Tin Works

2. Fragment Northern Cnap Llwyd Quarry Railway

   Most of the Landore Siemens Works Laboratory (until the 1980s)

4. Fragment of the South Wales Railway Viaduct

5. Part of the Landore Siemens Works (until the 1980s)

6. Most of the Steel Works Quay (until the 1980s)

7. Most of the Millbrook Cast-house

8. Fragment of the Copperhouse Chemical Works

Period 6 - c.1897-1902

1. Fragment Swansea-Morriston Railway

2. Entire South Cnap Llwyd Quarries

3. Entire North Cnap Llwyd Quarries

4. Fragment Lisbon Quarry

5. Entire Cwm Quarry

Full descriptive entries for the transport and water-power sites at Landore and references to the sources used can be found in S.R. Hughes, 'Landore: the evolution of an industrial landscape, c.1770-1897', *Bulletin of the South-west Wales Industrial Archaeology Society*, 23 & 24 (January & March 1980), 2-10; S.R. Hughes, 'Landore: a study of water-power during the Industrial Revolution', *Melin*, 3 (1987), 43-59.

References

1. For example the excavated brake engine located by digging down at the head of a self-acting tramroad incline (J. van Laun, 'Industrial Archaeology at Blaenavon', South East Wales Local History Newsletter 6 (September 1978), 29-30, and in Industrial Archaeology Review 3, No. 3). Other buried sites located include the weighbridge excavated at Banwen ironworks and the tramroad found intact at Abercrâf ironworks (see S.R. Hughes, *The Archaeology of an Early Railway System: The Brecon Forest Tramroads*, Aberystwyth, 1990).

2. The cornmill situated in the Anghiddy wireworks complex at Tintern was an intrusive element in the development of the surrounding works..

3. v.i. Rodney Law; a similar leat to convey water for steam-engineuse existed at Wanlockhead.

4. The intake of condenser-water required was twenty times that required for feeding the boiler.

5. The collieries at Tir Canol and Fforest took water from the Swansea Canal for the use of their steam-engines (Swansea Canal Minute Books, various dates (PRO).

6. See the section on the Llansamlet Railway for details of the Pwll Mawr leat.

7. It is known that Lockwood, Morris & Co. used mine water-wheels such as the 'Clyn-du Water Engine'. 'Clyndu Water Machine Pit' is marked on the Lockwood, Morris & Co. holdings (Collection of T. Bryn Richard, University of Wales Swansea Library. an undated copy of a probably early nineteenth-century map showing the underground workings between Morriston and Llangyfelach in relationto the surface road plan).

8. M. Tucker, 'The System of Watercourses to Lead Mines from the River Leri', *Ceredigion*, VIII, 2, (1977), 217-223.

9. W.G.R.O. William Kirkhouse's Letter Book, No. 466.

10. P.R. Reynolds, 'Clyndu and Pentremalwod Tramroads, Morriston', *South West Wales Industrial Archaeology Society Newsletter*, 13 (July 1976), 6-7.

11. R.O. Roberts, *'A Chronology of the copper, silver and lead smelting works in the Lower Swansea Valley'.*

12. Dates given in Charles Hadfield, *The Canals of South Wales and the Border* (Cardiff and London, 1960), 60-61.

13. Ibid, 68.

14. As 46, section 18.

15. Ibid.

16. Ibid, section 16.

CHAPTER 7
CONCLUSION

Fig. 327. Pictured from the air in the 1920s, Landore and the Morfa and Hafod Copperworks looking southwards to the White Rock Copperworks in the right background. (National Museums & Galleries of Wales – Department of Industry)

There were three main centres of the copper-smelting trade in the Severn Estuary area. These were Bristol, Neath and Swansea which were sited along tidal rivers and surrounded by hills which overlay large coal reserves. Of the three the lower Swansea Valley was the last to be developed. However, Swansea was the nearest smelting area to the main sources of copper-ore in Cornwall and Devon and developed at an intensity which far exceeded the other areas and which made it an industrial area of international significance.

Eighteenth- and early nineteenth-century Swansea had no less than 13 copper and other non-ferrous works almost continuous succession along three miles of river, which were generally navigable to sailing ships and with reserves of coal immediately available in the hillsides behind. Bristol had a far more dispersed industrial landscape with only five smelting-works spread out over 14 miles of navigable river, with two of them located over three miles from navigable water, to give access to fuel from the Bristol Coalfield that lay inland and to the east of the city. In addition to this, three of the five riverside works could not be reached by sailing ship but only by river-barge because of the restrictive headroom of Bristol Bridge. Of the 13 works at Swansea, all were sited alongside navigable water and only three were inaccessible to sea-going vessels. Because of these advantages, the primary activity of copper-refining came to be concentrated on Swansea. Bristol survived as a centre for the production and

Fig. 328. Swansea has retained many industrial sites and monuments of national and international importance. Key: 2 – Clyne Valley Colliery Tunnel; 3-4 – Areas of early shaft mounds ('bell-pits'); 6 – Ynys Collieries Leat; 7 – Ynys Leat Overflow; 8 – Ynys Colliery Shaft & Wheelpits; 9 – New Cornmill; 10 – Tirdeunaw Colliery site; 11 – Colt Pit & Waggonways; 12 – Six Pit Waggonway formation; 13 – Pwll Mawr Pumping Shaft; 14 – Pwll Mawr Winding Shaft, Engine house and Waggonway formation; 15 – Townsend's or Calland's Engine house & Shaft; 16 – Landore Pits; 18 – Morris Castle; 20 – Plas-y-Marl Pits; 24 – Double Pit Colliery Leat; 25 – Waggonway Tunnel site; 26 – Townsend's Wharf; 27 – Morfa Quay; 28 – Landore Quay; 29 – Landore River Docks; 30 – Birmingham Copperworks Dock site; 31 – Smith's Canal Tipping Stages; 32 – Llangyfelach Copperworks; 33 – 'The Great Workhouse' of the White Rock Copperworks; 34 – White Rock Riverdock; 35 – White Rock Canal Tunnel; 36 – White Rock Quays; 37 – Hafod Copperworks Engine Houses & steam-engine; 38 – Morfa Copperworks Rolling-mill; 39 – Clyne Wood Arsenic Works. (Based upon Ordnance Survey material with the permission of Ordnance Survey on behalf of the Controller of Her Majesty's Stationery Office ,, Crown Copyright)

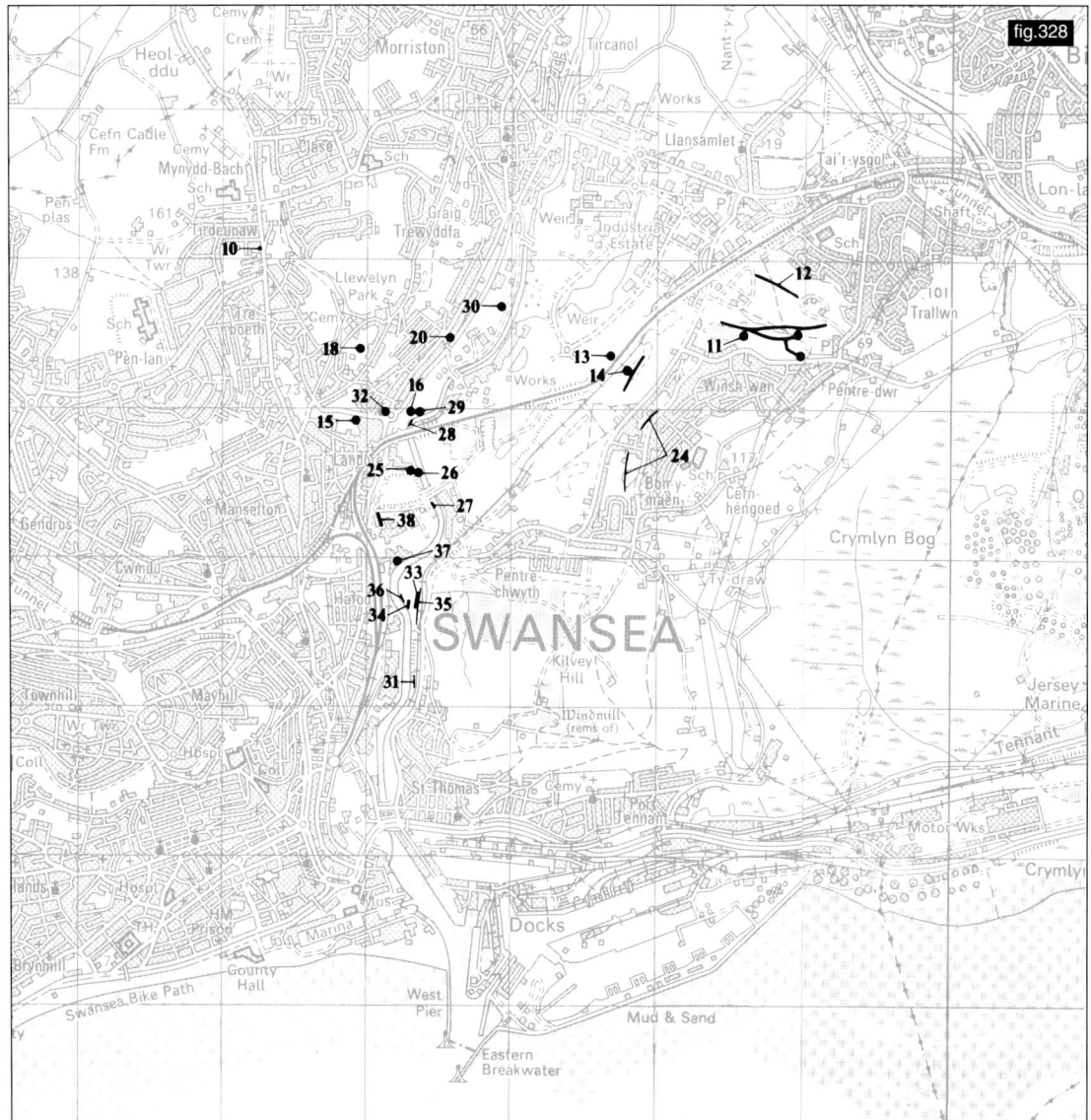

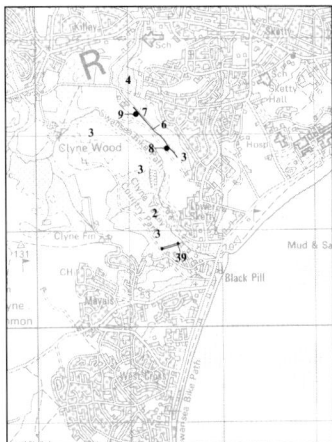

manufacture of brass, but only because of the calamine deposits (zinc ore) in the nearby Mendip Hills. Thus, even though Bristol was the first urban copper-smelting centre, it never experienced the impact that the concentration of heavy industry had on the lower Swansea Valley. Similarly, Neath was never industrialised to the same extent as Swansea and its landscape remained a mixture of the industrial and the picturesque, with parkland that was never submerged by intensive industrial activity. Even at Swansea this combination survived into the early years of the nineteenth century; the landscape, whilst busy, retained a picturesque element which attracted artists and the homes and parks of the entrepreneurs.

The character of this semi-industrialised landscape was captured by the American observer, Joshua Gilpin, who wrote in 1796 that he:

'...came to Swansea river where a wide scene of manufactory buildings displayed themselves, here being two very large copper smelting houses and a rolling and slitting copper works... The environs about it are very romantic and afford a number of fine [prospects] giving views of the bay, the Bristol Channel, the rocks called the Mumbles and a variety of country.'[1]

There is no doubt that the second generation of locally resident entrepreneurs who were active at this time, such as John Morris I, saw unity in a landscape that encompassed not only his Palladian mansion, with its banqueting house and park; but also the smelting-works, with its cupolas, rotundas and octagon; this next to the neat chequer-board layout of Morriston with its uniform houses, central island church and pedimented poor house; and finally the graceful arch

Fig. 329. The remaining buildings of Morfa Copperworks. The former large rolling-mill is the Swansea Museum Service stores. (935055/60)

of William Edwards' Wychtree Bridge. These were the views elegantly displayed on Rothwell's transfers for the Swansea Pottery. Sir Herbert Mackworth at Neath had a similar vision, as did George Haynes in the first decade of the nineteenth century with the creation of his garden-village at Clydach.

The third generation of the Morrises perceived that the large scale of nineteenth-century industry was enforcing a separation of the various elements of this copper-smelting landscape. The Vivians, arriving in 1808 to build a new and larger industrial unit run on scientific principles, did not even try to integrate their working and living areas. Nearly all the resident entrepreneurs in the first and second decades of the nineteenth century migrated to the sea views of Swansea Bay.

It would be difficult to claim that this early industrial landscape, with a busy but picturesque aspect, was radically different from other scenes of industry in the early phases of the world's first Industrial Revolution.

The lawns of Warmley House at Bristol, the home of the copper and brass smelter William Champion, sloped down to a large lake that served both as an ornamental feature and as a source of water power. In the centre was a huge copper-slag statue of Neptune, framed by a castellated cast copper-slag pavilion on one side and the intake to his copper-rolling mills on the other. Visitors to Matthew

Boulton's great pedimented Soho brass manufactory outside Birmingham likewise commented on the serpentine walks leading past the rustic hermitage and grotto, through the wooded grounds of his elegant Soho House to the ornamental lakes which also served as a source of water-power for the workshops, where copper refined in Swansea was transformed into beautiful objects of desire.

By the beginning of the nineteenth century, however, visitors to Swansea were beginning to note the grosser effects of the burgeoning number of copper-smelters erected in the 1780s and 1790s. The Reverend J. Evans, in his tour of south Wales published in 1804, noted that: 'The volume of smoke from the different manufactories contribute to make Swansea, if not unwholesome, a very disagreeable place of residence.'[2]

Tourists and artists alike found the completely novel and intriguing, but relatively small-scale, buildings and landscapes of the world's first Industrial Revolution a source of delight and fascination. Continental European tourists reporting on Great Britain's unique industrial process could, at this stage, still show an endless fascination with the special landscape that was evolving. The Swede, Eric Svedenstierna, observed in 1803-4:

'To return to Swansea, one finds near by the town on the River Tavey [Tawe] and up the Swansea Canal, as well as on the road to Neath

Fig. 330. Calland's Coalpit, which re-used the eighteenth-century shaft of Townsend's Engine Pit, was buried by redeposited copper-slag from the Hafod Copperworks tips in the mid twentieth century. The remaining end-walls of Morris Castle can be seen in the background. (University of Wales Swansea, Library & Information Services, Archives)

Fig. 331. The historic Landore Pit shafts revealed during reclamation works. (960203/2)

[i.e. through Morriston], such a profusion of copper works, coalmines, steam engines, ponds, canals, aqueducts, and railways, that the traveller, on arrival, becomes quite undecided as to where he should first direct his attention. Also a stranger would be able to go around here for several days without being able to make order out of this chaos ...

It is remarkable that these coal deposits were little worked up till thirty to forty years ago, and that on the unfruitful and bare heathland, which still twenty years ago was merely grazed by goats, there now live thousands of people, whose work brings millions into the country, and contributes to an extraordinarily high degree to its culture and growth.'[3]

However, the growing pall of pollution from the relatively unsophisticated methods of these early smelting-works began to repel the observers of this process. By contrast another world centre of metallurgy in Glamorgan - Merthyr Tydfil - might also be a crowded, chaotic scene of fevered activity but not one where the very atmosphere was becoming cloaked in a toxic fog. The population might by this creation of a scientific and industrially-based society, be freed from the vagaries of subsistence economy and the absolutism of a hierarchical regime, but in the short-term these beginnings were at some cost. The provision of sound housing, education for workers children and a communal and spiritual life in the countless nonconformist chapels, gave the workers some relief from such conditions and continued to suck labour in from the rural hovels of impoverished upland Wales.

Smelting on a huge scale switched to Swansea for economic reasons and the success of the industry created a landscape that became unloved by artists and where there was neither room, purity of air, nor inclination for the second and third generations of the entrepreneurs to linger - instead they migrated west to the still beautiful, 'sea-breeze-swept' slopes of Swansea Bay. Their workers still had to endure the effects of the industry that gave them the opportunity to earn wages and to survive. Contemporary engineering and processing knowledge was not capable of controlling the pollution caused by this huge concentration of booming heavy industry. From the picturesque landscape of the eighteenth-century copper-smelters evolved one of the first blighted landscapes of man-made despoliation and toxic pollution.

The smelting of copper and zinc released sulphur from their ores in the form of sulphur dioxide and droplets of sulphuric acid in suspension. By 1848, some 92,000 tons of sulphuric acid a year or 2,325,000 cubic ft a day, were released into the air by the copperworks of the Swansea Valley.

The prevailing winds were south-westerly and the mixed farming and coal-mining communities to the east of the river at Llansamlet and Bon-y-maen were worst hit. The brown arid slopes of Kilvey Hill, above the copperworkers' community of Foxhole, still vividly show the effect. The establishment of the huge industrial complex of the Hafod Works in 1810, and of the adjoining Morfa Works in 1828, intensified the effects of vast integrated works employing thousands rather than hundreds. Crops were ruined and cattle affected, and between 1822 and 1841 a series of indictments for nuisance were brought against individual copper firms by local farmers; but so essential was the industry to the economy of the town and the financial well-being of its growing population that none was successful.

However, the smelters seem to have been sensitive to the adverse criticism caused by the pollution from their works, and to have felt a sense of responsibility to their dependent communities that was also manifested in their provision of housing, schooling and religion. Under Swansea's Royal Institution of South Wales, a prize was raised to be awarded to whichever individual or company was deemed to have done most to reduce the nuisance.

The coppermasters experimented within the constraints of their technical knowledge. Both John H. Vivian of the Hafod Copperworks and Bevington Gibbin of the Rose Copperworks developed forms of water filter that were commended by the Institution but were not very effective. Some of the best scientists of the day, such as Sir Humphrey Davy and Michael Faraday, were also involved in the investigations.

The changing perceptions of this landscape can be vividly seen in the work of successive artists. Turner and De Loutherbourg, in the 1790s, were depicting small-scale industry as points of interest in a picturesque riverscape surrounded by rolling hills, with deliberate skyline eyecatchers such as the palladian mansion at Clasemont, the battlemented Morris Castle, or the later gothic Belvedere on the lip of the Clyne Valley.

By the 1860s the French artist who illustrated the geographical magazine *Le Tour du Monde* saw something very different. The natural element in what had been industry set in a semi-rural landscape had disappeared. Huge clouds of smoke were emitted from the vast Hafod Works towards Llansamlet and forests of chimneys, workers' housing and dwarfed canals and the River Tawe cowered beneath man-made mountainous tips.

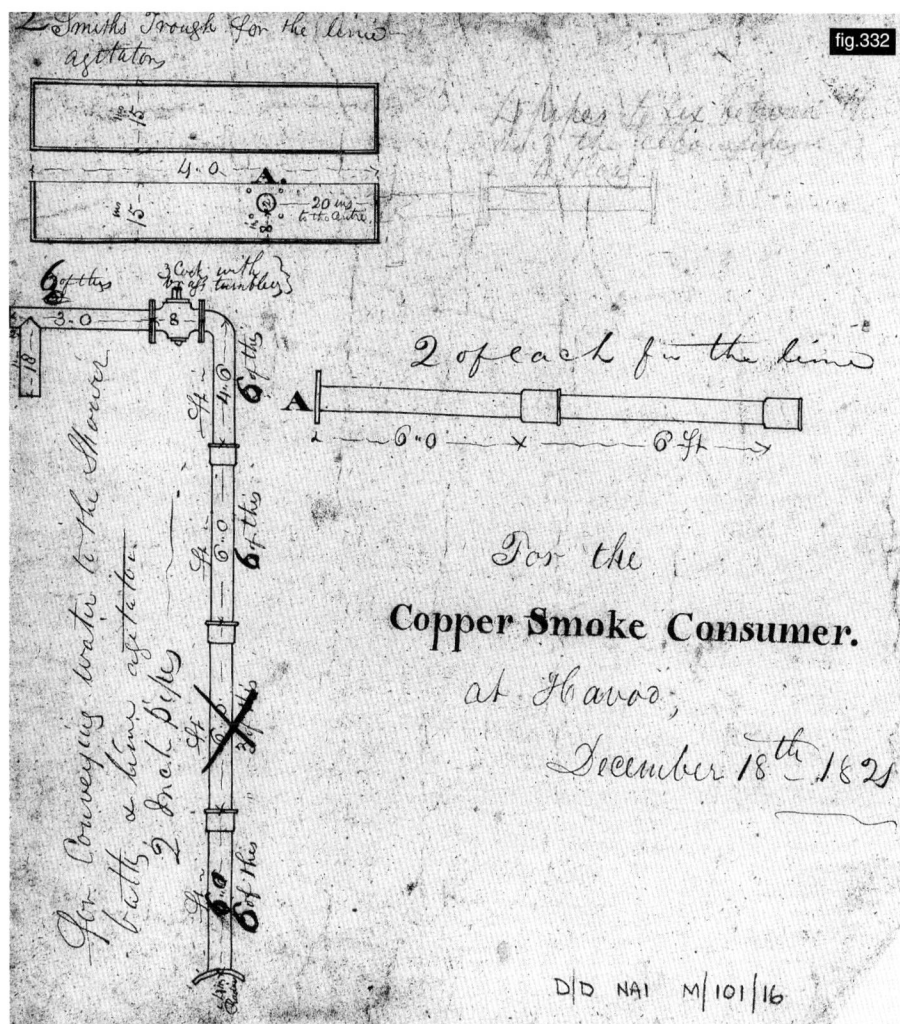

No effective control of the smoke could be made before a successful means of removing the sulphur from the furnace discharge was evolved. Successive members of the Vivian family had received a high standard of technical education in Germany and it was the adoption of Gerstenhofer's process in 1865 that effected a marked improvement in the situation. This was, in addition, an economically sustainable improvement, for the huge new chambers on top of the original tip at Hafod produced sulphuric acid to the value of £200,000 which now had a ready market in the burgeoning tinplate industry.

No sooner was the problem of pollution from the copperworks partially alleviated, however, than a

fig.332

Fig. 332. Attempts were made to harness contemporary technology to limit the pollution caused by the toxic smoke from the reverbatory smelting furnaces, initially with little success. In 1821, the Copper Smoke Consumer consisted of shower butts and lime agitators. (Held by the West Glamorgan Archives Service; Neath Abbey Ironworks Collection, D/D NAI M/101/16)

325

fig.333

Fig. 333. Gerstenhofer's process allowed the transformation of the sulphur by-product of refining into saleable sulphuric acid, significantly reducing pollution. A box-flue leads up to the acid chambers on top of the original riverside tip of the Hafod Copperworks (1909). The chimney that had originally released harmful fumes is visible together with newly-built aerial ropeway. (City & County of Swansea: Swansea Museum Collection)

new problem arose. The zinc industry flourished in Britain between 1840 and 1914 and in that period moved from use of the rarer calamine ore to the much commoner but more pollutant Blende. When demand intensified production in World War I, the effects on what crops were still being raised on the neighbouring farms at Llansamlet became clearly visible. Gardens nearby suffered damage until the surviving zinc-smelter switched to blast-furnace technology in 1961, shortly before the total demise of the industry in Swansea.

For nearly 100 years the copper industry centred on Swansea was one element in a picturesque landscape. For another hundred the copper smoke consumed the lower Swansea Valley. The majority of coppermasters went further to the west to the scenic delights of Swansea Bay where the castellated and gothic mansions of Singleton Abbey, Sketty Hall, Clyne Castle and Park Beck (formerly Parc Wern) still mark the presence of the Morrises and the Vivians.

The legacy to the east was not so kind and the indigenous sessile oak and the birch woodland of Kilvey Hill were destroyed. The topsoil, no longer held by plant roots, was washed off the valley sides and the subsoil was eroded by gullies. Much of these visible wounds have now been cloaked in conifers.

Trees and lakes, service industries and recreational facilities cover the valley floor. Most of the sores created by the uncontrolled excesses of past industry have been healed or disguised.

It might be asked what happened to the picturesque landscape that so attracted artists and fascinated industrial observers from all over the world. The answer is that there are remains but they have been submerged, both by Swansea's success as a centre of industry in the nineteenth century and by features of the late twentieth-century 'post-industrial' age - such as mass retailing, and leisure monuments like the tropical garden pyramid, the leisure centre, and the Marina development.

The latter has saved elements of the nineteenth-century industrial landscape of Swansea in the form of a dock which in the 1970s, was partly filled in. It is the sympathetic re-use of large-scale functional buildings that offers the only long-term prospect of effective conservation. Without this, preservation orders can be seen only as a short-term safeguard against unjustified demolition on cosmetic grounds, the motivation for which is especially strong in a community embittered by large-scale loss of jobs and livelihoods.

The Marina, formerly the South Dock, creates a new landscape of pleasure for all, which contrasts with the landscape of late eighteenth- and nineteenth-century leisure represented by the mansions of the coppermasters in west Swansea; the survivors of these latter are now often part of the educational infrastructure of the late twentieth century. This new landscape of leisure has provided a setting in which funds are generated that can safeguard the isolated structural survivals within an architectural framework that complements the remaining large buildings.

Thus, in the use of the South Dock, the infil was removed to expose the cyclopaedian dock walls and lock-chambers. The huge rock-faced bastions that once provided high-level access to the rows of coal-tipping stages have been retained as a visual border to the area. The hydraulic powerhouse is a restaurant; half of the opening-bridge has been retained; the general goods warehouse is now the Maritime Museum; the seaman's chapel is an art gallery and the pilot's house is re-used.

Just as interesting has been the regenerative effects on the Maritime Quarter, which is very largely made-up of buildings erected at the end of the 'picturesque period' at the beginning of the nineteenth century when Swansea still had aspirations to be a seaside resort. The Regency terraces erected by the Swansea architect, William Jernegan, have been refurbished and the fabric of his Assembly Rooms, where the great copper ticketings took place, has been retained.

On the valley floor alongside the River Tawe, once solidly filled with heavy industry, new areas of mixed entrepreneurial housing and industrial activity developed. Similar developments took place on the periphery of Swansea at Clydach and in the Clyne Valley.

The 'garden-village' developed by George Haynes at Clydach incorporated spoil-banks from the adjacent Swansea Canal; his ornamental gardens and fruit-walls enclosed, in part, by the colliery railways emerging from Cwm Clydach. His mansion of

fig.334

Ynys-Tanglwys directly faced the waterway, as did the model gothick-style houses of his workforce. Walks on the wooded meadows above led to his

Banqueting or Summer House. The mansion was de-gothicised upon conversion to further working-class housing and all was swept away, together with the gothic stables, as the Swansea Canal was filled in during the 1960s. All that has survived, generally unrecognised, is a rather plain secondary mansion built by George Haynes' successor Charles Parsons, and the crow-stepped Banqueting House (possibly by William Jernegan) with an open ground-floor that was converted to a dwelling-house in 1985.

John Morris I was largely responsible for creating a picturesque landscape at Morriston. At Sketty Park his son, John Morris II, was mainly responsible for the creation of a second. One of the reasons why the Oystermouth Railway was promoted was to develop the coal-seams running along the fringes of the Morrises newly acquired Sketty Park, on the eastern side of the Clyne Valley. Simultaneously, this part of the valley was provided with a large leat, initially tapped in two places, to provide water-power for corn-milling and colliery pumping. The new colliery lay just west of the site of Sketty Park House and a scenic drive zigzagged down the hill alongside the colliery and over the water leat as it passed through the woods fringing the park and valley. William Jernegan's castellated tower of the Belvedere, with a vaulted interior based on the Margam Chapter House, allowed observers to admire this mixed sylvan and industrial scene.

The coal resources in this valley, which was near the large estates of both the Morrises and Vivians, in turn attracted smelting that began to threaten, in the 1830s, the purity of air of the coppermasters' estates with a new type of toxicity. If it had been successful, the Clyne Wood Arsenic Works might well have turned western Swansea into another mixed area of gentlemen's parks and smelters, unless the coppermasters had in turn moved further out into less polluted territory. The matter was probably resolved when the steeply-inclined and faulted coal-seams of the Clyne Valley proved very difficult to work. The smelter had closed by the 1860s, but in the 1840s Sir John Morris II determined once again to exploit the coal-seams lying alongside the western side of his home and park and opened new collieries alongside his water-power leat. So, a third generation of a copper-smelting and coal-mining dynasty was prepared to move away from the greatest concentration of copper-fumes, but still exploit the colliery reserves on, and around, their new home park.

In many ways the Clyne Valley retains all the elements of an early industrial landscape that have been largely lost with the wholesale cleansing of the lower Swansea Valley. There are several reasons for this: first and foremost was the comparative

Fig. 334. Ynys-Tanglwys, one of the fine mansions built by eighteenth-and early nineteenth-century entrepreneurs in the Swansea area. It survived with the Gothic workers' housing alongside until the 1960s.
(Collection of J.C.B. Rye)

economic failure of the Clyne Valley and the arsenic works due to the heavily faulted coal-seams, thus the multiple shallow coal-pits have survived in the Clyne Valley when they have totally disappeared on the coal outcrops of the lower Swansea Valley. The next stage of colliery development, using water-powered pumping, has left a network of extensive embanked water-leats along the sides of the Clyne Valley which have largely disappeared from the lower Swansea Valley. In the latter, only short sections of leat remain between later areas of housing on the eastern slopes of Kilvey Hill and in the wooded slopes to the south of Glais. The only threat to the Clyne Valley leats, after initial recognition of their importance, was the continued tipping of domestic refuse on the floor of the valley, this has reduced the southern end of the Ynys Collieries leat into an unrecogniseable and impenetrable area of marshy bog which still contains the only remains known to survive of a water-powered colliery in the Swansea area.

In the Swansea Valley copper-smelting ceased in 1924. All the works were subsequently demolished except for part of the eighteenth-century Upper Bank Works and part of the older White Rock Copperworks (1737) which remains built into the hillside; the former was re-roofed for a wartime munitions factory and is now incorporated into the Addis Plastics Factory, while the latter is now part of the rolling greenery of the White Rock Industrial Archaeology Park. Hafod and Morfa Rollingmills continued until 1980 and the significant remains at Hafod are deteriorating badly.

By contrast, the Clyne Wood smelter retains a layout with extant high walls and the only visible remains of furnaces in the Swansea area. The many flues remaining on site are only paralleled in the Swansea area by the flues on and above the White Rock site. The Clyne Wood Works remained in use intermittently before its final demise in the 1860s and was subsequently used as hay sheds by the occupants of Clyne Castle. For that reason, the high upstanding walls survived into the early twentieth century, to become hidden by dense woodland and undergrowth. The base of the flue chimney at the top of the site has survived because it was later converted into a summer house.

The large-scale demise of the Swansea copper-smelting industry in 1924-5 left behind what was described as the largest tract of derelict land in Europe. *The Lower Swansea Valley Study* of the 1960s was headed by a keen amateur industrial archaeologist and suggested that significant remains should be conserved. However, the more aesthetically acceptable and historically significant remains disappeared, along with the greater remains of despoliation, in a campaign that cleared 14 acres in the years 1961-3. Swansea was the Headquarters of the 53rd Divisional Engineers (TA) and they were asked to clear designated sites with mechanical plant and explosives. Five sites were cleared as a training exercise by the Territorial Army, with no part of the sites deliberately retained for their historical, technical or archaeological value. A further four sites, comprising some 30 acres, were cleared by their owners between 1957 and 1962.[4]

The remains of the earliest copperworks in Swansea (from 1717 and 1720) have long been buried under later urban redevelopment and,

Fig. 335. The Hafod engine houses are one of the most significant survivals of the Swansea copper industry but are deteriorating significantly. (960204/2)

fig.336

Fig. 336. The 1910 copper rolling mill engine is a survival in deteriorating condition in the now largely roofless engine house at the former Hafod Copperworks. (960205/3)

fig.337

Fig. 337. The former electrical powerhouse of the Morfa Copperworks is one of the few functional buildings of the industry to remain partly intact. It was listed as a building of architectural interest before the works closed in 1980. No new use has been found for it and it has since been gutted by fire and vandalism on two occasions. (National Museums & Galleries of Wales – Department of Industry)

hopefully, the future demolition of buildings above them will be accompanied by some meaningful excavations, which would be capable of illuminating a crucial stage in the development of modern metallurgy.

Until 1963 one of the most visually impressive sights, amongst the large areas of dereliction, in the lower Swansea Valley were remains of the third copper-smelter founded in the area, the White Rock Copperworks of 1737, although the impressive long arcades that pierced the walls of the remaining smelting-halls largely dated from the nineteenth-century expansion of the works. In 1963, Kenneth Hilton, the administrator of the Lower Swansea Valley Project Team recognised that:

'Although with some exceptions the buildings were not of great importance, the site was one of the oldest industrial sites in the area and had been occupied more or less continuously since 1737 until the Works closed about 1930.

After a preliminary inspection of the site by staff of the Royal Commission on Ancient Monuments, it was decided to photograph the buildings and then to demolish them. Preliminary work was carried out in 1963, again by 48 Squadron RE but the main clearance was undertaken by 291 Plant Squadron RE (TA). This Squadron, with its headquarters in Walsall, held a Training Camp in Swansea in May and June 1965, during which the remaining walls were demolished.'[5]

In fact, the sequence of events were not as straightforward as that. The Investigators from the Royal Commission considered the buildings at White Rock were significant and impressive enough to merit long-term preservation.[6] However, the desire to quickly clear the whole of this area of derelict land did not allow the development of any significant body of opinion sympathetic to the retention of remains which, even though relatively graceful and aesthetically interesting in themselves, could not be separated in people's minds from a vast legacy of pollution and unsightly dereliction. Interestingly, by 1975, opinions had changed enough for such high and upstanding remains of other metalliferouss arcades to be conserved in the Swansea Valley.

The two cases were in many ways comparable,

329

except that these latter remains (of the 1860s rebuilding of the Ynysgedwyn Ironworks) were part of a much smaller 'island' of dereliction. In the popular imagination they did not attract the same hatred that the vast area of dereliction in the lower Swansea Valley of the 1960s did. In these circumstances, it was possible for the late Morgan Rees, Keeper of Industry at the National Museum of Wales, to persuade the promoters of this Land Reclamation Programme to retain these particular arcades which were reminiscent of the monastic arcades of Tintern or Llanthony, and so recognisably conformed to accepted notions of an 'ancient monument'.

A full ground-survey, by Royal Commission staff, also revealed that the large earthen bank on the Ynysgedwyn site incorporated the lower remains of the furnaces where George Crane and David Thomas had perfected the ability to smelt iron with anthracite by the use of hot blast in 1837 - a development with momentous implications for the iron industry of south-west Wales and the United States of America. The bank disappeared in the agreed plan of clearance and it was the aesthetically impressive, rather than the technologically significant, remains that survived. In the circumstances, the same would probably have happened at White Rock Copperworks in 1963.

Fieldworkers from the Royal Commission re-examined the White Rock site in the late 1970s and realised that significant remains, particularly from the earlier eighteenth-century period of activity, were still extant. Changed perceptions of the industrial past allowed Swansea City Council to designate this area as the White Rock Industrial archaeology Park. This present study has sought to identify these remains and those of other sites as precisely as possible, and it is likely that substantial parts of their substructures remain underground. It is important that development proposals on these and other sites recognise the likely existence of significant buried remains.

What did the Swansea industrial landscape mean in historical terms? What can a study of a complex infrastructure belonging to the world's first Industrial Revolution tell us that we do not already know from documentary sources? Two elements that can distort conventional historical synthesis are the patchiness and uneven nature of the surviving documentary record and the selectiveness with which this is studied. A historical landscape analysed in totality should advance objective historical analysis.

One significant example is that of the quantification of entrepreneurial philanthropy. One historical text by W.O. Henderson, on the industrialisation of Europe, in a standard popular series, offers the following generalisation:

'But for one enlightened employer like Robert Owen [of New Lanark textile mills, originally from Newtown in Montgomeryshire] or Titus Salt [founder of the model textile mill community at

Fig. 338. The establishment of the White Rock Industrial Archaeology Park in the late 1970s facilitated the excavation of one of the copperworks' river dock, consolidation of riverside quays and the preservation of the substructures of a series of eighteenth- and early nineteenth-century smelting-halls below the greensward. The white building in the background is former offices of the Hafod Copperworks reused as a social club. (960207/3)

Fig. 339. The settlement of Trevivian is an unusually complete survival with the housing and schools for the Hafod Copperworks. (9500055/30)

Saltaire, near Bradford, *c.*1853] there were a hundred who ignored the plight of their workers.'[7]

The accuracy of this statement in relation to other industries, such as that of copper-smelting, can be assessed from the fairly comprehensive topographical and archaeological evidence that has been studied in this volume.

There were 12 main copper-smelting works in operation in Swansea between 1717 and 1860. The nature of the industry meant that attempts were made, from the very first. to keep the foundation of large new works away from the existing urban core of Swansea. Therefore, all the works needed to make provision for some of their key workers. However, no less than six of the works made provision for housing a substantially greater number, starting with the White Rock Works in 1737. Five of these settlements were also provided with large gardens for the employees to keep pigs and grow vegetables. Beyond this, schooling provision started in 1806 and at least four of the works' owners are known to have made provision for the schooling of their workforce. Four of the works' owners provided, in full or in part, Anglican churches and one provided a nonconformist place of worship. One also built a secular meeting hall for the use of the employees of three of the local works and one provided a poorhouse.

Henderson, and many other historians besides, have assumed that the few very famous philanthropists were the exception and that only very few employers made provision for the welfare of their workforce. By contrast, the Swansea evidence suggests that some 50% of employers made substantial provision for the housing of their workforce and that 33% provided schooling for employees' children and places of worship. Indications from published works on Welsh workers' housing and works schools suggest that this level of provision in industrial communities may not have been exceptional.

It is hoped that some of the historical conclusions of this archaeological and architectural study will help to clarify perceptions of what was one of the most significant landscapes of the world's first Industrial Revolution.

References.

1. *Joshua Gilpin's Tour of Wales, 1796*, transcribed by Anthony Woolrich (Philadelphia University Library).

2. J. Evans, quoted in B.Dean, *Slum Living Conditions in 19th Century Swansea Part 1* (Swansea History Project, c.1980)

3. E.T. Svedenstierna, *Svedenstierna's Tour of Great Britain* (Newton Abbot, 1973), 42.

4. K.J. Hilton (ed.), *The Lower Swansea Valley Project* (London, 1967), 26.

5. Ibid., 237.

6. v.i. the late Douglas Hague.

7. W.O. Henderson, *The Industrialization of Europe: 1780-1914* (London, 1969), 138.

SELECT BIBLIOGRAPHY & ABBERVIATIONS

Abandoned Mines Map	Annotated 1/10.560 Ordnance Survey Map of 1921 on which most known abandoned coal mines have been plotted. (Abandoned Coal Mines Records Office, Ystrad Mynach, Mid Glamorgan).
B. Ferry estate plans	'Survey of Briton Ferry estate lands, plans and reference tables', 84 n.d., but before 1780 until after 1798. (West Glamorgan Archives Service, D/DBF E/1).
B.L	British Library.
Badminton MSS.	Badminton manuscript collection, Group II (National Library of Wales, Aberystwyth).
Barton, *Cornish Copper Mining*	D.B. Barton, *A History of Copper Mining in Cornwall and Devon* (Truro, 1961).
Baxter, *Tramroads*	B. Baxter, *Stone Blocks and Iron Rails (Tramroads),* (Newton Abbot, 1966).
Beaufort Plan & Case	Broadsheet entitled the 'Case of the Duke of Beaufort and the other noblemen and gentlemen on the proposed Swansea Canal' and headed by an excellent, detailed plan of the Landore-Morriston area. Not dated but probably 1793. (Swansea Reference Library).
Boat Check Book	'Cheque Book', Boat Check Book, 1812-22, of the upper Swansea Canal (National Library of Wales, Aberystwyth, MS 14098.D.).
Boorman, *The Brighton of Wales*	D.Boorman, *The Brighton of Wales: Swansea as a fashionable seaside resort, c.1730-c.1830* (Swansea, 1986).
Boulton Papers	Matthew Boulton Papers (Birmingham Reference Library).
Briggs, *John Johnson, Architect*	N. Briggs, *John Johnson, 1732-1814, Georgian Architect and County Surveyor of Essex* (Chelmsford, 1991).
Brooke, *Tin-plate Wks. Chronology*	E.H. Brooks, *Chronology of the Tin-plate Works of Great Britain* (Cardiff,1944).
C.G.S.	Court of Great Sessions documents (National Library of Wales).
C.L.	Cardiff Central Library Manuscripts Collection.
Canal Act	*An Act for Making and Maintaining a Navigable Canal from the Town of Swansea, in the county of Glamorgan, into the Parish of Ystradgunlais in the County of Brecon* 34 Geo. III, c.109. (23rd May 1794).
Canal C. Minutes	Swansea Canal Committee Minutes, (Public Record Office, London).
Canal Map of 1796-99	*G.W.R. Swansea Canal 20.* A large-scale strip map, entitled 'The Within Maps were surveyed and mapped by John Williams, Surveyor, Newland, Margam, Glamorganshire in 1796'. (The Estate Department, British Waterways Board, Dock Office, Gloucester). A similar map is *Swansea Canal 17.*

| *Canal Plan of 1832* | A large-scale strip map of G.W.R. *Swansea Canal 20046* the Swansea Canal surveyed by William Bevan and son in 1832. (The Estate Department, British Waterways Board, Dock Office, Gloucester). |

| *Canal Map of 1875-76* | *G.W.R. Swansea Canal 28072,* a large-scale strip map at a scale of 80' to 1". This is a printed copy of R.J. George's survey of 1875-1876. (The Estate Department, British Waterways Board, Dock Office, Gloucester). |

| Canal Minutes | Swansea Canal General Assembly Minutes, 2 volumes, RAIL 876. (Public Record Office, London). |

| Chamberlain, 'The Grenfells' | M. Chamberlain, 'The Grenfells of Kilvey', *Glamorgan Historian,* 9 (1973), 123-42. |

| Childs, *Penlle'r-gaer House* | J. Childs, 'Inside Penlle'r-gaer House Two Hundred Years Ago', *Gower,* 41, (1990), 22-38. |

| Childs, 'Morris Estate' | J. Childs, 'The Growth of the Morris Estate in Llangyfelach Parish, 1740-1850', *Gower,* 42, (1991), 50-69. |

| Clark, *Swansea Sanitary Inquiry* | 'Report to the General Board of Health, on a Preliminary Inquiry into the Sewerage, Drainage, and Supply of Water, and the Sanitary Condition of the Inhabitants of the Town and Borough of Swansea. By George Thomas Clark, Superintending Inspector,' London, 12 July 1849. |

| Cocks & Walters, *Zinc Smelting* | E.J. Cocks and B. Walters, *A History of the Zinc Smelting Industry in Britain* (London, 1968). |

| Collier, *S. Wales Coal* | L.B. Collier, 'A detailed survey of the history and development of the South Wales Coal Industry (from 1750-1850)', unpublished Ph.D. thesis, University of London, 1940. |

| *Collieries of Wales* | S. Hughes, B. Malaws, M. Parry & P. Wakelin, *Collieries of Wales: Engineering & Architecture* (Aberystwyth, 1994). |

| Cook, *Nonconformity & Reform* | L.A. Cook, *A Critical Investigation into Nonconformity & Social Reform in Swansea, 1851-1914* (University of Wales, M. Phil., 1992 Swansea Institute of Higher Education). |

| Cook, *Vivians Social Impact* | L.A. Cook, *An Examination of the Social Impact of the Vivians on Swansea, 1809-1894,* unpublished Ph.D thesis (University of the West of England, Bristol, 1997). |

| *Copper Through the Ages* | Copper Development Association, *Copper Through the Ages* (Radlett, Herts., 11th. Ed. 1954). |

| *Copper Committee Book* | 'Committee Book of the Copper Company' [i.e. John Freeman & Co. of White Rock] MS. B.4771 (Bristol Reference Library). |

| *County Rate, 1842* | J.E. Bicheno, 'Notes and Reports on the County Rate', 1842. (There is a copy in Glamorgan Record Office, Cardiff, D/DXgc 3). |

| Cowley, *Pentre Colliery* | F.G. Cowley, 'Pentre Colliery: A Short History'. *South West Wales Industrial Archaeology Society Newsletter.* XXI (March 1979) 3-7. |

| Crossley, *Sheffield Water-power* | D. Crossley (ed.), Water-power on the Sheffield Rivers (Sheffield, 1989). d. dated |

| Daunton, *Coal Metropolis* | M.J. Daunton, Coal Metropolis, Cardiff, 1870-1914 (Leicester, 1977). |

| Daunton, Miners Houses | M.J. Daunton, 'Miners' houses: South Wales and the Great Northern Coalfield, 1880-1914', *International Review of Social History,* 25 (1980). |

| Davies, *Hanes Cymru* | J. Davies, *Hanes Cymru* (London, 1992). |

| Davies, 'Morris Family' | J.M. Davies, 'The Morris Family and Swansea', *Gower,* 5, (1951), 26-30. |

Davies, *Pontardawe*   J.H. Davies, *History of Pontardawe and District* (Llandybie, 1967).

Davies, *Religion in the Ind. Rev.* E.T. Davies, *Religion in the Industrial Revolution in South Wales* (Cardiff, 1965).

Davies, *South Wales, 1814*   W. Davies, *A General View of the Agriculture and Domestic Economy of South Wales,* 2 volumes (London, 1814).

Davies, *Swansea Valley*   J.M. Davies, 'The Growth of Settlement in the Swansea Valley', unpublished M.A. thesis, University of Wales, 1942.

Davies & Williams, *Greenfield*   K. Davies and C.J. Williams, *The Greenfield Valley* (Holywell, 1977).

Day, *Bristol Brass*   J. Day, *Bristol Brass* (London, 1976).

Day & Tylecote, *Metals Revolution* J. Day and R.F. Tylecote (eds), *The Industrial Revolution in Metals* (London, 1991).

Dean, *Swansea Slums*   B. Dean, *Slums: Living Conditions in 19th Century Swansea, Parts 1-3* (Swansea History Project, c.1980).

Donnachie & Hewitt, *New Lanark*   Donnachie and G. Hewitt, *Historic New Lanark: The Dale and Owen Industrial Community since 1785* (Edinburgh, 1993).

Engineer's Report   Broadsheet of an engineering report on the canal by Charles Roberts entitled, 'To the Chairman of the Swansea Canal Committee' (Swansea Museum, in a scrap-book entitled 'Materials and Illustrations for a History of Swansea; collected by George Grant Francis Esq. 1862').

English Railways 1826   C. Von Oeynhausen and H. Von Decken, *Railways in England, 1826 and 1827* (Cambridge, 1971).

Evans, *Industrial Education*   L.W. Evans, *Education in Industrial Wales, 1700-1900; A Study of the Works Schools System* (Cardiff, 1971).

Fagg, *Llansamlet Mining*   B.C. Fagg, 'Coal Mining in Llansamlet Parish', *South West Wales Industrial Archaeology Society Newsletter,* 28 (July 1981) 3-9, 29 (November 1981) 8-10 and 30.

Farmer, *John Humphreys*   D. Farmer, *The Remarkable Life of John Humphrey, God's Own Architect* (Swansea, 1997).

Flinn, *British Coal*   M.W. Flinn, *The History of the British Coal Industry, Volume 2, 1700-1830 The Industrial Revolution,* (Oxford, 1984).

Francis, *Swansea Copper Smelting*   G.G. Francis, *The Smelting of Copper in the Swansea District* (Manchester, 2nd. edn.,1881).

Gabb, *Later Copper Works*   G. Gabb, *Later Copper Works,* Lower Swansea Valley Factsheet 6 (Swansea Museum, 1980s).

Gabb, *The Grenfells*   G. Gabb, *The Grenfells,* Lower Swansea Valley Factsheet 9 (Swansea Museum, 1980s).

Gabb, *The Morris Family*   G. Gabb, *The Morris Family,* Lower Swansea Valley Factsheet 10 (Swansea Museum, 1980s).

Gibbs & Roberts, *Copper 1723*   D.E. Gibbs and R.O. Roberts, 'The copper industry of Neath and Swansea: record of a suit in the Court of Exchaquer, 1723', *South Wales and Monmouth Record Society, Publications No. 4.*

Gilpin, *Tour of Wales 1796*   Joshua Gilpin, *Tour of Wales in 1796* (Philadelphia University Library).

Glamorgan Historian   *Stewart Williams' Glamorgan Historian,* ed. Stewart Williams *et al.,* 12 vols. (Cowbridge and Barry 1963-80).

Green, *Melincryddan Copperworks*   H. Green, 'Melincryddan Copperworks & Village', *Neath Antiquarian Society Transactions* (1978), 47-85.

Griffiths, *Clyne Castle*   R.A. Griffiths, *Clyne Castle, Swansea, a history of the building and its owners* (Swansea, 1977).

Griffiths, *Singleton & Vivians*

R.A. Griffiths, *Singleton Abbey and the Vivians of Swansea*, (Llandysul, 1988).

Griffiths, *Swansea, (1990)*

R.A. Griffiths (Ed.), The City of Swansea: Challenges & Change, (Stroud, 1990).

Gwernllwynchwith Colliery Plan, 1808

'PLAN of the GWERNLLWYNCHWITH colliery LANDS. The property of BENJAMIN MORGAN ESQ[R] under lease to Charles & Henry Smith Esqr[s]) referring to the PITS ENGINES and UNDERGROUND WORKINGS appertaining thereto', Measured, Plan'd & Copied by EDWARD MARTIN & D[D] DAVIES COLLIERY SURVEYORS MORRISTON near SWANSEA *Nov[r] 10th 1808* Copied by T.D. WILKS, July 1867 Copied by P.G. LLOYD. DEC 1981. A photocopy has been kindly sent to the Commission by Mr. B. Fagg.

Hadfield, 1960

C. Hadfield, *The Canals of South Wales and The Border,* (Cardiff & London, 1960).

Hadfield, *S. Wales*

C. Hadfield, *The Canals of South Wales and The Border,* (Newton Abbot, 2nd. edition, 1967).

Hadfield & Skempton, *William Jessop*

C. Hadfield and A.W. Skempton, *William Jessop, Engineer* (Newton Abbot, 1979).

Hafod: 1920's-1950's

Hafod History Society, *The Hafod: 1920's-1950's* (Swansea, 1996).

Hafod Copperworks' School

'W.G.W.', *A Brief History of the Hafod Copperworks' School From its Foundation* (Swansea, 1905) (supplement to *St. John's Parish Tidings,* copy in NLW, Vivian Papers, L46).

Hafod Works Manager Memos

'Memorandum Book of Mr. William Jones, for many years manager of the Hafod Works', (NLW MSS 15113B).

Hague, *Lighthouses of Wales*

D.B. Hague, *Lighthouses of Wales* (Aberystwyth, 1993).

Hallesy, *Glamorgan Pottery*

H.H. Hallesy, *The Glamorgan Pottery,* Swansea, 1814-38 (Llandysul, 1995).

Harris, *Copper King*

J.R. Harris, *The Copper King; a biography of Thomas Williams of Llanidan* (Liverpool, 1964).

Henderson, *Europe Industrialization*

W.O. Henderson, *The Industrialization of Europe: 1780-1914* (London, 1969).

'Hill Sketches'

Ordnance Surveyors' Drawings, Hill Sketches & Revisions for the first edition one inch to the mile map. Manuscript maps executed at a scale of two inches to the mile in *c.*1826. (British Library, London).

Hilton, *Swansea Valley Project*

K.J. Hilton (ed.), *The Lower Swansea Valley Project* (Swansea, 1967).

Hughes & Thomas, 'Chapel Architects'

V. Hughes and M. Seaborne, 'Welsh Chapel Architects', *Capel Newsletter,* 27, (Winter 1993/94), 13.

Hughes, *British Navigational Levels*

S.R. Hughes, 'The Development of British Navigational Levels', XXVII, 2 (July 1981).

Hughes, *Earthenware*

G.B. Hughes, *English & Scottish Earthenware* (London, c.1980).

Hughes, *I.A. Swansea Valley*

S.R. Hughes, 'The Industrial Archaeology of Water and Associated Rail Transport in the Swansea Valley', unpublished M. Phil. Thesis (University of Birmingham, 1984).

Hughes, *Montgomeryshire Canal*

S.R. Hughes, *The Archaeology of the Montgomeryshire Canal: A guide and study in waterways archaeology* (4th. ed., Aberystwyth, 1988).

Hughes, 'Landore'

S.R. Hughes, 'Landore: the evolution of an industrial landscape, c.1770-1897', *Bulletin of the South-west Wales Industrial Archaeology Society,* 23 & 24 (January & March 1980), 2-10.

Hughes, 'Landore Water-power'	S.R. Hughes, 'Landore: a study of water-power during the Industrial Revolution', Melin, 3 (1987), 43-59.
Hughes, *Railway System Archaeology*	S.R. Hughes, *The Archaeology of an Early Railway System: The Brecon Forest Tramroads* (Aberystwyth, 1990).
Hughes, *Swansea Canal*	S.R. Hughes, 'The Swansea Canal: Navigation and Power Supplier', *Industrial Archaeology Review,* IV, 1, Winter 979-80, 51-69.
Hughes & Reynolds, I.A. *Swansea*	S.R. Hughes & P.R. Reynolds, *A Guide to the Industrial Archaeology of the Swansea Region* (Aberystwyth, 1988).
Hunts	R. Hunt, *Memoirs of the Geological Survey of Great Britain, and of the Museum of Practical Geology, Mining Records, Mineral Statistics of the United Kingdom of Great Britain and Ireland* (London, annual volumes).
Ince, *South Wales Iron*	L. Ince, *The South Wales Iron Industry, 1750-1885* (Birmingham, 1993).
J.R. & C.H.S.	Journal of the Railway & Canal Historical Society.
J. Morris' *Commonplace Bk.*	Morris Manuscripts, John Morris' Commonplace Book, University of Wales Swansea Library.
Jenkins, *Tinplate Industry*	P. Jenkins, *Twenty by Fourteen: A History of the South Wales Tinplate Industry* (Llandysul, 1995).
Jervoise, *Ancient Bridges*	E. Jervoise, *The Ancient Bridges of Wales and Western England,* (London, 1936).
John, *Glamorgan 1700-1750*	A.H. John, 'Introduction: Glamorgan, 1700-1750', *Glamorgan County History, Volume 5, Industrial Glamorgan,* ed. A.H. John and G. Williams (Cardiff, 1980), 1-46, 27.
John, *Llansamlet Mining*	A.H. John, 'Iron and Coal on a Glamorgan Estate, 1700-1740', *Economic History Review,* 13 (1943), 93-103.
John, *South Wales 1750-1850*	A.H. John, *The Industrial Development of South Wales 1750-1850* (Cardiff, 1950).
Jones, *Charity Schools*	M.G. Jones, *The Charity School Movement: A Study of Eighteenth Puritanism in Action* (London, 1964).
Jones, *Swansea Port*	W.H. Jones, *History of the Port of Swansea* (Carmarthen, 1922).
Jones, *Welsh Chapels*	A. Jones, *Welsh Chapels* (Stroud, 1966).
Kirkhouse, 'Letter Book'	W. Kirkhouse, 'Letter Book', Swansea Reference Library, Local History Collection.
Lee, *Mumbles Railway*	C.E. Lee, *The Swansea & Mumbles Railway,* (South Godstone, Surrey, 3rd edn., 1970).
Lewin, *Early Railways*	H.G. Lewin, *Early British Railways 1801-44* (London, 1925).
Lewis, *Top. Dict.*	S. Lewis, *A Topographical Dictionary of Wales,* 2 vols. (London, 1833).
Lewis, *Swansea Guide 1851*	J. Lewis, *The Swansea Guide 1851* (Swansea, 1851).
Lewis, *Wooden Railways*	M.J.T. Lewis, *Early Wooden Railways,* (London, 1970).
Llansamlet Colliery	'Account Book, colliery Job Work Accounts. The collieries include *Great Pit, Charles Pit, Six Pit, Double Pit, Church Engine Pits,* in the *Llansamlet* area, and were owned by the *Smith Family* of Gwernllwynchwyth. Acc. no. 407176 (Swansea Reference Library, MSS. 1,046).
Lloyd, *Lost Houses of Wales*	T. Lloyd, *The Lost Houses of Wales* (London, 2nd. Ed. 1989).
Lloyd, *Swansea Regency Architects*	T. Lloyd, 'The Architects of Regency Swansea', *Gower,* 41, (1990), 56-69.

Logan, *S. Wales Coalfield 1837*

'Notes & Sections relating to the S. Wales Coalfield by Sir William E. Logan (then Mr. W.E. Logan). 1836 to 1842 (Geological Sciences Museum, Library, South Kensington, London, MSS. 1/218). Mostly datable by Logan's notebooks in the Ordnance Surveyors Notebooks collection to 1837.

Lowe, 'Industrial Housing Stock'

J.B. Lowe, 'Survey for Thematic-Based Research: An Industrial 'Housing Stock" in *Buildings Archaeology: Applications in Practice* ed. J. Wood (Oxford, 1994).

Lowe, *Welsh Workers Housing*

J.B. Lowe, *Welsh Industrial Workers Housing, 1775-1875* (Cardiff, 1977).

M.I.C.MSS

Millbrook Ironworks Company Manuscript collection (University of Wales Swansea Library).

M. MSS.

Morris manuscript collection (University of Wales Swansea Library).

Malkin, *South Wales Scenery*

B.H. Malkin, 'New bridge', *The Scenery, Antiquities and Biography of South Wales* (London, 1804).

Martin, *Mineral Bason*

E. Martin, 'Description of the Mineral Bason in the Counties of Monmouth, Glamorgan, Brecon, Carmarthen and Pembroke', *The Philosophical Transactions of the Royal Society of London,* (1806), 342-347.

Marshall, *British Railways*

C.D. Marshall, *History of British Railways down to the Year 1830* (London, 1938).

Minchinton, *Tinplate Industry*

W.E. Minchinton, *The British Tinplate Industry: A History* (Oxford, 1957).

Mining Districts, 1806

From an Appendix headed, 'I am undebted to Mr. Edmund Buckley, Iron-merchant, Manchester, for the following Tables:- 'Number of Furnaces and Make of Iron in England, Scotland and Wales, in the Year 1806', in *Report on the Mining Districts* or Report of the Commissioner (Appointed under the Provisions of the Act 5 & 6 Vict., c.99) to inquire into the operation of that Act, and into the state of the population in the mining districts, 1848. (London, 1848), 25.

Mining Districts, 1848

List of furnaces in 1848 from ibid., 27.

Mitchell & Deane, *Statistics*

B.R. Mitchell and P. Deane, *Abstract of British Historical Statistics* (Cambridge, 1962).

Morgan, *Upper Tawe History*

D. Watkin Morgan, 'An outline History of the Upper Reaches of the Tawe River', Brycheiniog, XII, 1966-67, 121-130.

Morris, *Copper History*

UWS, Morris MS., 'History of the Copper Concern' (1774).

Morris, *Morris Castle*

B. Morris, 'More Evidence for the Date of Morris Castle', *South West Wales Industrial Archaeology Society Bulletin,* (November 1994) 4-5, and 'Another Early Reference to Morris Castle', SWWIAS Newsletter, (September 1983) 9.

Morris, *Singleton Houses*

B. Morris, *The Houses of Singleton: A Swansea Landscape and its History* (Swansea, 1995).

Morris, *Swansea Houses*

B. Morris, 'Swansea Houses - Working Class Houses, 1800-1850', *Gower,* 26, 53-61.

Morris, *Thomas Rothwell*

B. Morris, *Thomas Rothwell; Views of Swansea in the 1790s* (Cardiff, 1991).

Morton, *Swansea Pottery*

E. Nance Morton, *The Pottery and Porcelain of Swansea and Nantgarw.*

Mullineux, *Worsley Canals*

F. Mullineux, 'The Duke of Bridgewater's Underground Canals at Worsley', *Transactions of the Lancashire and Cheshire Antiquarian Society,* 71 (1961).

N.L.W.   National Library of Wales, Aberystwyth.

N.M.R.W.   The National Monuments Record for Wales, Crown Building, Plas Crug, Aberystwyth, Ceredigion.

Neath Abbey Ironworks drawings   Neath Abbey Ironworks drawings (West Glamorgan Archives Service).

Newman, *Buildings of Glamorgan*   J. Newman, S. Hughes & A. Ward, *The Buildings of Wales: Glamorgan* (London, 1995).

O.R.P.   Oystermouth Railway Papers in the manuscript collections of the University of Wales Swansea Library.

O.S.   Ordnance Survey.

O.S. Drawings   Ordnance Surveyors' Drawings for the first edition one inch to the mile map. Manuscript maps executed at a scale of two inches to the mile in c.1813. (British Library, London).

O.S. Gl. Sh.   Ordnance Survey Glamorgan Sheet. (The dating given to each sheet is the date of survey and not the date of publication).

O.S.A.P   Ordnance Survey Aerial Photograph (followed by the flight number, e.g. 67/091 (i.e. a flight carried out in 1967) and then the frame number (e.g. 093). Copies of all these photographs may be consulted in the National Monuments Record for Wales at Aberystwyth.

Outram's Observations   Benjamin Outram's *Observations* on the construction of railways prepared for the shareholders of the Brecknock and Abergavenny Canal, printed for them in 1799 (one extant copy is N.L.W., Maybery MSS. 383).

p.c.   Personal communication.

P.C.B. MSS   Penlle'rgaer Manuscript collection, Group B (National Library of Wales).

P.R.O.   Public Record Office, London.

Phillips, *Vale of Neath*   Phillips, *History of the Vale of Neath.*

Percy, *Copper Metallurgy*   J. Percy, *Metallurgy, Copper* (London, 1861).

R.A.F.A.P.   Royal Air Force Aerial Photography (followed by flight details). Copies of all these photographs may be consulted in the National Monuments Record for Wales at Aberystwyth.

R.C.A.H.M. Wales   The Royal Commission for Ancient and Historical Monuments in Wales, Crown Building, Plas Crug, Aberystwyth, Ceredigion, SY23 1NJ. Tel. (01970) 621200.

Rees & Thomas, *Hanes Annibynol*   T. Rees and R. Thomas, *Hanes Eglwysi Annibynol Cymru* (Liverpool, 1871-75), Vols. 1-4.

Rees, *Industry before Ind. Rev.*   W. Rees, *Industry before the Industrial Revolution* (Cardiff, 1968).

Rees, *Quakers in Wales*   T.M. Rees, *A History of the Quakers in Wales and their Emigration to North America* (Carmarthen, 1925).

Rees's Manufacturing Industry   A. Rees, *Rees's Manufacturing Industry,* 2 Volumes (Newton Abbot, 1972): a selective reprint of Rees's *Cyclopaedia* of 1802-19.

Reynolds, *An Unusual Tramplate*   P.R. Reynolds, 'An Unusual Type of Tramplate', S.W.W.I.A.S.N. 20 (November 1978), 6-7.

Reynolds, *Brecon F.*   P.R. Reynolds, *The Brecon Tramroad Forest Tramroad* (Swansea, 1979).

Reynolds, *Industrial Development*   P.R. Reynolds, 'Industrial Development', *Swansea: an Illustrated History,* ed. G. Williams (Swansea, 1990), 29-56.

Reynolds, *Townsend's Waggonway*   P.R. Reynolds, 'Chauncey Townsend's Waggonway', *Morgannwg,* XXI (1977) 42-68.

Richards, *William Edwards*

H.P. Richards, *William Edwards: Architect, Builder, Minister* (Cowbridge, 1983).

Riden, *Charcoal Blast Furnaces*

P. Riden, *A Gazetteer of Charcoal-fired Blast Furnaces in Great Britain in use since 1660* (Cardiff, 2nd.ed., 1993).

Roberts, *Copper Development*

R.O. Roberts, 'The development and decline of the copper and other non-ferrous metal industries in South Wales', *The Transactions of the Honourable Society of Cymmrodorion* (1956).

Roberts, *Copperwks. Chronology*

R.O. Roberts, 'The Smelting of Non-ferrous Metals since 1750', *Glamorgan County History, Vol. V, Industrial Glamorgan from 1700 to 1970,* ed. A.H. John and G. Williams (Cardiff, 1980), 47-96, 90.

Roberts, *Gower Coalmining*

R.P. Roberts, 'History of Coalmining in Gower from 1700 to 1832', unpublished M.A. thesis, University of Wales (Cardiff), 1953.

Roberts, *Neath & Swansea Copper*

R.O. Roberts (ed.), 'The Copper Industry of Neath and Swansea; Record of a Suit in the Court of Exchequer, 1723', *South Wales and Monmouth Record Society, Publications No. 4* (1957), 123-64.

Roberts, *Non-ferrous Smelting*

R.O. Roberts, 'The Smelting of Non-ferrous Metals', in *Glamorgan County History, Vol. 5, Industrial Glamorgan* (ed. A.H. John & G. Williams, 1980).

Roberts, *The White Rock Works*

R.O. Roberts, 'The White Rock Copper and Brass Works, near Swansea, 1736-1806'. *Stewart Williams' Glamorgan Historian,* 12 (1981), 136-151.

Rogers, *History of Swansea*

W.C. Rogers, *A Pictorial History of Swansea* (Swansea, 1981).

Rogers, *Swansea Calendar*

MS. of W.C. Rogers, *The Swansea and Glamorgan Calendar* in two volumes (National Library of Wales).

Rogers, *Western Newcomen Engines*

K.H. Rogers, *The Newcomen Engine in the West of England* (Bradford-on-Avon, 1976).

Rolt, *The Potters' Field*

L.T.C. Rolt, *The Potters' Field: A History of the South Devon Ball Clay Industry* (Newton Abbot, 1974).

Rolt & Allen, *Newcomen's Engine*

L.T.C. Rolt and J.S. Allen, *The steam engine of Thomas Newcomen* (Hartington, 1977).

Ross, *Robert Morris Diaries*

J.E. Ross, *Radical Adventurer: the Diaries of Robert Morris, 1712-1744* (Bath, 1971).

Rowe, *Lead Manufacturing*

D.J. Rowe, *Lead Manufacturing in Britain* (London, 1983).

Royal Commission on Rel. Bodies

Report of the Royal Commission on the Church of England and Other Religious Bodies in Wales and Monmouthshire (London, 1911).

Ruddock, *Arch Bridge Builders*

T. Ruddock, *Arch Bridges and their Builders 1735-1835* (Cambridge, 1979).

S.C. Archives

Swansea City Archives Office.

S. Mus.

Swansea Museum (Royal Institution of South Wales).

S.R.Lib

Swansea Reference Library.

S.W.W.I.A.S.

South West Wales Industrial Archaeology Society.

S.W.W.I.A.S.N.

South West Wales Industrial Archaeology Society Newsletter.

S. Wales Coalfield Report, 1805

May 1805, 'Report of John Grieve and John B. Longmire on the South Wales Coalfield' (D/Lons/W/Collieries 138 in the Cumbria County Record Office, Carlisle). I am very grateful to the late Professor Michael Flynn for drawing this document to my attention.

Scoville, *Morriston's Past*

A. Scoville, *Morriston's Pictorial Past* (Cowbridge, 1988).

Seaborne, *Schools in Wales*

M. Seaborne, *Schools in Wales, 1500-1900; A Social and Architectural History* (Denbigh, 1992).

Shaw, *Cornwall Methodism*

T. Shaw, *A Methodist Guide to Cornwall* (London, 1991).

Sheasby, *Canal Plan & Section*

T. Sheasby, 'Plan of an Intended Navigable Canal from Swansea to Pentrecrybarth in the counties of Glamorgan and Brecon', 1793 + 'profile'. (Glamorgan Record Office, Cardiff, Q/DP 3 & 5).

Sheasby, *Hen-noyadd Plan*

T. Sheasby, 'Plan of an Intended Navigable Canal from Swansea to Hen-Noyadd in the Counties of Glamorgan and Brecon', 1793 (Glamorgan Record Office, Cardiff, Q/DP 4).

Smiles, *William Edwards*

S. Smiles, 'William Edwards, bridge builder', *Lives of the Engineers* (London, 1862), 266-75.

South Wales Coal Report, 1810

'Report from the Committee on the Petition of the Owners of Collieries in South Wales, House of Commons, 1810 (344) IV.

State of Llansamlet Colliery, 1771-72

Appendices entitled 'A State of Llansamlet Colliery taken by J. smith, Sept. 1771' and 'A State of Llansamlet Colliery taken by Jas. Townsend & T. Law, 1 Jan. 1772' transcribed on pages 248-254 of an article by H.K. Jordan, 'The South Wales Coalfield', *Proceedings of the South Wales Institute of Engineers,* 27 (1910-11), 172-254.

Svedenstierna's Tour

E.T. Svedenstierna, *Svedenstierna's Tour of Great Britain 1802-3. The Travel Diary of an Industrial Spy* (Newton Abbot, 1973). Transl. E.L. Dellow. Introd. M.W. Flinn.

Swansea Canal Case

Broadsheet entitled the 'Swansea Canal Case', not dated but probably 1794 (Swansea Reference Library).

Swansea Extension Rly., c.1845

The large-scale printed plans of a railway that was to have been built up the western side of the Swansea Valley and entitled the 'Swansea Extension Railway'. n.d. but coincident with 'The South Wales Railway' plans published in 1845. (Stored in the base of the geological cabinets in Swansea Museum).

Swansea Steam Power

'Statistical Table of Steam Power', n.d. but c.1840 (Morgan MSS., Swansea Museum (RISW)).

Swansea Univ.

Swansea University College Library.

Symons, *Llanelli Coal Mining*

M.V. Symons, *Coal Mining in the Llanelli Area: Volume One: 16th Century to 1829* (Llanelli, 1979).

Tithe

Tithe map and schedule relating to the parish mentioned (National Library of Wales).

Thomas, *Hanes Annibynol*

J. Thomas, *Hanes Eglwysi Annibynol Cymru,* Vol.5 (Liverpool, 1891).

Thomas, *I.A. Upper Tawe*

W.G. Thomas, 'Industrial Archaeology. A study of some remains of past mining activity in the Upper Tawe and Twrch Valleys', *Brycheiniog,* XIV, (1970), 67-77.

Thomas, *Industrial Llangiwg*

J.D.H. Thomas, unpublished M.A. thesis on 'The Industrial Development of Llangiwg Parish', University of Wales, 1974.

Thomas, *S. Wales Mines*

W.G. Thomas, *Coal Mines of S. Wales* (NMW Cardiff, manuscript).

Thomas, *Swansea Districts I*

N.L. Thomas, *The Story of Swansea's Districts & Villages, Volume I* (Swansea, 3rd. edn., 1969).

Thomas, *Swansea Districts II*

N.L. Thomas, *The Story of Swansea's Districts & Villages, Volume II,* (Swansea, 1969).

Tomlinson, *Cyclopaedia*

C. Tomlinson (ed.), *Cyclopaedia of Useful Arts, Mechanical and Chemical, Manufactures, Mining and Engineering, Volume I, Abbatoir to hair-pencils* (London and New York, 1854).

Toomey, *Vivian & Sons*

R.R. Toomey, 'Vivian and Sons, 1809-1924' (Ph.D. Thesis, Wales, 1979); subsequently published as *Vivian and Sons 1809-1924, A Study of the Firm in the Copper and Related Industries* (New York, 1985).

Trewyddfa Canal Traffic Book	Journals of the Trewyddfa Canal, 1829-33 (National Library of Wales, The Badminton manuscript collection, Group II, 816-817).
Trewyddfa Map	'A Map of the Fee of Trewyddfa in the Parish of Clace within the Parish of Langevelach in the County of Glamorgan Belonging to His Grace the DUKE OF BEAUFORT. This map was drawn by Lewis Thomas in 1761, from a survey made by Whitterley some years before, with some additions theretomade by Lewis Thomas in 1761.' (NLW Badminton MSS. 2136).
Trinder, *Industrial Landscape*	B. Trinder, *The Making of the Industrial Landscape,* (London, 1982).
Tylecote, R.F., *Metallurgy*	R.F. Tylecote, *A History of Metallurgy* (London, 1976).
v.i.	Verbal information.
Watkin, *C.R. Cockerell*	D. Watkin, *The Life and Work of C.R. Cockerell* (London, c.1980).
Williams, Coal Industry	J. Williams, 'The Coal Industry, 1750-1914' in A.H. John & G. Williams (ed.), Glamorgan County History, Volume V, Industrial Glamorgan from 1700 to 1970 (Cardiff, 1980), 155-210.
Williams, *Swansea, 1990*	G. Williams (Ed.), *Swansea: An Illustrated History,* (Swansea, 1990).
Williams, *Hanes Llansamlet*	W.S. Williams, 'Hanes a Hynafiaethau Llansamlet (History and Antiquities of Llansamlet)', (Dolgellau, 1908).
Wood, *Rivers of Wales*	J.G. Wood, *Rivers of Wales Illustrated* (London, 1813).
Yates Map, 1799	'A Map of the County of Glamorgan from an actual survey' Made by George Yates of Liverpool on Which was Delineated the Course of the Rivers, and Navigable Canals; with The Roads, Parks, Gentlemens Seats, Castles, Woods. Scale approximately 1 inch to a mile. 1799.

INDEX

346

Llanelli coalfield, 5, 6, 24; Genwen, 7; Llanelly
    Railway, 146; steam-powered collieries, 122;
    water-powered pits, 121, 122
Llanelli Copperworks Schools, 243
Llangyfelach: Welsh Trust School, 242
Llangyfelach aqueduct, *16*, 20, 40, 119, *296*, 313
Llangyfelach Copper Works (Landore), *16*, 17, 20-21,
    32, 40, 41, 54-55, 60, 83, 99, 109, 111, 115, 122,
    124, 133, 255, 295, *296*, 298, 302, *302*m, 313, *322*;
    accounts, 44, 59-60; battering-mill, 41; blast
    furnace, 25, 60; copper rod-mill, 41; labour force,
    163; Padmore (John) and, 20, 132, *134*; red-lead
    production, 61; silver and lead smelting, 59-60, 60,
    60-61, 296; smelting-hall, 21, 26, 27, 27-28, 32;
    workers' housing, 161
Llangyfelach Road: settlements, 168, 174-75, *174*m,
    175-76, 186
Llanidloes & Newtown Railway, 146
Llansamlet: air pollution, 325; arsenic and copper
    production, 63; canal, 62, 80, 84, 96, 193, 227;
    chapels, 268, 272, 280; Landore Siemens
    Steelworks, 308; lead smelter, 60-61; schools, 243,
    250; tinplate industry, 193; workers' housing, 165,
    199; Ysgoldy Capel-y-Cwm, 247; zinc smelters, 63,
    256, 326
Llansamlet coalfield, 4, 5, 24, 62, 80, 81, 82, *84*m, 90,
    108-09; atmospheric steam engines, 144, 151;
    Briton Ferry engine, 5; Double Pit, 11, 108, *108*,
    322; drainage, 107; Great Pit (Pwll Mawr), 5, *5*, 8,
    108, 311; Gwernllwynchwith, 5, *6*, 7, 82, *84*, 108,
    109, *109*; quays, 80; Scott's Pit, *9*, 93, *93*, *322*
Llewellyn, John Dillwyn, 168, 172, 186, 267
Llewelyn Park, 161, 186
Lloyd, Edward, 255, 256
Llwynypia: rents, 194
Llynfi Spelter Works, 268
locks: Neath River, 78; Swansea Canal, 79, 179;
    Tennant Canal, 78-79
Lockwood and Mawson (architects), 209
Lockwood, Morris and Co: accounts, 45, 60; Beaufort
    Bridge and, 74; collieries, 90, 95-96, 110, 122, 132,
    303, 305, 312; copper products, 37, 44, 45, 54, 55,
    60; copper-smelters, *2*, 3, 4, 5, *8*, *17*m, 21, 29, *42*,
    44, 47, *71*, 83, 98-99, 111, 200, 303, 305; dry dock,
    306; Edwards (William) and, 134, 136, 139;
    Morris's canal, 28, 29, 30, *71*, 84, 88, 96, *96*m, 97,
    98, 100, 101, 111, 112-13, 148, 167, 201, 214, 303,
    304; Padmore (John) and, 132, *134*; Mr Powell
    and, 83, 110, 142-44, *142*; railways, 95, 303,
    305-06, 312; steam power use, 110; water-power
    systems, 109, 110; wharves, 303-04; workers'
    housing, 158, 159, 162-63, 166, 167, 176, 177-78,
    178, 179, 190, 198
Lockwood, Thomas, 217, 312
London: copper product manufactures, 36, 39, 44
London Lead Company, 60
Loughor: tinplate production, 58; zinc works, 62, *62*
Loughor, River: quays, 76
Lower Forest Coppermill, *28*, 35, 41, 42-45, *42*,
    47-49, 48, 49, 52, 57, *59*, *71*, 113, 123, *136*;
    Edwards (William) and, 133; labour force, 163;
    Padmore (John) and, *28*, 42, 123, 133, *135*; Tawe
    navigation and, 74; water supply, 111; water-power
    availability, 80, 105, 123; water-wheels, 42, *42*,
    43-44, 44, 80, 105; workers' housing, 158, 162

Lower Forest Tinplate Works *see* Beaufort Tinplate
    Works
Lower Swansea Valley Study, The (1960s), 328-29
Lydney Canal and Dock, 151

Machen Upper Forge: workmen's dwellings, 157
McKinley, President William, 58
Mackworth, Sir Herbert, 4, 170, 183-84, 213, 214,
    222, 323
Mackworth, Sir Humphrey, 40, 50, 59, 60, 74, 82, 83,
    123, 140, 157, 183, 242
Maesteg House (Kilvey Hill), 210, 226, 228, 265
Malkin, Benjamin: on William Edwards, 133, 138
manillas, 45
Mannesmann Tube Works, 300
Mansel, Sir Bussy, 4th Baron, 22, 27, 46, 82, 121, 122,
    183; colliery pumping engine, 3-4; Great Coal
    Road, *23*, *80*, 82; Little Pit, 108; New Mill
    (Llansamlet), 46, 104, 106; water-power
    engineering, 107, 109, 112
Mansel, Rawleigh, 215
Mansel, Thomas, 2nd Baron, 76
Manselton, 4, 58, 83, 95-96, *95*, 196; churches, 266;
    colliers' housing, 160, 199
marble mills, 116
Margam Copperworks: workers' housing, 195
Margam Tinplate Works, 58
Marina development, 326-27
Marine Villas, *215*, 216
Marino (pleasure villa), 134-35, 147, *215*, 216, 217,
    221, *221*, 222, 223, *223*, *225*, 226, 233
Maritime Museum, 327
Martin, Edward, mining engineer, 90, 91, 93, 98, 104,
    105, 114, 132, 144, 146, 151
Martin, Sir Richard, 247
Mary, Queen-consort to George V, *155*
Mathews, John, 257
Melin-y-Frân Cornmill, 106
Melincryddan Copperworks (Neath), 20, 40, 45, 82,
    123, 140; lead smelting, 59, 60; school, 242; tidal
    pill, 96
Melingriffith Forge, 55
Melingriffith Pump, 138
Melingriffith Tinplate Works, 256, 292
Merthyr Tydfil: Cyfarthfa Ironworks, 31-32, 32, 34,
    55, 120; Dowlais Schools, 243, 248, 251; iron
    edge-rails, 91; ironworkers' townships, 163; lack of
    toxic fog, 324; Penydarren Ironworks, 50;
    population, 165; public health and mortality, 172;
    social amenities, 289; timber trough aqueduct, 119;
    workers' housing, 180-81, 196, 197, 234
Merton Copper Mills (Wimbledon), 44
Middle Bank Copperworks, 24, 27-28, *72*, 82, *84*, *85*,
    108, 112, 170, 173, 227, 228, 256; labour force, 163;
    melting furnaces, 25, 34; smelting-halls, 27-28, 34;
    and the Thames and Greenfield Valley Coppermills,
    47; use of copper-slag blocks, 53; workers' housing,
    162, 163, 182, 185, 190, 204-05, *204*m
Midland tinplate works, 58
Mile End, 166, 167, 206
Millbrook Ironworks, 221, *295*, 298-99, *298*, 313;
    cast-house, 316-18; internal railway, 308; Steel
    Furnaces Building, 318; turning and boring mill,
    299; water-wheels, 115, 173; workers' houses, 169,
    186, 206

and, 157; systems, 103-25; Tawe River and, 80, *105*, *115*; tinplate production and, 57

water-wheels, *14*, 80, 105-11, *106*, 115, 119, 120, 296, 299, 311; agricultural uses, 118; Blackpill Mill, 104; Cambrian Copperworks, 41-41, 105-06; colliery winding gear, 4, 5, 106-11, *106*, 108, 109, 110, 111, 142, 296, 298.*305*; Cwm Pit, 298, *305*; Landore Water-wheel Coal-pit, 296, 299; at Llangyfelach, *16*, 20, 41, 105-06, 115; at Lower Forest Coppermill, 42, *42*, 43-44, 44, 80; steam powered feeds, 6, 7, 46; Swansea Canal, *118*; water pumping, 4, 5, 6, 6-7, 108, *108*, 109, 121-22; White Rock Copperworks, 24, 46; Ynyspenllwch Forge, *105*

Watkins, A C (architect), 277

Watson, George: on Mr Powell, 142, 143-44

Watt, James, 6, 38, 144

Waun Gwn: workers' settlements, 168, 186

Wayne, Gabriel, 20

Wedgwood, Josiah: glazes, 14; pottery processes, 12, 13, 15, 229

Welsh language: education and, 245, 249, 250, 290; nonconformity and, 243, 249, 255, 261, 268; Sunday Schools, 243, 249

Wesley, John, 288

Wesleyan Methodists, 256, 257, 278, 288, 291; Bible Christian Chapel (Trevivian), 269, 287, *287*; Goat Street Chapel (Swansea), 257, 269, 275, 278, 279; Gothic architecture, 269; market building meeting (Morriston), 267, 286; Wern Road (Landore), 257

Westminster Bridge (London), 137

white lead, 61

White Rock canal tunnel, *322*

White Rock Copperworks, *21*, 22-24, *23*m, 27, *27*, 28, *28*, *29*, 30, *31*, *46*, *72*, 82, *84*, *85*, 106, 112, 170, 192, 256, 257, *321*, 328; battery-mill, 46; brass production, 55; and the Bristol Coppermills, 45-46; cast 'manillas', 45; clay mill, 10, 22, 115; copper products, 46; demolition, 329; docks, *24*, *72*, 76, 79, *80*, *322*; furnaces, 22, 27; housing leases, 197; John Smith's canal, *22*, *23*, *24*; labour force, 163; 'Manilla House', 46; recycled water, 121; rolling-mill, 24, 46, *47*; silverworks flues, 61; "The Great Workhouse", *21*, 22, *22*, *23*, *24*, 27, 28, *29*, 30, 32, 34, *322*; tips, *36*; waste gas flues, 32, 328; water supply, *23*, 46, 103, 106, *106*, 107, 118, 120; workers' housing, 162, 163, 166, 169, 171, 173, 183, 185, 203, 233, 331

White Rock House, 163

White Rock Industrial Archaeology Park, 22, *24*, 328, 330, *330*

Whitebrook Valley (Monmouthshire): leats and ponds, 125, 311; workers' houses, 157

Whitefield, George, 133

Whitehaven (Cumbria): iron rail tracks, 91

Whitehurst, Mr, of Derby, 142-43

William, Thomas, 140

Williams, Mr (architect), 284

Williams, Revd. David, 132-33, 138

Williams family, of Parys Mountain, 24, 36, 47

Williams family, of Scorrier (Cornwall), 48, 256

Williams, Foster & Co, 113, 256, 307

Williams, Foster and Pascoe Grenfell, 228

Williams, John ('King of Gwennap'), 256

Williams, John (Margam land-surveyor), 144

Williams, Michael, 35, 210, 235, 256

Williams, Owen, 47, 182, 203, 204, 227

Williams, Penry, 31

Williams, Thomas (the 'Copper King'), 36, 47, 51, 192-93, 211, 227, 256, 266

Williams, W Samuel (Independent Minister), 195

Williams, William, 257, 288

Wills, H O, 270, 284

Wills, John, 268

Wilson, J Buckley (architect), 266

Winch Wen Industrial Estate, *4*

windlasses, hand-operated, 4, 109, 122

windmills, 103

wiremills, 41, 124, 157

Woodlands (Clyne) Castle, 65, 147, 222-23, *222*, 224, 225, 326

Woods Mine (Hawarden, Flints), 4

Woodyer, Henry (architect), 224, 226, 253, *260*, *261*, 262, 263-64

Woollard (Somerset), 46

woollen-mills: water-powered, 117-18, 125

Worcester Tinplate Works, 57, 58, 288

workers: Belgian, 62, 157; German, 156, 157; housing and settlements *see* housing; New Cut construction, 78; Tawe navigation, 79-80

Worsley canal system, 8, 98, 99, 103

Wyatt, Benjamin (land agent), 275

Wychtree Bridge, 74, 137, 138, 139, *139*, 141, *142*, 200, 322-23

Wye Bridge (Brecon), 141

Yellow Tower of Gwent, 6

Ynys Collieries (Clyne Valley), *3*, 104, 110, 120, *123*, 219, *322*, 328

Ynys Copperworks, 38, 62, 166

Ynys-hir lead smelting works (Ceredigion), 60

Ynys-pen-llwch (Clydach): copper rolling-mill, 41, 105, 123; iron forge, 40, 55, 114-15; Tawe navigation and, 74; tinplate works, 57, 74, 114; water-power availability, 80, 105, *105*, 111, 114-15, *322*; wiremill, 41

Ynys-Tanglwys Mansion (Clydach), *230*m, 231, 232, *232*, *233*, 327, *327*

Ynys-Tawe: cornmill, 104; waste-water weir, *105*

Ynys-y-Gerwyn (Aberdulais): copper rolling-mill, 40, 57, 58, 123; tinplate works, 57, 58, 114

Ynysgedwyn Iron Furnace, 40, 55, 104, 115, 229, 232; arcade preservation, 330; workers' houses, 158-59, 192

Ynyshowell Copperworks, 115

Ynysmeudwy: aqueduct, 120; pottery, 16

Ynyspenllwch Forge: water-wheels, *105*; workers' housing, 157

York Minster, 19

Yspitty Tinplateworks, 288

Ystalyfera, 180, 192, 288

Ystradgynlais: workers' houses, 158, 180

zinc: in brass alloys, 39, 61, 63; coal and, 62-63; at Llangyfelach Works, 41; ores, 18, 22, 24, 61, 326; smelters, 16, 24-15, 55, 59, 61-63, *61*, *62*, 326